MATERIALS SCIENCE RESEARCH

Volume 12

BORATE GLASSES

STRUCTURE, PROPERTIES, APPLICATIONS

MATERIALS SCIENCE RESEARCH

Volume 1: Proceedings of the 1962 Research Conference on Structure and Properties of Engineering Materials
–edited by H. H. Stadelmaier and W. W. Austin

Volume 2: Proceedings of the 1964 Southern Metals/Materials Conference
–edited by H. M. Otte and S. R. Locke

Volume 3: Proceedings of the 1964 Conference on the Role of Grain Boundaries and Surfaces in Ceramics
–edited by W. Wurth Kriegel and Hayne Palmour III

Volume 4: Proceedings of the 1967 International Symposium on Kinetics and Reactions in Ionic Systems
–edited by T. J. Gray and V. D. Fréchette

Volume 5: Proceedings of the 1970 Conference on Ceramics in Severe Environments
–edited by W. Wurth Kriegel and Hayne Palmour III

Volume 6: Proceedings of the 1972 International Symposium on Sintering and Related Phenomena
–edited by G. C. Kuczynski

Volume 7: Proceedings of the 1973 International Symposium on Special Topics in Ceramics – Surfaces and Interfaces of Glass and Ceramics
–edited by V. D. Fréchette, W. C. LaCourse, and V. L. Burdick

Volume 8: Proceedings of the 1974 Conference on Emerging Priorities in Ceramic Engineering and Science
–edited by V. D. Fréchette, L. D. Pye, and J. S. Reed

Volume 9: Proceedings of the Eleventh University Conference on Ceramic Science devoted to Mass Transport Phenomena in Ceramics
–edited by A. R. Cooper and A. H. Heuer

Volume 10: Proceedings of the Fourth International Conference on Sintering and Related Phenomena
–edited by G. C. Kuczynski

Volume 11: Proceedings of the Fourteenth University Conference on Ceramic Science – Processing of Crystalline Ceramics
–edited by Hayne Palmour III, R. F. Davis, and T. M. Hare

Volume 12: Proceedings of the Conference on Boron in Glass and Glass Ceramics
–edited by L. D. Pye, V. D. Fréchette, and N. J. Kreidl

MATERIALS SCIENCE RESEARCH • Volume 12

BORATE GLASSES

STRUCTURE, PROPERTIES, APPLICATIONS

Edited by

L.D. Pye
and
V.D. Fréchette
Alfred University
and
N.J. Kreidl
Professor Emeritus
University of Missouri-Rolla

PLENUM PRESS · NEW YORK AND LONDON

Library of Congress Cataloging in Publication Data

Conference on Boron in Glass and Glass Ceramics, Alfred University, 1977.
Borate glasses.
(Materials science research; vol. 12)

"Proceedings of a Conference on Boron in Glass and Glass Ceramics, held at Alfred University, Alfred, New York, June 5–8, 1977."
Includes index.
1. Glass–Congresses. 2. Ceramics–Congresses. 3. Borates–Congresses. I. Pye, L. David. II. Frechette, Van Derck. III. Kreidl, N. J. IV. Title.
TA450.C63 1977 620.1'44 78-9108
ISBN-13: 978-1-4684-3359-3 e-ISBN-13: 978-1-4684-3357-9
DOI: 10.1007/978-1-4684-3357-9

Proceedings of a Conference on Boron in Glass and Glass Ceramics
held at Alfred University, Alfred, New York, June 5–8, 1977

A Division of Plenum Publishing Corporation
227 West 17th Street, New York, N.Y. 10011

PREFACE

Boron Oxide plays a key role in numerous glasses of high technological importance, yet its role in glass structure is far from clear. Indeed, in recent years there have been serious challenges to previous structure concepts for both crystalline and glassy borates. These challenges were sufficient to warrant a re-examination of the structure of borate glasses using the most powerful tools currently available. To provide a suitable forum for this undertaking, a four-day conference on "Boron in Glass and Glass Ceramics" was convened at Alfred University, June 3-8, 1977 to review the best scientific thinking on structure and to debate conflicting views and discuss properties and applications of borate glasses. This conference was also the first in a New University series on Glass Science to be rotated among Alfred University, The Pensylvania State University, Rensselaer Polytechnic Institute, and the University of Missouri-Rolla. The present volume represents the proceedings of the first conference in this series.

The volume begins with a review of the remarkable contribution of Jan Krogh-Moe to the understanding of the structure of Borate glasses. This review, authored by Professor N. J. Kreidl, concludes by dedicating the proceedings of this conference as a Krogh-Moe Festschrift. The volume continues with a historical review by D. L. Griscom, originally prepared for circulation to the contributors prior to the conference. An Epilogue to the opening chapter brings the survey up-to-date in light of the conference papers.

The structure section includes contributions from institutions specializing in various techniques suitable for glass structure analysis. Among these techniques are X-ray diffraction, nuclear magnetic resonance, vibrational analysis, Mössbauer absorption, electron spin resonance, laser-induced fluorescence line narrowing, and gas permeability studies. This section also includes important theoretical considerations of model borate glass electronic structure, topology of triangularly connected networks, and a critical examination of random vector statistics related to some of the above structural methods.

The third and fourth sections deal with the properties of borate melts and glasses. It was natural to arrange the many chapters in this section into high and low temperature properties; the former includes viscous flow, immiscibility and microstructure, crystal growth, physical properties, viscoelastic behavior, volatilization and glass formation. Among the low temperature properties are electrical and dielectric behavior, chemical durability, optical characterization and acoustic spectra.

The final section, dealing with applications, begins with a survey paper on industrial borate glasses and continues with discussions of their use for radioactive waste disposal and their utilization in electronics and thermal insulation.

The conference chairmen acknowledge with pleasure the sponsorship of the conference by the New York State College of Ceramics at Alfred University jointly with the U. S. Borax Research Foundation (through R. B. Bessey), the U. S. Bureau of Mines (through N. L. Jensen), and the New York State Science and Technology Foundation (through D. H. Davenport). The chairmen further acknowledge the splendid cooperation provided by all speakers and contributors to this volume and by numerous individuals who assisted in the planning and execution of the conference:

Members of the Program Committee - C. P. Ballard, A. M. Bishay, P. J. Bray, A. R. Cooper, G. H. Frischat, D. L. Griscom, D. R. Uhlmann, S. Urnes and J. Zarzycki,

D. L. Griscom - for the pre-conference survey on Borate Glass Structure,

Session Chairmen - C. P. Ballard, J. Blachere, D. Day, W. LaCourse, W. G. Lawrence, K. Murthy, G. Rindone, B. Schwartz, H. Smyth and R. Van Der Beck,

Members of the Concluding Panel - J. D. MacKenzie, G. H. Frischat, R. A. Weeks, J. F. MacDowell, D. Evans and R. Eagan.

Additional thanks are extended to the local arrangements committee composed of V. Burdick, W. Earl and D. Rase for their timely help throughout the conference. Finally, the editors acknowledge the help of the following individuals in preparing the pre-conference survey and certain portions of this volume: B. Aldrich, J. del Campo, P. Gignac, L. Hanks, C. Link, E. Richardson and D. Snowden.

L. D. Pye
V. D. Fréchette

January, 1977

CONTENTS

A DEDICATION TO KROGH-MOE: HIS CONTRIBUTION TO THE UNDERSTANDING OF BORATE GLASSES

N. J. Kreidl

University of Missouri - Rolla
and Institute Physical Chemistry
Vienna, Austria

ATTITUDE AND PRINCIPAL ACHIEVEMENTS

Our friend and colleague Jan Krogh-Moe left, after his untimely death in 1975, 92 publications, which have been collected and preserved for the glass community by Urnes et al. (1975) and are reproduced here. A large portion of this work was concerned with the clarification of the structure of crystalline and vitreous B_2O_3 and borates. As an introduction to the Conference on Boron in Glasses which was intended to become a memorial to Krogh-Moe's work an attempt is made to assess his contribution to the topic of our discussion.

For his inspection of the structural details of borate glasses Krogh-Moe mobilized among various methods principally the correlation of IR and X-ray spectra of glasses with those of crystalline compounds of comparable composition exploiting the confidence generally placed in the classical structural analysis of crystalline materials.

The work culminated in the listing in 1962 of possible groupings in the structure of borate glasses as a function of alkali content which proved to be by and large in agreement with the evidence of nuclear magnetic resonance provided principally by Bray and his collaborators and that of Raman spectroscopy provided by Konijnendijk et al. after Krogh-Moe's death.

This creative contribution of Krogh-Moe, the comparison of his models with the techniques just mentioned and various other approaches, and the appearance of more recent and critical views represent, even at the present moment, the backbone of any discussion of the topic and, therefore, of Griscom's extensive review as well as of the discussion of all contributions to this conference.

Krogh-Moe himself gave us an example of keen adherence to the tenet of continuous critical review of scientific concepts in his development of ideas about the structure of alkali borate glasses. He and his coworkers started out (Borgen et al. 1954)[9] venturing doubt about the conclusiveness of the X-ray evidence for the development of BO_4 groupings as alkali is added to B_2O_3, going as far as suspecting the reverse structural change! Yet, in his important paper in Physical Chemistry of Glasses in 1962 (Krogh-Moe 1962)[46] he clearly revises his early stand in the light of his own findings (Krogh-Moe 1960)[39]. At the same time his evidence as well as that of Bray's work on nuclear magnetic resonance made it explicitly clear that the so-called boron anomaly was neither an anomaly, nor associated with unusual maxima or minima of properties at 15% R_2O, nor allied to the evolution and decline of BO_4 groupings. It is in this vein that Griscom introduces to us, and the Conference will discuss, the maxima and minima of properties in borate systems as predictable, variable phenomena deriving from multiple causes.

The principal achievements of Krogh-Moe's studies of borate glass systems were the detailed and well-founded structural models based on documented crystalline structures with which he confronted spectroscopic evidence by himself and others arriving at convincing assignments such as those of boroxol, diborate, pentaborate etc. groupings.

EARLY STUDIES

As stated above, Krogh-Moe and his collaborators, started out by analyzing the information that could be derived from X-ray spectroscopy regarding the postulated evolution of BO_4 groups in the systems R_2O - B_2O_3 upon addition of R_2O. As also stated above he first came to the conclusion later retracted (1960)[39] that, on the contrary, BO_4 changed to BO_3 groupings upon the addition of R_2O. Characteristically, he did not stop at this point, but decided upon the careful determination by infrared spectroscopy of various crystalline borates, considering that at that time one might extract more easily and convincingly useful analogies with infrared spectra of vitreous borates than with X-ray spectra.

In this work (1958-1959)[25, 26, 27, 33, 34, 36] are contained statements of great interest to the present generation of students of noncrystalline solids. In line with Turnbull's or Uhlmann's contemporary concepts stressing kinetic criteria for glass formation, he illustrates a related viewpoint by the extension of glass formation in the binaries with PbO. How, he reasoned, could a network (structural) theory account for this extended range. Referring to Fajans et al.[19] critique, he concludes that the rate of devitrification governed by various factors, not the concentrations of a SiO_2 or B_2O_3-based network determines the vitrifiability of lead glasses[35]. At this time, even before having developed models for group development, he also exposed as a fallacy the attribution of "anomalous" property extrema to the BO_4 concentration. He rather postulated, as we may well do, the result of "contradictory effects": e.g. (a) the decrease of expansion from B_2O_3 to borate structure, (b) the increase of expansion by the alkali concentration itself. He suggested in these early papers the now generally accepted possibility of non-random even paired distribution of nonvalent constituents (alkali, silver)[34, 36] and correlation of such distribution to the mechanism of electrical conductivity.

GROUP ASSIGNMENTS IN BORATE SYSTEMS

By 1960[39] Krogh-Moe felt ready to conclude, from the comparison of spectroscopy of crystalline and vitreous boron compounds, the presence of groupings analogous to those in certain crystals in the broader ranges of borate glasses. In particular, in 1960, he recognized once more the classical concept of a conversion of BO_3 groupings prevalent in pure B_2O_3 to BO_4 groupings as R_2O is added. In addition he postulated:

a) the boroxol structure in pure B_2O_3
b) the pentaborate structure in certain alkali borate glasses
c) the hybridization as s^2p by the non-bridging oxygen produced by alkali, as well as in the process of viscous flow; this latter bold statement implies a coordination change in flow
d) non-random distribution of modifiers

He elaborated by undertaking numerous X-ray studies on (mostly crystalline) borates (1960-1962)[41-46, 48, 52, 53, 56, 57, 59-61, 65, 66, 68, 74, 84, 87-90].

In his most important (to us) 1962 paper[46] he clearly finds that BO_4 groupings increase beyond 15% R_2O as seemed soon confirmed by Bray's et al. spin resonance data. He correlated crystalline and glass groupings on the basis of infrared data. He consequently detached borate glass property extrema ("anomalies") from the formation of BO_4 groups, pointing out - as Uhlmann soon was to confirm by a large range of experiments and consonant interpretation - that for various properties minima do indeed occur at different concentrations for Li_2O, Na_2O, K_2O. In this same year Krogh-Moe listed[51] most of the various complex groupings such as diborate, triborate, pentaborate etc. groups which are at present considered as models when spin resonance or Raman (Konijnendijk et al.) interpretations are attempted. In this same study these groupings are used to interpret melting point depression data. He arrives, at times, at intriguing details such as relating the deduced pairing of Ag atoms to the prevalence of the $(B_8O_{13})^{2-}$ over the $(B_5O_8)^-$ arrangement. At any rate, the phase diagram in that interpretation, clearly rules out random distribution between unassociated BO_3 and BO_4 islets.

CULMINATION OF STRUCTURAL ANALYSIS OF BORATE GLASSES

In our field, Krogh-Moe's deservedly most cited study is his "Interpretation of the IR Spectra of Boron Oxide and Alkali Borate Glasses" (1965)[58]. Here the previously developed ideas are clarified and clearly presented:

(1) Pure B_2O_3 is built up from boroxol groups (Fig. 1[a]), perhaps in what may be approximately as the arrangement of two interlocking networks.

(2) On adding alkali up to 20% one triborate (Fig. 1[c]) and one pentaborate (Fig. 1[b]) in adjacent positions are obtained (Fig. 2).

(3) Up to 33% diborate (Fig. 1[d]) groups replace pentaborate groups.

The experimental evidence included the spectroscopy of crystalline $K_2O\cdot 5B_2O_3$, $K_2O\cdot 5B_2O_3\cdot 8H_2O$ containing penta groups, of $Li_2O\cdot 2B_2O_3$ and $Na_2O\cdot 2B_2O_3\cdot 10H_2O$ (borax) containing diborate groups and $Cs_2O\cdot 3B_2O_3$ containing triborate groups.

In 1969 Krogh-Moe presented a concise review of his work on B_2O_3[13]. Later papers in the field of borates concentrated on crystalline compounds[74, 84, 86-90].

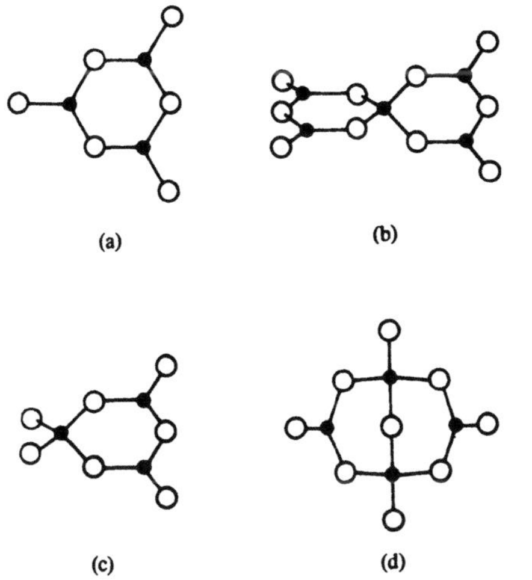

Figure 1. The borate groups found in borates with 33 1/3 mole % alkali or less. (a) The boroxol group. (b) The pentaborate group. (c) The triborate group. (d) The diborate group.

● Boron atoms O Oxygen atoms

DEDICATION

The proceedings of this conference are to be considered a Krogh-Moe Festschrift, following an informal resolution of some of us who heard about his death when we were together in Leningrad. May this conference honor one who has given us so much insight in our investigations.

REFERENCES

1. J. Krogh-Moe — "On the principle of thermal interaction" Acta. Chem. Scand. 7 (1953) 239.

2. J. Krogh-Moe — "A remark on some new methods of phase determinations" Acta. Cryst. 6 (1953) 568.

3. T. Forland, K. Grjotheim and J. Krogh-Moe — "Anvendelse av strukturmodeller ved beregning av de konsentrerte syreopplosninger fysikalsk-kjemiske egenskaper" Tidsskrift for Kjemi, Bergv. og Met. (1954) 1-7.

4. K. Grjotheim and J. Krogh-Moe — "Hydrogen bonding and the volume of water" Nature 173 (1954) 774.

5. K. Grjotheim og Krogh-Moe — "Elektrolytisk polering" Tidsskrift for Kjemi Bergv. og Met. (1954) 61.

6. K. Grjotheim and J. Krogh-Moe — "On the correlation between structure and some properties of water" Acta. Chem. Scand. 8 (1954) 1193.

7. J. Krogh-Moe — "The structure of liquid carbon disulphide" Acta. Chem. Scand. 8 (1954) 1949.

8. K. Grjotheim und J. Krogh-Moe — "Die borsäure-anomalie" Naturwiss 41 (1954) 526.

9. O. Borgen K. Grjotheim and J. Krogh-Moe — "A comment on the X-ray determination of glass structure" Kongelige Norske Vidensk. Selskabs forh. 27 (1954) Nr. 17.

10. K. Grjotheim and J. Krogh-Moe — "A structural explanation of the boron oxide anomaly" Krongelige Norske Vidensk. Selskabs forh. 26 (1954) Nr. 18.

11. A. Bjorhaug and J. Krogh-Moe — "A method for structure determination in simple centrosymmetrical systems" Acta. Cryst. 8 (1955) 441.

12. H. Holtan and J. Krogh-Moe — "On the correspondence between thermocells and isothermal cells" Acta. Chem. Scand. 9 (1955) 1022.

13. K. Grjotheim und J. Krogh-Moe — "Röntgenographische Strukturuntersuchung siner gestättigten Lösung von Schwefel in Schwefel-kolhlenstoff" Z. Phys. Chem. 5 (1955) 284.

14. K. Grjotheim F. Gronvold and J. Krogh-Moe — "The solution of cadmium in liquid cadmium chloride" J. Am. Chem. Soc. 77 (1955) 5824.

15. O. Borgen and J. Krogh-Moe — "The infrared spectra of some modifications of arsenic trioxide and antimony trioxide" Acta. Chem. Scand. 10 (1956) 265.

16. K. Grjotheim and J. Krogh-Moe — "On the structure of oxide glasses" Glasteknish tidskrift (1956) (2) 1.

17. J. Krogh-Moe — "On the proton conductivity in water" Acta. Chem. Scand. 10 (1956) 331.

18. K. Grjotheim and J. Krogh-Moe — "On the structure of vitreous lithium borates" Kongelige Norske Vidensk. Selskabs forh. 29 (1956) Nr. 6.

19. P. W. Schmidt, J. Krogh-Moe and H. D. Bale — "Small angle X-ray scattering of aluminum hydroxide gel" J. Phys. Chem. 60 (1956) 1589.

20. J. Krogh-Moe — "A method for converting experimental X-ray intensities to an absolute scale" Acta. Cryst. 9 (1956) 951.

21. T. Forland and J. Krogh-Moe — "A remark on some measurements of transference numbers" J. Phys. Chem. 61 (1957) 511.

22. J. Krogh-Moe — "Unit cell and space group for sodium tetraborate $Na_2O \cdot 4B_2O_3$" Acta. Cryst. 10 (1957) 435.

23. T. Forland and J. Krogh-Moe — "The structure of the high temperature modification of lithium sulfate" Acta. Chem. Scand. 11 (1957) 565.

24. T. Forland and J. Krogh-Moe — "Bemerkung zur Struktur von geschmolzenen Li_2SO_4" Z. für Elektrochem. 81 (1957) 1342.

25. J. Krogh-Moe — "Some new compounds in the system cesium oxide-boron oxide" Arkiv f. Kemi 12 Nr. 26 (1958) 247.

26. J. Krogh-Moe — "The infrared spectra of some vitreous and crystalline borates" Arkiv f. Kemi 12 Nr. 41 (1958) 475.

27. J. Krogh-Moe — "An X-ray investigation of lead silicate glass" Z. f. Phys. Chem. 18 (1958) 223.

28. T. Forland and J. Krogh-Moe — "The structure of the high temperature modification of sodium lithium sulfate" Acta. Cryst. 11 (1958) 224.

29. J. Krogh-Moe — "Electron microscope studies of borate glasses" Arkiv f. Kemi 14 (1959) 1.

30. J. Krogh-Moe — "An X-ray study of lithium borate glasses" Arkiv f. Kemi 14 (1959) 31.

31. T. Forland and J. Krogh-Moe — "On the connection between electrical potential and free energy for concentration cells with transference" Acta. Chem. Scand. 13 (1959) 520.

32. T. Forland and J. Krogh-Moe — "Transition point depression and its structural interpretation in the binary system sodium sulfate-potassium sulfate" Acta. Chem. Scand. 13 (1959) 1051.

33. J. Krogh-Moe — "The boron oxide anomaly" Glastechn. Ber. Sonderband. 32K (1959) VI 18.

34. J. Krogh-Moe — "The crystal structures of potassium pentaborate $K_2O \cdot 5B_2O_3$ and the isomorphous rubidium compound" Arkiv f. Kemi 14 (1959) 439.

35. J. Krogh-Moe — "The cation distribution in some crystalline and vitreous casium boartes" Arkiv f. Kemi 14 (1959) 451.

36. J. Krogh-Moe — "The relation of structure to some physical properties of vitreous and molten borates" Arkiv f. Kemi 14 (1959) 553.

37. J. Krogh-Moe — "On the structural relationship of vitreous potassium pentaborate to the crystalline modifications" Arkiv f. Kemi 14 (1959) 567.

38. J. Krogh-Moe "On the structure of boron oxide and alkali borate glasses" Doktoravhandling vid Chalmers Tekniska Högskola (1959).

39. J. Krogh-Moe "On the structure of boron oxide and alkali borate glasses" Phys. Chem. Glasses 1 (1960) 26.

40. J. Krogh-Moe "The transition of cesium chloride" J. Am. Chem. Soc. 82 (1960) 2399.

41. J. Krogh-Moe "A note on the structure of barium tetraborate" Acta. Chem. Scand. 14 (1960) 1229.

42. J. Krogh-Moe "The crystal structure of cesium triborate, $Cs_2O \cdot 3B_2O_3$" Acta. Cryst. 13 (1960) 889.

43. J. Krogh-Moe "Transition point depression of cesium chloride by rubidium chloride" J. Am. Chem. Soc. 82 (1960) 6196.

44. J. Krogh-Moe "Unit cell data for some anhydrous potassium borates" Acta. Cryst. 14 (1961) 68.

45. J. Krogh-Moe "Co-ordination and homogeneity in hydrated boron oxide glasses" Phys. Chem. Glasses 2 (1961) 24.

46. J. Krogh-Moe "New evidence on the boron coordination in alkali borate glasses" Phys. Chem. Glasses 3 (1962) 1.

47. S. E. Svansen, E. Forslind and J. Krogh-Moe "NMR-study of boron coordination in potassium borate glasses" J. Phys. Chem. 66 (1962) 174.

48. J. Krogh-Moe "The crystal structure of lithium diborate, $Li_2O \cdot 2B_2O_3$" Acta. Cryst. 15 (1962) 190.

49. W. D. Hand and J. Krogh-Moe "New data on the system $CdO-B_2O_3$" J. Am. Ceram. Soc. 45 (1962) 197.

50. J. Krogh-Moe "The KBr briquette method for studying the infrared spectra of anhydr.borates" Phys. Chem. Glasses 3 (1962) 61.

51. J. Krogh-Moe "Structural interpretation of melting point depression in the sodium borate system" Phys. Chem. Glasses 3 (1962) 101.

52. J. Krogh-Moe "The infrared spectra and the structure of some anhydrous zinc borates" Z. f. Kristallogr. 117 (1962) 166.

53. J. Krogh-Moe "An X-ray study of barium borate glass" Phys. Chem. Glasses 3 (1962) 208.

54. J. Krogh-Moe "Energy and length of the boron-oxygen bond" Acta. Chem. Scand. 17 (1963) 843.

55. J. Krogh-Moe "Crosslinking of borate polymer chains with silicates" J. Am. Ceram. Soc. 47 (1964) 307.

56. J. Krogh-Moe "The crystal structure of strontium diborate, $SrO \cdot 2B_2O_3$" Acta. Chem. Scand. 18 (1964) 2055.

57. J. Krogh-Moe "The crystal structure of silver tetraborate $Ag_2O \cdot 4B_2O_3$" Acta. Cryst. 18 (1965) 77.

58. J. Krogh-Moe "Interpretation of the infrared spectra of boron oxide and alkali borate glasses" Phys. Chem. Glasses 6 (1965) 46.

59. J. Krogh-Moe — "Least squares refinement of the crystal structure of potassium pentaborate" Acta. Cryst. 18 (1965) 1088.

60. J. Krogh-Moe and H. Jürine — "An X-ray study of thallium borate glasses" Phys. Chem. Glasses 6 (1965) 30.

61. M. Ihara and J. Krogh-Moe — "The crystal structure of cadmium diborate, $CdO \cdot 2B_2O_3$" Acta. Cryst. 20 (1966) 132.

62. J. Krogh-Moe — "Crystalline salts with a cubic body-centered defect structure". In: "selected topics in high temperature chemistry". Universitetsforlaget, Oslo (1966).

63. J. Krogh-Moe — "The structure of glass" Glass Industry 47 (1966) 306.

64. J. Krogh-Moe — "A method for the resolution of composite radial pair distribution functions" Acta. Chem. Scand. 20 (1966) 2890.

65. J. Krogh Moe and M. Ihara — "The crystal structure of caesium enneaborate $Cs_2O \cdot 9B_2O_3$" Acta. Cryst. 23 (1967) 427.

66. J. Krogh-Moe — "A note on the structure of pinneite" Acta. Cryst. 23 (1967) 500.

67. J. Krogh-Moe M. Vikan and C. Krohn — "K_2BaCl_4, another case of extreme ionic conductivity in a solid" Acta. Chem. Scand. 21 (1967) 309.

68. J. Krogh-Moe — "Refinement of the crystal structure of lithium diborate, $Li_2O \cdot 2B_2O_3$" Acta. Cryst. B24 (1968) 179.

69. L. A. Kristiansen and J. Krogh-Moe — "Vibrational assignment and valence force field of the boroxol skeleton" Phys. Chem. Glasses 9 (1968) 96.

70. K. Grjotheim H. G. Nebell and J. Krogh-Moe — "The solution of alkaline earth metals in their molten halides. I. On the phase diagram of the system barium-barium chloride" Acta. Chem. Scand. 22 (1968) 1159.

71. T. Forland T. Ostvold and J. Krogh-Moe — "Monte Carlo studies on fused salts. I. Calculations for a two-dimensional ionic model liquid" Acta. Chem. Scand. 22 (1968) 2415.

72. J. Krogh-Moe T. Ostvold and T. Forland — "Monte Carlo studies on fused salts. II. Calculations on a model of fused lithium chloride at 1073°K" Acta. Chem. Scand. 23 (1969) 2421.

73. J. Krogh-Moe — "The structure of vitreous and liquid boron oxide" J. Non-cryst. Solids 1 (1969) 269.

74. J. Krogh-Moe and M. Ihara — "On the crystal structure of barium tetraborate, $BaO \cdot 4B_2O_3$" Acta. Cryst. B25 (1969) 2153.

75. K. Grjotheim H. Ikeuchi H. G. Nebell and J. Krogh-Moe — "The solution of alkaline earth metals in their molten halides. II. The magnetic susceptibility of melts in the system barium-barium chloride" Acta. Chem. Scand. 24 (1970) 985.

76. S. Lyng J. Markali J. Krogh-Moe and N. H. Lundberg — "On the crystallization in aluminosilicate glasses containing fluoride and magnesia" Phys. Chem. Glasses 11 (1970) 6.

77. K. Jenssen and J. Krogh-Moe — "A modified design of a metal ribbon furnace for high-temperature X-ray diffraction" J. Appl. Cryst. 4 (1971) 334.

78. K. Grjotheim, H. Ikeuchi, S. Dhabanandana and J. Krogh-Moe — "The solution of alkaline earth metals in their molten halides. III. The densities of melts in the systems barium-barium chloride, barium-barium bromide and strontium-strontium chloride" Acta. Chem. Scand. 25 (1971) 3415.

79. K. Grjotheim, H. Ikeuchi, J. Krogh-Moe and Z. Moser — "The solution of alkaline earth metals in their molten halides. IV. EMF measurements of concentration cells containing solutions of barium metal in fused barium chloride" Kungl. Tekniska Hogskolans Handlinger, Stockholm (1972) nr 295.

80. J. Krogh-Moe — "The crystal structure of the high-temperature modification of potassium pentaborate" Acta. Cryst. B28 (1972) 168.

81. J. Krogh-Moe — "The crystal structure of a sodium triborate modification, β-$Na_2O \cdot 3B_2O_3$" Acta. Cryst. B28 (1972) 1571.

82. A. H. Schulz, B. Bieker and J. Krogh-Moe — "Phase equilibrium in the system BaF_2-AlF_3" Acta. Chem. Scand. 26 (1972) 2623.

83. K. Grjotheim, S. Dhabanandana and J. Krogh-Moe — "The solution of alkaline earth metals in their molten halides. V. The magnetic susceptibility of melts in the system strontium-strontium chloride" Acta. Chem. Scand. 26 (1972) 3427.

84. J. Krogh-Moe — "The crystal structure of potassium diborate, $K_2O \cdot 2B_2O_3$" Acta. Cryst. B26 (1972) 3089.

85. E. Rytter, B.E.D. Rytter, H. A. Oye and J. Krogh-Moe — "The crystal structure of potassium heptabromo-dialuminate, KAl_2Br_7" Acta. Cryst. B29 (1973) 1541.

86. J. Krogh-Moe, P.S. Wold-Hansen — "The crystal structure of hexalead pentaborate, $6PbO \cdot 5B_2O_3$" Acta. Cryst. B29 (1973) 2242.

87. J. Krogh-Moe — "The crystal structure of sodium diborate, $Na_2O \cdot 2B_2O_3$" Acta. Cryst. B30 (1974) 578.

88. J. Krogh-Moe — "The crystal structure of alpha sodium triborate, α-$Na_2O \cdot 3B_2O_3$" Acta. Cryst. B30 (1974) 747.

89. J. Krogh-Moe — "Refinement of the structure of caesium triborate, $Cs_2O \cdot 3B_2O_3$" Acta. Cryst. B30 (1974) 1178.

90. J. Krogh-Moe — "The crystal structure of pentapotassium ennea-kaidekaborate, $5K_2O \cdot 19B_2O_3$" Acta. Cryst. B30 (1974) 1827.

91. E. Rytter, B. E. Rytter, H. A. Oye and J. Krogh-Moe — "The crystal structure of ammonium heptabrom-dialuminate, $NH_4Al_2Br_7$" Klar for publisering i Acta. Cryst.

92. J. Krogh-Moe and K. Jenssen — "The crystal structure of beta caesium ennea-borate, β-$Cs_2O \cdot 9B_2O_3$" Forelopig ikke publisert

BORATE GLASS STRUCTURE

David L. Griscom

U. S. Naval Research Laboratory
Code 644 3
Washington, D. C. 20375

INTRODUCTION

Early in 1975, Norbert Kreidl approached me for my opinion on the topic and novel format which he was considering for the Fifth "Rolla" Ceramic Materials Conference.* The theme was to be borate glass structures, and part of the motivation was to produce a volume of original papers on the subject which could be dedicated as a "festschrift" in memory of Prof. Jan Krogh-Moe. The format he had in mind was to commission "a one-author monograph draft which would be distributed to, and exposed before a (the V Rolla) Conference." He further envisioned that "short contributed papers and extensive discussion would be tied to the monograph draft which subsequently would be edited by its author for publication together with the conference proceedings." I was then and I remain tremendously impressed by this concept, though I initially balked at Norbert's suggestion that I be the author of the central paper--which surely will be taken under fire by all of the true experts in each of the relevant subfields. Nevertheless, Norbert refused to accept "no"

---

*Renamed - The Alfred Conference on "Boron in Glass and Glass Ceramics," 10/1/76.

for an answer, and accordingly I have assembled the attached material for the consideration of the Conference attendees.

With regard to my personal style and point of view in this writeup, a few words of explanation are in order. First, I have been instructed that the present draft is to be informal and will not be published in its present form. Taking advantage of this freedom, I have arranged this manuscript in a rather unorthodox form, beginning with a somewhat editoralized chronology in the chapter titled, "Milestones." Before the reader begins to trace his finger down the list to see if his own work has been properly recorded, I should explain that I had no intention of making this tabulation exhaustive. Much good work has been excluded on the basis of some arbitrary criteria which I shall describe. These criterea fall into three classes: (1) those concerning glass composition, (2) those concerning types of structural information, and (3) those concerning experimental techniques. As regards composition, studies of pure B_2O_3 glass are given top weighting--with the alkali borate and isomophous silver borate systems following next on my (arbitrary) scale of emphasis. I have decided not to list studies of hydrated borates except in a few cases where such studies have a direct bearing on the structure of anhydrous borate glasses. Turning to the question of structural information, those types of data which shed light on the boron-oxgen framework are considered "king;" data concerning modifier coordination are rated one notch down, and many excellent works dealing with modifier diffusion, mixed-alkali effects, etc. are regrettably excluded. The criterion regarding experimental techniques is simple: Greatest consideration is given to those techniques which yield relatively unambiguous information regarding glass structure <u>on a molecular scale</u> (~1 - 20Å). Bulk property measurements which have not yet been connected in any direct or obvious way with glass structure are for this reason largely left out. Regardless of technique, comparative studies of both glasses and compounds of a given glass-forming system invariably result in more structural information than similar investigations of the glasses alone. These criteria are illustrated schematically in Fig. 1.

My first idea in preparing the list of "Milestones" was to keep it short and confined only to the most stunning breakthroughs. But after much agonizing, I began to include minor breakthroughs and important supplemental work as well. This process eventually snowballed to the point where I have now listed no less than 75 "events" covering 100 references. In order that the major landmarks should not become lost in this panorama, I opted for the "Michelin Guide" system of assigning stars in the lefthand margin to those accomplishments which I personally consider to be the most important. Two stars is the maximum that I went to, and there turn out to be 9 achievements to which I have assigned this highest rating. I cannot emphasize too strongly that I have awarded stars on the basis of the criteria which I have defined above and not on the basis of some abstract notion of "quality." I believe all of the works that I have listed are of the highest quality and many are of considerable importance from other points of view besides that of glass structure. The same can be said for most of the papers I consciously excluded (not to mention a number of which I am undoubtedly unaware). Stars usually represent some important "first," and two stars generally are given to novel works which provide in relatively unambiguous form some badly needed or long awaited piece of structural information; papers dealing with a wide range of glass compositions naturally tend to rate higher than those treating just a few. It should be clear that many entries on this list could only have been made with the benefit of hindsight. For example, Krogh-Moe's structural determinations for some seemingly obscure borate compounds would not have been listed, much less rated a star, 15 years ago at a time predating the recent torrent of data supporting the presence of structural units of these compounds in glasses of similar compositions. On the other side of the coin, structural determinations for compounds

(a) Relative emphasis on glass composition

max

pure B_2O_3

alkali and silver borates

alkaline-earth, lead, and other binary borates

important ternary systems, especially the borosilicates

min

other ternary, quaternary and more complex glasses, hydrated glasses

(b) Relative emphasis on glass structure

max

boron-oxygen network

modifier cation-cation separations

min

modifier diffusion, mixed alkali effects, other

(c) Relative emphasis on experimental techniques

max

techniques which yield maximum structural information

min

techniques which do not yield information readily related to glass structure

Figure 1. Criteria used in this paper for determining the relative emphasis to be accorded various published studies of the structure and properties of borate glasses. These criteria are arbitrary and have been imposed as a means of reducing the scope of this review to manageable proportions.

now known to bear little structural relationship to the glasses are not listed unless they should be of considerable academic interest.

It is reasonable to expect some disagreement as to which works should and should not have been included and which of these should and should not have been starred. Indeed, some may see little value in even attempting to produce such a chronology. The advantage of the present conference format as conceived by Prof. Kreidl is that my preliminary ideas concerning the master document can be read and criticized in advance of the actual meeting. If my "Milestones" chapter should be deemed at all useful, there should be ample opportunity to draw up a revised version which will more closely represent a concensus of the experts in attendance.

In addition to the "Milestones," the present draft comprises further chapters discussing nomenclature, the "boron-oxide anomaly," phase separation, and the information which can be derived from a selected number of the more structure-sensitive experimental techniques. The actual order in which the various experimental chapters appear is partially arbitrary, although I have made an attempt to begin with those techniques which have yielded the most or the best structural information. Here again I have exercised some judgements which may be disputed at the conference. Perhaps there will be little argument with my view that x-ray diffraction and nuclear magnetic resonance (NMR) should receive top billing; I've placed the x-ray work first mainly out of historical considerations. The ordering and content of the remaining chapters is apt to be more controversial. As an example, one may consider the practice of using defects as "probes" of glass structure. This is certainly an idea with a great deal of potential. However, to be more specific, if one uses optical techniques alone to study radiation-induced defects, it is unlikely that he can unambiguously identify the defects much less their relationship to glass structure. If electron spin resonance (ESR) is brought into the picture, the possibility of making an impeccable identification of the defect exists providing the proper concepts and mathematical tools can be mobilized. But even assuming this is done, it remains to show in what way the defect structure is sensitive to the structure of the "perfect" glass. Parallel ESR studies of both the glasses and polycrystalline compounds of a given system may resolve the issue, but single-crystal studies would be preferable whereas none have yet been carried out. The situation worsens when paramagnetic transition-group ions are used as "probes." Much evidence suggests that such ions dictate their own environments to an equal or greater degree than does the glass itself. Although there may exist individual exceptions to this rule (see Chapt. XIII), the value of ESR and optical studies of foreign ions in borate glasses as structure-sensitive tools is downgraded in this draft and no such investigations have been listed as "Milestones." This in no way implies a belief that such downgraded areas of research are at a deadend; on the contrary, a conference which may be held 15 years from now might well include with stars some of the very works I have deliberately excluded--this is the meaning of my earlier reference to hindsight. The present examples could be further elaborated and others given, but my main point here is simply to re-emphasize that unambiguous information relating to the nature of the boron-oxygen skeleton in pure borate glasses has been the overriding criterion in preparing the present draft. Every experimental technique to be discussed will be criticized in this light.

Finally, it is appropriate to remark upon the dual motivations of the 5th Rolla Conference, namely, to summarize the present state of our understanding of borate glass structures and to commemorate Jan Krogh-Moe. On the surface this might appear to be a mild conflict of interests, but on reviewing the literature I have personally reached the conclusion that these two themes represent a nearly ideal wedding of purposes. I don't believe that I have been biased in the large number of listings and stars which I have accorded to Krogh-Moe's works in the

"Milestones" chapter. But even assuming that I have erred in this direction, there is absolutely no question that Krogh-Moe was the chief advocate, if not sole author, of the basic structural theory of borate glasses which appears now to be accepted in some measure by virtually all workers in the field. The theory has not been simple to prove and Krogh-Moe had himself emphasized time and again that many forms of evidence are necessary to support each aspect of the theory. I think that the "Milestones" tabulation clearly shows that in the span of about 12 years beginning in 1958 it was Krogh-Moe alone who provided, or brought together for the first time, virtually all of the evidence upon which the theory originally rested. Recent Raman and NMR studies tend to prove its correctness. This is not to say that the "book" on borate glass structures is now closed. Interesting new "twists" are discovered, it seems, almost daily. The remarkable fact is that many of these "twists" (three-coordinated oxygens, for one example) were already anticipated by Krogh-Moe as a consequence of his vast familiarity with the structures of borate compounds.

I. MILESTONES

★1932 Zachariasen's classic paper[1] introduces the "random network theory", destined to exert a strong influence on most theories of borate glass structure for at least four decades.

1934 Gooding and Turner[2] report an "anomalous" minimum near 15 mol% Na_2O in the thermal expansion coefficient of sodium borate glasses.

1935 Hägg[3], dissenting with Zachariasen, publishes a view suggesting the possible importance of larger structural groupings in glass and presaging the later theories due to Krogh-Moe.

★1936 Warren, Krutter, and Morningstar[4] give X-ray evidence for BO_3 triangles in B_2O_3 glass.

1936 Morey and Merwin report a careful study[5] of phase relationships in the system sodium oxide-boron oxide.

★1937 Zachariasen[6] determines the crystal structure of potassium metaborate, finding all borons to be trigonally coordinated -- a fact which strongly influences subsequent theories of the compositional dependence of the fraction of borons in 4-coordination in alkali borate glasses.

★1938 Biscoe and Warren[7] present a theory explaining the "borate anomaly" in terms of boron coordination changes, as supported by an analysis of their X-ray diffraction data for B_2O_3 glass and a series of sodium borate glasses.

1938 Kracek, Morey, and Merwin[8] prepare crystalline B_2O_3.

★1938 Hibben[9] notes that the Raman spectrum of borax ($Na_2O \cdot 2B_2O_3 \cdot 10H_2O$) is very similar to that of sodium diborate glass. His conclusion that the structures of these two phases must be related is evidently forgotten until Krogh-Moe (1958) reaches the same conclusion based on his infrared studies.

1942 Green[10] carries out X-ray diffraction and other studies of a series of potassium borate glasses. His conclusions are generally in accord with those of Biscoe and Warren (1938).

1942 Warren[11] presents a geometrical explanation of the "borate anomaly".

1952 Abe[12] proposes a specific set of rules to describe the structure of borate glasses. These rules and several others subsequently proposed[13,14] are eventually disproved by NMR measurements.

1952 Fajans and Barber,[15] remarking on the extremely low viscosity of boron oxide melts relative to silica, propose a structure of B_2O_3 glass composed of B_4O_6 molecular units. NMR data subsequently militate against this model of glass structure, while alternative explanations of the melt viscosity are proposed (e.g. Mackenzie 1959).[16]

★1953 Goubeau and Keller[17] give evidence from Raman spectroscopy for the existence of boroxol groups in B_2O_3 glass.

★1953 Willis and Hennessy[18] discuss the activity of silver oxide in silver borate melts, demonstrating that Ag_2O enters the network as an entity $Ag_2O \cdot nB_2O_3$.

1953 Shartsis, Capps, and Spinner[19] carry out extensive bulk physical property studies of alkali borate glasses.

1953 Berger[20] publishes a crystal structure of boron oxide, initiating a period of erroneous belief that boron may be tetrahedrally coordinated in various forms of B_2O_3.

1954 Borgen, Grjotheim, and Krogh-Moe[21] caution against reliance on X-ray diffraction data for inferring coordination numbers in borate glasses, pointing out a number of sources of substantial errors in earlier works.

1955 Anderson, Bohon, and Kimpton[22] attempt to assign infrared absorption bands of alkali borate glasses to fundamental modes of vibration.

1955 Jellyman and Procter[23] study infrared reflectance spectra of some alkali borate glasses.

1956 Lotkova, Obukov-Denisov and Sobolev[24] publish a complete Raman spectrum of boron oxide glass.

1958 Sidorov and Sobolev[25] publish an infrared study of boron oxide glass. Subsequent works by Parsons and Milberg (1960)[26] and Borrelli _et al_ (1963)[27] confirm and extend the data of Sidorov and Sobolev.

★ 1958 Silver and Bray[28] demonstrate by NMR that boron does change from 3- to 4-coordination upon addition of alkali oxide to a B_2O_3 melt and that this coordination change continues to a higher alkali content than predicted by Warren and others.

1958 Skatula, Vogel and Wessel[29] give evidence for phase separation in borate glasses.

1958 Sastry and Hummel[30] further explore phase relationships in the lithium borate system, cataloging X-ray powder diffraction spectra for a large number of polycrystalline compounds.

★ 1958 Krogh-Moe[31] demonstrates that infrared data indicate a structural relationship between some borate crystalline and glass phases of the same composition.

★ 1959 Krogh-Moe[32] performs the first structural determination of a complex anhydrous borate structure having >50 mole % B_2O_3, viz, $K_2O\cdot 5B_2O_3$.

1959 Krogh-Moe[33] gives first X-ray evidence for non-random distribution of modifier cations (Cs^+) in borate glasses. Subsequent works by Krogh-Moe (1962)[34], Block and Piermarini (1964)[35], and Krogh-Moe and Jürine (1965)[36] provide similar evidence in the cases of Ba^{++}, Sr^+ and Tl^+.

1959 Adams and Douglas[37] attempt to assign specific infrared peaks to fundamental vibrations of BO_3 and BO_4 units. (Krogh-Moe (1958) cautions, however, that the majority of bands in alkali borate compounds cannot be looked upon as characteristic of simple BO_3 triangles or BO_4 tetrahedra).

1960 Parsons[38] discusses in detail the vibrational spectrum of orthorhombic metaboric acid, making use of Raman data obtained for various boron isotopic ratios.

1960 Krogh-Moe, in a review article[39] on the structure of borate glasses, suggests a mechanism of continuous coordination change to explain the low viscosity of B_2O_3 melts relative to that of SiO_2. Although somewhat influenced by Berger's (1953) erroneous structural determination for crystalline B_2O_3, this model remains plausible today.

1960 Edwards and Ross[40] propose a rule for the fraction of borons in 4-coordination in hydrated borates, $x\, Me_2O \cdot (1-x)\, B_2O_3 \cdot yH_2O$:

$$N_{BO_4} = \frac{x}{x-1} \quad \text{for} \quad x \leq 0.5$$

1961 Bray, Edwards, O'Keefe, Ross and Tatsuzaki[41] examine and classify the NMR quadrupole coupling constants for simple trigonal BO_3 units and tetrahedral BO_4 groups in a host of crystalline borates. By comparison with these, boron coordination in a number of polyborates and glasses is inferred and the basis is laid for future NMR studies of borate glasses.

1962 Svanson, Forslind, and Krogh-Moe[42] carry out NMR studies of a series of potassium borate glasses and using a method of data analysis superior to that of Silver and Bray (1958) show that N_{BO_4} closely follows the theoretical relation $x/(1-x)$ up to 30 mol % K_2O. Similarities in the NMR spectra of glassy and crystalline B_2O_3 are also noted, suggesting similar boron oxygen configurations in both.

★1962 Krogh-Moe elucidates the crystal structure of lithium diborate[43] and based on infrared data, postulates the importance of this structural grouping in borate glasses[44]. In particular, he concludes that Edwards and Ross' rule (1960) for boron coordination should be equally valid for anhydrous borate compounds and glasses containing $\leq$ 30 mol % modifier oxide.

★★1962 Krogh-Moe's studies[45] of melting point depression provide strong new evidence for the prime importance of boroxol, tetraborate, and diborate groupings in glasses of the sodium borate system.

1962 Becherer, Brümmer, and Herms[46] published improved X-ray diffraction data on sodium borate glasses, showing an increase in N_{BO_4} up to at least 26 mol % Na_2O in agreement with the work of Biscoe and Warren (1938) and taking into account the objections of Borgen, Grjotheim and Krogh-Moe (1954). As was the case for Biscoe and Warren's work, the estimated N_{BO_4} values are much higher than either the theoretical maximum $(\frac{x}{x-1})$ or the more reliable NMR data of Bray and O'Keefe (1963).

1963 Mc Swain, Borelli and Su[47] study the displacement of the UV absorption edge of alkali borate glasses as a function of alkali oxide content. A sharp increase in absorption in the composition range $x = 0.15 \rightarrow 0.20$ is attributed to nonbridging oxygen (NBO) formation. This conclusion is not supported by NMR measurements (e.g. Bray and O'Keefe, 1963) although an NBO content $\sim$1 mol % could cause the edge shift while escaping NMR detection.

1963 Kumar[48] using the free-volume theory for viscosity calculates the average size of a flow unit in sodium borate melts for $x \approx 0.33$ to be approximately that of the diborate group.

★★1963 Bray and O'Keefe[49] publish N_{BO_4} determinations for the entire glass forming ranges of all five alkali borate systems. NMR data analysis is improved to a new high standard, and deviations from the "ideal" $N_{BO_4} = \frac{x}{x-1}$ behavior are noted above $x \approx 0.3$, but N_{BO_4} is found not to return to zero until $x \approx 0.7$.

A companion paper by Bray, Leventhal and Hooper[50] dealing with lead borate glasses shows N_{BO_4} to depart from the "ideal" curve about 20 mole % PbO and gives evidence for Pb entering the network at high lead contents.

1963 Lee and Bray[51] show that the electron spin resonance (ESR) spectra of irradiated borate glasses can be attributed to holes undergoing a hyperfine interaction with boron. A follow-on paper[52] by the same authors shows that these spectra are sensitive to glass composition.

1964 Krogh-Moe's crystal structure determination[53] for strontium diborate provides the first example of a borate compound of x <0.5 which violates Edwards and Ross' 1960 rule. Subsequent studies by Park and Bray (1972) confirm that all borons in this compound are 4-coordinated.

★1965 Krogh-Moe[54] determines the crystal structure of silver tetraborate and demonstrates a likely interrelationship with the structures of sodium tetraborate compound and glass. Hyman, Perloff, Mauer, and Block (1967)[55] subsequently determine the structure of sodium tetraborate, confirming its similarity to the silver compound.

1965 Beekenkamp[56] proposes a modification of Abe's (1952) rules for the structure of borate glasses, essentially sticking with Zachariasen's "random network model" even as further evidence is being amassed by Krogh-Moe for the importance of larger groupings.

★★1965 Krogh-Moe[57] demonstrates that the characteristic vibrations of fundamental polyborate groups can be identified in the infrared spectra of hydrated borate compounds and from there traced to the corresponding anhydrous compounds and finally to alkali borate glasses of similar compositions, despite the broadening effects of polymerization and vitreous disorder. A similar comparison with the infrared spectrum of orthorhombic metaboric acid supports the conclusion of Goubeau and Keller (1953) that B_2O_3 glass contains mainly boroxol groups. These results lend strong new support to the group model of borate structure advocated by Krogh-Moe since 1958.

1965 Leventhal and Bray[58] present NMR data for lead borate crystalline compounds which, when compared with similar data for the glasses, are in accord with but do not prove, the theory of borate glass structure advanced by Krogh-Moe.

1966 Bishay and Maklad[59] lend some support to Krogh-Moe's concept of lead-borate glass structure, based on changes in radiation induced optical bands at compositions corresponding to crystalline compounds.

1966 Bishop and Bray[60] carry out NMR investigations of a tenary borate glass system, the calcium boroaluminates (CaBAl). Contrary to previous belief, the evidence shows that the 4-coordination of Al and the 4-coordination of B are processes which compete on nearly equal terms for the additional oxygens introduced with the modifier cations.

1966 Bray, Kline, and Poch[61] demonstrate that application of pressure (25 kbar) sufficient to produce permanent densification does not produce coordination changes 3 → 4 in B_2O_3 glass, contrary to some previous suggestions based on Berger's (1953) structural determination for crystalline B_2O_3.

1967 Zarzycki and Naudin[62] study the kinetics of metastable phase separation in the PbO-B_2O_3 system by small-angle X-ray scattering.

1967 Riebling[63] concludes from volume relations in sodium-borate melts at 1300°C that, at 40 mol % Na_2O, N_{BO_4} $\sim$0.5 -- contrary to previous suggestions that BO_4 tetrahedra may be unstable at high temperatures. Nagel and Bergeron (1974)[64] subsequently obtain similar results for the composition $Na_2O \cdot B_2O_3$ at temperatures near the transition range.

1968 Strong and Kaplow[65] show by X-ray diffraction studies that all borons in crystalline B_2O_3 are in planer trigonal configurations, thereby correcting a misimpression that had persisted since the 1953 publication by Berger. Gurr, Montgomery, Knutson, Gorres (1970)[66], studying a single-crystal sample, subsequently correct the structure of Strong and Kaplow although upholding the basic planar trigonal configuration.

1968 Kline, Bray and Kriz[67] simultaneously with Strong and Kaplow's X-ray work, show by NMR that crystalline B_2O_3 cannot have the structure proposed by Berger (1953).

★1968 Kriz, Bishop and Bray [68]demonstrate from a study of polycrystalline calcium metaborate that NMR is capable of distinguishing BO_3 units with one nonbridging oxygen from BO_3 units with three bridging oxygens and from tetrahedral BO_4 units. Kriz, Park and Bray (1971)[69] subsequently expand on the generality of this result, noting however that BO_3 units with one or two nonbridging oxygens cannot be distinguished one from another.

1968 Griscom, Taylor, Ware and Bray[70] execute an ESR study comparing radiation induced trapped-hole centers in lithium borate glasses with analogous defects in the polycrystalline compounds of the system, thereby providing a new form of support for Krogh-Moe's conclusion that tetraborate groups are the most important elements in binary borate glasses containing ∿20 mol. % modifier oxide. Computer simulation of the glass spectra provide the first explicit evidence for statistical distributions of spin Hamiltonian parameters deriving from vitreous disorder.

1968 Kristiansen and Krogh-Moe[71] calculate the vibrational spectra of boroxol groups with different groups attached and make comparisons with observed IR and Raman spectra.

1968 Shaw and Uhlmann[72] demonstrate the existence of metastable subliquidus miscibility gaps in all five alkali borate systems, thereby bringing to light a possible complication in relating physical and chemical properties of these glasses to glass structure.

★1969 Uhlmann and Shaw[73] show conclusively that the oft-quoted "borate anomaly" in the compositional variation of the thermal expansion coefficient of alkali borate glasses is not anomalous at all but instead presents a broad flat minimum over a wide composition range which is amenable to simple explanation.

★1969 Baugher and Bray[74] studying thalium borate glasses, produce the first evidence for a glass system in which N_{BO_4} unequivocally exceeds the theoretical maximum value of $\frac{x}{x-1}$. It is inferred that these glasses must contain 3-coordinated oxygens, thus violating one of Zachariasen's 1932 rules for oxide glass formation.

1969 DeWaal[75] interprets his internal friction data for CaBAl and NaBAl glasses in terms of nonbridging oxygen formation as alkali disrupts an Al_2O_3-B_2O_3 network. Subsequently, Boulos and Kreidl (1971) draw a similar conclusion in the case of silver borate glasses containing >30 mol. % Ag_2O.

1969 Krogh-Moe, in a compelling review paper[76], summarizes a great deal of diverse evidence for the existence of important numbers of boroxol groups in boron oxide glass.

★★1970 Mozzi and Warren[77], performing X-ray diffraction studies by the fluoresence excitation method and eliminating the "proportionality of scattering factors" approximation, confirm the prevalence of boroxol groups in B_2O_3 glass, in agreement with a great deal of previous evidence summarized in Krogh-Moe's (1969) review article.

★★1970 Taylor and Bray publish details[78] of their method for computer simulating the magnetic resonance spectra of glasses. In particular, it is shown how simulation of NMR spectra as a superposition of three well-definied components can lead to an acurrate measure not only of N_{BO_4} but also of $N^B{}_{BO_3}$ and $N^{NB}{}_{BO_3}$, the fractions of boron in 3-coordination with all bridging and one nonbridging oxygens, respectively. Taylor and Friebele (1974)[79] later expand on the theme of the uniqueness of these boron sites, countering a suggestion by Peterson, Kurkjian and Carnevale (1974)[80] that the data might be equally consistent with a "random coordination" model.

1971 Kriz and Bray[81], by means of computer line shape simulations, refine the information derivable from ^{11}B NMR of B_2O_3 glass. Planer trigonal BO_3 units with three bridging oxygens are concluded. An estimated distribution in quadrupole coupling constants due to vitreous disorder turns out to be smaller than earlier predicted.

1971 Griscom[82], in an ESR study uses an intrinsic trapped electron defect as a probe of the structure of Li, Na and K borate glasses, noting a rather sharp break in the boron hyperfine coupling constant between 15 and 25 mol. % alkali oxide which is best interpreted in terms of Krogh-Moe's structural model for these glasses.

★★1971 Rhee and Bray[83] studying caesium borates, perform the first correlated NMR study of glasses and crystalline compounds in an alkali borate system. Agreement in quadrupolar coupling constants and N_{BO_4} values over a wide range of compound and glass compositions lends strong, if indirect, support to Krogh-Moe's theory for the existence in alkali borate glasses of structural groupings characteristic of the compounds. Essentially identical conclusions in this regard are reached by Rhee (1971)[84] and Kim and Bray (1974) for the sodium borate and silver borate systems, respectively. In particular, these studies offer strong evidence for the pairing of BO_4 tetrahedra (as in diborate groups) in glasses containing ∿30-40 mol. % modifier oxide, in contradiction to the first of Abe's 1952 rules as well as the rules proposed by Beekenkamp (1965).

1971 Taylor and Griscom[85] show by computer simulation of the ESR spectrum that a radiation-induced defect center in lithium and strontium diborate compounds must be a hole trapped on an oxygen bridging between two BO_4 units. Soon after, Taylor and Bray (1972)[86] show by computer simulation of the spectra of irradiated lithium borate glasses that the same defect is likely to be present in glasses near the diborate composition, in complete agreement with the NMR evidence of Rhee and Bray (1971) and the earlier evidence mustered by Krogh-Moe.

★1971 Boulos and Kreidl[87] publish their multifaceted study of structure and properties of silver borate glasses. Additional support for Krogh-Moe's structural model is provided, while internal friction data provide evidence for nonbridging oxygen formation above 30 mol. % Ag_2O.

★1972 Milberg, O'Keefe, Verhelst, and Hooper[88] publish NMR studies of boron coordination in a technologically important ternary system, the sodium borosilicates. It is shown that N_{BO_4} can approach unity when the soda:boron-oxide ratio exceeds 1.0 and the silica content is sufficiently high. Similar conclusions were reached nearly simultaneously by Zvyagin, Kalinin, Kaplun and Shevelevich (1971)[89] and by Scheerer, Müller-Warmuth and Dutz (1973)[90]. Milberg et al find the effect of SiO_2 on N_{BO_4} to be independent of phase separation, while Scheerer et al give evidence that on the average borate polymers do not occur in the as-quenched glasses of high SiO_2 content.

★1972 Park and Bray[91] publish their extensive NMR study of glasses and compounds of the strontium borate system. Among their numerous conclusions is the observation that structural units of the diborate compound do not appear in substantial numbers in the glass of the same composition, in accord with the possibility anticipated by Krogh-Moe (1962) that thermodynamic and kinetic considerations may favor the appearance of some structural groupings while precluding others.

★1974 Kim and Bray's conclusions from their NMR studies[92] of glasses and compounds of the silver borate system are in perfect accord with Krogh-Moe's (1962) findings for the sodium borate system, i.e., supporting the prime importance of boroxol, tetraborate, and diborate groupings. Some evidence agrees with the presence of nonbridging oxygens above 30 mol. % Ag_2O, tending to cooroborate the internal friction work of Boulos and Kreidl (1971).

1974 Kim and Bray[93] complete their studies of the MgO-Na_2O-B_2O_3 system, finding among other things that magnesium is evidently incorporated into the network for $\gtrsim$ 15 mol. % MgO. This appears to be another exception to the rules of Zachariasen (1932).

★★1974 Bril[94] undertakes Raman investigations of a number of metaborate compounds and alkali borate glasses, carrying out a normal coordinate analysis of the compound spectra which permits an interpretation of the characteristic lines at 770 cm^{-1} and 806 cm^{-1} in the glass spectra in terms of Krogh-Moe's structural model. The calculations for the metaborate group give results superior to those of Kristiansen and Krogh-Moe (1968), who performed their analyses in the absence of Raman data. New Raman data obtained as a function of temperature show that structural groupings are unchanged in passing from the glass to the melt, thus substantiating a basic premise of Krogh-Moe's (1962) study of glass structure via melting point depression.

★★1975 Konijnendijk and Stevels[95,96] unveil the results of their comprehensive Raman scattering and IR investigations of borate and borosilicate glasses. Their conclusions vis-à-vis the presence of boroxol, tetraborate and diborate groups in alkali borate glasses are in total accord with Krogh-Moe's (1962) predictions based on melting-point depression studies. The Raman data can be interpreted as indicating that small numbers of "loose" BO_3 and BO_4 units are also present in the glasses (which is analogous to Krogh-Moe's (1974)[97] structural determination for crystalline $K_2O \cdot 3.8\ B_2O_3$). Conclusions based on Raman spectroscopy are also in agreement with the NMR work of Park and Bray (1972) to the effect that structural groupings of the $SrO \cdot 2B_2O_3$ compound are not present in the glass of that composition. However, there is a possible disagreement between the Raman and NMR (Milberg et al, 1972; Scheerer et al, 1973) measurements on borosilicate glasses of high silica content; the former data suggest the presence of agglomerated ring-type metaborate groups, while the latter show the presence mostly of "loose" BO_4 tetrahedra.

1975 Jenssen and Krogh-Moe[98] study the crystal structure of β-$Cs_2O \cdot 9B_2O_3$, finding evidence for structural disorder possibly relating to the relatively low energy required for boron to switch from 3- to 4-coordination. This provides some indirect support for Krogh-Moe's (1960) hypothesis explaining the low viscosity of molten B_2O_3.

★1976 Snyder, Peterson, and Kurkjian[99] perform the first ab initio theoretical calculations of boron quadrupole coupling constants for trigonally coordinated borons. Strong evidence is given that out-of-plane distortions of BO_3 units are probably not responsible for the observed[81] distortion of quadrupole coupling constants in B_2O_3 glass. This demonstrates a weakness in the Townes and Dailey theory as employed by Taylor and Friebele[79] (but does not affect the latter authors' demonstration of the uniqueness of BO_3 and BO_4 sites in glass).

★★1976/77 Panek, Jellison and Bray[100] carry out ^{17}O, ^{10}B, and ^{11}B NMR studies of glassy B_2O_3, finding explicit evidence for the presence of boroxol groups as manifested in two distinct oxygen sites. A means of independently determining distributions in boron quadrupole coupling constants and asymmetry parameters is also presented, and the physical basis for previous ^{11}B NMR measurements of N_{BO4} is unambiguously confirmed by the ^{10}B work. Furthermore, it is shown to be possible to discern five distinct boron sites in the alkali borate system. Quantitative measurements of the numbers and characteristics of these sites show excellent correlations with the two tetrahedral sites (tetraborate and diborate) and three trigonal sites (boroxol, tetraborate and diborate) predicted by Krogh-Moe's structural theory.

II. NOMENCLATURE

A certain amount of nomenclature and a number of acronyms have already been introduced in the preceding "Milestones." For purposes of convenience the most common acronyms are listed and defined in Table I.

TABLE I

A Partial Listing of Frequently Used Symbols and Acronyms

N_{BO_4}	fraction of borons which are four-coordinated by oxygens
N_{BO_3}	fraction of borons which are three coordinated by oxygens
$N^{B}_{BO_3}$	fraction of borons which are three coordinated by oxygens with all three oxygens bridging

TABLE I cont'd

$N_{BO_3}^{NB}$	fraction of borons which are three coordinated by oxygens with at least one oxygen non-bridging
NBO	nonbridging oxygen
x	mole fraction of modifier oxide in binary borate systems
RDF	radial distribution function
PFD	pair function distribution
$\vec{k}$	X-ray wavevector
EFG	electric field gradient
NMR	nuclear magnetic resonance
μ_o	Bohr nuclear magneton
g	electron or nuclear gyromagnetic factor
$\bar{H}$	laboratory magnetic field rector
$\bar{I}$	nuclear spin vector
m	nuclear magnetic quantum number
e	electronic charge
h	Planck's constant
ν_o	Larmour frequency
Q	nuclear quadrupole moment

TABLE I cont'd

q	zz component of electric quadrupole tensor
Q_{cc}	nuclear quadrupole coupling constant
η	nuclear quadrupole interaction asymmetry parameter
σ_G	linewidth of Gaussian convolution function
σ_L	linewidth of Lorentzian convolution function
$\sigma_{Q_{cc}}$	width of Gaussian distribution of quadrupole coupling constants
SCF	self consistent field (calculation)
LCAO	linear combination of atomic orbitals
MO	molecular orbital
T-D	Townes and Dailey (theory)
IR	infra red
ESR	electron spin resonance
$\mathcal{H}$	spin Hamiltonian
β	Bohr magneton
$\bar{S}$	electron spin vector
M	electron magnetic quantum number
A	hyperfine coupling constant

TABLE I cont'd

A_{ns}	atomic ns-state hyperfine coupling constant
A_{np}	atomic np-state hyperfine coupling constant
D	axial fine structure parameter
E	off-axial fine structure parameter
BOHC	boron-oxygen hole center
BEC	boron electron center

Throughout this writeup, the terminologies adopted by Bray (58, 101) and by Konijnendijk and Stevels (95, 96) will be applied interchangeably when referring to boron-oxygen atomic arrangements. Thus, BO_3 triangles and BO_4 tetrahedra will be referred to either as "configurations" or as "units." Indentifiable structural elements containing more than one boron will be termed either "groups" or "structural groupings."

The notation to be used in denoting anhydrous borate compounds will be that adopted by Krogh-Moe (102) (Table II). The same nomenclature will be used for the basic structural groupings of which the compounds are constructed. These groupings have been summarized nicely by Konijnendijk (96) in a figure which is reproduced as Fig. 2 of the present paper (see pages 24, 25.) An alternative notation proposed by Konijnendijk may provide a useful shorthand; the interested reader is referred to Ref. 96 for details.

TABLE II

Nomenclature for Crystalline Binary Borates

Formula	Notation[a]
$3\ MgO \cdot B_2O_3$	magnesium orthoborate
$2\ MgO \cdot B_2O_3$	magnesium pyroborate
$Na_2O \cdot B_2O_3$	sodium metaborate
$Na_2O \cdot 2B_2O_3$	sodium diborate
$Na_2O \cdot 3B_2O_3$	sodium triborate
$Na_2O \cdot 4B_2O_3$	sodium tetraborate
$K_2O \cdot 5B_2O_3$	potassium pentaborate
$Cs_2O \cdot 9B_2O_3$	caesium enneaborate

[a] After Krogh-Moe, refs. 102, 103.

The term "network" will from time to time be used in a general sense to mean a polymerized structure generally composed of boron-oxygen units or structural groupings joined at the corners; a completely random network in the sense proposed by Zachariasen[1] is not implied. Similarly, the notion of the interstitial "modifier cation" will be retained but with the understanding that certain of the traditional modifiers may enter the network under some circumstances.

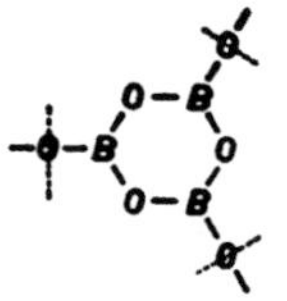

The boroxol ring (a_3), observed in vitreous B_2O_3.

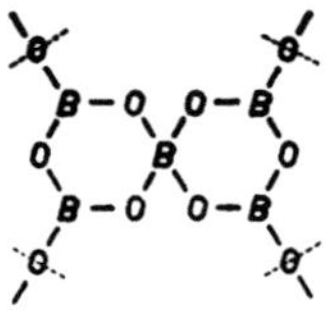

The pentaborate group (a_4c), observed in the compounds α-$K_2O \cdot 5\,B_2O_3$ and β-$K_2O \cdot 5\,B_2O_3$.

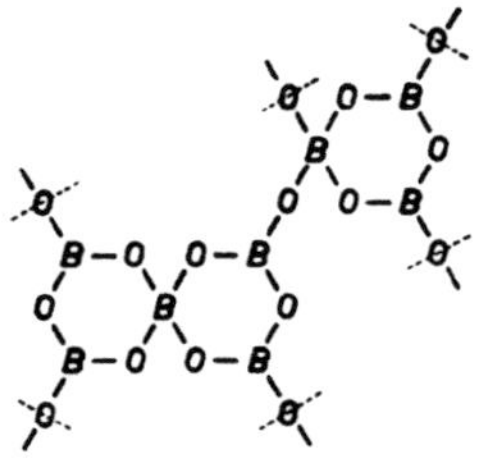

The tetraborate group (a_6c_2), observed in the compound $Na_2O \cdot 4\,B_2O_3$.

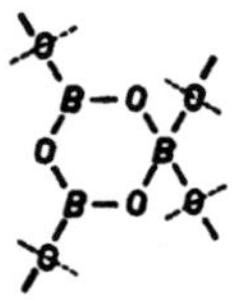

The triborate group (a_2c), observed in the compound $Cs_2O \cdot 3\,B_2O_3$.

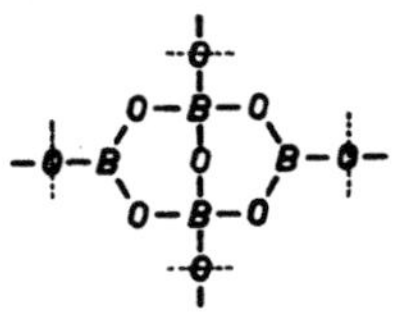

The diborate group (a_2c_2), observed in the compound $Li_2O \cdot 2\,B_2O_3$.

Figure 2. Boron-oxygen structural groupings. (Figure taken from ref. 96). A shorthand notation[96] designates the number of four-coordinated borons (a), of three-coordinated borons with all

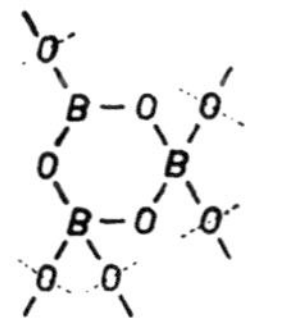

The di-triborate group (ac_2), observed in the compound $K_2O\ .\ 2\ B_2O_3$.

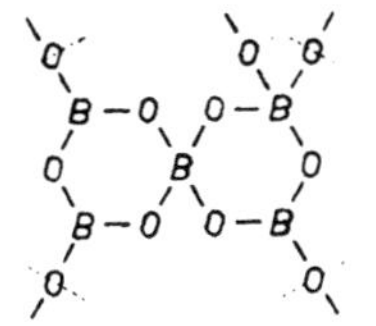

The di-pentaborate group (a_3c_2), observed in the compound $Na_2O\ .\ 2\ B_2O_3$.

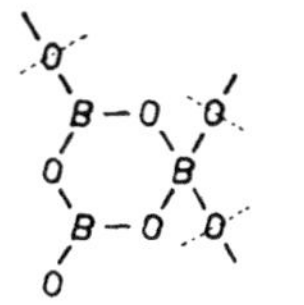

The triborate group with one non-bridging oxygen ion (abc), observed in the compound $Na_2O\ .\ 2\ B_2O_3$.

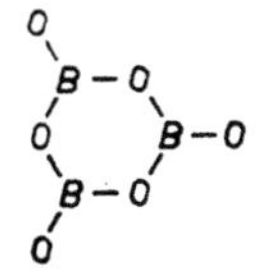

The ring-type metaborate group (b_3), observed in the compounds $Na_2O\ .\ B_2O_3$ and $K_2O\ .\ B_2O_3$.

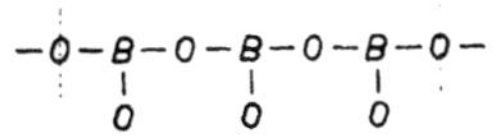

The chain-type metaborate group (b_{∞}), observed in the compounds $Li_2O\ .\ B_2O_3$ and $CaO\ .\ B_2O_3$.

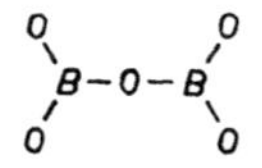

The pyroborate group (b_2''), observed in the compounds $2\ MgO\ .\ B_2O_3$ and $2\ CaO\ .\ B_2O_3$.

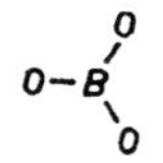

The orthoborate group (b'''), observed in the compounds $3\ MgO\ .\ B_2O_3$ and $3\ CaO\ .\ B_2O_3$.

bridging oxygens (c), and of three-coordinated borons with one, two, or three nonbridging oxygens (b, b",b"').

III. THE BORON-OXIDE ANOMALY

As attested by the "Milestones" chapter, much of the history of borate-glass structural studies has been strongly influenced--if not dominated--by the observed existence of maxima, minima, or inflections in the property-versus-composition curves for complex borate glasses. In particular, a number of abrupt property changes in binary borate glasses at compositions near 15 mol % modifier oxide became known as the "boron-oxide anomaly" (2, 7, 10). Such maxima and minima were not observed in most common boron-free glasses and they were further termed "anomalous" because they could not be readily explained by any obvious structural mechanism. The experimental data nevertheless stimulated the publication of a number of imaginative structural models, all of which were built around the relatively unique ability of boron to exist in two distinct coordination states. The supreme test of many of the earlier models therefore lay in obtaining an independent measure of the fractions of borons which occurred as BO_3 triangles or BO_4 tetrahedra. This test was ultimately provided by NMR, with the result that many models were promptly proven wrong. (The ability of NMR to quantitatively measure N_{BO_4} will be presumed throughout the present discussion but will be given further consideration in the NMR chapter to follow.) When it was demonstrated (42, 49) that N_{BO_4} varies smoothly as $x/(1-x)$ from 0 to ~30mol% modifier oxide -- with no unusual behavior in the critical 15-20 mol% composition range--the suspicion grew that the boron-oxide anomaly may be the result of incorrect interpretations of the experimental data. Such would now appear to be the case.

For historical reasons, then, some of the earlier evidence for the existence of a boron-oxide anomaly will be briefly touched upon, and this will be followed by a recounting of those data which were principally responsible for the demise of this concept. Figure 3 shows two examples of "boron-oxide anomalies" cited by

Warren (7, 11) in the course of presenting his theory for the structure of alkali borate glasses. It can be seen that there is a distinct minimum in the curve of thermal expansion coefficient at ~16 mol% soda content. Assuming all boron in pure B_2O_3 glass is triangularly coordinated, Biscoe and Warren argued that, up to a certain composition, additions of soda to the glass would result in some borons changing to tetrahedral coordination. The rule, simply expressed, is

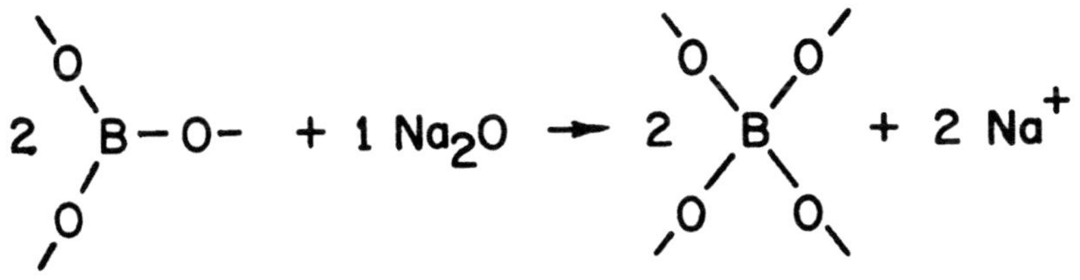

(1)

These authors went on to assert that "the BO_4 groups are bonded to the rest of the structure in four directions, and the structure is therefore tied together in three dimensions rather than in two," and "this will produce a marked increase in the strength and tightness of the structure." Thus they concluded that the rapid decrease in the coefficient of thermal expansion from 0 to 16% Na_2O is due to the progressive change of borons to tetrahedral coordination. In this regard Uhlmann and Shaw (73) have remarked that "while there is no simple relation between thermal expansion coefficient and cohesion, it does seem highly plausible to expect distances normal to planar BO_3 groups to increase readily with thermal motion of atoms, and to expect the resulting large thermal expansion coefficient to decrease as the boron-oxygen configuration changes from triangles to tetrahedra on adding alkali." The bottoming out of the thermal expansion coefficient at ~16 mol% modifier oxide (Fig. 3a) was therefore attributed to a cessation of the reaction represented by Eq. (1). The roll-off of the density curve near the same composition (Fig. 3b) was similarly explained.

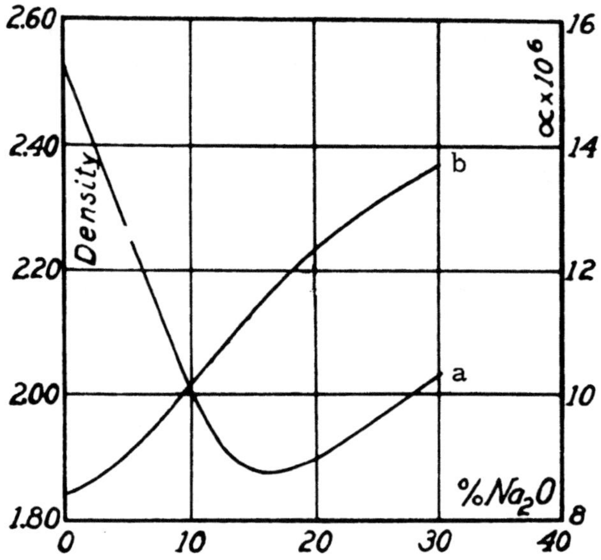

Figure 3. Thermal expansion coefficient (a) and density (b) of soda boric oxide glasses. (Figure taken from ref. 7).

Following Biscoe and Warren's work, virtually all workers accepted the conclusion that the fraction of four-coordinated borons in glasses of composition $x\,Me_2O\cdot(1-x)\,B_2O_3$ followed the relation

$$N_{BO_4} = \frac{x}{1-x}\;, \qquad (2)$$

at least up to the "anomalies" near 16 mol% Me_2O (Me=Li, Na, K, Rb, Cs). What happened at higher modifier contents was the subject of a number of hypotheses involving numerical rules (12-14). Many of these rules dropped by the wayside when the NMR data showed Eq. (2) to hold up to $x \approx 0.3$ (see Fig. 4) (49). At this point the boron-oxide anomaly could no longer be explained in terms of boron coordination changes alone.

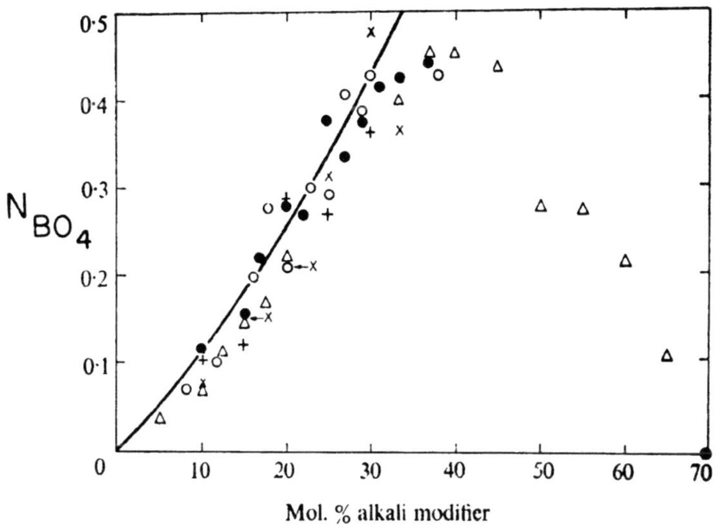

Figure 4. The fraction N_{BO_4} of boron atoms in four-coordination in alkali borate glasses. Figure taken from ref. 49). ●Na_2O ○K_2O △Li_2O ✚Rb_2O ✖CS_2O

Other examples of putative boron-oxide anomalies have included the dependences of viscosity (10, 19) and molar oxygen volume (13) on glass composition. But even before the availability of NMR data, Krogh-Moe (44, 105) had criticized the latter "anomalies" as not being acceptable proof of a saturation of N_{BO_4} in the 10-20 mol% alkali oxide range. The viscosity of boron oxide at 500°C was found in increase with additions of sodium oxide up to 10 mol% where a peak in this property was noted (19). This, of course, invited the same kind of interpretation as the thermal expansion data, namely, an association of increased cohesion with increasing numbers of tetrahedral structures. However, Krogh-Moe pointed out (105) that the opposite effect is noted when (tetrahedral) SiO_2 is added to the melt. Moreover the viscosity data for Na_2O- and NaF-B_2O_3 glasses at 700°C did not conform to such a simple picture either. Minima in oxygen molar volume-composition curves were explained by

Krogh-Moe as resulting from an improper neglect of the volume requirements of the alkali ions. (44, 105)

Against this background, Uhlmann and Shaw (73) undertook to re-examine the thermal expansion properties of alkali borate glasses as functions of glass composition. On reviewing the literature, they noted that there was no general agreement among various authors on the location of the minimum in the thermal expansion coefficient-composition curves. Indeed, in many cases the minima were not nearly so well defined as that of Fig. 3a and often they were positioned by but a single data point. Uhlmann and Shaw rectified this situation by obtaining new data for all five alkali borate systems using glasses prepared at composition intervals of 2 mol%. As seen in Fig. 5, one of the most striking features of their data was the complete absence of sharp minima such as portrayed in Fig. 3a. The general features of Fig. 5 were explained (73) in qualitative terms as possibly reflecting a competition between two processes: the formation of BO_4 tetrahedra, tending to decrease the expansion coefficient, and the introduction of modifying cations, tending to increase it. The pronounced increase in thermal expansion above about 30 mol% alkali oxide could be correlated with the NMR data (49) (Fig. 4) showing that fewer BO_4 configurations are created per unit of alkali oxide addition in this range. Thus, the thermal expansion characteristics of alkali borate glasses are not "anomalous" at all, but vary with structural changes in a generally predictable manner.

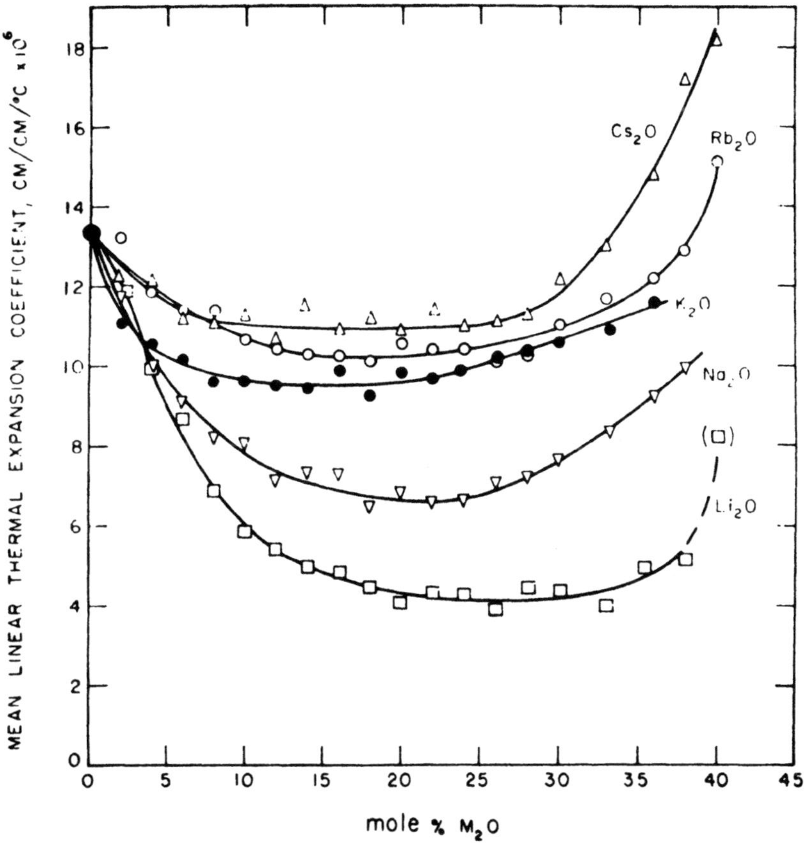

Figure 5. Thermal expansion coefficients of alkali borate glasses as a function of composition. (Figure taken from ref. 73).

IV. PHASE SEPARATION

The subject of phase separation will not be extensively treated in the present draft because of the author's decision to concentrate on borate glass structure on a scale $\lesssim$ 20 Å. Nevertheless, the occurrence of phase separation in glasses assumed to be homogeneous could obviously lead to incorrect interpretations of other structure-sensitive measurements. Since the existence of stable liquid-liquid miscibility gaps in many borate systems (e.g., $PbO-B_2O_3$, $CaO-B_2O_3$) has been well known for many decades (and can frequently be determined by visual inspection), no possible confusion could result in such cases. However, the occurrence of metastable (subliquidus) immiscibility on a scale of 50-5000 Å in the heavily studied alkali-borate system has only been recognized since the 1958 work of Skatulla et al (29). The phenomenon has subsequently been investigated in the lead borate system (62, 106), in the $SiO_2-B_2O_3$ system (107), in the sodium borosilicate system (108), and in all five alkali borate systems. (72) A recent review of the miscibility gap in the alkali borate systems has also appeared (109). The subliquidus miscibility gaps determined by Shaw and Uhlmann (72) for the lithium and sodium borate systems are shown in Fig. 6.

At present, there does not appear to be any well documented cases in which metastable immiscibility has confused the interpretation of other measurements. Uhlmann and Shaw (73) have evaluated the effect of phase separation on some properties of simple borate glasses and have concluded that the "anomalous" thermal expansion behavior is not associated with this phenomenon, Milberg et al (88), carrying out NMR studies of sodium borosilicate glasses, found the effect of silica content on N_{BO_4} to be independent of phase separation as monitored by electron microscopy. Nevertheless, continued caution is advisable, since many experimental techniques yield results which would tend to average together the properties of separated phases which in all liklihood have quite different local structures.

Phase separation is also of some interest from the point of view of glass structure on a scale $\lesssim 20$ Å. Modifier ion clustering now appears to be an intrinsic structural property of many homogeneous glasses (33-36). Thus, metastable phase separation can be viewed as an advanced stage of this intrinsic clustering phenomenon.

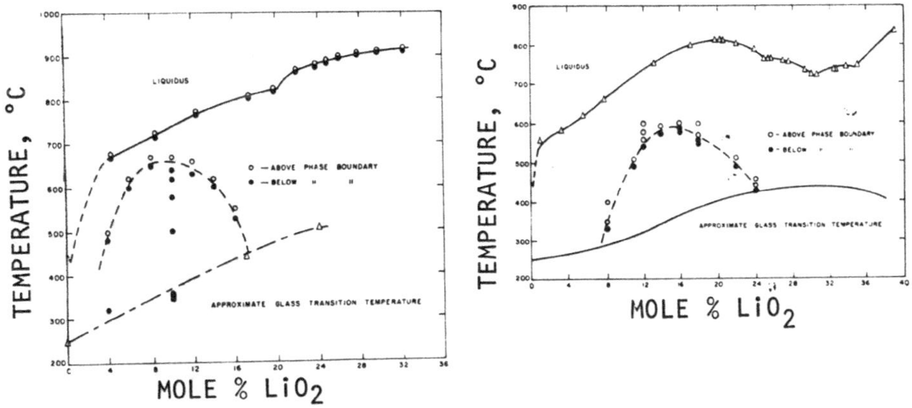

Figure 6. Phase diagrams showing subliquidus miscibility gaps in the systems $Li_2O-B_2O_3$ and $Na_2O-B_2O_3$. Respective liquidus curves are from refs. 30 and 5; approximate glass transition curves are from refs. 110 and 111 and 112. (Figure taken from ref. 72).

V. KROGH-MOE'S STRUCTURAL THEORY

The theory of borate glass structure due to Jan Krogh-Moe is by now widely, if not universally, accepted. Although many of the details are still arguable, it is clear from the "Milestones" chapter above that Krogh-Moe's theory is the one against which all new data and concepts are currently compared. For this reason, it is appropriate to attempt a concise statement of that theory, separated from the specific contexts of the individual experimental techniques used to test its validity (following chapters).

Krogh-Moe has asserted that borate glasses are not merely a random network of BO_3 triangles and BO_4

tetrahedra joined at the corners but, instead, they "actually contain well-defined and stable borate groups as segments of the disordered framework." (45) Those borate groups which are included in the glass structure should be identical with groupings which occur in crystalline borates of compositions bordering those of the glasses. Due to possible thermodynamic or kinetic considerations, not all polyborate anions existing in the crystalline materials need be found in the glasses, however. Perhaps the most general statement of Krogh-Moe's theory is that borate glasses will contain sizeable numbers of at least some of the polyborate groupings which occur in related crystalline materials.

With regard to the alkali and silver borate glass systems, Krogh-Moe's theory is more specific as a result of his pioneering thermodynamic (45) and infra-red (57) studies. That is, it has been possible to theorize (57) that only four different kinds of structural groupings exist in the composition interval up to 34 mol% alkali (or silver) oxide. These are the boroxol, pentaborate, triborate, and diborate groups -- re-illustrated for quick reference in Fig. 7. The more specific theory goes further, however, in making the assertion that pentaborate and triborate groups always occur in pairs; such pentaborate-triborate pairs are collectively referred to as "tetraborate groups" even though they might not always be connected as diagrammed in Fig. 2. Thus, to recapitulate, Krogh-Moe's theory of alkali and silver borate glasses states that the primary structural groupings in the range $\lesssim$ 34 mol% modifier oxide are boroxol rings, tetraborate groups, and diborate groups.

The relative numbers of these groups as a function of glass composition is not quantitatively predicted by Krogh-Moe's theory. Indeed, the probable existence of a few "loose" BO_3 and BO_4 configurations in the alkali borate glasses can be anticipated on the basis of an x-ray determination of their presence in the $K_2O \cdot 3.8B_2O_3$ compound. (97) Nevertheless, it is sometimes helpful to

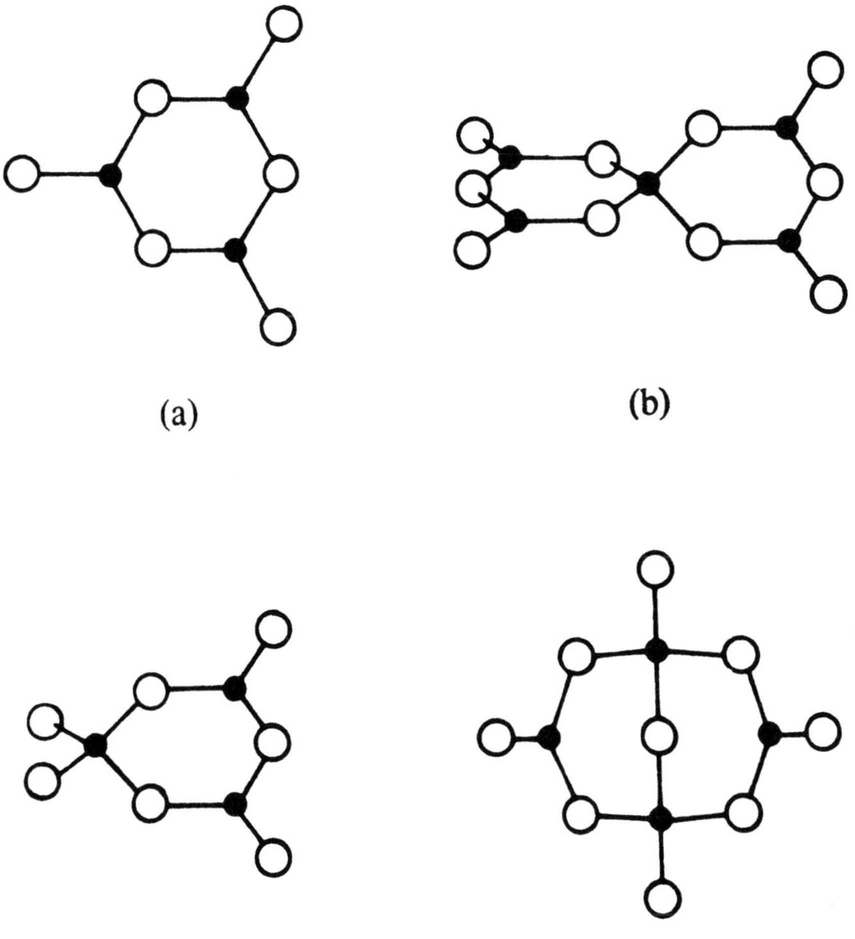

Figure 7. The borate groups postulated to exist in alkali and silver borate glasses with 34 mol % modifier or less. (a) The boroxol group. (b) The pentaborate group. (c) The triborate group. (d) The diborate group. (After Krogh-Moe, ref. 57).

compare various experimental results against what we shall call an idealized version of Krogh-Moe's theory. In this hypothetical picture, the types of polyborate groupings present in a given glass composition are just those characteristic of the nearest compounds on the phase-equilibrium diagram which have structures related to the glass structure. The relative numbers of these groupings in the glass would then be given by the lever rule. Thus, in an idealized alkali borate glass, $xMe_2O\cdot(1-x)B_2O_3$, the number of boroxol rings per mole would be (94) $(2-10x)/3$ and the number of tetraborate groups per mole would be x in the range $0 \leq x \leq 20$ mol%. At values of $x > 20$ mol%, the number of tetraborate groups per mole would decrease linearly to zero at x=33 mol%, while the number of diborate groups would increase linearly to a maximum of 1/3 diborate group per formula unit at the latter composition. Of course, NMR determinations of N_{BO_4} (Fig. 4) (49) have shown that the theoretical maximum number of diborate groups is impossible in many real glasses, while Krogh-Moe's 1962 thermodynamic studies (45) suggested likewise that fewer than x tetraborate units are formed per mole of real glass. Therefore, the idealized version of Krogh-Moe's theory described here was known from the start to be no better than a zero-order approximation. It does, however, illustrate the essential conceptual framework which may be tested, modified, refined (or possibly rejected) on the basis of the results of further careful experimentation.

VI. INTRODUCTION TO THE STRUCTURE-SENSITIVE EXPERIMENTAL METHODS

The following chapters attempt to review one-by-one the various structure-sensitive experimental techniques which have been applied to borate glasses, carefully discussing the kinds and qualities of information deriving from each. The treatment is deliberately non-mathematical, but adequate literature references are given to enable the reader to track down the theoretical foundations.

The total number of techniques covered is limited here by the early deadline for the present preliminary draft. Hopefully, guided by the recommendations of the Program Committee and other contributors, a more comprehensive manuscript may be compiled following the Conference.

Virtually all structure-sensitive experiments can be described by some variation on the following sequence of steps:

(1) A sample is prepared which contains a large number of atoms ($\sim 10^{23}$ usually), which are grouped in certain definable ways which the experimenter hopes to determine.

(2) An experiment is performed which usually results in a spectrum or pattern of some sort, i.e., a graph of some experimental intensity versus some experimental probe parameter(s), e.g., angle, wavelength, temperature, etc.

(3) One or more mathematical operations (e.g., base-line subtractions, integrations, multiplication by "sharpening" functions, Fourier transformations, etc.) is applied to the original data to convert them into a more convenient graphical form (e.g., a pair-function distribution curve) or a theoretically defined set of parameters (e.g., spin Hamiltonian parameters).

(4) A structural model is formulated and the predictions of this model are tested via the appropriate theory against the "distilled" experimental results of step (3). Step (3) is sometimes run in reverse so that the theoretical model is transformed in the necessary way for comparisons to be made with the raw experimental data of step (2).

The successful model is usually offered as possibly being a reasonably accurate portrayal of the atomic arrangements in the material selected for study. Whether or not it really is depends on many things, e.g., how well the sample is characterized, how big are the errors in the raw experimental data, how reasonable is the mathematical data processing (and to what degree this processing avoids magnification of the experimental errors), and how rigorous is the theory used to relate the model with the refined data. Most of the experiments to be described are actually quite well developed so that experimental errors are usually small and the requisite data processing is reliable. Many of the theories relating model to experiment are likewise highly rigorous, although there exist exceptions and these will be pointed out as they arise in context. Even assuming that a rigorous theory is available, one further question always remains: How unique is the model? Stated another way, is it possible that a completely different model might fit the data equally well? Such questions crop up over and over again in the history of almost any scientific endeavor, and the study of borate glass structures is no exception. The question of model uniqueness will be carefully considered in each of the following chapters.

VII. X-RAY DIFFRACTION STUDIES

The use of X-ray diffraction techniques for determining crystal structures is a highly developed art with many variations of both experimental technique and data analysis spread about in a truly enormous literature. A basic level discussion within a few pages is therefore next to impossible, even when the subject matter is limited to borate structures. Nevertheless an attempt will be made here to review and remark upon those studies which have contributed the most to our understanding of borate glass structure.

The basic X-ray diffraction experiment carried out on a single-crystal specimen yields as raw data a map in reciprocal space giving the change in X-ray wavevector $\Delta \bar{k}$ represented by each diffracted beam. This would comprise a complete map of the reciprocal lattice of the crystal at a radius $|\bar{k}|$ in the improbable event that the geometric structure factor for each diffracted beam is nonzero. For such an ideal case, the real crystal lattice could be constructed immediately. In most real cases, certain reflections are absent and the investigator is forced to "guess" the crystal structure aided by the symmetry of the diffraction pattern. One then refines the trial model until the calculated reciprocal lattice includes all of the experimental diffraction spots and until those spots which are weak or absent can be accounted for by theoretical form and structure factors. (See, for example, ref. 113.) The model lattice is compared with the experimental diffraction pattern (or some mathematical processed version thereof) by means of Fourier transforms and, by means of statistical methods, a reliability index can be calculated which allows the uniqueness of the model to be evaluated objectively.

Since single-crystal specimens are not always available, it is sometimes profitable to work with polycrystalline samples. It has been shown (114) that a one-dimensional Fourier transform can be used in deducing structures for polycrystalline data. However, when this is done the question of uniqueness becomes more serious, as illustrated by an example of great importance to the present theme, namely, the structure determination for crystalline B_2O_3I. It is recalled from the "Milestones" chapter that an incorrect structural determination for B_2O_3I by Berger (20) greatly confused structural interpretations of borate glasses for a period of about 15 years. Berger's work certainly was not all bad, since later workers (65, 66) have agreed with his identification of the space group and his measurement of unit-cell dimensions. The problem had to do with the nature of the boron-oxygen atomic arrangements within the unit cell. Berger determined by trial-and-error methods 15 parameters which allowed a qualitative fit of the positions and intensities of the observed diffraction lines. His final model involved two different sets of tetrahedra, one of them highly distorted. As shown by Strong and Kaplow (65), this model was not unique and moreover it did not give as good a fit to the experimentally derived radial distribution function (RDF) as did a model involving all planar trigonal BO_3 units. It was refinement by a random walk procedure of computing the RDF which permitted Strong and Kaplow to conclude rather firmly that B_2O_3I is built up of interconnecting infinite ribbons of BO_3 triangles. (It will be noted throughout the present write-up that modern computer methods greatly expand one's ability to test proposed structural models against the form of the experimental data.)

Even with the advantage of advanced computational tools, some caution must be exercised in accepting models which are founded on polycrystalline data. Gurr et al. (66), working with single-crystal samples, showed that the crystal structure of B_2O_3 inferred by Strong and Kaplow was itself incorrect. Gurr et al. noted, however, that Strong and Kaplow did infer essentially the correct ribbon as the structural basis and they remarked further on the source of the two false solutions (Berger's and Strong and Kaplow's): "The true structure has essentially the same radial distribution of interatomic distances as the model proposed by Strong and Kaplow, and since those authors did not have the advantage of single-crystal data they had no grounds for rejecting their model."

Against the rather sobering backdrop of the above studies of a simple crystalline borate, we will now turn to the use of X-ray diffraction methods for studying a more complicated situation, namely, borate glasses. Since a conscious effort is being made in the present discussion to avoid lengthy mathematical developments, the reader may be assisted in understanding the qualitative steps in data acquisition and analysis by reference to

Fig. 8 (see page 42). Step (1), of course, is to prepare the sample. In Step (2) one acquires the scattered X-ray intensity I_{obs} as a function of

$$k = \frac{4\pi \operatorname{Sin} \theta}{\lambda} ,$$

where 2θ is the scattering angle and λ is the wavelength of the (monochromatic) X-ray source. In Step (3) this raw data is processed: From I_{obs} is subtracted a quantity B, the theoretical background resulting from the same atoms (or ions--a choice must be made) in an unbound gaseous state. The difference (I_{obs} - B) is then multiplied by some factor (in the example of Fig. 8 that factor is k/B.) to obtain a so-called reduced intensity curve. Finally, the reduced intensity curve is converted to an RDF through a Fourier transform. Throughout these steps a number of approximations and corrections are usually made. The elements of Fig. 8 through Step (3) were taken from Borgen, Grjotheim and Krogh-Moe (21), who particularly wished to emphasize the errors that may creep in and even be amplified in the course of processing; the broken curve shown along with the RDF is a spurious sine-wave component which may result from a number of sources (e.g., finite termination of the Fourier integral). In Step (4) a model involving the presence of planar BO_3 triangles is formulated and tested against the RDF of Step (3).

The RDF, as the name implies, is (or should be) simply a superposition of the probability distributions of the radial distances to all atoms surrounding each crystallographically distinct atom in the structure. (Actually, a term quadratic in r must be added to the RDF Fig. 8 to obtain a true probability distribution.) For the example of B_2O_3 glass, the RDF includes the sum of three individual pair distributions, namely, B-O, O-O, and B-B. These three components overlap and cannot be separated except by assuming some specific model. More than one model must be tested before any strong conclu-

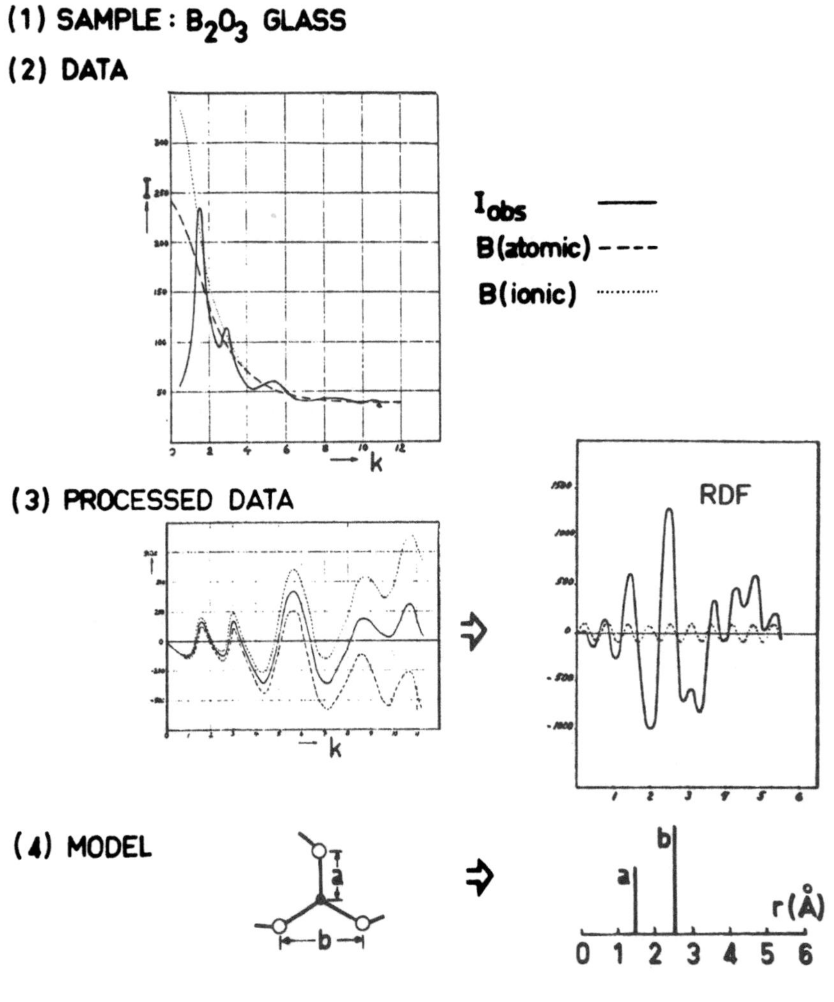

Figure 8. Typical X-ray diffraction study of a borate glass dating before 1970. Dashed lines in initially processed data demark error limits. Dashed sinusoid superimposed on RDF represents another type of error. (Data after ref. 4; processed data and error estimates due to ref.21).

sions may be drawn. It can be noted however that, assuming all spurious components are suppressed, the first peak at small r must correspond to the basic B-O bond(s). The position of that first peak would therefore give the average B-O bond length and the area under the peak the average coordination number of the boron. Indeed, the positional data are usually sufficiently accurate to infer reasonable B-O bond lengths: typically ≈ 1.37Å for triangular BO_3 units and ≈ 1.48Å for tetrahedral BO_4 units (115). Until the recent work of Mozzi and Warren (77) the areas under the peaks have not been accurate enough to measure coordination numbers directly; values of N_{BO_4} so obtained tended (7, 46) invariably to be vastly larger than the more accurate values derived from NMR (next chapter). The improved methods of Mozzi and Warren (see below) have not yet been applied to multicomponent borate glasses containing both BO_3 and BO_4 units, however.

The second peak in the experimental RDF arises primarily from O-O vectors, although a small B-B component is present in pure B_2O_3 glass (77) and an additional small Na-O component likewise contributes in soda-boric oxide glasses (7). Assuming the underlying components do not interfere too greatly, the average O-O distance can be determined. This can serve as an independent check of the boron coordination number inferred from the first peak, since the O-O distance is related to the B-O distance by a trigonometric factor: $\sqrt{3}$ for BO_3 triangles and $\sqrt{8/3}$ for BO_4 tetrahedra. Thus taking the experimental B-O bond lengths cited above, one calculates the O-O distance for BO_3 units

$$1.37 \sqrt{3} = 2.37\text{Å}$$

and for BO_4 units

$$1.48 \sqrt{8/3} = 2.42\text{Å}$$

Since it has been estimated (116) that peak positions can be determined within an uncertainty of 0.02-0.03Å, it is clear that under favorable circumstances (e.g., a single-component, glass) boron coordination can be inferred from the positions alone of the first two peaks in the RDF. In practice, however, this type of analysis has failed to detect differences between glasses for which N_{BO_4} varies between 0 and 0.3 (116).

Turning briefly to multicomponent glasses, we note that use can be made of the fact that heavy atoms make a much greater contribution to the diffracted X-ray intensity than do light atoms such as boron and oxygen. As pointed out by Krogh-Moe (33, 34), the RDFs for such glasses are consequently dominated by inter-atomic distances involving at least one heavy atom. At high modifier concentrations the peaks corresponding to heavy atom pairs are orders of magnitude more intense than B-O, O-O, and B-B peaks. Thus, it has been possible to conclude that Cs ions tend to pair in caesium borate glasses with average separations of 4.6Å(33), Ba ions tend to pair in barium borate glasses ($d_{ave} \approx 6.9$ Å) (34), Sr ions tend to pair in strontium borate glasses ($d_{ave} \approx 6.6$ Å) (35), and Tl ions tend to pair in thalium borate glasses ($d_{ave} \approx 7.1$ Å) (36). Krogh-Moe (36) has emphasized, however that these results add little to our knowledge of the glassy network on an extended scale, i.e., they tell us little about the nature of any boron-oxygen structural groupings which happen to be present. A similar caution was noted by Block and Piermarini (35).

It is correct to infer from the foregoing discussion that X-ray diffraction studies of borate glasses dating before 1970 have produced very little information relating to the boron-oxygen network other than reasonably firm evidence for the existence of BO_3 triangles in B_2O_3 glass and complex glasses of low modifier content. A quantum improvement in this situation has been manifested in the 1970 work of Mozzi and Warren (77) on pure glassy B_2O_3. At the core of this advancement is the use of fluorescence excitation (117), making possible measurements to high values of $\sin\theta/\lambda$ with Compton modified scattering removed, and the replacement of the "proportionality of scattering factors" approximation by an exact formulation in terms of pair functions (118). In analogy with Fig. 8, the essential steps in Mozzi and Warren's work are graphically presented in Fig. 9 (see page 46). Here it can be seen that Step (4) includes a new substep, namely, the synthesis of the RDF by means of a superposition of rigorously calculated pair function distribution curves (PFD). The particular case shown in Fig. 9 pertains to PFDs calculated for a model of planar BO_3 units joined in a "random network" wherein all B-O-B bonds are taken to be 130° and total orientational randomness about the B-O bonds is assumed. (A small randomness in B-O bond length $\sim\pm 0.05$Å was also built into this and the other model which will be described.) It is seen that the model of randomly oriented BO_3 units leads to peaks in the calculated PFDs extending no further than about 4.6Å. Since the experimental RDF (or PFD in the notation of Mozzi and Warren) shows well-defined peaks at distances out to about 6Å, a model incorporating some kind of longer-range order is evidently required.

A suitable model involving six-membered boroxol rings had already been suggested on the basis of Goubeau & Keller's Raman data (17) and supported by much additional evidence summarized by Krogh-Moe (57, 76). Accordingly, Mozzi and Warren calculated the PFDs for the model of randomly oriented boroxol groups (Fig. 10), (see page 47) where B-O-B angles within the ring were taken to be 120° and the B-O-B angle between rings was again taken as $\alpha=130^{o}$, with complete randomness in orientation about extra-ring B-O bonds. The final results of this calculation are shown in Fig. 10b, curve (B). Curve (C) is the difference between the experimental PFD and curve (B); this difference is reasonably ascribed to contributions to the PFD curve from vectors joining atoms on one group with atoms on another group to which it is not directly linked. The calculated PFDs for the first two peaks (B-O and O-O, respectively) are seen to be a little too sharp. This was attributed (77) to either a too-narrow B-O bond length distribution in the calculation or

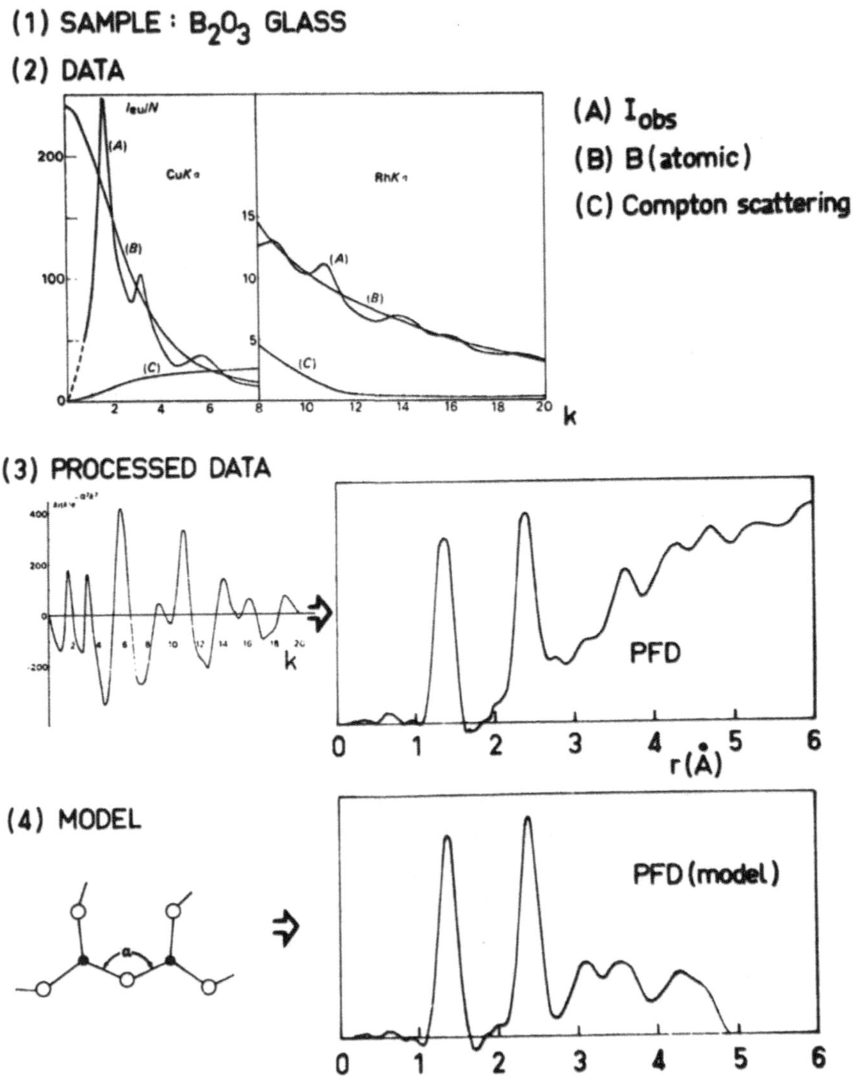

<u>Figure 9</u>. Modern X-ray diffraction study of a borate glass. Sample is glassy B_2O_3. Sources of error noted in Fig. 8 have been minimized or eliminated. Trial model in Step (4) is a random network of BO_3 triangles. A more successful model is illustrated in Fig. 10. (After ref. 77).

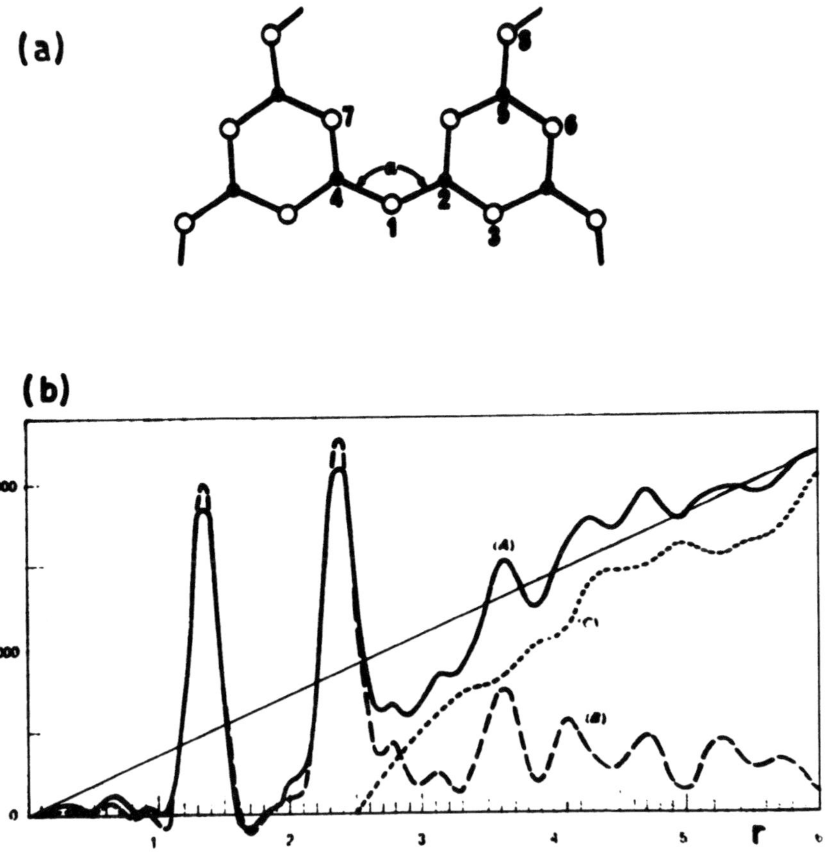

Figure 10. (a) Boroxol structural model for B_2O_3 glass. (b) Analysis of X-ray diffraction data under boroxol model: Curve (A), the measured PFD for glassy B_2O_3; curve (B), the sum of computed contributions to the PFD for a model of randomly linked boroxol groups; curve (C) is the difference (A)-(B). (Figures taken from ref. 77).

to the presence of a few randomly linked BO_3 units in the glass. Another possible explanation could conceivably be random variations in O-B-O bond angles. In any event the fit exhibited in Fig. 10b is quite good and clearly superior to that of the "random BO_3 unit" model, particularly as regards the small peak near 2.8Å which indentifies with the vector 2-6 in Fig. 10a.

It is thus seen how two plausible models for glass structure can be objectively compared against one another by means of careful curve fitting via a rigorous theory. In the present case, this process has led to the selection of the boroxol model of B_2O_3 glass as more suitable than a model consisting of a random network of BO_3 triangles. Of course, other models could be tried. Mozzi and Warren themselves remarked that a possibly improved fit might be achieved if it were to be assumed that a small fraction of the glass is in the form of BO_3 configurations linked in a random way to the boroxol groups. However, they did not believe that the computed PFD for this case would be sufficiently unique to clearly favor such a model over any other similarly subtle variation. (As described in the following chapter, recent ^{17}O NMR data do support the presence of ~20% of the borons outside of boroxol rings.) To the extent that no alternative to the "mostly boroxol" model has been proposed or supported by detailed model PFD calculations, Mozzi and Warren's study (77) by itself is powerful evidence that a major part of B_2O_3 glass is made up of boroxol groups. Taken together with NMR and Raman results (following chapters) the X-ray case for boroxol groups becomes essentially conclusive.

VIII. NMR STUDIES

Nuclei with integer or half integer spin I have nuclear magnetic moments given by $\vec{\mu} = g\vec{\mu}_0 I$, where g is the nuclear g factor (a dimensionless number), μ_0 is the Bohr nuclear magneton, and $\vec{I}$ is a dimensionless vector of magnitude $\sqrt{I(I+1)}$. This magnetic moment interacts with an externally applied magnetic field $\vec{H}$ to set up Zeeman energy levels given by

$$E = \vec{\mu} \cdot \vec{H} = -g\mu_o \; |\vec{H}| \, m, \qquad (3)$$

where m is the nuclear magnetic quantum number which may take on 2I + 1 values I, I - 1, . . . , -I. Transitions between adjacent levels ($\Delta m = 1$) can be stimulated and detected by bathing the sample in radio-frequency radiation of frequency ν_o such that

$$h\nu_o = g\mu_o \; |\vec{H}| \, . \qquad (4)$$

The value of ν_o which satisfies this resonance condition is known as the Larmour frequency.

In real solids, the Zeeman energy levels are shifted by interactions of the nucleus with its immediate surroundings. Because of the present emphasis on boron-oxygen structural arrangements, it is possible to limit discussion to only those interactions relevant to this problem. Thus, the principal local interaction of interest is the interaction of the nuclear quadrupole moment Q with the electric field gradient (EFG) tensor at the nuclear site. Only nuclei with $I > 1/2$ have quadrupole moments; we shall be concerned with $^{11}B(I = 3/2)$, $^{10}B(I = 3)$, and $^{17}O(I = 5/2)$. It is customary to define the quadrupole coupling constant $Q_{cc} = e^2qQ/h$, where e is the charge of an electron, h is Planck's constant, and q ($\equiv q_{zz}$) is the largest principal component of the EFG tensor. If the EFG departs from axial symetry, an asymmetry parameter is defined

$$\eta = \frac{q_{xx} - q_{yy}}{q_{zz}}$$

which may take on values ranging from 0 (axial symmetry) to 1 (complete rhombic symmetry).

Figure 11 (see page 51) illustrates the energy levels arising from (a) the Zeeman interaction with a nucleus with spin I = 3/2 and (b) the shifts in these levels due to a small quadrupole interaction. Three distinct resonance frequencies are apparent: ν_a, ν_b, and ν_c. These three frequencies are each functions of the angles between the applied magnetic field $\vec{H}$ and the principal axes of the EFG tensor. In a powder or glass--where all orientations are present with equal probability--the resonance spectrum corresponding to each frequency is smeared out in a well defined way between well defined limits. ν_a and ν_c are spread out over a far wider frequency range than ν_b and, as a consequence, are seldom observed at spectrometer settings appropriate for optimal acquisition of the narrower ν_b spectrum. Thus, in the following discussion we shall be concerned only with the "central" transition ν_b, i.e., the transition (m = 1/2) $\rightleftarrows$ (m = -1/2), for the two half-integer spin nuclei ^{11}B and ^{17}O. In the case of ^{10}B, of course, there is no "central" transition but, following the same reasoning, recent authors (100) have considered the two least-spread-out transitions, (m = 1) $\rightleftarrows$ (m = 0) and (m = 0) $\rightleftarrows$ (m = -1).

At this point we turn our attention to the actual shapes of the angularly averaged NMR spectra. To obtain these shapes theoretically, one must first write down the complete resonance condition giving the resonance frequency ν as a function of $|\vec{H}|$, Q_{cc}, η, and the angles (call them ϕ and θ) between $\vec{H}$ and the EFG. This resonance condition can be found in the literature (e.g., ref. 119) and will not be reproduced here. The angularly averaged absorption spectrum can then be generated by calculating ν for a large number of equal solid angle elements ($d\Omega = \sin\theta \, d\theta d\phi$) spanning the entire solid angle sphere and then histograming these results on a frequency scale. The so-called "powder patterns" calculated in this way have a number of singular features whose positions can be calculated exactly (120, 121, 122); see Fig. 12 for the results in the cases of (m = 1/2) $\rightleftarrows$

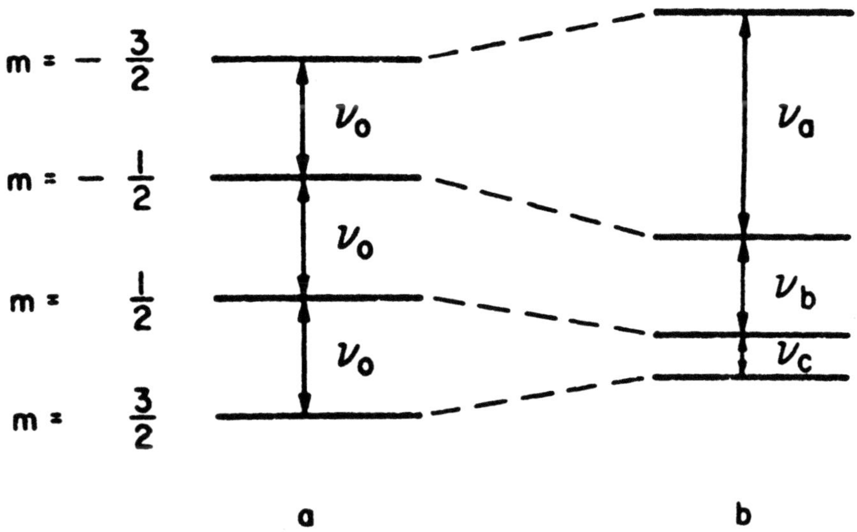

Figure 11. Energy levels arising from the interaction of the nuclear magnetic dipole moment with a magnetic field: (a) no electrical quadrupole interaction present; (b) small quadrupole interaction present. The levels shown are appropriate to the ^{11}B nucleus with spin I = 3/2. (Figure taken from ref. 101).

(m= -1/2) transitions where $Q_{cc} << \nu_0$ and the quadruple interaction has been treated to second-order in perturbation theory. Powder patterns such as those of Fig. 12 give the NMR line shape for the hypothetical case of an infinitesimally narrow single-crystal linewidth. The effect of finite single-crystal broadening due to dipole-dipole and/or relaxation effects can be simulated by convoluting the powder patterns with Gaussian and/or Lorentzian functions of appropriate widths (78, 121).

The various steps of carrying out and interpreting an NMR experiment in a powder or glass can now be set

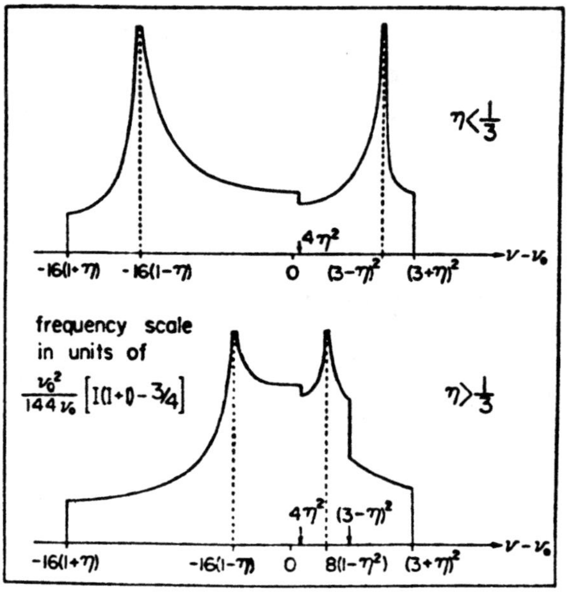

Figure 12. NMR powder patterns for second-order quadupole broadening of the central, $(m = 1/2) \leftrightarrow (m = -1/2)$, transition. These are appropriate to the ^{11}B and ^{17}O nuclei, but not to ^{10}B. (Figure taken from ref. 121).

down. Step (1), as always, is to prepare the sample and characterize it if necessary. Step (2) entails obtaining the experimental spectrum, which usually (though not always) consists of the first derivative of the absorption curve as a function of magnetic field at fixed requency or as a function of frequency at fixed field. (In most cases it is the field which is varied, but it is customary nevertheless to convert all data to a frequency scale.) In Step (3), numerical integrations may be performed to obtain the areas under the absorption curves, e.g., if an N_{BO_4} measurement is desired. Otherwise, the data themselves are rarely subjected to mathematical processing. Also in Step (3) the experimental spectrum is computer simulated (or compared with the powder patterns of Fig. 12) to obtain accurate Hamiltonian parameters, Q_{cc}, η, and the Gaussian or Lorenzian linewidth σ_G or σ_L. In certain circumstances, further information may be extracted from glass spectra, namely, the statistical distributions which may exist in the parameters Q_{cc} and η as consequences of vitreous disorder (79, 81, 100, 123). In Step (4), a structural model is formulated. At present there is only an empirical relationship between boron-oxygen atomic arrangements and the observed values of Q_{cc} and η; this relationship has been summarized by Kriz, Park, and Bray (69) whose results are reproduced in Table III. Figure 13 (see page 54) graphically illustrates the 4 steps alluded to above in the NMR context.

TABLE III

^{11}B quadrupole coupling parameters for various boron-oxygen configurations, after Kriz, Park and Bray (69)

Boron Coordination	$Q_{cc} = e^2qQ/h^a$		η^a
Tetrahedral	0 - 855 kHz		0 - 1
Trigonal	2.45-2.81 MHz	symmetric group (contains either 3 bridging or else 3 non-bridging oxygens)	0 - 0.23
		Asymmetric group (contains either 1 or 2 nonbridging oxygens)	0.47-0.75

[a]Experimental values determined for a large number of crystalline borates of known structures. NMR and structure references are given by Kriz and Bray. (69)

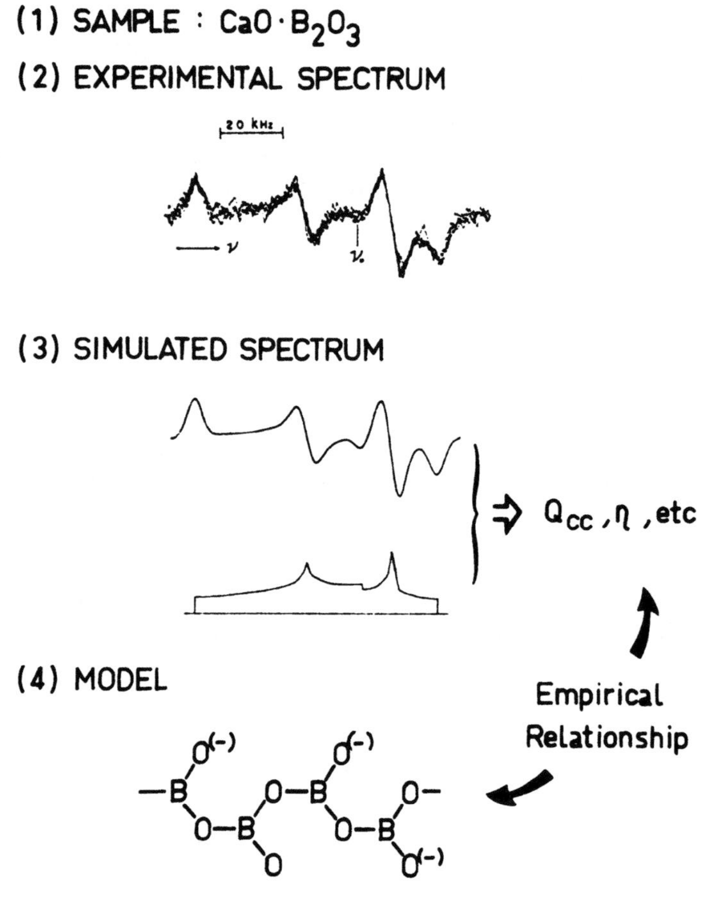

Figure 13. Typical NMR study of a borate glass or (in this example) polycrystalline compound. Hamiltonian parameters derived by computer simulation are empirically associated with certain structural configurations according to Table III. (After ref. 68).

As mentioned in the chapter dealing with the "boron oxide anomaly," one of the truly outstanding achievements of NMR has been the measurement of the fraction of borons which are in four-coordination, N_{BO_4} (28, 41, 49, 101). From a phenomenological standpoint, this has been possible because borate minerals containing solely tetrahedral BO_4 units have exhibited ^{11}B Q_{cc} values $\lesssim 0.85$ MHz, while compounds containing only triangular BO_3 units are characterized by values in the range of 2.4 - 3 MHz (41). Theoretically, the magnitude of the quadrupole coupling constant can be understood in terms of covalent bonding effects as well as the EFGs set up by ionic point charges. For perfect tetrahedral geometry, theory predicts zero EFG at the central boron regardless of ionicity. For trigonal geometry and zero ionicity, the coupling constant should be ~5.4 MHz; the lower values observed for BO_3 groups in various borates is presumed to be due to ionic resonance effects (28, 41). In practice, the large disparity between the coupling constants for trigonal and tetrahedral boron permit the resonances of each to be separated and quantitatively measured even in materials containing both types of units. This is in dramatic contrast to the X-ray diffraction results cited in the preceding chapter. It has been common to refer to the NMR signal of four-coordinated boron as the "narrow line" and that due to three coordinated boron as the "broad line." N_{BO_4} can be obtained by measuring the integrated intensity of the narrow line and comparing it to an absolute standard (49, 101) or by ratioing the narrow-line intensity to the total (88, 90, 100).

Peterson, Kurkjian, and Carnevale (80, 123) have recently raised the question of whether narrow lines in glasses might arise from sources other than four-coordinated borons. As an alternative they suggested a "random model" whereby an extremely broad distribution of Q_{cc} values (presumably arising from trigonal or irregularly coordinated borons) could conceivably give rise to a narrow "spike" near ν_0. They computer simulated such an effect using theoretical machinery which has been shown (124) to be equivalent to that already in use by Taylor and Bray (78). Peterson et al. did not attempt any detailed curve fitting, even though they stipulated that "detailed curve fitting must...be the ultimate test of the validity of a given distribution..." (80). One particular line shape calculated (123) by these workers did appear somewhat convincing when not directly overlaid on an actual experimental spectrum. That simulation has been reproduced in Fig. 14 (see page 57) along with an actual comparison of the computed absorption curve with an experimental spectrum. It can be seen in Fig. 14 that Peterson et al's "random model" does not in fact give a very good fit to the experimental data in the regions marked by arrows. By contrast, Taylor and Friebele (79) have carefully curve fitted spectra of glasses where N_{BO_4} ranges from 0 to ~0.4 using a sharp bimodal distribution of coupling constants (Fig. 15; see page 58). In analogy with borate compounds and consistent with theoretical EFGs, the smaller-coupling-constant peak in $P(Q_{cc})$ in Fig. 15 is presumed to represent BO_4 units while the peak at larger coupling constants is taken to represent BO_3 units. Finally, and most conclusively, ^{10}B NMR (which is not subject to "spiking," due to the absence of a central transition) demonstrates normal behavior vis-a-vis the appearance of a "narrow line" when alkali oxide is added to a pure B_2O_3 glass. (100) The ability of NMR to measure N_{BO_4} is thus upheld beyond all doubt.

Referring back to Step (4) of Fig. 13, we comment on the obvious need to make a theoretical (rather than simply empirical) connection between the measured values of Q_{cc} and η and the actual geometry and electronic structure of the BO_3 configurations giving rise to the broad resonances. One way to make such a connection is by means of model calculations. The first attempt at model calculation was that of Taylor and Friebele (79), whose model related experimentally observed (81) variations in Q_{cc} to out-of-plane distortions of the basic BO_3 configuration via Townes and Dailey (T-D) theory. (125) The success of the fits already noted in Fig. 15 was

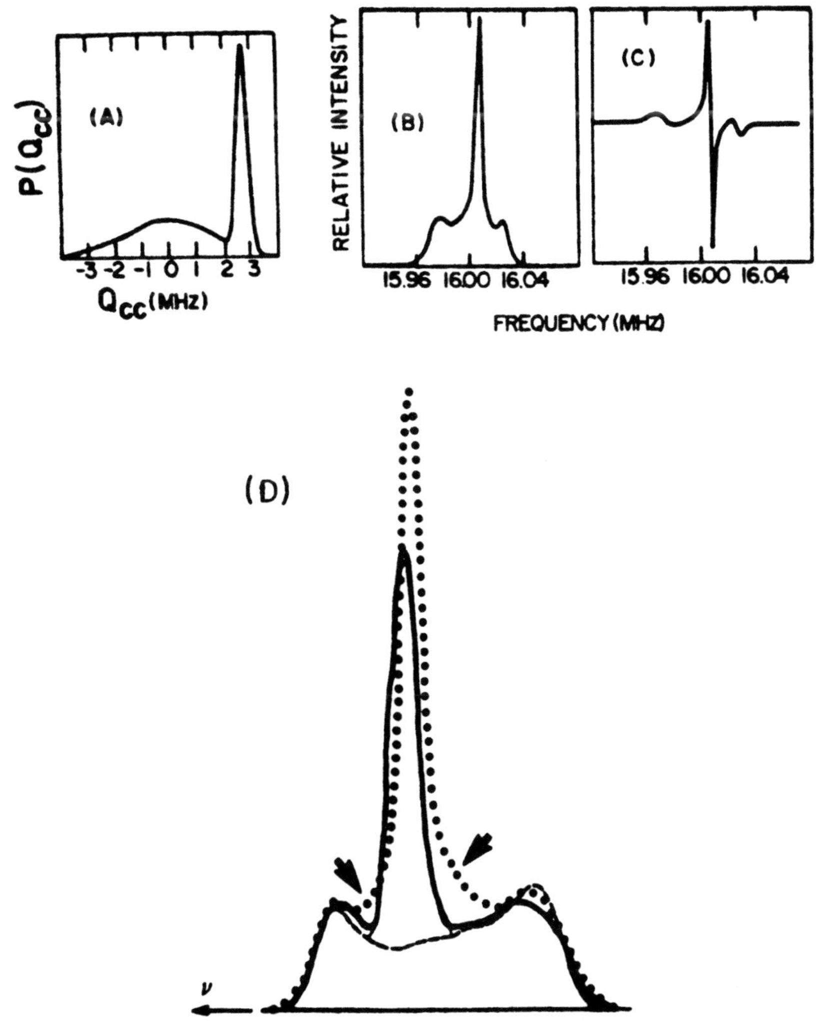

Figure 14. Calculated NMR lineshape for a hypothetical distribution of quadrupole coupling constants. (A) Hypothetical density function for Q_{cc}. (B) Calculated absorption curve. (C) Calculated derivative curve. (D) Curve of (B) shown superimposed (dotted) on experimental ^{11}B NMR spectrum for a borosilicate glass. The model of (A) yields a poor fit to the experimental spectrum in the regions marked by arrows. The shift in the computed "spike" to the low-frequency side of the experimental "narrow line" may also be real, due to the intrinsically asymmetric nature of the "spiking" phenomenon described in refs. 80 and 123. (Calculated spectra taken from ref. 123; experimental spectrum from ref. 90).

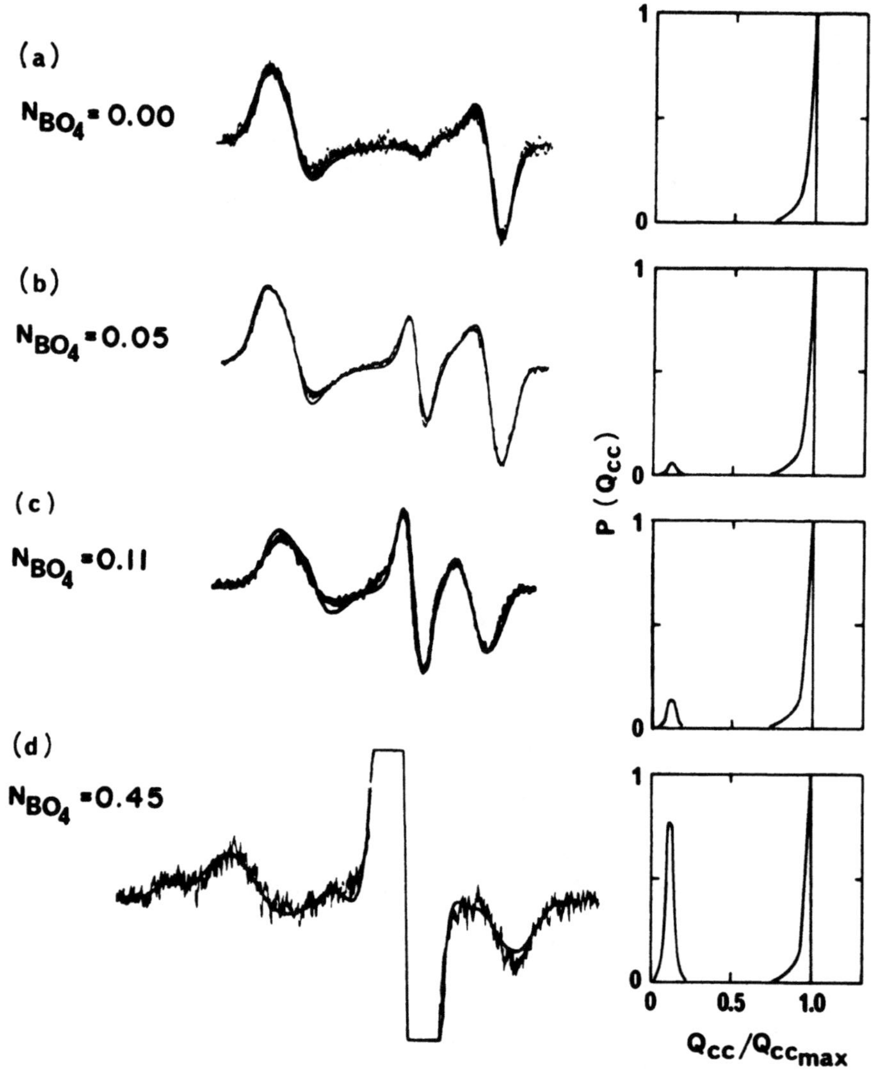
(a)
NBO4 = 0.00
(b)
NBO4 = 0.05
(c)
NBO4 = 0.11
(d)
NBO4 = 0.45
P (Qcc)
1
0
0
0.5
1.0
Qcc/Qccmax

Fig. 15.

encouraging, but the authors readily admitted that their T-D model was not unique. Moreover, there is now a substantial basis for the belief that the T-D approach is misleading with regard to the physical origin of the Q_{cc} variations: Snyder, et al (99) carried out much more sophisticated SCF model calculations on trigonal $B(OH)_3$ and BH_3 configurations, coming to the conclusion that Q_{cc} is almost totally insensitive to out-of-plane distortions and that such distortions are probably energetically prohibited. The actual source of the experimentally observed coupling constant distributions in B_2O_3 glass was ascribed by Snyder et al (99) to "charges in the molecular environment of boron in glass." In their model calculations they achieved variations ~0.4 MHz in Q_{cc} by placing charges ~+1 at distances 3.12Å above and below the basic

Figure 15. Experimental and computer-simulated ^{11}B NMR derivative spectra at ν_0 = 16 MHz for various borate glasses whose fractions of boron atoms in four-coordination, N_{BO_4}, vary from 0 to to 0.45; (a) B_2O_3 glass; (b) $(Ag_2O)_{0.05}$ $(B_2O_3)_{0.95}$ glass; (c) $(Cs_2O)_{0.1}$ $(B_2O_3)_{0.9}$ glass; (c) $(Cs_2O)_{0.1}$ $(B_2O_3)_{0.9}$ glass; (d) $(Cs_2O)_{0.4}$ $(B_2O_3)_{0.6}$ glass. Quadrupole-coupling-constant density functions used for the respective simulations are shown at the right. For convenience, the density function for (d) is drawn as though there were only a single large-coupling constant site; actually, two such sites were required for the simulation. This figure demonstrates that a relatively sharp bimodal distribution in Q_{cc} values (corresponding to three-and four-coordinated borons) is implied by the experimental spectra. The detailed shape of $P(Q_{cc})$ for the large-coupling constant site is probably incorrect, however (see text and Fig. 16). Simulation parameters and sources of experimental spectra are given in refs. 79 and 127. (After ref. 79).

planar trigonal boron unit. On the other hand, in-plane displacement of one hydrogen on a basic $B(OH)_3$ unit caused a large change in asymmetry parameter η but a variation in Q_{cc} of only ~0.07 MHz. The latter calculation was believed to model the metaborate configuration, where each boron is bonded to one nonbridging and two bridging oxygens. Initially, it seems prudent to accept all these calculations at face value, while searching out further experimental data which may test the various predictions.

Some relevant experimental results (81, 91, 100, 126), plus the model calculation of Taylor and Friebele, have been gathered together in Fig. 16. The dots represent the distribution in Q_{cc} values estimated by Kriz and Bray (81) for pure B_2O_3 glass under the simplifying assumption that the asymmetry parameter is undistributed. It can be seen that Kriz and Bray's distribution agrees rather well with that determined by Jellison and Bray (100) from ^{10}B NMR, where Q_{cc} and η and their distributions can be measured independently* (bell shaped curve in Fig. 16). Note that the shape of the curve calculated from T-D theory bears no resemblance to those determined by careful curve fitting. This disagreement tends to confirm the conclusion of Snyder et al (99) that out-of-plane distortions of BO_3 configurations either do not occur in glass or at least they do not determine the distribution of

---

*Jellison and Bray performed their analysis by assuming a Gaussian distribution in Q_{cc} and then showing that excellent computer simulations of both the ^{10}B and ^{11}B spectra can be achieved by taking $^{\sigma}Q_{cc}$ = 0.10 MHz, where $^{\sigma}Q_{cc}$ = 0.85 x (half-width of distribution). Simulations of the ^{11}B spectrum alone were found to be relatively insensitive to changes in $^{\sigma}Q_{cc} \sim \pm$ 50%. Peterson et al (123) have also computer simulated the ^{11}B spectrum of B_2O_3 glass, citing $^{\sigma}Q_{cc}$ = 0.2 MHz as yielding a "satisfactory fit." It is possible that Peterson et al use a different definition of σ, however.

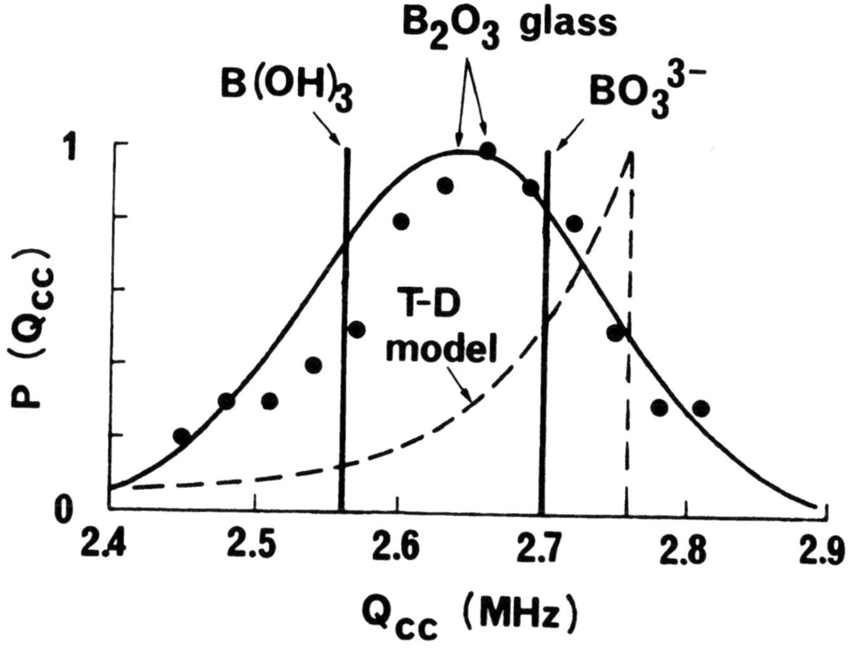

Figure 16. Distributions in ^{11}B quadrupole coupling constants experimentally estimated for BO_3 triangular configurations in B_2O_3 glass, with comparisons to measured values for $B(OH)_3$ triangles in orthoboric acid and $BO_3{}^{3-}$ triangles in strontium orthoborate and to a calculated distribution function based on Townes and Dailey (T-D) theory. Here, the T-D curve has been arbitrarily shifted to make its peak coincide with the measured (67) coupling constant for ^{11}B in crystalline B_2O_3I. (See text for sources of the plotted quantities).

Q_{cc} values. In fact, both of these statements are probably true, since the T-D distribution of Fig. 16 corresponds to mean out-of-plane distortions $\sim 8^o$, whereas the maximum out-of-plane distortion determined by X-ray diffraction for any borate crystalline compound--even highly polymerized and distorted polyborates--has been $\sim 1^o$ (e.g., the results of Gurr et al (66) for B_2O_3I or those of Krogh-Moe (97) for $K_2O \cdot 3.8\ B_2O_3$). Presumably, $B(OH)_3$ in orthoboric acid and BO_3^{3-} in strontium orthoborate are examples of almost perfect planar trigonal structures, and it is seen in Fig. 16 that the measured quadrupole coupling constants (91, 126) for the latter units differ from one another by an amount (0.14 MHz) equivalent to ~60% of the full width of the Q_{cc} distribution in B_2O_3 glass. Thus, chemical bonding effects exterior to essentially planar BO_3 triangles could account for the whole of the Q_{cc} distribution in B_2O_3 glass. As pointed out by Taylor (127), there is now a need for further model calculations which take into account larger structural groupings, e.g., boroxol rings, pentaborate groups, etc.

In the absence of ab initio model calculations for large polyborate structural groupings, major advances in understanding the structure of borate glasses have still been achieved by imaginative new NMR experimentation. The ^{10}B and ^{17}O work (100) of Jellison, Panek and Bray provides a good example. The ^{17}O spectrum of pure glassy B_2O_3 revealed the presence of two basic sites, one with a narrow distribution of η values and one with a substantially broader distribution in η. Since the existence of two distinct oxygen sites cannot be reconciled with a "random network of BO_3 triangles model," the boroxol model was considered. In the latter picture, it was logical to associate the narrower η distribution with the in-ring site O(R) and the broader distribution with the connecting sites O(C). Computer simulation of the spectra revealed an O(C)/O(R) ratio of 1.2, which implied (100) that ~18% of the borons must reside outside of boroxol rings. This means that on the average there must be almost one "loose" BO_3 triangle for each boroxol ring. Eventually, this may provide a new impetus to try to fit the X-ray diffraction data of Mozzi and Warren (77) with a PFD based on a "mixed boroxol-loose triangle model" (see preceding chapter). Beyond that, the availability of Q_{cc} and η values for all oxygen and boron sites has already permitted a semiempirical T-D calculation for the boroxol ring in glass. The results of Jellison and Bray (100) imputed a slight aromatic character to the ring, with a boron π orbital occupancy of 0.360. This result is in excellent (though perhaps fortuitous) agreement with the value of 0.376 calculated in an ab initio SCF-MO treatment of the trigonal $B(OH)_3$ unit (a non-ring structure) by Snyder et al (99). A self-consistent calculation (100) of the electronic charge distributions on the boroxol group in glass based on observed quadrupole parameters gave an ionicity ~0.4, which is about the same as that predicted by the Pauli electronegativity criterion. The ab initio SCF calculation (99) on $B(OH)_3$ gives an ionicity about 50% smaller.

A further result of the ^{10}B NMR studies of Jellison and Bray (100) has been the recognition of five distinct boron sites in sodium borate glasses containing up to 35 mol% Na_2O. Two of these sites are four-coordinated borons and correspond to the two "narrow-line" sites already discerned by Rhee (84) by ^{11}B NMR and attributed to BO_4 units in tetraborate and diborate structural groupings. The other three sites are three-coordinated borons and are distinguished one from another mainly on the basis of their asymmetry parameters η, which are more accurately determined from ^{10}B NMR than from ^{11}B spectra. Jellison and Bray obtained quantitative estimates of the relative prevalence of the five sites by careful trial-and-error computer synthesis of the experimental ^{10}B lines. The results shown in Fig. 17 are seen to be in excellent agreement with the predictions of Krogh-Moe's structural theory. In fact, these data--taken in conjunction with the Raman, IR, and melting-point-depression results of the following chapters--seem to comprise as

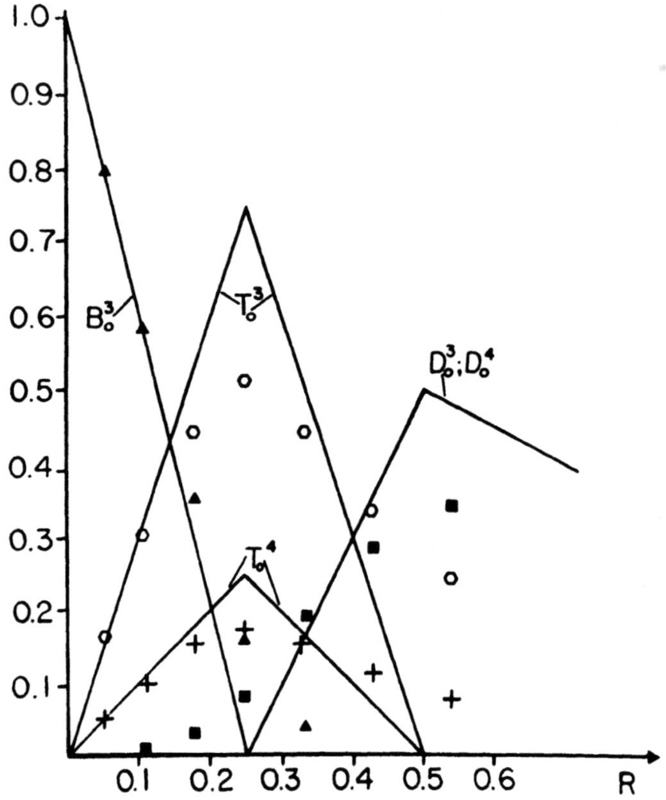

Figure 17. ^{10}B NMR measurement of the fractions of borons in five distinguishable sites, plotted as a function of R = x/(1-x), where x is the molar fraction of Na_2O in the sodium borate glass. Boron coordination numbers are unambiguously determined and the individual sites are tentatively ascribed as follows:

▲ three-coordinated borons in boroxol groups and "loose" BO_3 triangles

○ three-coordinated borons in tetraborate groups

✚ four-coordinated borons in tetraborate groups

■ three- and four-coordinated borons in diborate groups

(overlapping data points)

Straight lines show predicted values based on an idealized version of Krogh-Moe's theory (see Chapt. V). (Figure from the Thesis of G. E. Jellison, ref. 100).

good a proof of the theory as could be hoped for. The NMR experiment, of course, cannot "see" out beyond the basic BO_3 triangle or BO_4 tetrahedron to identify a specific polyborate unit such as those of Fig. 7, but the compositional dependencies of the five sites in Fig. 17 now seem unlikely to have any explanation other than that proposed by Krogh-Moe (45, 57).

IX. RAMAN SCATTERING

When light passes through a transparent medium the local electronic charge distributions are set into oscillation in response to the electric field vector of the light waves. These oscillating electrons act as dipole emittors causing the light to be scattered. Most of the scattered light is of the same frequency as the incident radiation, and this is known as Rayleigh scattering. However, if there should be a coupling between the electronic polarization tensor and the vibronic modes of the medium, some of the scattered photons may be augmented (or diminished) by absorption (or creation) of a quantum of vibrational energy. The latter phenomenon is known as the Raman effect.

Raman scattering in glasses was observed (128) not long after the discovery of the effect in the 1920's. However, the technique could never be fully exploited as a probe of glass structure before the advent of intense monochromatic ion-gas laser sources. Since 1970 a number of important Raman studies of glass structure have appeared in the literature (see, Ref. 94, for an extensive listing of references). In the context of borate glasses, the most important contributions have come out of Stevel's group at Eindhoven, viz, the theses of Bril (94) and Konijnendijk (96). The latter have been the primary sources for the following discussion.

It must be remarked that the Raman frequency shifts span the same energy range as the infra-red (IR) spectrum and in fact both forms of spectroscopy look at crystal or molecular vibrations. Ideally, both experiments are carried out together, since, due to selection rules some vibrations are active in Raman but not IR, while for other vibrations the reverse is true. Indeed, the success of any theoretical treatment (see below) is critically dependent on identifying reasonably complete sets of both Raman- and IR-active modes. Nevertheless, the two techniques will be discussed separately here because they have on occasion been applied individually to good advantage and because the phenomenological properties of the two types of spectra differ greatly one from another. Raman spectra of glasses are found to have the following advantageous properties (94):

(1) The observed peaks are generally well defined, limited in number, and often polarized (in cases of totally symmetric vibrations).

(2) The spectra look reasonably simple by comparison to IR spectra for the same materials (see Fig. 18, page 67).

(3) The spectra are markedly dependent on glass composition.

(4) Bulk, rather than thin-film or matrix-dispersed, samples can be employed.

(5) Raman spectra are less sensitive to surface effects.

(6) The occurrence of small amounts of water has little influence on the Raman spectra, in contrast to the IR case.

(7) High temperature measurements are more easily carried out.

(8) Lower frequencies ($\sim 200\ cm^{-1}$) are easier to measure.

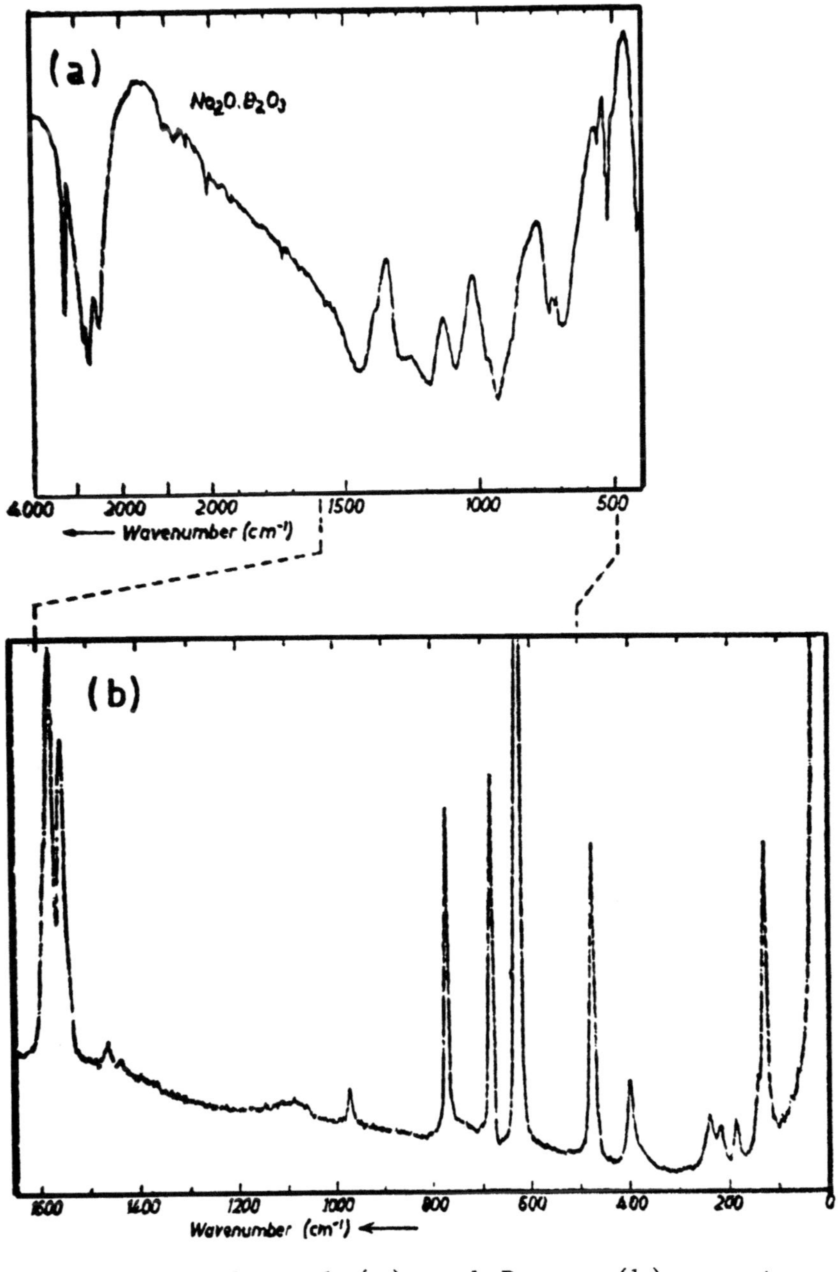

Figure 18. Infrared (a) and Raman (b) spectra for polycrystalline sodium metaborate with boron in its natural isotopic abundance (81% ^{11}B, 19% ^{10}B). (Figures taken respectively from refs. 96 and 94).

In the manner of the preceding experimental chapters, we shall want to outline what is entailed in carrying out a complete Raman spectroscopic study including theoretical model calculations. For this it will be necessary to specialize to the case a crystalline material, since the randomness in most glassy structures complicates vibrational assignments to the point where model calculations, tedious in any event, become impossible. Application to borate glasses continues to be the motivation, however. Reasonably complete studies of well characterized crystalline materials having the same compositions as the glasses are almost essential to the successful use of the "fingerprint method" in interpreting the glass spectra (see Appendix A). Specific reference in the following example is made to the work of Bril (94), where a number of ring-type metaborate compounds were selected for study.

Step (1), as always, is to prepare the sample(s). It is also essential from the standpoint of vibrational analysis that complete X-ray structural determinations have been performed on at least some of the compounds selected for study and that the structures of the Raman samples themselves be verified by X-ray diffraction. Moreover, it is wise to commence (as did Bril (94)) with a crystal structure with certain symmetries which may facilitate vibrational analysis. In Step (2), one obtains Raman spectra not only of the targeted compound, but several isomorphous materials (in order to disentangle boron-oxygen network vibrations from vibrations involving "modifier" cations). Beyond that, one also carries out whatever isotopic enrichments are possible (usually ^{10}B for the borates) and notes the effect on the experimental spectra. The polarizations of the Raman lines are determined and usually the IR spectra of the same materials are likewise obtained. In Step (3), model calculations are performed and compared with the results of Step (2). The models generated in Step (4) all involve the same fundamental molecular units already known from X-ray structural determinations. Thus, a set of vibrational frequencies are defined and classified according to the symmetry properties of the hypothetical free molecular unit, e.g., the metaborate ring $B_3O_3(O^-)_3$. The various trial models differ one from another only as regards which theoretical vibration is assigned to which experimental peak(s). Assignments are facilitated by group theoretical concepts which categorize all possible vibrations in to subsets called symmetry species, for which the relative IR activity, Raman activity, polarization, and isotope effects can be predicted. For example, totally symmetric species (e.g., the metaborate ring modes shown in Fig. 19, page 69) are known to be active in Raman but not IR, to be

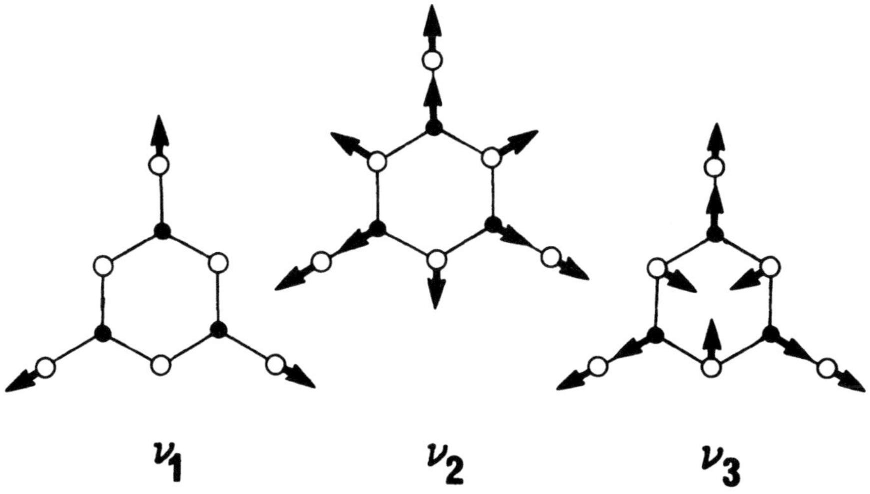

Figure 19. Symmetric vibrational modes of the "free" metaborate anion, $B_3O_6^{3-}$. All three are Raman active and polarized.

polarized, to have high oscillator strengths, and to be sensitive to isotopic substitution of ring atoms. Thus, the vibrational frequencies ν_1, ν_2, ν_3 of Fig. 19 are rather easily assigned in our example and were in fact taken the same in each trial model (94). Other peak assignments must be made on the basis of more subtle reasoning and it is at this point that several trial models must be considered. It should be mentioned that further complications arise because the vibrations of the solid compound are altered from those of the free molecular unit; the consequence changes in symmetry classification must ultimately be taken into account. Assuming all of this has been done, the model calculations are carried out by rather highly involved matrix methods (c.f., ref. 94). The end product of each such calculation is a set of force constants governing relative motions within the basic molecular unit. Of the several models tried, the best one is selected on how well it predicts the observed isotope effects and on the question, "How reasonable are the calculated force constants?" For example, shorter B - O bonds must have larger force constants for stretching modes than longer B - O bonds (129). The four basic steps described above are summarized diagrammatically in Fig. 20 (see page 71).

By comparison to the rich structure evident in the Raman spectra of borate crystalline compounds (e.g., Fig. 18b) the Raman spectra of alkali borate glasses are vastly less complicated (Fig. 21, see page 72). However, those spectral features which remain in the glass spectra can be understood by analogy with the results obtained for the crystalline compounds. Goubeau and Keller (17) originally suggested that the strongly polarized peak at 806 cm^{-1} in pure glassy B_2O_3 (Fig. 21, top) probably arises from a symmetric vibration (ν_2 in Fig. 19) of the boroxol group, in analogy with similar vibrations in several boroxol derivatives. Krogh-Moe (57) expanded on this argument by comparison of the B_2O_3 glass spectra with the IR and Raman spectra of orthorhombic metaboric acid (38). Kristiansen and Krogh-Moe (71) then carried

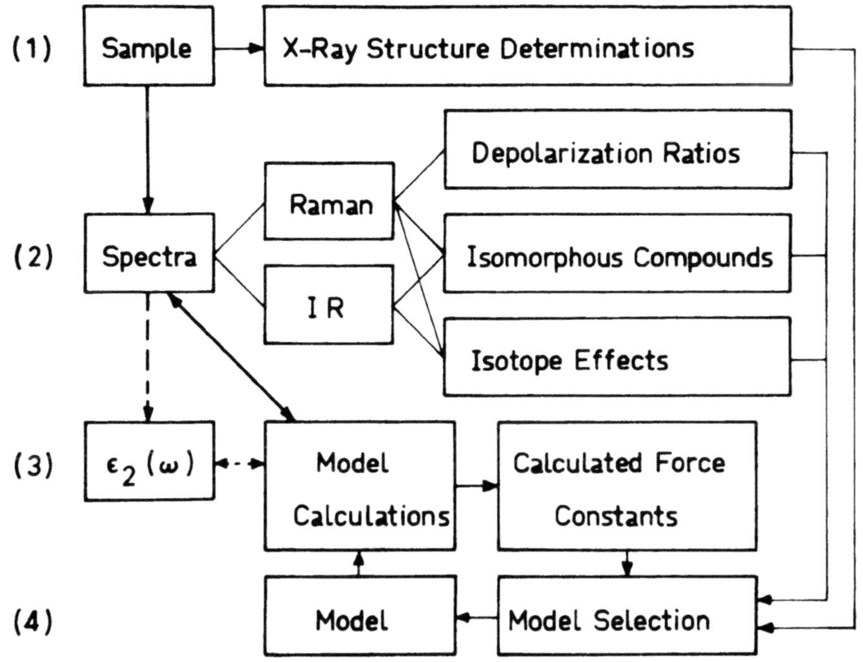

Figure 20. Block diagram illustrating the steps of a typical Raman scattering study of a polycrystalline material. The Raman spectra of glasses are best interpreted by comparison to the results obtained in this manner for compounds of similar compositions.

out normal-coordinate model calculations on the metaboric acid ion (a six-membered boron-oxygen ring with OH groups attached) and extended these to the metaborate anion (the same ring with the hydrogens removed) and to hypothetical structures where the hydrogens were substituted by more massive atoms. They were able to

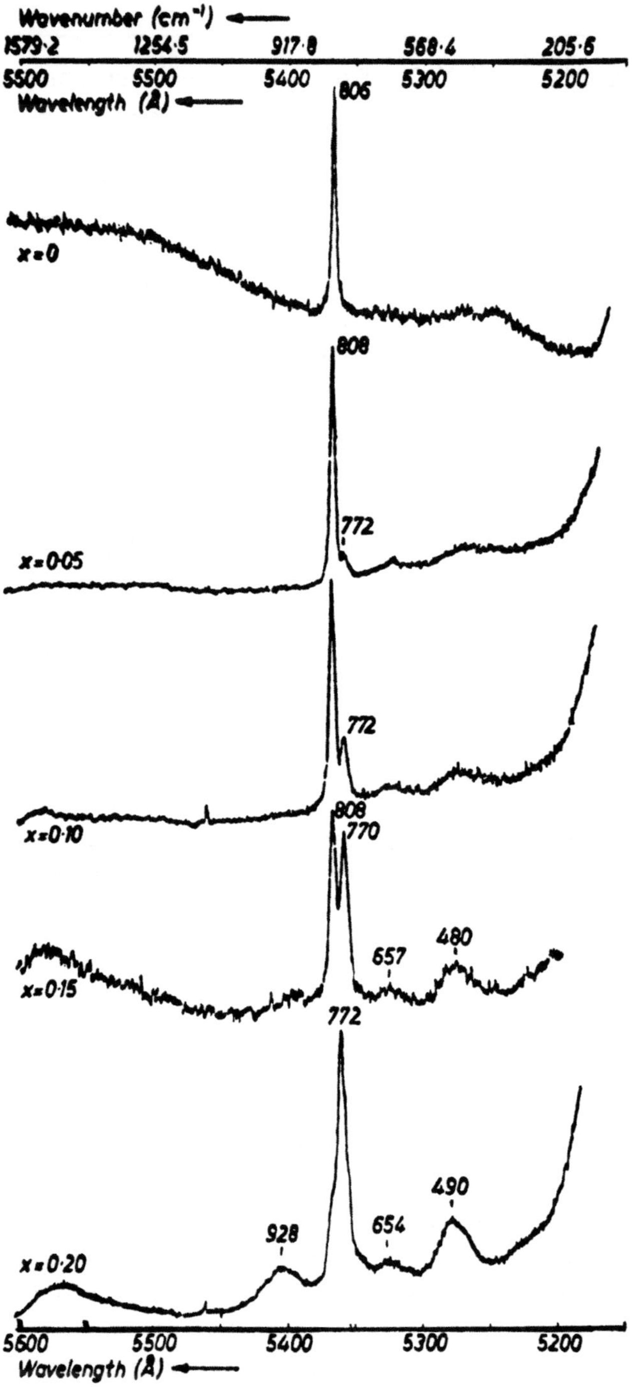

Figure 21. Raman spectra of sodium borate glasses: $xNa_2O\cdot(1-x)\ B_2O_3$. (After ref. 96).

show that while many frequencies are quite senstitive to the mass of the outer atoms, the symmetric ring breathing mode ν_2 is not (Fig. 22, page 73). Thus it was possible to understand why the sharp 806 cm^{-1} peak is characteristic of boroxol-type rings in so many different matrix environments, while other peaks (e.g., those corresponding to ν_1 and ν_3) vary from compound to compound. The best fit to the spectra for B_2O_3 glass was achieved with an outer-atom mass of 5, rather than 11 as might have been expected for boron. Enough approximations and assumptions were involved in the calculation of ref. 71 that this kind of discrepancy was not unexpected. Indeed, in an

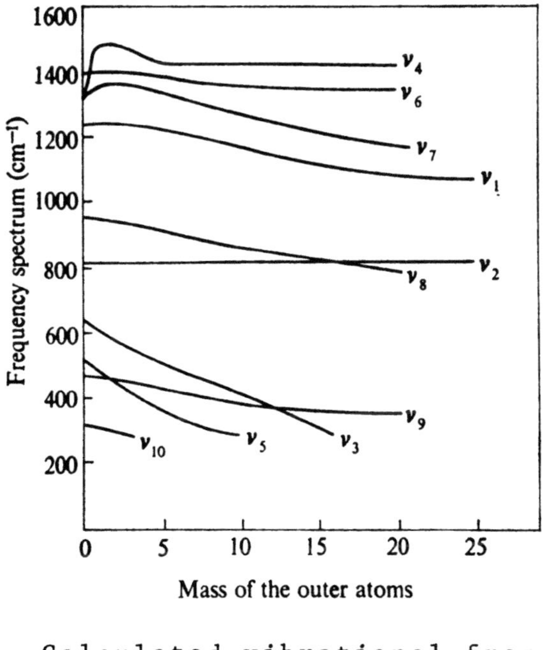

Figure 22. Calculated vibrational frequencies for a boroxol ring with atoms of various masses attached to the outer oxygens. (Figure taken from ref. 71).

extended network of boroxol rings--as has been proposed for the structure of vitreous B_2O_3 --couplings between vibronic modes of neighboring groupings must be considered (130). One may speculate that to a crude approximation such couplings may produce the same effect as a variable outer-atom mass in Kristiansen and Krogh-Moe's calculation. It is quite instructive, though not necessarily physically accurate, to imagine further that the randomness of the vitreous state alters the vibronic modes in the same way as a statistical distribution in outer-atom masses. Referring to Fig. 22, it can be seen that such a distribution would smear out all peaks but the one corresponding to ν_2 (810 cm^{-1}). It is almost certain that some effect of this nature is responsible for the simple one-sharp-peaked Raman spectrum of pure B_2O_3 glass (Fig. 21, top).

Bril (94) has emphasized that there are important differences between the boroxol group and the isolated metaborate ring. Indeed, Kristiansen and Krogh-Moe overlooked these differences since they did not have Raman data for any metaborate compounds and hence could not know that ν_2 for the metaborate ring must be $\approx$ 770 cm^{-1} (94) rather than 823 cm^{-1} as guessed from IR data. This is an important point when one turns to interpreting the Raman spectra of alkali borate glasses. Bril (94) was able to argue quite cogently that higher values of ν_2 are to be associated with higher intra-annular force constants. Higher intra-annular force constants are in turn associated with a degree of intra-annular π bonding, or aromatic character. It can be anticipated that boroxol groups with their outer oxygens bridging, as in B_2O_3 glass, may be somewhat aromatic --and this is supported by recent NMR studies (previous chapter). On the other hand, the metaborate anion, with its large intra-annular B - O distances, evidently undergoes π bonding only within the short extra-annular B - O bonds. Thus, ν_2 is lower for the metaborate ring (770 cm^{-1}) than for the boroxol ring in glass (806 cm^{-1}).

It can be seen in Fig. 21 that as alkali oxide is added to a B_2O_3 melt a new Raman peak at 770 cm^{-1} arises (96). Since small alkali oxide additions are known from NMR to result in four-coordinated borons (see Chapt. VIII), it would be inappropriate to ascribe the 770 cm^{-1} line in low-alkali glasses to a metaborate ring. However, Krogh-Moe (45, 57) had postulated the existence of pentaborate, tetraborate, and triborate structural groupings in this composition range. And, as aptly pointed out by Bril (94), no π bonding is possible with four-coordinated borons, since all valence orbitals are used up in forming σ bonds. Thus, Bril suggested that six-membered ring structures containing one or more BO_4 configurations may have breathing-mode frequencies similar to that of the metaborate ring, where aromatic character is also suppressed. This gives weight to a similar conclusion reached by Konijnendijk (96) on purely empirical grounds.

It is evidently not possible to discriminate among pentaborate, tetraborate, triborate, or metaborate rings on the basis of the position of the principal Raman line (96). However, Bril pointed out (94) that the amplitude ratio of the 806 cm^{-1} line to the 770 cm^{-1} line should be proportional to the ratio of the number of boroxol groups to the total number of rings involving four-coordinated borons if only boroxol, pentaborate, tetraborate and triborate groups are present. He further demonstrates that h_{806}/h_{770} is <u>equal</u> to the ratio of boroxol groups to triborate rings in the caesium enneaborate compound which is known (103) to contain only these two groups. Thus it seems possible that the amplitude ratios will be equal to the ring-group ratios in the glasses as well. Bril therefore measured h_{806}/h_{770} for a range of alkali borate glasses and compared the results to the theoretical boroxol-to-other-ring ratios for the three hypothetical cases in which all four-coordinated borons are incorporated as (1) triborate rings, (2) tetraborate groups, or (3) pentaborate groups. His results, reproduced as Fig. 23 (see page 76) of this review, suggest that some diborate groups (or other groups characteristic of high alkali-to-boron ratios) must be present at all compositions above 5 mol% alkali

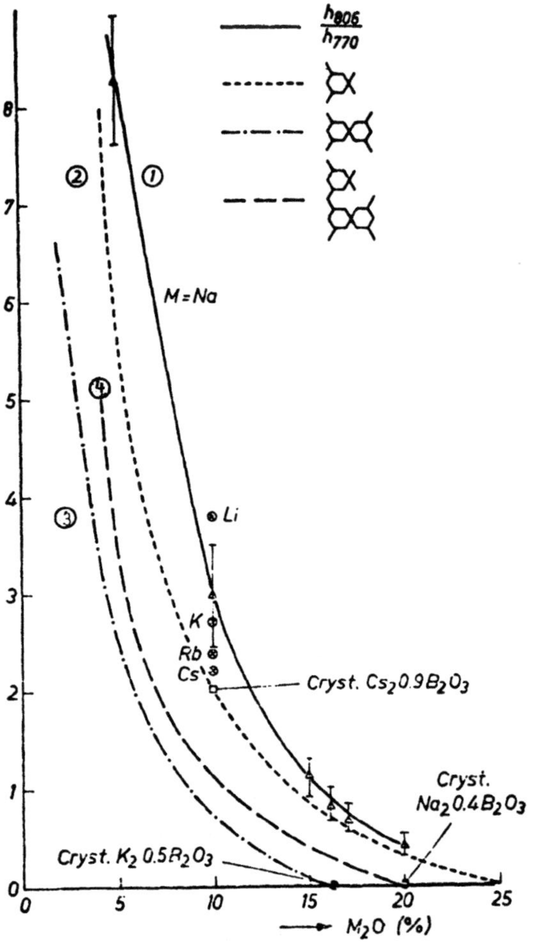

Curve 1: the ratio of the peak heights of the peaks at 806 cm^{-1} and 770 cm^{-1} for sodium borate glasses. The ratio of the peak heights of the peaks at 806 cm^{-1} and 770 cm^{-1} is also given for some other alkali borate glasses of the composition 10% M_2O . 90% B_2O_3.
Curve 2: the calculated ratio of the maximum number of boroxol rings to the number of triborate rings (all BO_4 units are situated *in* the rings).
Curve 3: the calculated ratio of the maximum number of boroxol rings to twice the number of pentaborates groups.
Curve 4: the calculated ratio of the maximum number of boroxol rings to three times the number of tetraborate groups.

Figure 23. Ratios of various hypothetical ring structure populations in alkali borate glasses compared to the observed peak-height ratio of the Raman lines at 806 and 770 cm^{-1}. Theoretical curves are based on various idealized versions of Krogh-Moe's structural theory (see Chapt. V). (Figure taken from ref. 94).

oxide. This is consistent with a slight dissociation of the melt into boroxol plus diborate groups--a possibility envisioned by Krogh-Moe (45, 57). His conclusion (45, 57) that tetraborate groups predominate in alkali borate glasses of 20 mol% alkali oxide content is in no way contradicted by the results of Fig. 2. Konijnendijk (96) arrived at similar conclusions by semiquantitative comparisons of the Raman spectra (including peaks other than the principal 806 and 770 cm^{-1} lines) of a vast range of borate glasses. Indeed, considering a number of possible models for glass structure and, by a process of elimination subject to the constraints of NMR data, Konijnendijk was able to support Krogh-Moe's conclusions in considerable detail.

In summary, Raman scattering is a powerful technique for probing the structure of borate glasses. To date, however, the Raman spectra do not provide unambiguous signatures of various ring structures, e.g., the observation of a Raman peak at 806 cm^{-1} in B_2O_3 glass alone does not prove the presence of boroxol rings. However, interpretation of this peak within the random network theory (1) is unsatisfactory (96). If the 806 cm^{-1} line were due simply to BO_3 triangles, NMR evidence (Fig. 4) implies a possible decrease in intensity by no more than 50% in glasses containing ~35 mol% modifier oxide. Yet experimentally (94-96) the 806 cm^{-1} peak disappears when the modifier oxide exceeds ~25 mol%. For the two decades prior to 1970 it was chiefly Raman data which supported the existence of boroxol groups in B_2O_3 glass; the best present evidence has come from X-ray diffraction (77). Thus, Raman and IR studies of borate glasses have historically depended upon the results of other techniques for final conclusions to be drawn. This dependence seems in the process of being overcome in the case of pure SiO_2 glass (131) and ultimately may be similarly surmounted in the case of the borates.

X. INFRA-RED

Infra-red spectroscopy, like Raman, is a probe of vibrational states. As alluded above, the two techniques belong together but are being discussed separately mainly out of historical considerations. The present chapter comprises essentially a synopsis and critique of the IR studies of borate glasses and compounds published in 1965 by Krogh-Moe (57). Many other important IR studies of glassy boron oxide and alkali borate glasses have appeared in the literature (we mention particularly refs. 23, 25-27, and 132), but in retrospect it has been Krogh-Moe's work which first converged upon what appears to be the correct structural interpretations.

By 1965, X-ray studies of a number of acid crystalline anhydrous borates (principally due to Krogh-Moe, e.g., refs. 32, 43, and 54) had revealed the presence of three-dimensional networks based on characteristic borate groups (Fig. 7, page 35). These same groups were known to exist as isolated polynuclear anions in several hydrated borates. Thus it appeared that dehydration of the latter compounds was equivalent to the polymerization of borate groups. Melting-point depression studies (45) (following chapter) and preliminary IR work (31) suggested that similar, if more random, polymerization of groups may take place in borate glasses. Krogh-Moe (57) reasoned that IR spectroscopy could be a valuable tool to further test this notion, provided it could be shown (1) that various isolated polyborate anions could be discriminated among on the basis of their spectra and (2) that they could still be recognized on this basis following polymerization. To do this, it was necessary to obtain IR spectra not only for the hydrated and anhydrous compounds but also for deuterated equivalents, in order that the boron-oxygen skeletal motions of the hydrated structures could be discriminated from OH vibrations. The deuteration experiment demonstrated (57) that most peaks in the wavelength regime above 5 microns were due to skeletal vibrations. Comparisons of the normal hydrated compounds with their anhydrous counterparts are illustrated in Fig. 24 (see page 80). In this figure, considerable similarity can be noted between the various pairs of spectra, suggesting that polyborate structural groupings might indeed be recognized by their IR spectra in spite of obvious broadening effects due to polymerization. Certain new peaks near 1250 cm^{-1} and 870 cm^{-1} in the spectra of the anhydrous compounds were singled out by Krogh-Moe as probably arising not from the borate groups themselves but from their interlinkages.

Comparisons of the IR spectra of some alkali borate glasses with those of the corresponding anhydrous compounds are shown in Fig. 25 (see page 81) (44, 57). The agreement seen here is rather good on first glance, but a substantial case for the existence of the same structural groupings in both the glassy and crystalline forms can only be built upon a detailed peak-by-peak discussion of the discrepancies as well as the similarities in the spectra. Krogh-Moe, for example, pointed out that the tetraborate compound is composed of interpenetrating twin networks each comprised of pentaborate and triborate groups (54) and that these two groups alone (not counting their interlinkages) would account for 38 IR-active absorption bands. This is clearly far more than are experimentally resolved. Moreover, it would be expected that the spectrum of the tetraborate compound would be a superposition of the spectra of the pentaborate and triborates compounds. Krogh-Moe (57) noted that "a complete accordance is not achieved, however, showing that the simple approach has its limitations." Another illustration of these limitations is the fact that the IR spectra of crystalline and glassy B_2O_3 bear strong resemblances to one another (31), whereas the actual structures of these two phases are evidently distinctly different (66, 77).

Many of the discrepancies between the crystal and glass spectra of Fig. 25 were explained by Krogh-Moe (57) in terms of two effects: (1) a general broadening of the glass peaks due to disorder and (2) the effects of partial dissociation in the melt. That is, a tetraborate melt

(a)

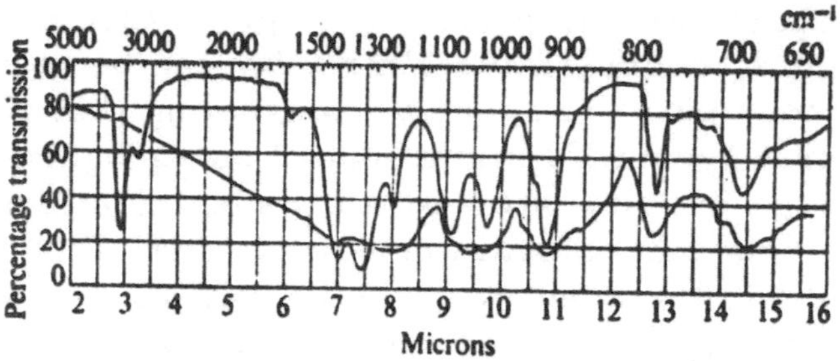

The infra-red spectra of anhydrous crystalline potassium pentaborate, $K_2O.5B_2O_3$ (lower curve) and potassium pentaborate octahydrate, $K_2O.5B_2O_3.8H_2O$ (upper curve)

(b)

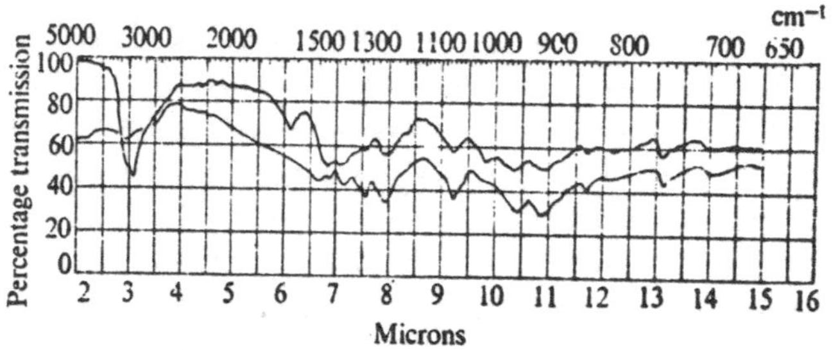

The infra-red spectra of anhydrous crystalline caesium triborate, $Cs_2O.3B_2O_3$ (lower curve) and the same phase exposed to the humidity of the air for a short time (upper curve)

(c)

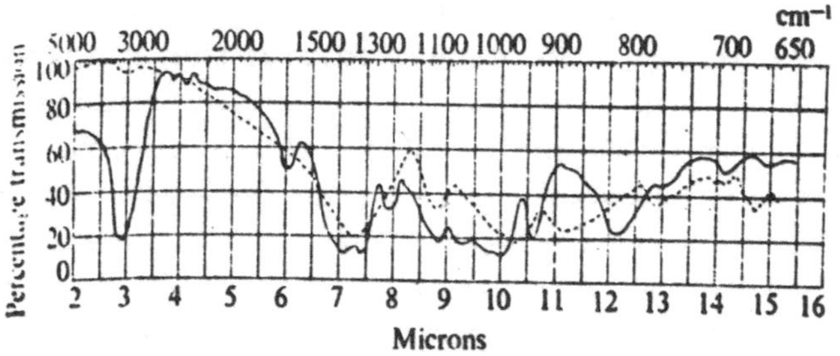

The infra-red spectra of anhydrous crystalline lithium diborate, $Li_2O.2B_2O_3$ (dashed curve) and sodium diborate decahydrate, $Na_2O.2B_2O_3.10H_2O$ (fully drawn curve)

Figure 24. Infrared spectra of three anhydrous borate compounds with comparison to spectra of hydrated compounds known to contain the same boron-oxygen ring structures. (Figures taken from ref. 57).

(a)

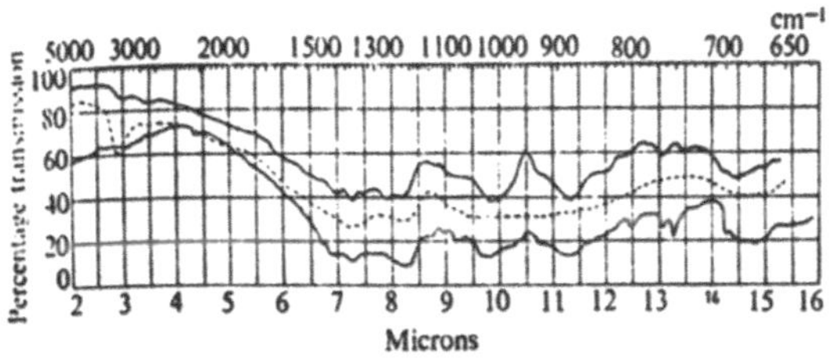

The infra-red spectra of anhydrous crystalline sodium tetraborate, $Na_2O.4B_2O_3$ (upper solid curve) and anhydrous crystalline silver tetraborate, $Ag_2O.4B_2O_3$ (lower solid curve). The spectrum of sodium tetraborate glass is given as a dashed curve

(b)

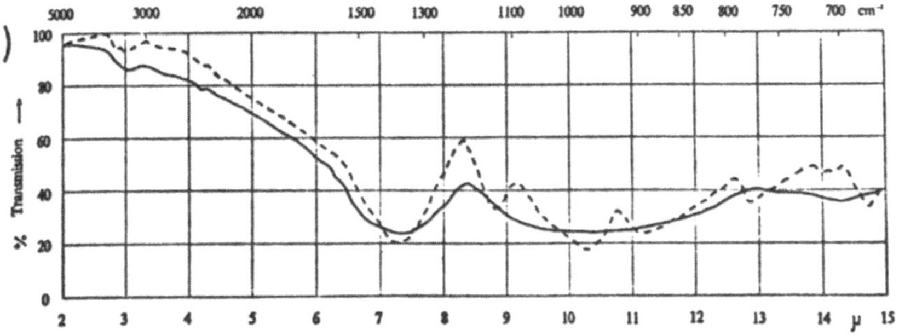

The infra-red absorption spectra of lithium diborate glass (solid line) and anhydrous crystalline lithium diborate (dashed line)

Figure 25. Infrared spectra of two alkali borate glasses with comparison to the spectra of polymerized anhydrous compounds of the same boron-to-oxygen ratios. (Figures taken from refs. 44 and 57).

might consist mainly of pentaborate-triborate pairs, but a certain fraction of these might be expected to dissociate into boroxol plus diborate groups (45). (Recall from the previous chapter that recent Raman results (94) give strong evidence that this does in fact occur to a measurable degree). Krogh-Moe (57) remarked that the presence of some diborate groups could very well account for certain specific discrepancies between the IR spectrum of the tetraborate glass and that of the corresponding compound.

Moreover, lowering the alkali oxide content of the glass would certainly move the melt equilibrium in the direction of the boroxol group and away from the diborate group, thus accounting for the better observed agreement between the IR spectrum of a 15 mol % sodium borate glass (27) and that of the sodium tetraborate compound.

Krogh-Moe (57) admitted that his interpretation of the infra-red absorption curves of borate glasses was not the only possible one, so far as the spectra themselves were concerned. But he did feel that when taken together with other forms of evidence the structural picture advanced on the basis of the above IR data was fairly compelling. Even without the hindsight of recent developments, the author would be inclined to agree with that judgement. Krogh-Moe's 1965 IR studies are perhaps among the all-time most successful applications of the "fingerprint method," a procedure against which some cautions have been mounted in Appendix A.

The full potentialities for IR spectroscopy in borate glass research are yet to be realized--and may be unrealizable in the near future due to the immense complexities involved. However, progress could be made by carrying out concurrent IR and Raman studies of both glassy and crystalline materials, as has been done recently (96), but with a carefully developed program for presenting IR band shapes in terms of $\epsilon_2(\omega)$, the imaginary part of the dielectric constant. Further processing of the data and model calculations would then be carried out much in the manner as has been done recently for crystalline α-quartz and glassy silica by Gaskell and Johnson (131). In general overview, the outline of a complete experiment would be the same as described in the previous chapter on Raman scattering (see especially Fig. 20).

XI. MELTING POINT DEPRESSION

Thermodynamic data can often be used quite effectively in discussing the structures of molten salts and ceramic oxides (18, 45, 133). The information derived from such "phase diagram methods" is not of the same precise nature as that flowing from the experimental techniques described in the preceding chapters; bond angles, bond lengths, and other geometric factors, for example, cannot be determined. Nevertheless, some very specific conclusions can be drawn regarding the sizes and atomic compositions of the identifiable atomic groupings present in the melt as functions of composition. If the melt should be highly viscous near the liquidus (as is true for the alkali borates), it can be reasonably inferred that the same atomic groupings are present as well in the corresponding glasses (supercooled liquids). Furthermore, it is entirely probable that the geometric arrangements of atoms within these groupings are the same as those in at least some of the compounds which exist on the phase diagram. Thus, the "phase diagram method" can in principle yield very specific information about the structure of glasses. The fact that this is evidently the case in practice is illustrated by Krogh-Moe's classic study (45) of melting point depression in the sodium borate system, to be discussed below.

We begin by presenting a minimal sketch of the theoretical background (see ref. 45 for additional details). Consider a compound A melting at the temperature T_f with a heat of fusion ΔH_f. In molten A, let a given proportion of a compound B be dissolved and assume B to be insoluble in solid crystallized A. From the molten mixture of A and B, pure crystallized A will start to separate at the temperature T_m. We define $\Delta \overline{S}_A$ as the partial entropy of molten A in the mixture relative to pure A at the same temperature. Krogh-Moe (45) gives the derivation of an expression for $\Delta \overline{S}_A$ involving several reasonable approximations; the result can be expressed

$$\Delta \overline{S}_A = \frac{\Delta H_f}{T_f}\left(\frac{T_f - T_m}{T_m}\right) - \frac{\Delta c_p}{2}\left(\frac{T_f - T_m}{T_m}\right)^2, \qquad (5)$$

where Δc_p is the difference in specific heat of crystallized A and supercooled molten A at the temperature T. Values of ΔH_f and Δc_p have been determined for the sodium diborate and tetraborate compounds by Smith and Rindone (134). Phase diagrams for the system sodium oxide - boron oxide have been reported by Morey and Merwin (5), who found no evidence for solid solubility of any of the compounds therein. The liquidus curves provide T_m as a function of melt composition and, of course, $T_m = T_f$ at the composition of a pure compound such as the diborate or tetraborate. Thus, at least for the sodium borate system, experimental estimates of $\Delta\overline{S}_A$ are possible over ranges of compositions centered on $Na_2O \cdot 2B_2O_3$ and $Na_2O \cdot 4B_2O_3$. (Values of ΔH_f and Δc_p are not generally available for many other glass-forming systems).

The essence of the "phase diagram method" is to derive <u>theoretical</u> expressions for $\Delta\overline{S}_A$ as functions of melt composition under a wide variety of assumptions regarding the structural elements which comprise the melt. That is, one performs a number of model calculations for comparison against the experimental data--in exactly the same spirit as such model calculations are performed in interpreting X-ray or NMR data (see respective chapters above). If the melt composition is expressed as $xNa_2O \cdot (1-x)B_2O_3$, each model calculation yields $\Delta\overline{S}_A(x)$. Krogh-Moe's practice was to insert these expressions into Eq. (5) and solve for T_m, so that the predictions of each model could be compared directly against the experimental liquidus curve. While a good fit to the data does not necessarily imply the correctness of a given model (just as in the X-ray and NMR cases), a poor fit is conclusive grounds for rejecting it. The ability to unambiguously reject some models elevates the "phase diagram method" substantially above the usual "fingerprint method" employed in some spectroscopic approaches (Appendix A).

It is appropriate now to outline the method by which $\Delta\overline{S}_A(x)$ can be calculated under various model assumptions. We will specifically consider only the case of binary mixtures of structural elements, although Krogh-Moe (45) has shown how ternary mixtures may be treated as well.

Suppose that n_A mol of compound A are mixed with n_B mol of a compound B. The ensuing mixture is taken to consist of two structural elements, α and β, in amounts n_α and n_β respectively. For any specific model postulated for the mixture, it is possible to express n_α and n_β as functions of n_A and n_B. Usually, linear relationships can be found:

$$n_\alpha = s_1 n_A + t_1 n_B, \tag{6}$$

$$n_\beta = s_2 n_B + t_2 n_B. \tag{7}$$

The constants s and t will vary with the model. Following experience for a wide variety of molten salts, it is assumed that the entropy of mixing is mainly configurational (133). Therefore an expression for the entropy of mixing $\Delta \bar{S}$ can be written down in terms of the statistical weights associated with the interchange of equivalent structural elements (see ref. 45 for the explicit formulation). Then the partial molar entropy of A in the mixture, $\Delta \bar{S}_A$, is obtained as

$$\begin{aligned} \Delta \bar{S}_A &= \frac{\partial \Delta \bar{S}}{\partial n_A} \\ &= -R \left[\frac{\partial n_\alpha}{\partial n_A} \ln \frac{n_\alpha}{n_\alpha + n_\beta} + \frac{\partial n_\beta}{\partial n_A} \ln \frac{n_\beta}{n_\alpha + n_\beta} \right] - S_A^{o}, \end{aligned} \tag{8}$$

where R is the gas constant and S_A^{o} is a constant which can be evaluated (45) but is usually zero for cases of interest. The use of Eq. (8) to obtain $\Delta \bar{S}_A(x)$ under various model assumptions can be illustrated by several simple examples provided by Krogh-Moe (45):

Example 1. Suppose that the structural elements α and β are molecular A ($=B_2O_3$) and B ($=Me_2O$) respectively.

$$n_\alpha = n_A = 1 - x$$

$$n_\beta = n_B = x$$

Hence $\Delta \bar{S}_{B_2O_3} = -R \ln(1 - x)$.

Example 2. A mixture of n_A mol B_2O_3 with n_B mol of metal oxide, Me_2O, is taken to be a completely random distribution of boron atoms in four-fold coordination among boron atoms in three-fold coordination. $n_\beta = n_{BO_4}$ and $n_\alpha = n_{BO_3}$ are obtained from Eq. (2), giving:

$$n_\alpha = 2n_A - 2n_B = 2(1 - 2x)$$

$$n_\beta = 2n_B = 2x$$

Hence $\Delta\bar{S}_{B_2O_3} = -2R \ln \frac{1 - 2x}{1 - x}$.

Example 3. Suppose that in a mixture of n_A mol of $Me_2O \cdot 4B_2O_3$ with n_B mol of $Me_2O \cdot 2B_2O_3$ there are n_α mol of the tetraborate group and n_β mol of the diborate group (see Fig. 2). Eqs. (6) and (7) in this case simply become:

$$n_\alpha = n_A = \frac{1}{2}(1 - 3x)$$

$$n_\beta = n_B = \frac{1}{2}(5x - 1)$$

So that $\Delta\bar{S}_{Me_2O \cdot 2B_2O_3} = -R \ln \frac{n_B}{n_A + n_B}$,

or, converting to the mol fraction scale of the system $Me_2O - B_2O_3$

$$\Delta\bar{S}_{Me_2O \cdot 2B_2O_3} = -R \ln \frac{5x - 1}{2x} .$$

The specific results of Krogh-Moe's melting point depression analysis of the sodium borate system will now be recounted. First to be considered was the depression resulting from additions of B_2O_3 to molten sodium diborate (Fig. 26a). The circles in this figure represent the experimental liquidus curve, while the fully drawn curves are the theoretical predictions based on Eqs. (5) through (8) under various model assumptions. Curve 1 in Fig. 26a is the predicted behavior if the melt dissociates completely into BO_4 tetrahedra and BO_3 triangles in a ratio given by Eq. (2). Still flatter depressions would result for lower values of N_{BO_4}. Other models of the same type, such as BO_4 pairs mixed with BO_3 triangles, offered no improvement in the fit over Curve 1. Hence it was firmly concluded that sodium diborate is not dissociated in the melt. The other models that were tried were then based on mixtures of the diborate group (Fig. 7) with other entities. Mixtures of the diborate group with triborate groups (curve 4) and of the diborate group with a hypothetical "small tetraborate group" $B_4O_{6.5}^-$ (curve 3) gave depressions which were too severe, and hence these models also were rejected. On the other hand, a mixture of diborate groups with normal tetraborate groups (Fig. 7) gave excellent agreement (curve 2 of Fig. 26a). Krogh-Moe noted that a mixture of diborate and boroxol groups also gave an acceptable fit, demonstrating the kind of nonuniqueness that can occur in this type of model calculation. However, the existence of substantial numbers of boroxol groups above 20 mol % Na_2O could be ruled out by calculating melting point depressions of $Na_2O \cdot 4B_2O_3$ melt by addition of Na_2O (Fig. 26b). Curve 1 of Fig. 26b pertains to complete dissociation into BO_4 and BO_3 units, but an equally unsatisfactory curve was obtained by assuming complete dissociation into boroxol and diborate groups:

$$3 \text{ tetraborate} \rightleftarrows 3 \text{ diborate} + 4 \text{ boroxol}. \qquad (9)$$

Thus, Krogh-Moe (45) concluded that the equilibrium in Eq. (9) must be displaced to the left. Whence, there remains only one known interpretation of the melting point depression of sodium diborate (Fig. 26a), namely, the units of the melt are diborate and tetraborate groups.

Continuing with the analysis of Fig. 26b, a mixture of tetraborate and diborate groups (curve 3) gave a fair fit for the depression of $Na_2O \cdot 4B_2O_3$ by sodium oxide additions. An excellent fit was achieved (curve 2) by allowing 10% dissociation according to Eq. (9). It can be recalled that a slight degree of dissociation into diborate and boroxol groups is now strongly suggested by recent Raman studies (94-96) (see especially Fig. 23). The alternative interpretation of the data of Fig. 23 in terms of the sole presence of boroxol and triborate groups is emphatically ruled out by the melting point depression studies under discussion, as witnessed by the poor fit of curve 4 of Fig. 26b. Finally, the very same assumptions which went into calculating curve 2 of Fig. 26b also led to the excellent agreement between the calculated and experimental depression of $Na_2O \cdot 4B_2O_3$ by additions of B_2O_3 (Fig. 26c, curve 2).

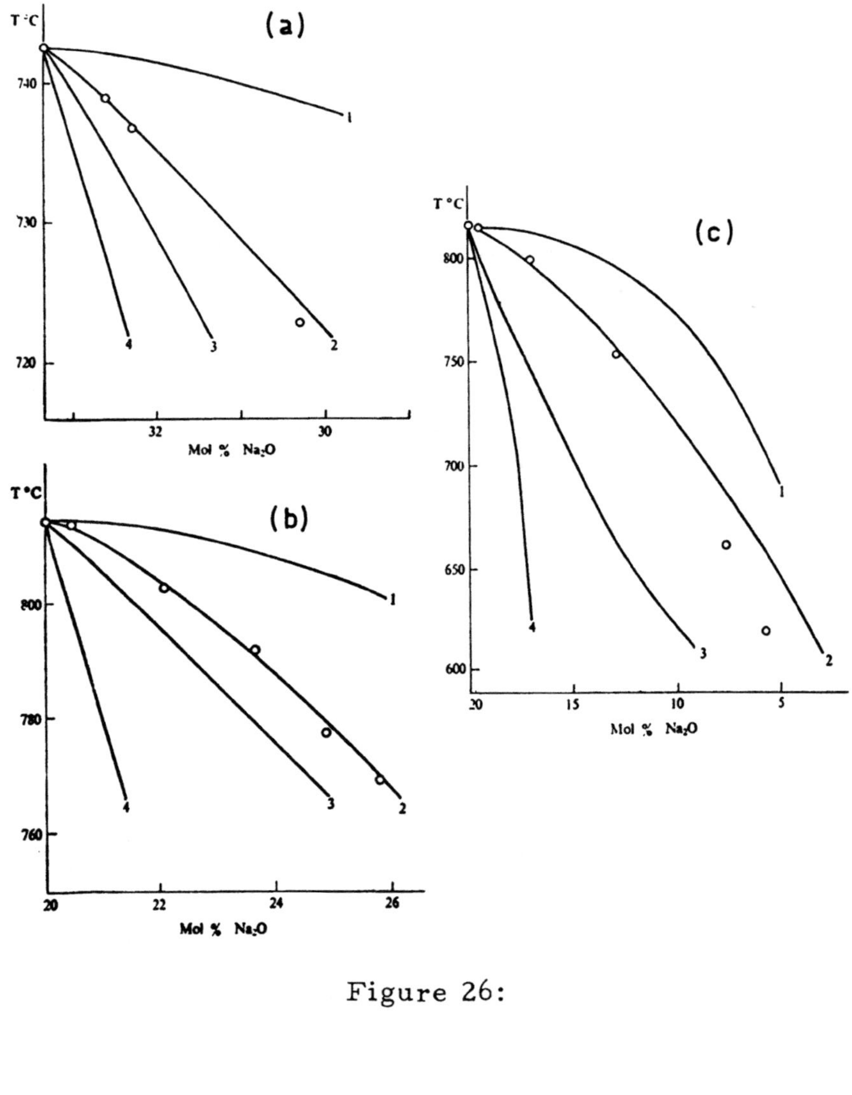

Figure 26:

Melting point depression studies provide perhaps the strongest evidence of any type for the existence and relative prevalence of tetraborate and diborate groups in alkali borate glasses. As has been mentioned, the Raman data (94-96) alone could have been interpreted in several different ways; IR evidence (57) was suggestive but still sketchy. Recent NMR studies (100) are now capable of showing five different boron sites, but this finding alone says nothing about the sizes of the polyborate groupings in which the respective sites are situated. X-ray evidence for pairing of modifier cations (33-36) does tend to support the existence of tetraborate groups over pentaborate and triborate groups, however, as pointed out by Krogh-Moe (45). All considered, the "phase diagram

Figure 26. Parts of the phase diagram for the system Na_2O-B_2O_3, showing the melting point depression (a) of $Na_2O \cdot 2B_2O_3$ by B_2O_3, (b) of $Na_2O \cdot 4B_2O_3$ by Na_2O, and (c) of $Na_2O \cdot 4B_2O_3$ by B_2O_3. Experimental values according to Morey and Merwin(5) are shown as circles. Curves were calculated by Krogh-Moe (45) on the basis of various structural models for the melt, assuming random distributions of the following structural groupings:

In (a): Curve 1: BO_4 tetrahedra and BO_3 triangles: Curve 2: diborate and tetraborate groups: Curve 3: diborate and $B_4O^-_{6.5}$ groups: Curve 4: diborate and triborate groups.

In (b): Curve 1: BO_4 tetrahedra and BO_3 triangles: Curve 2: diborate, boroxol, and tetraborate (with 10% dissociation): Curve 3: tetraborate and diborate groups: Curve 4: tetraborate and triborate groups.

In (c): Curve 1: BO_4 tetrahedra and BO_3 triangles: Curve 2: boroxol, diborate, and tetraborate (with 10% dissociation): Curve 3: tetraborate and BO_3 triangles: Curve 4: tetraborate and pentaborate groups. (Figures taken from ref. 45).

method" deserves to rank high among the structure sensitive techniques which have been successfully applied to borate glasses.

XII. ESR : DEFECTS

Radiation-induced paramagnetic defect centers are potentially excellent probes of glass structure when studied by the technique of electron spin resonance (ESR). In order for this potential to be realized, however, one must accomplish two things: first, to correctly identify and characterize in detail the defect itself and, second, to recognize in what ways the structure of the defect is related to the structure of the "perfect" glass. In many cases, the first task has been performed with enormous success, as will be illustrated below by examples from the alkali borate and the borosilicate glass systems. The second task is generally much more difficult than the first, and the present chapter will focus on the nature of these difficulties as well as the successes which have been achieved.

The ESR phenomenon has much in common with that of NMR, described in a previous chapter. Whereas NMR dealt with nuclei having quantum mechanical spin I, ESR primarily concerns electrons with spin S (= 1/2 for most paramagnetic defects of interest). The electron like the nucleus has a magnetic moment, this given by $\vec{\mu} = -g_e\beta\vec{S}$, where β is the electron Bohr magneton and g_e is the free electron g value (= 2.0023). An externally applied magnetic field $\vec{H}$ will then set up electronic Zeeman energy levels in complete analogy to Eq. (3), where m must now be replaced by M, the electron-spin magnetic quantum number (which may take on values $M = \pm 1/2$ for $S = 1/2$). Transitions between the two Zeeman sublevels may be stimulated by <u>microwave</u> radiation of frequency ν_0 such that, in analogy to Eq. (4),

$$h\nu_0 = g_e\beta\left|\vec{H}\right|. \qquad (10)$$

For a given magnetic field, the ratio of the ESR to NMR frequencies is of the order of $\beta/\mu_0 = 1836$, which is the ratio of the proton mass to the electron mass.

As in the NMR case, the simple one-peaked spectrum described by Eq. (10) is of minimal interest. What is of interest are measurements which reflect the unpaired electron's interaction with its environment. In the NMR case, it was the nuclear electric quadrupole interaction which was of the most value; for ESR of paramagnetic defects, we shall be concerned with the spin-orbit, orbit-Zeeman, and hyperfine interactions. It is convenient to describe these effects in terms of an effective spin Hamiltonian

$$\mathcal{H} = \vec{H} \cdot \vec{\vec{g}} \cdot \vec{S} + \vec{I} \cdot \vec{\vec{A}} \cdot \vec{S}, \tag{11}$$

where the spin-orbit and orbit-Zeeman interactions (whose names we can now safely forget) are subsumed into the first term and the hyperfine interaction is represented by the second term. The experimental quantities which one seeks to measure are the principal-axis components of the g tensor, g_{xx}, g_{yy}, and g_{zz}, and of the hyperfine tensor A_{xx}, A_{yy}, and A_{zz}.

We shall touch only briefly on the means by which the g factors and hyperfine coupling constants are extracted from experimental glass spectra; more detailed discussions are available in numerous places, e.g. Refs. 70, 78, 122, and 135. First, a resonance condition is derived by diagonalization of Eq. (11), giving the magnetic field at which resonance occurs as a function of ν_0, g_{xx}, g_{yy}, g_{zz}, A_{xx}, A_{yy}, A_{zz}, m, and θ and φ, the polar or Euler angles describing orientation of $\vec{H}$ with respect to the principal axis system of $\vec{\vec{g}}$ and $\vec{\vec{A}}$. Here, m is the magnetic quantum number of the nucleus with which the unpaired electron spin is undergoing the hyperfine interaction. For powdered crystals or glasses, all solid angle orientations are equally probable and--just as in the NMR case--the resonance condition must be angularly averaged, resulting in the calculation of "powder patterns" (122). In glasses, it has been known since the 1968 work of Griscom, Taylor, Ware, and Bray (70) that vitreous disorder has a tendency to impose rather broad statistical distributions on at least some of the spin Hamiltonian parameters. The effects of angular averaging and of distributions in spin Hamiltonian parameters can both be computer simulated. The program of Taylor and Bray (70, 78) performs the angular averaging first and handles the distribution effects second, while the formulation due to Peterson, Kurkjian, and Carnevale (136) performs both functions in one step. The approach of Peterson et al is more elegant but generally consumes more computer time; as pointed out in the NMR context, the results of both algorithms are equivalent (124). It should be noted that, whereas angular averaging is carried out "automatically" in these programs, the distributions of spin Hamiltonian parameters must be estimated and confirmed by cut-and-try procedures. The end product of computer simulation is usually a rather precise set of g values and A values and their distributions. The uses to which this information can be applied are partially summarized in the next two paragraphs.

The defects of primary interest in borate glasses are mostly of the type wherein the wavefunction ψ_o of the unpaired spin is highly localized on a single atom in the glass matrix. If this particular atom happens to have a nuclear magnetic moment ($I \neq 0$), the hyperfine coupling constants, A_{xx}, A_{yy}, and A_{zz}, may be measured (generally subject to confirmation by isotopic substitution). If so, and the symmetry is axial ($A_{xx} = A_{yy} = A_\perp$, $A_{zz} = A_\parallel$), it is immediately known that

$$\psi_o = \alpha|ns\rangle + \beta|np\rangle + \gamma(\text{other}), \qquad (12)$$

where $|ns\rangle$ and $|np\rangle$ are atomic orbitals of the magnetic nucleus and

$$\alpha^2 = \frac{A_{iso}}{A_{ns}} \equiv \frac{(A_\parallel + 2A_\perp)}{3A_{ns}}, \qquad (13)$$

$$\beta^2 = \frac{A_{aniso}}{A_{np}} \equiv \frac{(A_{\parallel} - A_{\perp})}{3A_{np}} \qquad (14)$$

In Eqs. (13) and (14), A_{ns} and A_{np} are atomic hyperfine coupling constants available in the literature from atomic beam data or theoretical calculations (137). From this information, one knows not only what kind of atom the unpaired spin is located upon, but often something about the geometry of the complex into which it is bonded. The sensitivity of these data to geometry is particularly well illustrated in the cases of isolated AX_2 and AX_3 molecules, where bond angles can be calculated with reasonable accuracy simply from the ratio $\xi^2 = \beta^2/\alpha^2$ (Fig. 27, page 94) (138). It will be seen later on that trapping sites analogous to the molecular structures of Fig. 27a have been identified in borate glasses (where A = B and X = a bridging oxygen).

The three principal values of the g tensor can also be related to the structure of the defect and thence to the atomic arrangements surrounding it, but interpretations are considerably more involved than for the hyperfine data. Indeed to use the g value information properly, one must perform model calculations which involve assumptions which can only be tested by optical data and/or theoretical calculations. The latter are not always available. Up-to-date discussions of g value calculations for defects in borate glasses can be found in the recent literature (139-141).

It is now apropos to turn to some examples of what has been learned about defects in boron-containing glasses and what they tell us about glass structure. The most commonly observed defect in irradiated borate glasses is one that exhibits five sharp lines and a low-field shoulder for boron in its natural isotopic abundance (81% ^{11}B, 19% ^{10}B) and $\nu_0 \approx 9$GHz (51, 56, 70). The character of this spectrum changes markedly when the sample is enriched to 95% ^{10}B (Fig. 28, page 95) (139), thus suggesting that the sharp lines are due to a hyperfine interaction with boron (51). This has been confirmed by computer simulations (70, 139) (dotted curves in Fig. 28) which take rigorous

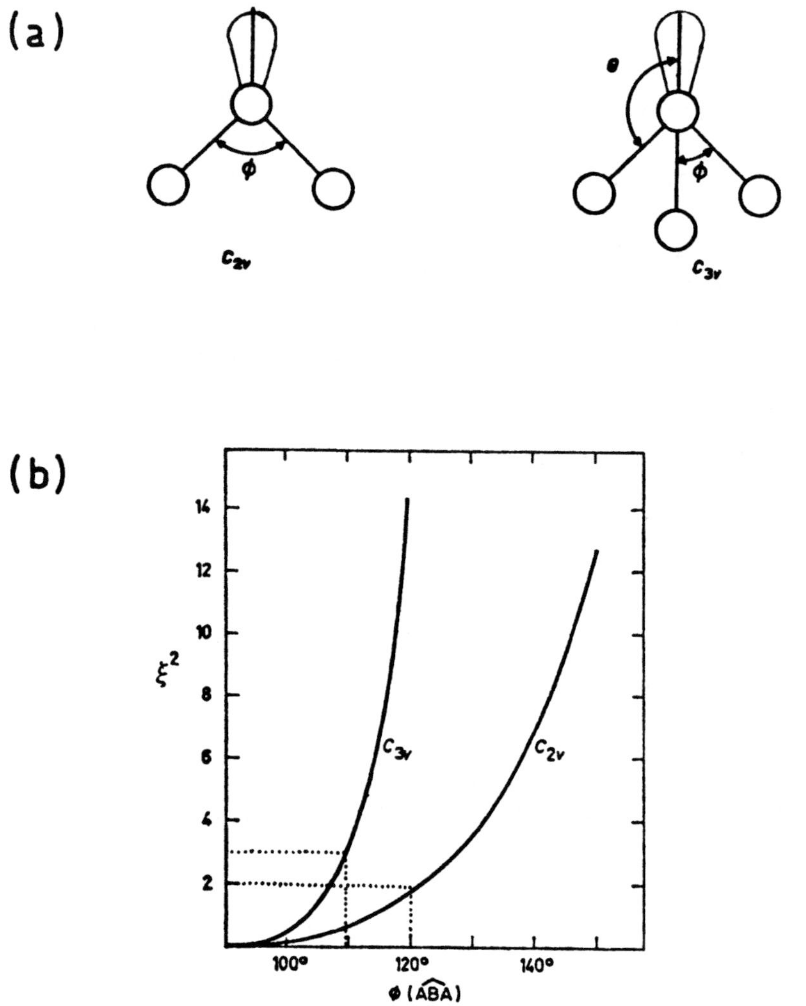

Figure 27. Relationship between ESR parameters and geometry for AX_2 and AX_3 molecules. In (a), the unpaired spin is located in the dangling orbital ψ_o projecting along the respective symmetry axes. In (b), $\xi^2 = |\langle\psi_o|np\rangle|^2/|\langle\psi_o|ns\rangle|^2$ can be determined by ESR. Dotted lines indicate the normal situations for sp^3 (tetrahedral) and sp^2 (trigonal) hybridization. Note that a measurement of ξ^2 does not determine whether one is dealing with C_{2v} or C_{3v} molecules; this must be inferred on another basis. (Figures adapted from ref. 138).

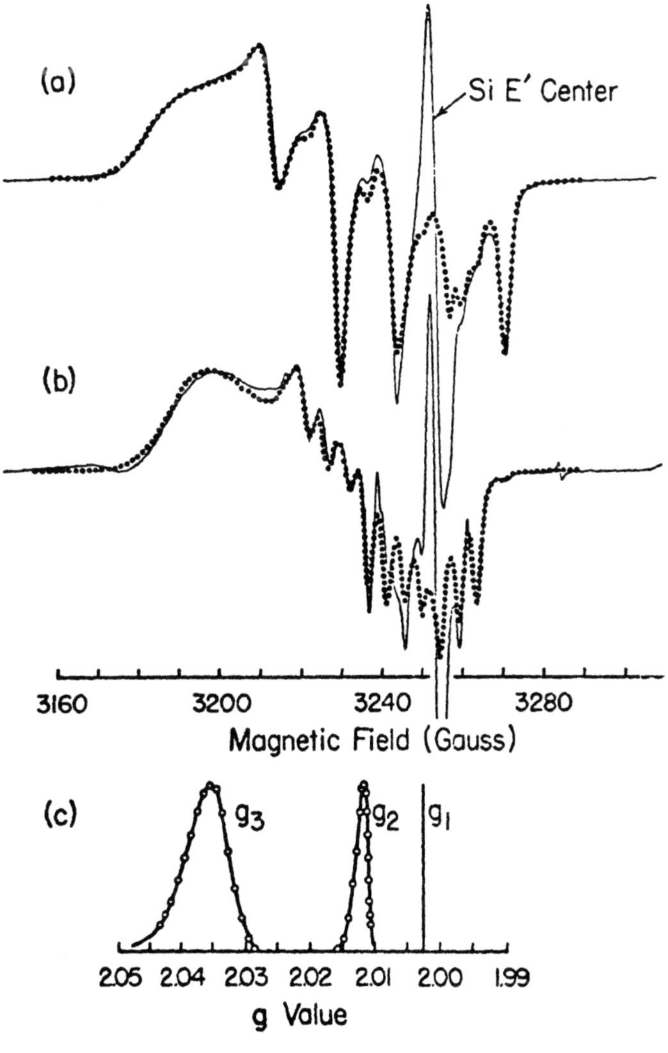

Figure 28. X-band (9GHz) ESR spectra of the boron-oxygen hole center (BOHC) in a γ-irradiated binary borosilicate glass. (a) Sample with normal boron isotopic abundance (81% ^{11}B, 19% ^{10}B). (b) Sample enriched to 95% in the ^{10}B isotope. Unbroken curves are experimental; dotted curves are computer simulations based on standard theory with 10 adjustable parameters, including the positions and widths of the g value distributions shown in (c). (Figure taken from ref. 139).

account of both the prerecognized isotopic abundances and the known ratio of the nuclear magnetic moments of ^{10}B and ^{11}B. Note the distributions in g values (Fig. 28c) required to achieve these simulations. In the present example, the hyperfine interaction is very weak and application of Eqs. (12) - (14) tells us that the unpaired spin is only about 1% localized on a boron (70, 139). Localizations this small are essentially "no localization at all" and it can be firmly concluded that the unpaired spin is located away from, though adjacent to, a single boron nucleus. (Other boron nuclei could be nearby, but these must be in a different bonding arrangement since the observed hyperfine splitting is shown to be due to only <u>one</u> boron). The spin cannot be located on a modifier ion or else an additional hyperfine interaction would be seen; (in the borosilicate glass of Fig. 28 no modifier oxide is involved, anyway). Thus, by a rigorous process of elimination we conclude that the unpaired spin must be on a (nonmagnetic) oxygen. That this unpaired spin is a trapped hole rather than a trapped electron has been supported by competitive trapping experiments (70, 142). Whence, it has been convenient to refer to the defect of Fig. 28 as the boron-oxygen hole center (BOHC). It can be remarked that a g-tensor analysis is consistent with the hole being located in a pure O_{2p} orbital (139-141). The best model for this defect in alkali borate glasses would seem to be that of a hole on an oxygen bridging between a three-coordinated boron and one in four coordination (85). A full discussion of the various models and their relative merits is beyond the scope of this paper, but the involvement of a four-coordinated boron is supported by the near absence of the BOHC in "pure" B_2O_3 glass. Beekenkamp (56) has shown that the BOHC concentrations in glassy B_2O_3 for fixed radiation dose can be correlated with impurity and water contents sufficient to account for the required number of coordination changes.

The BOHC of Fig. 28 has also been observed in the lithium tetraborate (1:4) and triborate (1:3) crystalline compounds, but not in any of the more alkaline compounds (70). Since it has been shown (54, 55) that the tetraborate compound is comprised of pentaborate and triborate structural groupings, Taylor and Griscom (85) have adopted the shorthand notation of calling this particular BOHC the "1:3 center." One may picture a hole trapped in an out-of-plane orbital of an oxygen bridging between a three-coordinated boron and a four-coordinated boron in a triborate ring structure (refer to Fig. 7). This does not preclude, however, the existence of holes on pentaborate groups or on oxygens which link triborate groups with their neighbors; indeed, computer simulations of the spectrum of the tetraborate compound clearly showed the existence of at least two distinct sites differing mainly in their values of g_3 (70). The prime motivation for Taylor and Griscom's shorthand notation was the discovery (85) in the lithium and strontium diborate compounds of BOHC's with nearly identical g tensors but weak hyperfine interactions with two equivalent borons. Since all the borons in $SrO \cdot 2B_2O_3$ are four coordinated (53, 91), this new "1:2" defect is identified as a hole trapped on an oxygen bridging between two four-coordinated borons (85). It is thus logically deduced that in the lithium diborate compound the 1:2 center must be associated with the paired four-coordinated borons in the diborate structural grouping (Fig. 7).

Having thus identified and characterized the 1:3 and 1:2 BOHC's in borate compounds of known structures, the implications of observing the same defects in borate glasses can be examined. Taylor and Bray (86) have made tentative estimates of 1:3 and 1:2 defect concentrations in lithium borate glasses by carrying out computer simulations of the observed spectra. Their results are shown in Fig. 29 (see pages 98, 99) where comparison is made with the total BOHC concentrations after S. Lee (quoted in ref. 142) and with the qualitative predictions of Krogh-Moe's melting point depression studies (45) (preceding chapter). The reasonable qualitative agreement between the estimated 1:3 and 1:2 defect concentrations and the relative numbers of tetraborate and diborate groups, respectively, predicted by Krogh-Moe's structural theory serves as an additional confirmation of the latter and a demonstration of how paramagnetic defects can sometimes serve as useful probes of glass structure.

A second example of a radiation-induced defect serving as a structural probe is the boron electron center (BEC) (82). This center was identified by boron isotopic enrichment and computer simulation studies of the ESR spectra of alkali borate glasses irradiated at 77°K. However, careful analysis of the boron hyperfine splittings were incapable of distinguishing between four models,

Figure 29. Comparison of ESR intensity data for radiation-induced hole centers (a) with predicted numbers of structural groupings (b) in alkali borate glasses. Total intensity data for room-temperature irradiated samples are due to S. Lee (quoted in ref. 142). Break-down into relative numbers of "1:3" and "1:2" centers is the result of a computer simulation analysis by Taylor and Bray (86). Concentration of "1:3" centers at 33 1/3% alkali oxide is probably overestimated due to the latter authors' admitted neglect of the well known "four-line" spectrum, which is probably associated with non-bridging oxygens (NBOs). The "1:3" center is typical of both the triborate and tetraborate compounds. Dashed lines in (b) represent an idealized version of Krogh-Moe's structural theory. Fully drawn curves qualitatively take into account Krogh-Moe's conclusion (45) that the tetraborate groups are ∿10% dissociated, plus NMR evidence (49) for the onset of NBO formation near 30 mol % alkali oxide and the persistence of four-coordinated borons (assumed here to be in diborate groups) up to ∿66 mol % modifier. Perfect correlations between defect concentrations (a) and actual numbers of structural groupings (b) are not to be expected. For example, low "NBO center" concentrations at high alkali contents are probably due to efficient electron-hole recombination due to electron hopping between alkali ions.

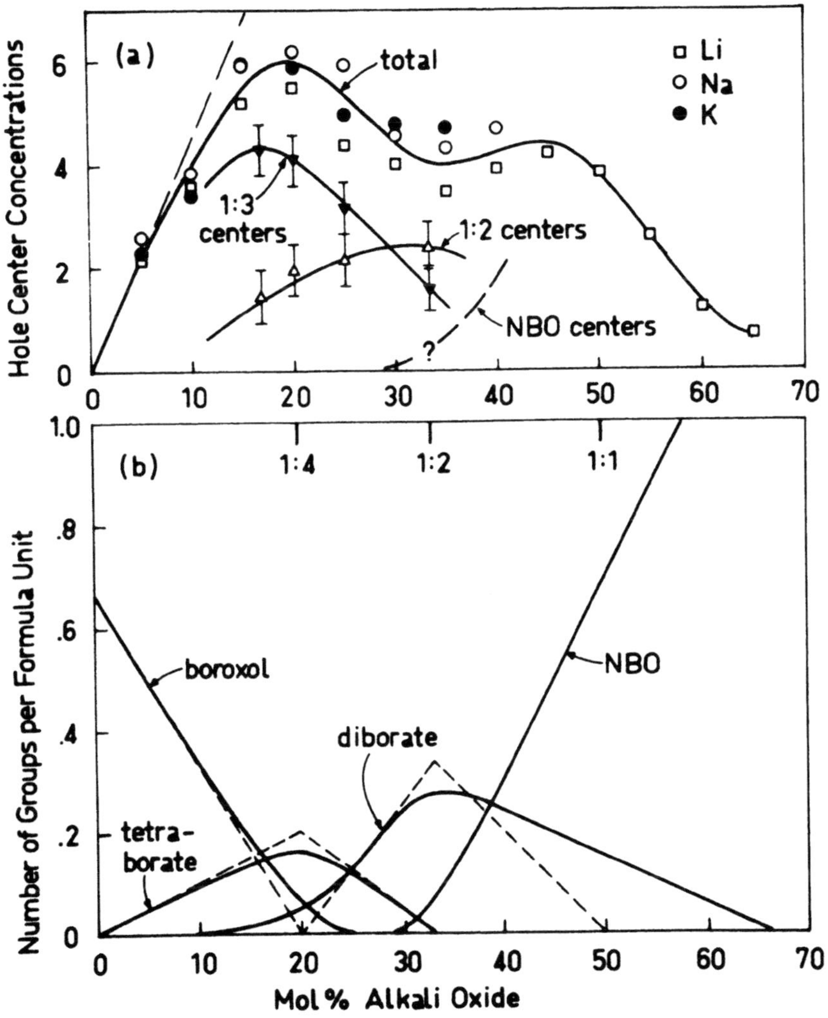
(a)
total
□ Li
○ Na
● K
1:3 centers
1:2 centers
NBO centers
?
Hole Center Concentrations
0
2
4
6
0
10
20
30
40
50
60
70
(b)
1:4
1:2
1:1
boroxol
NBO
diborate
tetra-borate
Number of Groups per Formula Unit
0
.2
.4
.6
.8
1.0
Mol % Alkali Oxide

Figure 29.

namely, the two structures of Fig. 27a (with the apex atom = boron, and the basal atoms = bridging oxygens) plus the same two structures with alkali ions weakly bonded to the apex borons (82). By Eqs. (12) - (14), the BEC unpaired spin is known to be ~50% localized in a boron sp^n hybrid orbital (82). More recently a very similar defect, the boron E' center, has been identified in alkali-free borosilicate glass, whereby contrast the localization in a boron sp^n orbital was shown to be ~100% (139). (See Fig. 30, page 101, for experimental and computed boron E' center spectra for two different isotopic abundance ratios). The borosilicate glass study favored the pyramidal structure of Fig. 27a for the boron E' center, and by inference this structure plus an alkali ion now appears to be the best model for the BEC. It might be noted that the four original models for the BEC led to a probability of 1/4 that any given one was correct. Had the isotope and computer simulation studies not been performed, the possible models would have run in the dozens and the probability of guessing the correct answer would have approached zero (Appendix A). But even assuming the accumulated evidence is now sufficient to identify the BEC as an electron trapped in an sp^n orbital of a trigonal boron weakly bonding to an alkali ion, there remains the question of whether the trigonal boron is involved in a boroxol group, a pentaborate group, a triborate group, or some other. Obviously, when one gets this specific, the probability of being wrong rises again--unless still further evidence can be brought to bear.

The compositional dependence of A_{iso} for the BEC has been studied for the lithium, sodium, and potassium borate glass systems, with the results displayed in Fig. 31 (see page 102). The dependence of these data on type of alkali is understandable (82) on the basis of the BEC model favored above. The dependence on quantity of alkali is of prime importance from the structural point of view. The sharp drop in coupling constant between 15 and 25 mol % alkali oxide does not correlate with N_{BO_4} as known from NMR nor with most of the purported boron-oxide anomalies (see Boron-Oxide Anomaly chapter above). On the other hand, one can make some sense out of the curves of Fig. 31 in terms of Krogh-Moe's structural theories. One may picture the data as representing averages among the spectra of BECs associated with several structural groupings. In this view, the change between ~15 and ~25 mol% alkali

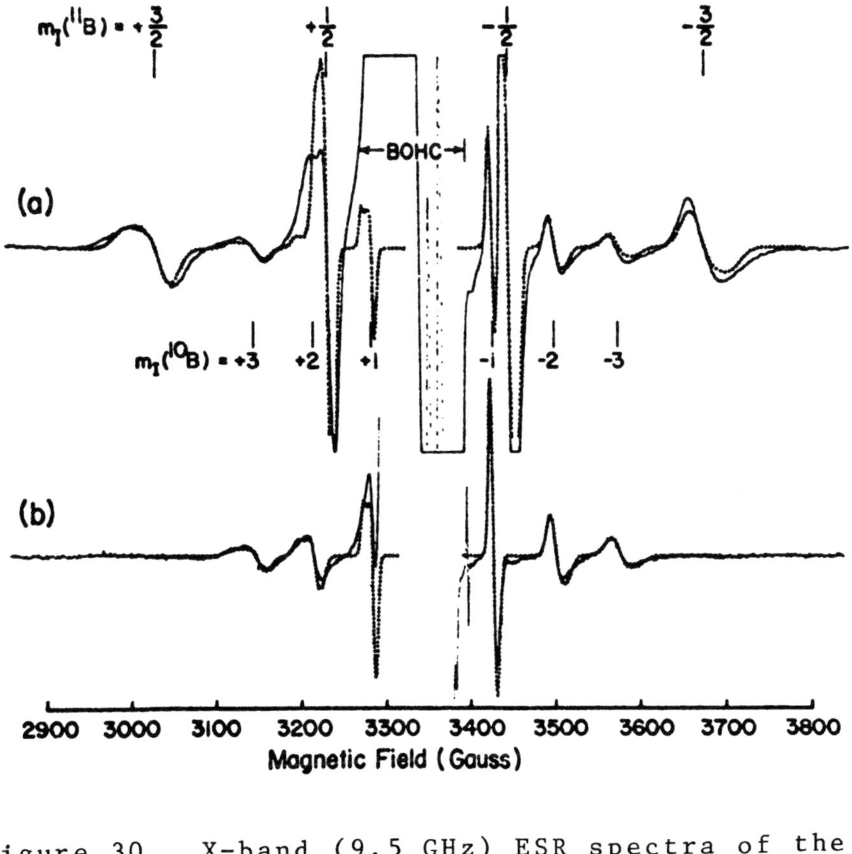

Figure 30. X-band (9.5 GHz) ESR spectra of the boron E' center in irradiated $B_2O_3 \cdot 3SiO_2$ glass (a) for boron in its natural isotopic abundance and (b) for boron enriched to 95% in ^{10}B. Dotted curves are computer simulations of the respective boron E' center spectra (neglecting the m = 0 component in the ^{10}B manifold). The BOHC dominates the central parts of the spectra. (Figure taken from ref. 139).

oxide represents a decline in numbers of the site prevalent below 15 mol % and the subsequent dominance of sites characteristic of more alkaline compositions. Since, according to Krogh-Moe (45, 57) the pentaborate and triborate groups occur in pairs (= the tetraborate group), it is reasonable to suggest that the large-coupling-constant BEC site in the acid glasses can only be attributed to

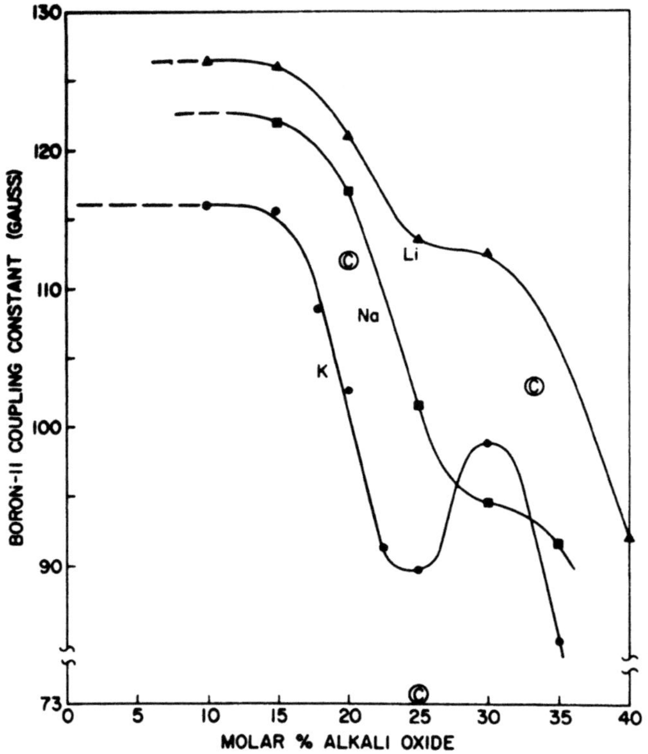

Figure 31. Variation with glass composition of the isotropic ^{11}B hyperfine coupling constant for the boron electron center (BEC) in alkali borate glasses. This defect is evidently sensitive to the type of structural grouping on which it is located. Circles with inscribed 'C' are tentative data for irradiated lithium borate compounds. (Figure taken from ref. 82).

boroxol rings, which evidently (45) disappear just above 20 mol % alkali oxide. BEC's located on triborate groups may well dominate near 25 mol % alkali oxide, but pentaborate or diborate groups cannot be eliminated. At still higher alkali contents the situation is even more problematical because of the onset of a new type of defect center associated with clusters of alkali ions (143).

The examples discussed above demonstrate some of the ways in which paramagnetic defects studied by ESR can serve as useful probes of borate glass structure. The coverage of source material has by no means been exhaustive. A review of numerous types of defect centers in alkali borate glasses has been given (135). Also, recent work in the silver borate system (144) is of major interest because of the explicit evidence from ^{107}Ag and ^{109}Ag hyperfine splittings for the pairing of Ag^+ ions at Ag_2O concentrations exceeding ~1 mol %. The work of Assabghy et al (144) thereby confirms the thermodynamic evidence for silver pairing developed by Willis and Hennessy (18) and adds yet further support to Krogh-Moe's contention (45, 57) that tetraborate groups are formed when silver or alkali oxide is added in amounts $\lesssim$ 20 mol % to B_2O_3 melts.

In keeping with the spirit of the previous experimental chapters a diagrammatic outline of the performance of a typical experiment is provided in Fig. 32 (see page 104). It should be apparent that great care is required just to identify defects in glasses, but that when sufficient care is taken an impeccable identification is often possible. Such identifications can sometimes be so specific as to include quantitative estimates of the bond angles at the defect sites (Fig. 27). It should be borne in mind, however, that these bond angles are often different before and after electron or hole trapping has occurred (145). Beyond that aspect, it should also be remembered that most defects are sensitive only to their first coordination sphere of neighbors. Thus, unlike X-ray and Raman techniques, ESR studies of defects cannot by themselves demonstrate the existence of boron-oxygen ring structures, for

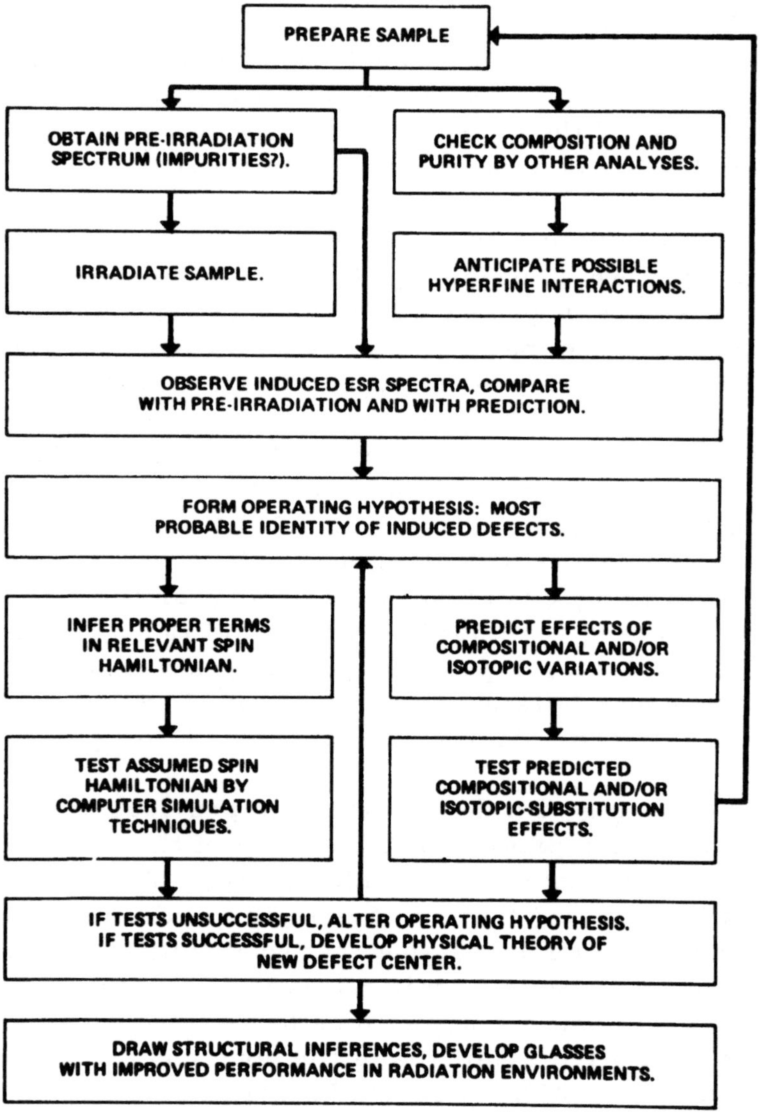

Figure 32. Flow diagram illustrating a typical approach to the study of radiation damage and structure in glassy materials by means of electron spin resonance techniques. (Figure taken from ref. 135).

example. But on the other side of the coin, analysis of ESR (and NMR) spectra of glasses is a far more highly developed art than the analyses of IR and Raman spectra as currently applied to borate glasses. Thus, as further experimental and theoretical innovations are exploited, ESR structural studies promise to at least keep pace with the other experimental techniques described in the preceding chapters.

XIII. ESR : FOREIGN IONS

The suggestion has long been around that paramagnetic transition-group ions studied by ESR (or optically) may serve as useful probes of glass structure. The introduction of as little as ~0.1 mol % of many such iron-group or rare earth ions is sufficient to observe a strong ESR signal. Whether or not that signal will tell us something about the atomic arrangements in the undoped pure glass cannot be predicted in advance, but in the final analysis the answer will be mirrored in whether or not the spectrum is sensitive to glass composition. The present chapter reviews the relevant physical principles and considers the results of a reasonably representative group of ESR studies of paramagnetic foreign ions in borate glasses.

The basic ESR phenomenon has already been outlined in the preceding chapter in the context of radiation-induced defects. It will be recalled that all defect centers of interest were characterized by electronic spin $S = 1/2$. By contrast, the paramagnetic electrons in transition-group ions are located in partially filled d or f shells. According to Hund's rules (see, e.g., Pake (146)) these shells fill up in such a way that all electron spins are parallel up to and including the situation where the shell is half filled. Thus, many transition-group ions of interest will have total electronic spin S = (integer or half integer) $> 1/2$. When this is true, additional terms must be added to the effective spin Hamiltonian of Eq. (11):

$$\mathcal{H}_{\text{fine structure}} = D[S_z^2 - (1/3)S(S+1)] + E(S_x^2 - S_y^2) \quad (15)$$

The physical origins of these terms are complicated (see, e. g., ref. 146), but can be loosely described as follows. In free space transition group ions having orbital angular momentum may become "quenched" in a solid matrix as a consequence of ligand electric fields or covalency effects which constrain the electronic charge distributions of the unfilled shells to point in certain directions--thus preventing them from circulating about an axis. The details depend in an important way on the nature of the ions, so we shall specialize the remaining discussion to iron-group ions, which are the only ones to have been extensively studied in borate glasses. For these cases, "quenching" is rather complete but the relative importance of ligand-field and covalency effects vary from case to case and must be tested for by highly involved model calculations (see, e.g., refs. 147, 148). If quenching is due to electric field effects alone, the terms of Eq. (15) can be thought of as arising from the spin-orbit and orbit-Zeeman interactions which remix a small amount of angular momentum into the quenched ground state. In addition to the components of the g and hyperfine tensors (preceding chapter), the axial and off-axial principal components of the fine-structure tensor (D and E, respectively) are also parameters which must be determined experimentally.

To solve an actual problem for an iron-group ion in a glass, one must diagonalize the effective spin Hamiltonian to obtain the resonance condition and then use it to construct powder patterns, just as described in the "NMR" and "ESR : Defects" chapters. The only difference is that for $S > 1/2$ the terms of Eq. (15) must be added to the effective spin Hamiltonian of Eq. (11). A special case of particular academic and practical interest concerns the eventuality when the terms of Eq. (15) are much larger than those of Eq. (11). This situation was originally worked out by Castner, Newell, Holton, and Slichter (149) to explain the ESR spectrum of Fe^{3+} ($S = 5/2$)

in glass. Castner et al showed that the observed spectrum could be explained by assuming that $D = 0$ and $E >> g\beta H$. (To prevent confusion later on, we should point out that by redefining one's coordinate system, the situation, $D = 0$, $E \neq 0$, is perfectly equivalent to $|E/D| = 1/3$) (150). By treating the Zeeman interaction as a perturbation on the fine structure term, Castner et al demonstrated how the $S = 5/2$ manifold of Fe^{3+} in glass is split into three very widely spaced Kramer's doublets, each of which can be described by a new pseudo-spin Hamiltonian comprising only the first term on the right-hand side of Eq. (11) with a ficticious spin $S' = 1/2$. It was shown further that two of these doublets have identical highly anisotropic effective g tensors with principal values ranging from zero to ten; these two doublets therefore accounted for an extremely smeared out background spectrum with a small shoulder at $g_{eff} = 10$. The third doublet was demonstrated to have an isotropic effective g value of 4.3. This accounted for the sharp spectral feature observed at $g = 4.3$ for Fe^{3+} in glass since the 1955 work of Sands (151). (See Fig. 33, page 108). The compositional dependence (or lack thereof) of the Fe^{3+} spectrum in alkali borate glasses will be mentioned later on in this chapter.

Mn^{2+}, which is isoelectronic with Fe^{3+}, is among the most thoroughly studied iron-group ions in borate glasses. As far as ESR parameters are concerned, however, there are two big differences between these two $S = 5/2$ ions. First, Mn^{2+} has a nuclear spin $I = 5/2$ giving rise to a 6-line hyperfine multiplet, whereas the abundant isotope of iron has $I = 0$. Second, whereas the fine-structure terms for Fe^{3+} in glass are very large with respect to the Zeeman interaction ($E >> g\beta H/h \approx 9$GHz), this is not true of Mn^{2+}. De Wijn and van Balderen (152) and Griscom and Griscom (153) inauguarated the first detailed attempts to explain and parameterize the spectra of Mn^{2+} in alkali borate glasses. The former authors concentrated on the 6-line hyperfine spectrum centered near the free electron g value, calculating all lines including the observable $\Delta m \neq 0$ "forbidden" transitions by treating

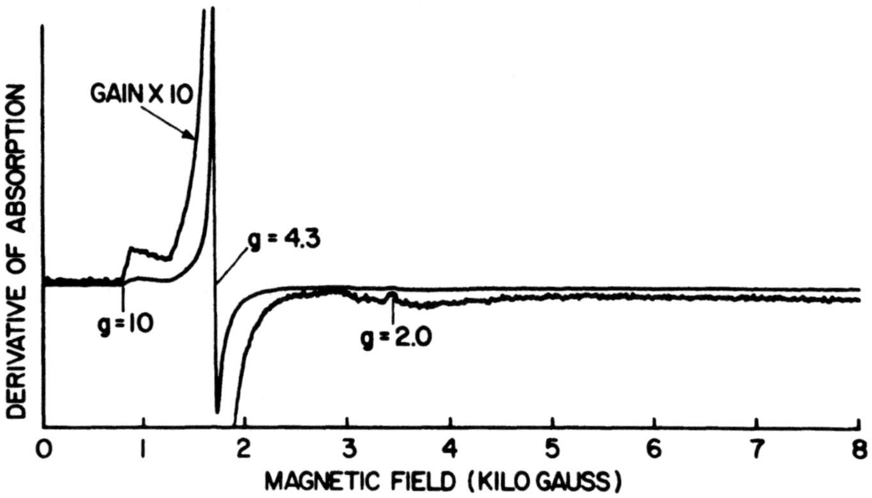

Figure 33. X-band (9 GHz) ESR spectrum of Fe^{3+} in an alkali borate glass of composition $Li_2O \cdot 4B_2O_3$.

the hyperfine and fine structure interactions to third order in perturbation theory. It was concluded (152) that $E \approx 0$ and that the axial fine structure parameter $|D|/h$ was quite small, i.e., ~ 0.26 GHz. In complete contrast, Griscom and Griscom (153) concluded that the Mn^{2+} sites in alkali borate glasses of $\lesssim 15$ mol % alkali oxide are characterized by $|E/D| \sim 1/3$ and $|D|/h \sim 2$GHz. This is a rather dramatic illustration of a major theme woven into the present review, namely, the question, "Is it possible that two different models can fit the data equally well?" As will be seen below, the answer in this case turns out to be in the negative. One of the aforementioned models clearly explains many more features of the data than the other.

By studying the frequency dependence of the glass spectra and also that of the spectrum of Mn^{2+} doped into the $Li_2O \cdot 4B_2O_3$ compound, Griscom and Griscom came to the realization that neither of the special cases previously treated ($g\beta H << |D|$, 149; $g\beta H >> |D|$, 152) can explain all details of the behavior of Mn^{2+} in these materials. If indeed $g\beta H$ and $|D|$ were of the same order of magnitude, the appropriate model calculation could not employ perturbation theory; thus an exact diagonalization of Eq. (11) plus Eq. (15) was necessary. This had not been done before in the context of powders or glasses (except for unpublished work by Barry; 154), so Griscom and Griscom performed the necessary computations for the special case of $|E/D| = 1/3$ but with $|D|$ values ranging from 0 to about two times the Zeeman energy. The results are illustrated in Fig. 34 (see page 110). (More extensive calculations for other values of $|E/D|$ have subsequently become available; 154-156). The graph to the upper right in Fig. 34 comprises a theoretical plot of normalized resonance field ($g\beta H/h\nu_o$) versus normalized axial fine structure constant ($|D|/h\nu_o$). The integrated spectra obtained at two frequencies for Mn^{2+} in a $5Li_2O \cdot 95B_2O_3$ glass are displayed to the left of Fig. 34. To account for the shapes and frequency dependence of the glass spectra, Griscom and Griscom (153) were compelled to postulate the existence of statistical distributions in D values in the glasses. Although no specific shape function was suggested at that time, the bell-shaped curves at the bottom of Fig. 34 are qualitatively capable of explaining the observed glass spectra. Clearly, as pointed out by Griscom and Griscom, any transition will be smeared out which is not stationary with respect to variations in D in the region where D is distributed. Note how the model fits the observed fact that the X-band spectrum extends from zero field to well above $g\beta H/h\nu_o = 1$, while the Ka-band spectrum collapses on $g\beta H/h\nu_o = 1$ (Fig. 34, left). The model of de Wijn and van Balderin (152), who did not consider the broad parts of the resonance, yields a spectrum which is collapsed upon $g\beta H/h\nu_o = 1$ at both frequencies. Hence the latter model can be rejected, and there remains but a single model which can account for the principal aspects of the data.

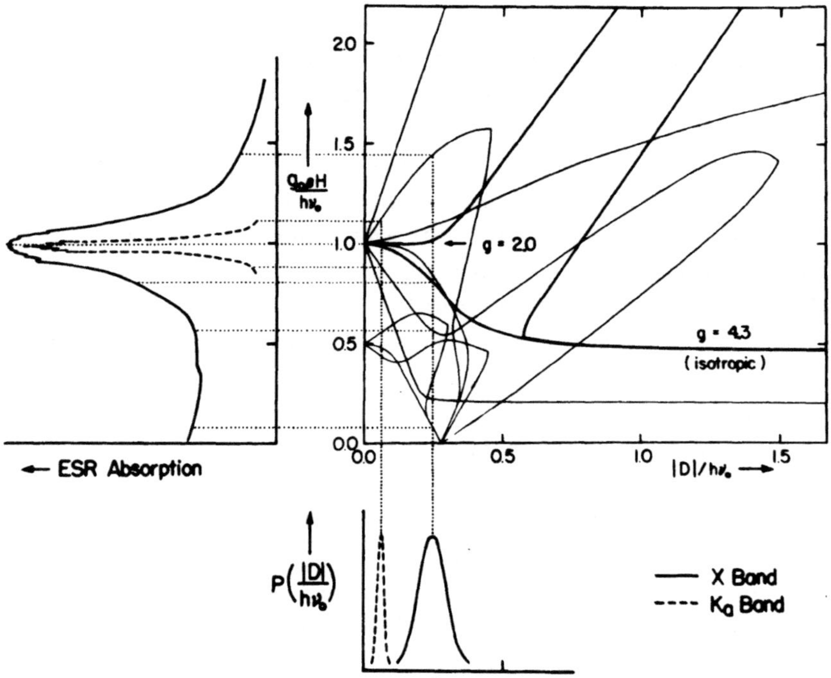

Figure 34. Computed resonance fields (upper right) for the sharp features in the ESR powder spectra of Fe^{3+} or Mn^{2+} as a function of $|D|/h\nu_o$ for the special case $|E/D| = 1/3$. At the left are shown the integrated experimental spectra of Mn^{2+} in a $5Li_2O$-$95B_2O_3$ glass at two frequencies ν_o, X-band (9 GHz) and Ka band (35 GHz). Note that the magnetic field axis is normalized by dividing by $h\nu_o$. At the bottom is indicated a possible distribution in $|D|$ values (two distributions in $|D|/h\nu_o$, one for each frequency) which may account for the observed behavior of the experimental spectra. (Figures adapted from ref. 153, where additional details may be found).

Griscom and Griscom concluded that the predominant Mn^{2+} site in alkali borate glasses of low alkali content is a randomly distorted version of the site in the lithium tetraborate compound. However, lacking single crystal samples they were unable to decide if the manganese substituted for lithium or for boron. In light of the success of Krogh-Moe's structural model, some educated speculations are now possible on the basis of the compositional dependence of the Mn^{2+} glass spectra. In Fig. 35a (see page 112) is shown a plot of certain amplitude ratios which, it has been argued (153), are probably indicative of the relative prevalence of at least two types of sites. The one type of site (A/B) is evidently characteristic of glasses containing $\lesssim$ 25 mol % alkali oxide. It seems noteworthy that this site (we shall call it the "1:4" site) appears to be obliterated when the alkali oxide content becomes comparable to or lower than the MnO doping level. This is compelling evidence that, whatever the 1:4 site, it demands extra oxygens to satisfy its coordination requirements. Thus, a manganese substituting for a four-coordinated boron in a tetraborate group must be considered. Such a model is not irreconcilable with other features of the ESR data (153) nor is it inconsistent with the conclusions of fluorescence studies (157). If this model for the 1:4 site is accepted, however, it becomes necessary to explain why the curve A/B begins to decline above ~10 mol % Li_2O instead of holding constant up to ~20 mol % modifier oxide as might be predicted from Krogh-Moe's theory (45, 57). A possible explanation of this "anomaly" will be tendered in conjunction with the discussion below of the ESR spectra of Cu^{2+} in alkali borate glasses.

Before leaving the Mn^{2+} problem, it is appropriate to mention two other related works. First, Taylor and Bray (158) studied the ESR spectra of Mn^{2+} in glasses and compounds of the strontium borate system. These authors concluded that a large fraction of the glass sites were characterized by $|E/D|$ ratios substantially less than 1/3. They went on to suggest that the ubiquitous nature of the ESR spectra in many diverse glass types is due to certain resonance transitions which are stationary not only with

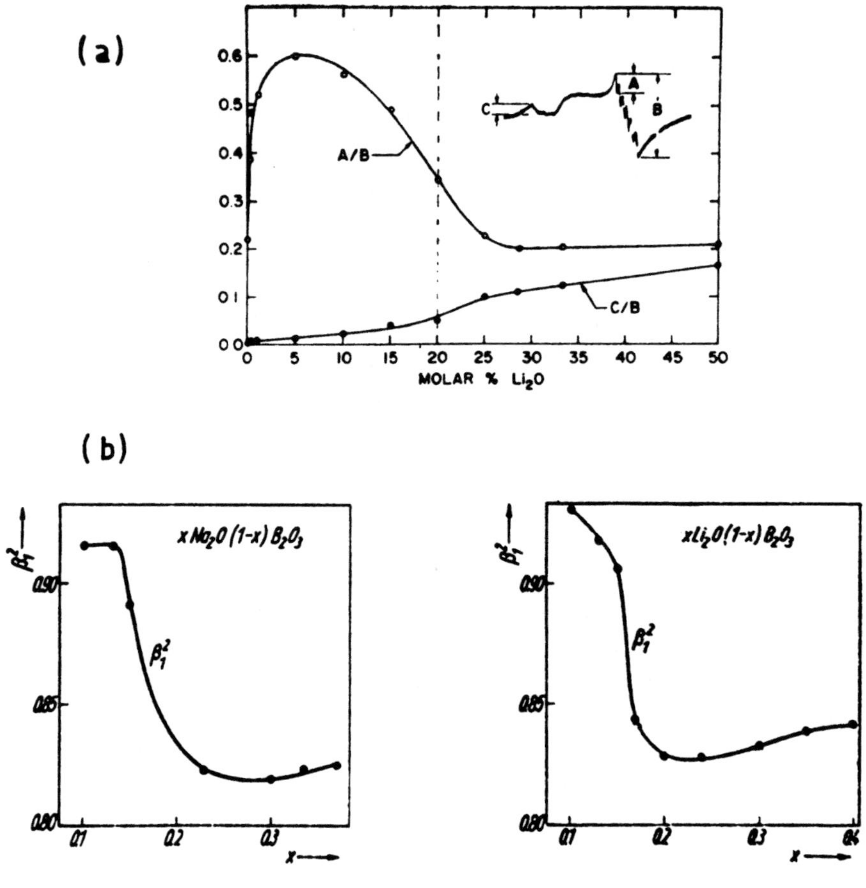

Figure 35. ESR property-versus-composition curves for (a) Mn^{2+} and (b) Cu^{2+} doped into alkali borate glasses. (Figures taken from refs. 153 and 148, respectively).

respect variations in D, but also with respect to variations in |E/D|. This is a highly reasonable suggestion and is certainly well supported by comparisons of the glass and compound spectra for the strontium borates (158). However, homogeneous glasses can only be prepared for SrO contents exceeding 20 mol %. Thus, Taylor and Bray were only able to investigate glass compositions in a range where the 1:4 site discussed above is essentially absent. The detailed arguments of Griscom and Griscom (153) attributing |E/D| ratios of 1/3 to the 1:4 sites are therefore not contradicted. In the second related work, Peterson, Kurkjian, and Carnavale (136) intimated the possibility that the spin Hamiltonian proposed by Castner et al (149) to explain the Fe^{3+} spectrum in glass (Fig. 33) may not be unique. Peterson et al showed that an axial effective g tensor with $g_{\parallel} = 2$ and $g_{\perp} = 6$ could yield a resonance at $g \approx 4$ if there existed equally broad distributions in both $g_{\parallel}$ and $g_{\perp}$ with a correlation coefficient of -1. It is pointed out here that such a model is purely mathematical. To test its applicability one must refer to a diagram such as Fig. 34, which contains the relevant physical constraints (diagrams appropriate to axial symmetry can be found in refs. 154-156). For the case E/D = 0 considered by Peterson et al (136), a distribution of D values could result in distributions in both $g_{\perp}$ and $g_{\parallel}$, but the computed diagrams indicate that the specific correlation required to produce an "accidental" resonance at $g \approx 4.3$ is not present. Therefore, the model of Castner et al (149) should not be regarded as impugned. Note in Fig. 34 that for $|E/D| \approx 1/3$ and $|D|/h\nu_0 \gtrsim 1$ a broad distribution in D values has no effect whatever on the principal resonance transition, which remains pinned at g = 4.3. Thus, under the Hamiltonian proposed by Castner et al no special distributions need be invoked.

We turn now to the study Cu^{2+} in alkali borate glasses as carried out by Imagawa (148). Cu^{2+} is a d^9 ion, which means there is a single hole in the otherwise filled 3d shell; thus, S = 1/2. The naturally occurring isotopes of copper both have nuclear spin I = 3/2 and nearly identical nuclear g factors; thus, the resonance spectra of Cu^{2+} can be described by Eq. (11) alone and are expected to consist of hyperfine quartets. The powder pattern analysis upon which Imagawa relied was originally expounded by Sands (151). Imagawa, like Griscom and Griscom, cited evidence for statistical distributions in spin Hamiltonian parameters as apparent consequences of the intrinsic disorder in the vitreous state; in the Cu^{2+} case, it was the ^{63}Cu and ^{65}Cu hyperfine coupling constants which were most sensitive to such effects. By comparison to the results for Cu^{2+} in crystals of well known structure, it could be inferred from the measured g values that the cupric ion in alkali borate glasses is coordinated by six ligands which form an octahedron elongated along one axis. Thus Cu^{2+} is located at network-modifier sites in the glass structure. By means of a linear combination of atomic orbitals (LCAO) analysis, Imagawa was able to convert the observed g values and hyperfine coupling constants into parameters relating to the degree of covalency in the Cu - O bonds. The parameter β_1^2 in Fig. 35b reflects the competition between the Cu^{2+} ion and its neighboring network-forming cations for attracting the π electrons of the intervening oxygens (148). An increase in the strength of the B - O bond causes β_1^2 to increase towards 1.0. Thus, Imagawa attributed the results of Fig. 35b to the gradual weakening of the average B - O bond as the number of four-coordinated borons increases with increasing alkali content. Although this explanation is undoubtedly correct in the qualitative sense, it fails to explain the sharp break in the curves near x = 15 mol % modifier oxide. A possible answer to this puzzle and also to that surrounding the Mn^{2+} data of Fig. 35a can now be posed.

Krogh-Moe has emphasized in a number of papers (33, 39) that the coordination requirements of modifer cations are fairly strict: "Firstly, these ions will always be located in the neighborhood of those borate groups with an excess negative charge. Secondly, they will tend to attract and share other such negatively charged groups in order to screen themselves more efficiently (39)." Indeed, it is evident from the above discussion that Cu^{2+} demands and achieves something approximating octahedral coordination by oxygens at all alkali borate glass compositions. However, Krogh-Moe (54) has also remarked that "the cation coordination sphere is to a considerable extent governed by the requirements of the borate polymer network and not only by the cation itself." Thus we have a situation in which the foreign modifier cations under study may demand and achieve a certain coordination sphere subject only to the constraints of what negatively charged polyborate structural groupings are available in the melt. Below 20 mol % alkali oxide, the tetraborate group is, of course, the most prevalent polyborate anion (45, 57). However, Imagawa's studies concerned only 0.2 wt % CuO doping--which means that if as much as ~0.5 mol % diborate groups were available the Cu^{2+} could improve its screening by searching out these compact doubly charged anionic structures. And, indeed, Krogh-Moe (45, 57) has cited several pieces of evidence for a partial dissociation of tetraborate groups into diborate groups plus boroxol groups (Eq. (9)). The degree of dissociation estimated by Krogh-Moe was ~10% at 10 mol % alkali oxide (45). Accepting this estimate, ~1 mol % diborate groups are available at 10 mol % alkali oxide and more become available at higher alkali contents. Assuming that Cu^{2+} scavanges these diborate groups with something less than 100% efficiency, the $\beta_1{}^2$ curves of Fig. 35b can be understood in terms of a change-over at the earliest possible moment of Cu^{2+} coordination by tetraborate groups to coordination by diborate groups. Thus we are concerned not with a "boron-oxide anomaly" but instead with the efficiency of a tiny number of foreign ions attempting to optimize their screening by seeking an almost equally small number of polyborate groupings. This hypothesis might well be tested by repeating Imagawa's studies with differing Cu^{2+} doping levels. In the meantime, one must resist the temptation to fall back upon a belief in "boron-oxide anomalies" merely because of the behavior of a minuscule number of foreign ions.

The above explanation of the Cu^{2+} ESR may also be adapted to understand the Mn^{2+} results of Fig. 35a. Assuming the correctness of our speculation that Mn^{2+}

substitutes for a tetrahedral boron at low alkali oxide content, we can speculate further that the driving force for Mn^{2+} to become coordinated by diborate groups is smaller than in the Cu^{2+} case. Whence the more gradual decline of the Mn^{2+} 1:4 site (Fig. 35a, curve A/B) vis-a-vis the Cu^{2+} tetraborate site (Fig. 35b, $\beta_1{}^2$). Other explanations are conceivable, of course.

If the ESR spectra of doped-in Mn^{2+} and Cu^{2+} ions were difficult to interpret in terms of borate glass structure, the spectra of other doped-in transition-group ions leave little at all to interpret. Fig. 36 (see page 117) illustrates the compositional dependences of several ESR parameters for dilute Fe^{3+} and V^{4+} in alkali borate glasses. The approximate 20% variation of linewidth ΔH observed by Loveridge and Parke (159) for Fe^{3+} between 15 and 25 mol % sodium oxide (Fig. 36a) is reminiscent of the larger effect noted for the C/B curve of Fig. 35a, but both effects are too cryptic to warrant any speculations regarding a possible relationship to glass structure. The data of Fig. 36b, pertaining to V^{4+} in sodium borate glasses, were obtained by Hecht and Johnson (147). The latter authors were able to argue from their ESR and optical results that they were dealing with a vanadyl-type ion (VO^{2+}) occupying a well-defined site at all glass compositions. The same conclusion was reached earlier by Hochstrasser (160) in carrying out a computer simulation of the ESR spectrum of vanadium in a borosilicate glass. Hecht and Johnson carried out a splendid LCAO-MO calculation supporting an octahedral site and relating the small compositional variations of both ESR and optical data to slight dilations or contractions of the coordination sphere about that site. The ill-defined minimum in $g_{\parallel}$ near $x/(1 - x) = 0.15$ (Fig. 36b) was thought to be correlated with the "boron-oxide anomaly." However, more recent data gathered by Toyuki and Akagi (161) for VO^{2+} in lithium, sodium, potassium, and caesium borate glasses do not show such a minimum; rather, both $g_{\parallel}$ and $g_{\perp}$ remain constant up to ~18 mol % alkali oxide and approach each other slightly at higher modifier levels. It is evident that the V^{4+} (in the form of VO^{2+}) dictates its own environment in borate glasses and does so with little

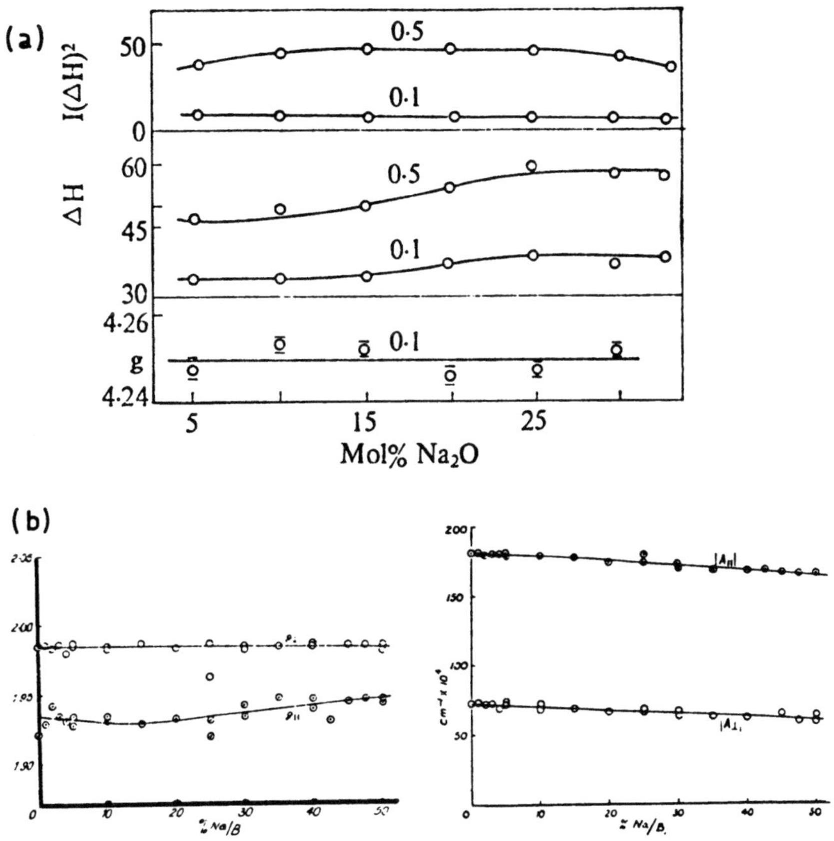

Figure 36. ESR property-versus-composition curves for (a) Fe^{3+} and (b) V^{4+} doped into alkali borate glasses. Note the data of (b) are plotted versus x/(1-x) where x is the molar fraction of sodium oxide. (Figures taken from refs. 159 and 147, respectively).

interference from the various types of boron-oxygen structural groupings which may be present. Vanadium is therefore a poor structural probe. For reasons which are as yet undetermined, Fe^{3+} is also a poor probe of borate glass structure, at least at X-band frequencies. Conceivably, the fine structure parameter $|D|$ is sensitive to glass composition but, for $|D| > h\nu_o$, Fig. 34 shows that variations in $|D|$ cannot affect the observed isotropic resonance at g = 4.3. Studies of Fe^{3+}-doped glasses at higher frequencies may therefore be worthwhile--just as low-frequency (0.5 GHz) studies were important in identifying the resonance of Ti^{3+} in a calcium borate glass (162).

The examples given above have included some of the most successful and some of the least successful examples of the use of transition-group ions as probes of borate glass structure. Many more studies have been carried out and the reader is referred to the recent book by Wong and Angell (163) for a more comprehensive listing. The present chapter, as it is, has exceeded the average length of the other experimental chapters. This was quite necessary in order to convey a proper feeling for the many complexities and pitfalls involved in performing and interpreting this kind of experiment. Still, the tangible results have been few. Clearly, ESR studies of doped-in foreign ions is a terribly hard way to learn about the structure of the undoped glass.

APPENDIX A: THE FINGERPRINT METHOD

In the literature, one frequently comes across what some authors refer to as the "fingerprint method." This method is usually invoked in cases when identical spectra are observed in two different materials, A and B, but the spectral shapes, positions, and widths are not explainable on the basis of existing information. The central, but usually unproved, premises of the fingerprint method are (1) that identical spectra must arise from identical sources and (2) that there exists a one-to-one correspondence between the observed spectrum and some intrinsic property (e.g., the existence of a particular atomic arrangement) in the materials examined. Thus, if the structure of B happens to be known, rather specific inferences are drawn regarding the structure of A, e.g., "A contains some of the same atomic groupings as B."

The probability of any experimental analysis yielding a wrong answer can in principle be estimated. Let n_c be the number of conceivable models which can explain the observed experimental results. Let n_r be the number of models which can be rejected on some basis or other. As an example, in an ESR experiment the observation of a hyperfine interaction with a given element (checked by isotopic substitution) precludes all conceivable models which do not allow the unpaired spin the appropriate degree of localization on a nucleus of that element. In other cases, detailed model calculations and curve fitting may be necessary to eliminate some models. Then the probability of a wrong answer is

$$P_{wrong} = \frac{n_c - n_r - 1}{n_c} .$$

In the fingerprint method as described above, the observed spectral lines are not understood in detail and consequently curve fitting or model calculations are ruled out. In this worst case $n_r = 0$ and

$$P_{wrong} \text{ (fingerprint method, worst case)} = \frac{n_c - 1}{n_c} .$$

Given a fertile mind, a large value can usually be placed on n_c, in which situation

$$P_{wrong} \text{ (fingerprint method, worst case)} \rightarrow 1.$$

On the other hand, if a great number of well characterized materials, B_1, B_2, ... B_q, are studied and a consistent pattern emergies (including, perhaps, a partial theoretical understanding of the spectra of materials B_i), some models can be rejected and $n_r \neq 0$. Thus every conclusion based on simple spectral comparisons is not necessarily wrong. It is important, however, for the reader to develop within himself a caution mechanism which automatically trigers on reading the words "fingerprint method." Any article bearing those words should be carefully scanned for the extent of the data base, what model calculations (if any) have been brought to bear, what constraints (if any) on the theoretical interpretation are provided by the spectral line shapes and, above all, how many conceivable alternative models could equally well explain the observed data. In the worst cases of simple A-versus-B comparisons where the two premises cited above have not been proved (e.g., if impurity effects are not ruled out), it is indeed very probable that any hypothesis advanced regarding the structure of A is false.

REFERENCES

1. W.H. Zachariasen, J.Am.Chem.Soc. 54 (1932) 3841.
2. E.J. Gooding and W.E.S. Turner, J.Soc.Glass.Techn. 18 (1934) 32.
3. G. Hägg, J.Chem.Phys. 3 (1935) 42; ibid 363.
4. B.E. Warren, H. Krutter and O. Morningstar, J.Am.Ceram. Soc. 19 (1936) 202.
5. G.W. Morey and H.E. Merwin, J.Am.Chem.Soc. 58 (1936) 2248.
6. W.H. Zachariasen, J.Chem.Phys. 5 (1937) 919.
7. J. Biscoe and B.E. Warren, J.Am.Ceram.Soc. 21 (1938) 287.
8. F.C. Kracek, G.W. Morey, and H.E. Merwin, Am.J.Sci. 35A (1938) 143.
9. J.H. Hibben, Amer.J.Sci. 35A (1938) 113.
10. R.L. Green, J.Amer.Ceram.Soc. 25 (1942) 83.
11. B.E. Warren, J.Appl.Phys. 13 (1942) 602.
12. T. Abe, J.Am.Ceram.Soc. 35 (1952) 284.
13. F.E. Eversteijn, J.M. Stevels and H.I. Waterman, Phys. Chem.Glasses 1 (1960) 123.
14. M.L. Huggins and T. Abe, J.Amer.Ceram.Soc. 40 (1957) 287.
15. K. Fajans and S.W. Barber, J.Amer.Chem.Soc. 74 (1952) 2761.
16. J.D. Mackenzie, J.Phys.Chem. 63 (1959) 1875.
17. J. Goubeau and H. Keller, Z.Anorg.Chem. 272 (1953) 303.
18. G.M. Willis and F.L. Hennessy, J.Metals (1953) 1367.
19. L. Shartsis, W. Capps and S. Spinner, J.Amer.Ceram. Soc. 36 (1953) 35; ibid 319.
20. S.V. Berger, Acta Chem.Scand. 7 (1953) 611.
21. O. Borgen, K. Grjotheim, and J. Krogh-Moe, Kgl. N.V.S. Forh 27 (1954) 1; K. Grjotheim and J. Krogh-Moe, Glasteknisk Tidskrift 2 (1956) 1.

22. S. Anderson, R.L. Bohon and D.D. Kimpton, J.Amer. Ceram.Soc. 38 (1955) 370.

23. P.E. Jellyman and J.P. Procter, J.Soc.Glass Technol. 39 (1955) 173.

24. E.N. Lotkova, V.V. Obukhov-Denisov and N.N. Sobolev, Opt. Spectrosc. 1 (1956) 772.

25. T.A. Sidorov and N.N. Sobolev, Opt. Spectrosc. 4 (1958) 9.

26. J.L. Parsons and M.E. Milberg, J.Amer.Ceram.Soc. 43 (1960) 326.

27. N.F. Borrelli, B.D. Mc Swain, and G.-J. Su, Phys.Chem. Glasses 4 (1963) 11.

28. A.H. Silver and P.J. Bray, J.Chem.Phys. 29 (1958) 984.

29. W. Skatulla, W. Vogel, and H. Wessel, Silikat Tech. 9 (1958) 51.

30. B.S.R. Sastry and F.A. Hummel, J.Am.Ceram.Soc. 41 (1958) 7.

31. J. Krogh-Moe, Ark.Kemi 12 (1958) 475.

32. J. Krogh-Moe, Ark.Kemi 14 (1959) 439.

33. J. Krogh-Moe, Ark.Kemi 14 (1959) 451.

34. J. Krogh-Moe, Phys.Chem.Glasses 3 (1962) 208.

35. S. Block and G.J. Piermarini, Phys.Chem.Glasses 5 (1964) 138.

36. J. Krogh-Moe and H. Jürine, Phys.Chem.Glasses 6 (1965) 30.

37. R.V. Adams and R.W. Douglas, Glastech.Ber. 32K (1959) VII/12.

38. J.L. Parsons, J.Chem.Phys. 33 (1960) 1860.

39. J. Krogh-Moe, Phys.Chem.Glasses 1 (1960) 26.

40. J.O. Edwards and V.F. Ross, J.Inorg.Nucl.Chem. 15 (1960) 329.

41. P.J. Bray, J.O. Edwards, J.G. O'Keefe, V.F. Ross, and I. Tatsuzaki, J.Chem.Phys. 35 (1961) 435.

42. S.-E. Svanson, E. Forslind, and J. Krogh-Moe, J.Phys. Chem. 66 (1962) 174.

43. J. Krogh-Moe, Acta Cryst. 15 (1962) 190.

44. J. Krogh-Moe, Phys.Chem.Glasses 3 (1962) 1.

45. J. Krogh-Moe, Phys.Chem.Glasses 3 (1962) 101.

46. G. Becherer, O. Brümmer and G. Herms, Silikat Tech. 13 (1962) 339.

47. B.D. Mc Swain, N.F. Borelli and G.-J. Su, Phys.Chem. Glasses 4 (1963) 1.

48. S. Kumar, Phys.Chem.Glasses 4 (1963) 106.

49. P.J. Bray and J.G. O'Keefe, Phys.Chem.Glasses 4 (1963) 37.

50. P.J. Bray, M. Leventhal, and H.O. Hooper, Phys.Chem. Glasses 4 (1963) 47.

51. S. Lee and P.J. Bray, J.Chem.Phys. 39 (1963) 2863.

52. S. Lee and P.J. Bray, J.Chem.Phys. 40 (1964) 2982.

53. J. Krogh-Moe, Acta Chem.Scand. 18 (1964) 2055.

54. J. Krogh-Moe, Acta Cryst. 18 (1965) 77.

55. A. Hyman, A. Perloff, F. Mauer, and S. Block, Acta Cryst. 22 (1967) 815.

56. P. Beekenkamp, Philips Res.Repts.Suppl. 4 (1966).

57. J. Krogh-Moe, Phys.Chem.Glasses 6 (1965) 46.

58. M. Leventhal and P.J. Bray, Phys.Chem.Glasses 6 (1965) 113.

59. A.M. Bishay and M. Maklad, Phys.Chem.Glasses 7 (1966) 149.

60. S.G. Bishop and P.J. Bray, Phys.Chem.Glasses 7 (1966) 73.

61. P.J. Bray, D. Kline, and W. Poch, Glastechnische Berichte 39 (1966) 175.

62. J. Zarzycki and F. Naudin, Phys.Chem.Glasses 8 (1967) 11.

63. E.F. Riebling, J.Amer.Ceram.Soc. 50 (1967) 46.

64. S.R. Nagel and C.G. Bergeron, J.Amer.Ceram.Soc. 57 (1974) 129.

65. S.L. Strong and R. Kaplow, Acta Cryst. B24 (1968) 1032.

66. G.E. Gurr, P.W. Montgomery, C.D. Knutson, and B.T. Gorres, Acta Cryst. B26 (1970) 906.

67. D. Kline, P.J. Bray, and H.M. Kriz, J.Chem.Phys. 48 (1968) 5277.

68. H.M. Kriz, S.G. Bishop, and P.J. Bray, J.Chem.Phys. 49 (1968) 557.

69. H.M. Kriz and P.J. Bray, J.Magn.Res. 4 (1971) 76; H.M. Kriz, M.J. Park, and P.J. Bray, Phys.Chem.Glasses 12 (1971) 45.

70. D.L. Griscom, P.C. Taylor, D.A. Ware, and P.J. Bray, J.Chem.Phys. 48 (1968) 5158.

71. L.A. Kristiansen and J. Krogh-Moe, Phys.Chem.Glasses 9 (1968) 96.

72. R.R. Shaw and D.R. Uhlmann, J.Amer.Ceram.Soc. 51 (1968) 377.

73. D.R. Uhlmann and R.R. Shaw, J.Non-Cryst.Solids 1 (1969) 347.

74. J.F. Baugher and P.J. Bray, Phys.Chem.Glasses 10 (1969) 77.

75. H. de Wall, Phys.Chem.Glasses 10 (1969) 101.

76. J. Krogh-Moe, J.Non-Cryst.Solids 1 (1969) 269.

77. R.L. Mozzi and B.E. Warren, J.Appl.Cryst. 3 (1970) 251.

78. P.C. Taylor and P.J. Bray, J.Magn.Res. 2 (1970) 305.

79. P.C. Taylor and E.J. Friebele, J.Non-Cryst.Solids 16 (1974) 375.

80. G.E. Peterson, C.R. Kurkjian, A. Carnevale, Phys.Chem. Glasses 15 (1974) 59.

81. H.M. Kriz and P.J. Bray, J.Non-Cryst.Solids 6 (1971) 27.

82. D.L. Griscom, J.Chem.Phys. 55 (1971) 1113.

83. C. Rhee and P.J. Bray, Phys.Chem.Glasses 12 (1971) 165.

84. C. Rhee, J. Korean Phys.Soc. 4 (1971) 51.

85. P.C. Taylor and D.L. Griscom, J.Chem.Phys. 55 (1971) 3610.

86. P.C. Taylor and P.J. Bray, J.Amer.Ceram.Soc. 51 (1972) 234.

87. E.N. Boulos and N.J. Kreidl, J.Amer.Ceram.Soc. 54 (1971) 368.

88. M.E. Milberg, J.G. O'Keefe, R.A. Verhelst, and H.O. Hooper, Phys.Chem.Glasses 13 (1972) 79.

89. A.I. Zvyagin, P.S. Kalinin, V.A. Kaplun and R.S. Shevelevich, Isvestiya Akademi Nauk SSSR 7 (1971) 350.

90. J. Scheerer, W. Müller-Warmuth and H. Dutz, Glastechn. Ber. 46 (1973) 109.

91. M.J. Park and P.J. Bray, Phys.Chem.Glasses 13 (1972) 50.

92. K.S. Kim and P.J. Bray, J.Nonmetals 2 (1974) 95.

93. K.S. Kim and P.J. Bray, Phys.Chem.Glasses 15 (1974) 47.

94. T.W. Bril, Thesis, Technological University of Eindhoven, The Netherlands (1975).

95. W.L. Konijnendijk and J.M. Stevels, J.Non-Cryst.Solids 18 (1975) 307.

96. W.L. Konijnendijk, Philips Res.Repts.Suppl. 1 (1975).

97. J. Krogh-Moe, Acta Cryst. B30 (1974) 1827.

98. K. Jenssen and J. Krogh-Moe (publication pending).

99. L.C. Snyder, G.E. Peterson, and C.R. Kurkjian, J.Chem. Phys. 64 (1976) 1569.

100. G.E. Jellison, Jr. and P.J. Bray, Solid State Comm. 19 (1976) 517; G.E. Jellison, Jr., Ph.D. Thesis, Brown University, 1977; L.W. Panek, G.E. Jellison, Jr., and P.J. Bray, in Structure and Excitation of Amorphous Solids, edited by G. Lucovsky and F.L. Galeener (AIP Conf.Proc., 1976).

101. P.J. Bray, in Magnetic Resonance, edited by C.K. Coogan et al , Plenum Press, New York (1970) p. 13-1.

102. J. Krogh-Moe, Thesis, Chalmers Technical University, Göteborg (1959).

103. J. Krogh-Moe and M. Ihara, Acta Cryst. 23 (1967) 427.

104. J. Krogh-Moe, Glastech.Ber. 32K (1959) VI/18.

105. J. Krogh-Moe, Ark.Kemi 14 (1959) 553.

106. D.L. Liedberg, C.G. Ruderer, and C.G. Bergeron, J.Am. Ceram. Soc. 48 (1965) 440.

107. R.J. Charles and F.E. Wagstaff, J.Am.Ceram.Soc. 51 (1968) 16.

108. W. Haller, D.H. Blackburn, R.E. Wagstaff, and R.J. Charles, J.Am.Ceram.Soc. 53 (1970) 34.

109. P.B. Macedo and J.H. Simmons, J.Res.Natl.Bur.Standards 78A (1974) 53.

110. K.H. Karsch, Glastech.Ber. 35 (1962) 234.

111. E. Jenckel, Z.Elektrochem. 41 (1935) 211.

112. M.O. Samsoen, Ann.Phys. 9 (1928) 35.

113. C. Kittel, Introduction to Solid State Physics, 3rd ed. John Wiley, New York (1968) Chapt. 2.

114. B.E. Warren and N.S. Gingrich, Phys.Rev. 46 (1934) 368.

115. H. Zachariasen, Acta Cryst. 16 (1963) 385.

116. J. Krogh-Moe, Ark.Kemi 14 (1959) 31.

117. B.E. Warren and G. Mavel, Rev.Sci. Instrum. 36 (1965) 59.

118. B.E. Warren, X-ray Diffraction, Addison-Wesley, Reading (1969) p. 135.

119. M.H. Cohen and F. Reif, in Solid State Physics, F. Seitz and D. Turnbull, Eds., Vol. 5, Academic Press, New York (1957) p. 321.

120. G.H. Stauss, J.Chem.Phys. 40 (1964) 1988.

121. J.F. Baugher, H.M. Kriz, P.C. Taylor, and P.J. Bray, J.Magn.Res. 3 (1970) 415.

122. P.C. Taylor, J.F. Baugher, and H.M. Kriz, Chem.Rev. 75 (1975) 203.

123. G.E. Peterson, C.R. Kurkjian and A. Carnevale, Phys. Chem.Glasses 16 (1975) 63.

124. G.E. Jellison, P.J. Bray, and P.C. Taylor, Phys.Chem. Glasses 17 (1976) 35.

125. C.H. Townes and B.P. Dailey, J.Chem.Phys. 17 (1949) 782.

126. A.H. Silver, J.Chem.Phys. 32 (1960) 959.

127. P.C. Taylor, in Glass, (to be published).

128. J. Kujumzelis, Z.Phys. 100 (1936) 221.

129. C.A. Coulson and T.W. Dingle, Acta Cryst. B24 (1968) 153.

130. S. Brawer, Phys.Rev. B11 (1975) 3173.

131. P.H. Gaskell and D.W. Johnson, J.Non-Cryst.Solids 20 (1976) 153, 171.

132. N.F. Borrelli and G.-J Su, Phys.Chem. Glasses 4 (1963) 206.

133. W.J. Knapp and H. Flood, J.Amer.Ceram.Soc. 40 (1957) 246; S. Urnes, Trans.Brit.Ceram.Soc. 60 (1961) 85.

134. G.S. Smith and G. Rindone, J.Am.Ceram.Soc. 44 (1961) 72.

135. D.L. Griscom, J.Non-Cryst. Solids 13 (1973/74) 251.

136. G.E. Peterson, C.R. Kurkjian, and A. Carnevale, Phys. Chem.Glasses 15 (1974) 52.

137. C.M. Hurd and P. Coodin, J.Phys.Chem.Solids 28 (1967) 523.

138. P.W. Atkins and M.C.R. Symons, The Structure of Inorganic Radicals, Elsevier, Amsterdam (1967).

139. D.L. Griscom, G.H.Sigel, Jr., and R.J. Ginther, J. Appl. Phys. 47 (1976) 960.

140. D.L. Griscom, in Defects and their Structure in Non-metallic Solids, B. Henderson and A.E. Hughes, eds., Plenum Press, New York (1976) p. 323.

141. D.L. Griscom, E.J. Friebele, G.H. Sigel, Jr., and R.J. Ginther, in Proceedings of Symposium on the Structure of Non-Crystalline Materials, Cambridge, England, Sept. 20-23, 1976 (to be published by Society of Glass Technology).

142. D.L. Griscom, P.C. Taylor, and P.J. Bray, J.Chem.Phys. 50 (1969) 977.

143. D.L. Griscom, J.Non-Cryst.Solids 6 (1971) 275.

144. F. Assabgby, S. Arafa, E. Boulos, A. Bishay, and N.J. Kreidl, Proc. X Int.Cong. on Glass 5 (1975) 30: submitted to J. Non-Cryst. Solids, 1976.

145. F.J. Feigl, W.B. Fowler, and K.L. Yip, Solid State Comm. 14 (1974) 225.

146. G.E. Pake, Paramagnetic Resonance, Benjamin, New York (1962).

147. H.G. Hecht and T.S. Johnson, J.Chem.Phys. 46 (1967) 23.

148. H. Imagawa, Phys.Stat.Sol. 30 (1968) 469.

149. T. Castner, G.S. Newell, W.C. Holton, and C.P. Slichter, J.Chem.Phys. 32 (1960) 668.

150. H.H. Wickman, M.P. Klein, and D.A. Shirley, J.Chem. Phys. 42 (1965) 2113.

151. R.H. Sands, Phys.Rev. 99 (1955) 1222.

152. H.W. de Wijn and R.F. van Balderen, J.Chem.Phys. 46 (1967) 1381.

153. D.L. Griscom and R.E. Griscom, J.Chem.Phys. 47 (1967) 2711.

154. T.I. Barry, National Laboratory Report, Teddington, Middlesex, England (1967), unpublished.

155. R.D. Dowsing and J.F. Gibson, J.Chem.Phys. 50 (1969) 294.

156. R. Aäsa, J.Chem.Phys. 52 (1970) 3919.

157. K. Bingham and S. Parke, Phys.Chem.Glasses 6 (1965) 224.

158. P.C. Taylor and P.J. Bray, J.Phys.Chem.Solids 33 (1972) 43.

159. D. Loveridge and S. Parke, Phys. Chem. Glasses 12 (1971) 19.

160. G. Hochstrasser, Phys.Chem.Glasses 7 (1966) 178.

161. H. Toyuki and S. Akagi, Phys.Chem.Glasses 13 (1972) 15.

162. G.E. Peterson and C.R. Kurkjian, Solid State Comm. 11 (1972) 1105.

163. J. Wong and C.A. Angell, Vitreous State Spectroscopy, Marcell Dekker (1976).

EPILOGUE

David L. Griscom

U. S. Naval Research Lab

Washington, D. C.

Much has changed since the preparation of my Pre-Conference Manuscript "Borate Glass Structures" in the summer of 1976. Most visibly, the title of the actual conference was changed from "borate glass structures" to "Boron in Glass and Glass Ceramics" and the location was moved from Santa Fe to Alfred. Beyond that, the decision was made to publish my position paper in its original form without editing, revision, or expansion. There were two very good reasons for this: (1) the desire for rapid publication and (2) the fact that many of the other authors contributing papers to the Conference Proceedings have undoubtedly included discussions based on the original version. Indeed, the Pre-Conference Manuscript has been too widely circulated to ever hope to cover up whatever mistakes it may contain, so it is just as well that it be printed "as was". But as a consequence, I must amplify my apology for the informal style and limited scope of this document.

It was neither expected nor intended that my review would be all things to all people. I made the arbitrary decision to focus on those experiments which I believed told us that most about boron-oxygen structural arrangements in defect- and impurity-free boron-oxide and alkali-borate glasses. Even so, time did not permit me to cover all techniques that yield useful information of this nature. For example, optical spectra of foreign ions in glass are often more readily interpreted than ESR spectra of the same systems. However, as the deadline for submission of my first draft approached, I felt I could make a more useful contribution to the literature by summarizing the ESR work, with which I was intimately familiar, rather than the optical area for which I had not yet amassed an adequate bibliography. In any event, the Pre-Conference Manuscript took shape as a critical commentary, as opposed to an exhaustive source of references. It is hoped that the discussions surrounding those experimental techniques singled out for emphasis will serve as examples for assessing the reliability and uniqueness of other methods of probing borate glass structure.

Another thing that has changed since the writing of the original manuscript has been the holding of the Conference itself. The various papers which were presented, together with the ensuing discussions and private conversations, have made a decided impression on my view of our current understanding of borate glass structures. In as much as my Pre-Conference Manuscript is reproduced in this volume in a state which became frozen in late 1976, it seems appropriate to append a brief up-date indicating how my own thinking has changed since the Conference at Alfred. Due to the very short time alotted for its preparation, as well as obvious page restrictions, this Epilogue will be even more informal than the preceding manuscript; for this I must also apologize. Most of the subheadings in the following sequence correspond to consecutive chapters in my Pre-Conference Manuscript. Where I quote individuals by name without a specific literature citation, it should be understood that I am referring to remarks which were made (either formally or informally) at the Alfred Conference.

Milestones. If I were now assigned the task of rewriting the "Milestones" chapter of my Pre-Conference Manuscript, I would make very few changes notwithstanding a number of arguments aired at the Conference itself. However, it has occurred to me that our current understanding of borate glass structures is equally well described as a "modified random-network theory" or a "modified crystallite theory." Therefore it would be appropriate to add to the list of milestones the crystallite theory of glasses as proposed by Lebedev in 1921 (164) and further argued by Randall, Rooksby and Cooper in 1930 (165) and Valenkov and Porai-Koshits in 1936 (166). As will be explained in separate paragraphs below, one might also advocate the removal of a star from Krogh-Moe's 1962 melting-point depression study (45) and/or from Uhlmann and Shaw's apparent disproof of the "boron-oxide anomaly" (73). However, it will be left to the reader to make his own judgement on these and other possibilities.

The Boron-Oxide Anomaly. In Chapter III of the Pre-Conference manuscript, following arguments of Krogh-Moe (44, 105) and of Uhlmann and Shaw (73), the "boron-oxide anomaly" was dismissed as being no longer extant. It was surmised that all truly sharp property changes in alkali borate glasses in the vicinity of 16 mole % alkali oxide could be attributed to poor data or incorrect interpretations, since the relatively flat thermal expansion coefficient data of Uhlmann and Shaw (73) were amenable to rather straight forward explanation. However, this position must definitely be revised in light of the results presented at the Alfred Conference by Mader and Loretz.

The latter authors prepared large batchs of optically homogeneous sodium borate glasses sharing a common thermal history; on these they performed careful measurements of some eight properties as functions of glass composition. Anomalous characteristics were observed in at least five of these properties in the range 16 to 19 mole % Na_2O. This included a thermal expansion coefficient curve virtually duplicating that shown by Biscoe and Warren (7) (Fig. 3a of the Pre-Conference Manuscript). The sharp minimum observed by Mader and Loretz at ~ 16.5 mole % Na_2O contrasts with the broad flat minima determined by Uhlmann and Shaw (73) (Fig. 4 of the Pre-Conference Manuscript). This apparent conflict does not imply that one of the data sets need be erroneous, however, since Mader and Loretz measured the expansion coefficient in a temperature range well above room temperature while Uhlmann and Shaw's data were for a range below ambient. It seems reasonable that the high-temperature and low-temperature expansion coefficients may be sensitive to different aspects of the glass structure. Thus, while Uhlmann and Shaw did not observe a "boron-oxide anomaly," they were evidently wrong in inferring that such phenomena do not occur in other property-versus-composition curves.

What then is the origin of the "boron-oxide anomaly"? Possibly, some or all such effects can be related to Krogh-Moe's structural theory (Chapter V.), but it is not immediately obvious in what ways. Stevels, Kreidl, and others were quick to point out, as they have been doing for years, that the occurence of non-bridging oxygens may provide the explanation. This proposition raises the question: How many nonbridging oxygens would be required to account for the observed anomalies? Beekenkamp (56) explored this question from a thermodynamic point of view by postulating that the log of the equilibrium constant governing the ratio N_{BO_4}: N_{NBO} is a simple linear function of x (x=mole fraction alkali oxide). This postulate was qualitatively supported by NMR data (49), at least in the approximate composition range $0.25 < x < 0.4$. Beekenkamp (56) assumed, however, that the experimental N_{BO_4} values of Bray and O'Keefe (49) may be in error by a small scale factor and went on to calculate a quasi-theoretical N_{BO_4} curve based essentially on the above postulate and the first two of Abe's rules (12). However, more precise ^{10}B and ^{11}B NMR data (100) have since shown that Beekenkamp's theoretical curve predicts far too many nonbridging oxygens. Moreover, Abe's first rule (BO_4 tetrahedra cannot be bound to each other) violates Krogh-Moe's structural model as well as explicit NMR evidence (84, 100) that paired BO_4 units do occur in alkali borate glasses (presumably in diborate groups). More recently, Arafa (167) adopted Beekenkamp's thermodynamic approach without regard to any particular structural rules, showing that the experimental N_{BO_4} curve can be fit rather well in the approximate range $x < 0.4$ by suitable adjustment of parameters. In this way the fraction of 3-coordinated borons with non-bridging oxygens at 20 mole % alkali oxide could be predicted to lie in the range ~0.002-0.004. These numbers are too small to be confirmed or disproved by present NMR techniques. Although a complete thermodynamic treatment would presumably take into account structural groupings of the type advanced by Krogh-Moe (45, 57), Arafa's calculation may serve as an interim basis for evaluating the possible influence of nonbridging oxygens on the "boron-oxide anomaly".

Mader and Loretz essentially duplicated the results of McSwain et al. (47) for the shift in the UV absorption edge from ~175 to ~210 nm with increasing Na_2O additions from ~15 to ~20 mole % Na_2O. An estimate of the minimum number of NBO's required to produce this shift could be made by subtracting the absorption spectrum of, say, the 15 mole % Na_2O glass from that obtained for a 20 mole % glass of the same thickness. By integrating the area under this difference spectrum and assuming an oscillator strength of unity, a lower limit on the number of absorbing centers could be obtained by means of Smakula's formula (168). When such a calculation is performed and compared, for example, with Arafa's calculation, it should be possible to render a less speculative judgement on the possible relationship between the "boron-oxide anomaly" and the presence of nonbridging oxygens.

Phase Separation. An undeniable source of controversy at the Alfred Conference concerned the question of phase separation in alkali borate glasses. One school of thought, strongly argued by Macedo, is that all alkali borate glasses of compositions falling within the subliquidus immiscibility gaps (Chapter IV of the Pre-Conference Manuscript) are phase separated instantly on quenching, irrespective of cooling rate. The other school, championed by Porai-Koshits, Golubkov, and Titov (paper read by Kreidl), hold that there are no subliquidus immiscibility gaps in any alkali borate glass systems except for a very narrow region at low LiO_2 content in the lithium borate system.

Porai-Koshits et al. based their conclusions on small angle X-ray scattering (SAXS), which theoretically should reveal the presence of phase separation with a high degree of resolution. Leaving aside the question of theoretical resolving power of the technique, an irreversible change in SAXS intensity with prolonged heating near the softening point is sufficient evidence of phase separation. The latter phenomenon was observed for the low-lithia glasses but not for the other alkali-borate glass systems. Porai-Koshits et al. (conference preprint) concluded, "Thus, no immiscibility gaps are present in these systems, and structures observed by electron microscopy are not related to the processes of phase separation." On the other hand, it would be difficult to imagine the origin of the submicrostructure observed electron microscopically by Shaw and Uhlmann (72) if it were not related to phase separation. The variation with temperature of the scale of this structure, together with the results of clearing experiments, gave strong indication that the miscibility gaps determined by the latter authors are indeed true phase boundaries. One possible way out of this dilemma would be to assume with Macedo that in all alkali borate glasses (save possibly the low lithia glasses) initial phase separation takes place spontaneously on a time scale short compared to the performance of any physical measurements. Increases of scale with prolonged heating near the phase boundaries are then observable by electron microscopy but not via SAXS for reasons yet to be elucidated. Seward has remarked on the variable suitability of different experimental probes of phase-separated glass microstructure depending on the particular glass forming system under study; his review in the present volume should be consulted.

Krogh-Moe's Theory. In the wake of the Alfred Conference, it still appears that major portions of Krogh-Moe's structural theory (Chapter V) are accepted by virtually all researchers in the field of borate glasses. Nevertheless, many individuals have continued to challenge one aspect or another of the theory, although such challenges appeared to be relatively low keyed. For example, Cooper pointed out that there are topological problems involved in trying to construct a B_2O_3 glass of the proper density using nothing but boroxol groups. He further noted that the 2.8Å peak attributed to the boroxol ring in the experimental X-ray diffraction PFD of vitreous B_2O_3 (77) can also be found in the diffraction pattern of crystalline B_2O_3I (66). A model for glassy B_2O_3 based on 2-dimensional ribbons of BO_3 units (similar to B_2O_3I) has been advanced by Dunlevey and Cooper (169). Cooper conceded, however, that the experimental difficulty in crystallizing B_2O_3I from glassy B_2O_3 suggests that the two polymorphs may have different structures and that the mixed boroxol-BO_3 triangle model evinced by ^{17}O NMR (100) might circumvent the topological problems he alluded to.

The resurection of the "boron-oxide anomaly" (see above) itself presents a new challenge. It is not immediately obvious how Krogh-Moe's structural model can account for well-defined maxima, minima, or inflections in property-composition curves in the range 16-19 mole % modifier oxide. Perhaps the answer lies in the topological considerations in fitting together the indicated polyborate groupings to form a continuous network. On the other hand, it could turn out that the various anomalies are completely unrelated to Krogh-Moe's theory (be it right or wrong), as, for example if NBO's were the primary culprits.

X-ray Diffraction. There does not appear to be any need to modify Chapter VII of the Pre-Conference Manuscript summarizing the present state of large-angle X-ray diffraction studies of borate glasses. However, the reader is referred to Zarzycki's article in this volume, particularly with regard to the SAXS technique, and also to the recent review "Diffraction Studies of Glass Structure" by Wright and Leadbetter (170).

NMR Studies. The Alfred Conference witnessed substantial agreement among all workers that NMR can be used to provide an unambiguous measure of the fraction of borons in four coordination (c.f. Chapter VIII). On the other hand, Kurkjian, Peterson, and Carnevale presented a convincing demonstration that unique distributions in boron quadrupole coupling constants Q_{cc} cannot be extracted beyond the aforementioned division into three-fold and four-fold coordination. Attention is called to the recent paper on this subject by Peterson, Carnevale, and Kurkjian (171). Jellison and Bray agreed with the latter premise but wished to emphasize that their own estimate of the "resolving power" of ^{10}B NMR for determining distributions in asymmetry parameter η is about 30 times that for Q_{cc}. Thus, various trigonal boron sites co-present in glass which cannot be distinguished one from another on the basis of their Q_{cc} values can be distinguished on the basis of their η values (c.f. Chapter VIII and ref. 100).

N. B. --Jellison and Bray informed us that there now exist much better sets of N_{BO_4} data than those portrayed in Fig. 4 of the Pre-Conference Manuscript. These data, which show considerably less scatter and follow the theoretical curve $x/(1-x)$ quite closely below $x \approx 0.28$, are being readied for publication.

Raman Scattering. The interpretations of Bril (94) and Konijnendijk (96) vis-a-vis the Raman spectra of borate glasses (Chapter IX) were not strongly disputed at the Alfred Conference, although considerable discussion focused on possible alternatives. White suggested that the Raman spectra might turn out to be sensitive to structures (clusters) somewhat larger than the structural groupings theorized by Krogh-Moe. Still, compared to the Raman spectra of other glasses, the 808 and 770 cm^{-1} lines are exceedingly sharp, and this sharpness keeps leading one back to the simple ring structure interpretation argued so lucidly by Bril (94).

Jellison pointed out that the peak height ratio h_{806}/h_{770} in Fig. 23 may deviate from theoretical expectation as a result of the two peaks having different (composition-dependent) widths, rather than being a consequence of the dissociation of tetraborate groups as suggested by Bril (94). This ambiguity could be eliminated by plotting I_{806}/I_{770}, where I_λ denotes the areas under the respective peaks as might be determined by a Gaussian resolution.

Melting Point Depression. In Chapter XI of the Pre-Conference Manuscript the author gave high marks to Krogh-Moe's 1962 study of the structure of alkali borate melts by the "phase diagram method." The particularly attractive feature of this method was its seeming ability to unambiguously eliminate certain structural models, e.g., the random network of BO_3 triangles and BO_4 tetrahedra. Unfortunately, I had overlooked a 1970 publication by

Ostvold and Kleppa (172) which throws some doubt on the reliability of this procedure. (A. Paul is thanked for reopening this issue.)

Krogh-Moe's derivation of Eq. (5) of the Pre-Conference Manuscript involved what was termed "several reasonable approximations." One of these approximations was that the partial molar enthalpy $\overline{\Delta H}_A$ of A in a molten mixture with B can be neglected relative to the term $-T_m\overline{\Delta S}_A$ in the expression for the partial free energy of A at small concentrations of B. The real question is, how small a concentration of B is necessary for the approximation to be valid? In the case of the melting point depression of $Na_2O.4B_2O_3$ by B_2O_3, Krogh-Moe (45) assumed (in effect) that additions of $B_2O_3 \sim 10$ mole % were "small." However, Ostvold and Kleppa (172) measured the partial molar enthalpy of B_2O_3, $\overline{\Delta H}_{B_2O_3}$ in a wide range of sodium borate melts. From these data they obtained the partial molar enthalpy of $Me_2O.4B_2O_3$ as a function of B_2O_3 additions by means of Gibbs-Duhem integrations. The results showed that $\Delta H_{tetraborate}$ could not be neglected in calculating melting point depressions for additions of more than ~ 2 mole % B_2O_3 to tetraborate melts. When additions ~ 10 mole % are examined, Ostvold and Kleppa showed what all structural models considered by Krogh-Moe could be eliminated by comparing the theoretical melting point depressions of $Me_2O.4B_2O_3$ by B_2O_3 with the actual liquidus curves. At face value, this would appear to be a serious blow to the "phase diagram method" as employed by Krogh-Moe (45).

One may consider three possibilities: (1) Krogh-Moe's theory is wrong, (2) Krogh-Moe's theory is right but his case built on melting point depressions is invalid, or (3) Krogh-Moe's theory is right and his melting point depression studies were somehow valid in spite of Ostvold and Kleppa's objections. In view of the recent support of Krogh-Moe's theory from NMR and Raman studies (Chapters VIII and IX, respectively), the author is inclined to reject the first possibility. In choosing between possibilities (2) and (3), it should be remarked that if one confines himself to the true region of validity of the approximation (additions approximately <2 mole %) Krogh-Moe's study still supports the models he postulated while eliminating alternative models such as a random network of BO_3 triangles and BO_4 tetrahedra. (Note in Fig. 26 of the Pre-Conference Manuscript that the curves of melting point depressions of $Na_2O.2B_2O_3$ by B_2O_3 and of $Na_2O.4B_2O_3$ by Na_2O cover ranges of only 3 and 6 mole %, respectively). On the other hand, if one extends the range of comparison by introducing Ostvold and Kleppa's enthalpy data, all conceivable models are seemingly rejected. This simply means that some other approximation breaks down when an attempt is made to fit a broad portion of the liquidus curve using the enthalpy data; Ostvold and Kleppa suggest that neglect of vibrational entropy is a probable source of the difficulty (172). It remains conceivable therefore that Krogh-Moe's neglect of both the enthalpy of mixing and vibrational entropy--poor approximations if made individually--had an offsetting effect which fortuitously extended the range over which he could calculate liquidus curves on the basis of configurational entropy alone. Clearly, these matters are deserving of additional study before any final conclusions can be drawn.

Foreign Ions in Borate Glasses. In the Pre-Conference Manuscript, studies of doped-in foreign ions were downgraded as a tool for probing the structure of the undoped glasses. The reason for this was that transition-group ions in particular frequently dictate their own atomic surroundings and, even when they don't, it is difficult to know this without performing a separate experiment of a more direct nature. Paul presented a pretty example wherein NMR verified that the optically determined coordination changes of Co^{2+} in the CaBAℓ glass system mimics those of the boron almost exactly. But the usual experience is different. For example, in the abstract of the paper presented by Weber, Hegarty, and Blackburn, it was concluded on the basis of laser-induced fluorescence line narrowing that the oxygen ligand coordination and bonding rather than the network forming cation appear to be predominant in determining the spectroscopic properties of Eu^{3+} in the lithium borate glass system. On the other hand, Weber emphasized that when the ultimate object of one's research is the development of a better laser glass, the study of foreign ions in glass is quite naturally upgraded. This again underscores the very narrow viewpoint of the Pre-Conference Manuscript, which was preoccupied mainly with boron-oxygen atomic arrangements in glass. Apropos, the laser spectroscopy method is clearly very powerful and would have been appropriately included in the Pre-Conference Manuscript but for the fact that no such work had been performed on borate glasses as of fall 1976.

Quantum Chemical Calculations. As was learned from Snyder's presentation at the Alfred Conference, considerable advances have been made in the past year in ab initio calculations of the electronic structure and bonding of polyborate structural groupings. Complexes as large as a boroxol ring with one attached BO_3 unit have now been treated. The results should be of particular value in interpreting the structural significance of the quadrupolar coupling constants and asymmetry parameters obtained from ^{11}B, ^{10}B, and ^{17}O NMR of borate glasses (c.f. Chapter VIII and ref. 100). For example, Jellison and Bray's empirical calculation of electronic charge distributions on the boroxol group from the NMR parameters required the making of some assumptions regarding the principal axes of the EFG tensor. While Snyder's preliminary calculations of Q_{cc} and η for ^{17}O in this group did not accurately reproduce the experimental values, his findings did suggest that the q_{zz} axis for a ring oxygen should be parallel to the line joining the adjacent boron nuclei, rather than parallel to the lone pair axis as surmised by Jellison (100). Additional details will undoubtedly be found in Snyder's contribution to this volume.

Chromatography. Filter paper chromatographic methods have been applied with good advantage in studies of the structure of phosphate glasses (e.g., ref. 173). At the Alfred Conference, Paul made an appeal for the application of similar techniques to vitreous borates, as this would seem to be a natural way to test the validity of Krogh-Moe's structural theories. The problem lies in finding a suitable solvent, but Paul felt it should be possible to do so if a determined search is mounted.

Acknowledgments. The first draft of my Pre-Conference Manuscript was prepared while I was on sabbatical leave at the Départment de Physique des Matériaux, Université Claude Bernard-Lyon I. I thank Professors R. Uzan and C. H. S. Dupuy for their hospitality and encouragement during this period. A lengthy exchange of correspondence with Dr. P. C. Taylor was a great help in trying to understand what was at issue in some controversies concerning the NMR results. Many individuals too numerous to list were very kind in mailing me reprints, but E. J. Friebele stood out for continually responding to my requests for copies of obscure articles, figures, etc. An indispensible aid in preparing my review was a bound volume of Krogh-Moe's collected works; I am deeply grateful to Professor S. Urnes for providing me with it. I want to thank also the secretarial staffs at Lyon and Rolla for typing rough drafts and, especially, the staff at Alfred for the excellent job in preparing the camera-ready copy. Finally I thank Professors L. D. Pye and V. D. Fréchette for their expenditure of time and resources in assembling and distributing the Pre-Conference Manuscript while simultaneously burdened with organizing the Conference itself.

ADDITIONAL REFERENCES

164. A. A. Lebedev, Trudy Gosud. Optichesk. Inst. 2 (1921) 10.

165. J. T. Randall, H. P. Rooksby, and B. S. Cooper, Z. Krist. 75 (1930) 196.

166. N. Valenkov and E. A. Porai-Koshits, Z. Krist. 95 (1936) 195.

167. S. Arafa, in Recent Advances in Science and Technology of Materials, A. Bishay, Ed., Vol. 2, Plenum Press, New York (1974) p. 107.

168. D. L. Dexter, Solid State Phys. 6 (1958) 353.

169. F. M. Dunlevey and A. R. Cooper, Bull. Am. Ceram. Soc. 51 (1972) 374.

170. A. C. Wright and A. J. Leadbetter, Phys. Chem. Glasses 17 (1976) 122.

171. G. E. Peterson, A. Carnevale, and C. R. Kurkjian, J. Non-Cryst. Solids 23 (1977) 243.

172. T. Ostvold and O. J. Kleppa, Inorg. Chem. 9 (1970) 1395.

173. A. E. R. Westman and J. T. Crowther, J. Amer. Ceram. Soc. 37 (1954) 420; A. E. R. Westman and P. A. Gartaganis, J. Amer. Ceram. Soc. 40 (1957) 293.

QUANTUM CHEMICAL CALCULATIONS TO MODEL BORATE GLASS ELECTRONIC STRUCTURE AND PROPERTIES

Lawrence C. Snyder

Bell Laboratories
600 Mountain Avenue
Murray Hill, New Jersey 07974

INTRODUCTION

I consider myself to be a quantum chemist, having worked in that area for twenty years. I know little about borate glasses and have depended upon my Bell Laboratories colleagues Chuck Kurkjian and George Peterson for guidance as to the significant questions of glass chemistry. The overview of recent borate glass science provided by the preprint of David Griscom's talk[1] has been quite valuable.

In contemplating this talk I planned to show how some of the methods of quantum chemistry might be applied questions of borate glass structure and properties. I concurrently undertook computations to provide a few example applications. Although I have not completed everything planned, the bulk of this talk will be a sketch of the quantum methods used and my findings in the several applications.

This talk will attempt to accomplish three objectives:

a) To describe the methods used,
b) To provide evidence of the credibility of these methods for the phenomena to which they are applied.
c) To outline my findings relative to borate glass science.

Perhaps the first application of quantum chemical methods to molecules important to understanding borate glass structure, was a Huckel calculation by Coulson to explain the difference in length of interior and exterior B-O bonds of metaboric acid.[2] Subsequently

Coulson and Dingle made Huckel molecular orbital calculations to explain the variation of B-O bond lengths in a variety of borate ions.[3] More recently Dill, Schleyer and Pople made self-consistent field molecular orbital calculations, of the type which are used in this work, to predict the geometry and energy of small boron compounds.[4] Of these, hydroxyborane (H_2BOH) is of most interest here. In no quantum chemical study was information on borate glass structure or properties a primary objective.

THE QUANTUM CHEMICAL METHODS

Quantum chemistry may be thought to have come into existence with the discovery in 1926 of wave mechanics and the Schrödinger equation:[5]

$$H\bar{\Psi} = E\bar{\Psi} \tag{1}$$

This is a partial differential eigenvalue equation which states that the Hamiltonian operator H operating upon the wave function $\bar{\Psi}$ is equal to a constant E, the energy, times the wavefunction. The Hamiltonian operator is a sum of four major terms, here written in atomic units:

$$H = -\frac{1}{2}\sum_{i}\nabla^2(i) - \sum_{\mu}\sum_{i}\frac{Z_\mu}{r_{\mu i}} + \sum_{i>j}\frac{1}{r_{ij}} + \sum_{\mu>\nu}\frac{Z_\mu Z_\nu}{r_{\mu\nu}} \tag{2}$$

The first sum on the right is the kinetic energy of the electrons, a sum of second derivatives with respect to electron positions. The second sum is the attraction of the electrons to the nuclei. The third sum is the repulsion of the electrons for one another. The last is the repulsion of the nuclei for one another. The wave function is a function of the positions of the n-electrons of the molecular system;

$$\Psi = \Psi(r_1, r_2, r_3 \ldots r_n) \tag{3}$$

For systems of more than one electron, it has not been possible to solve the Schrödinger equation. This is because of the mathematical difficulties introduced by the mutual repulsion of the electrons. We will reserve the symbol Ψ of equation (3) for an approximate or "trial" wave function. It can be shown that the expectation value of the energy $\bar{E}$, which may be computed as an integral over the position space of all electrons dτ,

$$\bar{E} = \int \Psi^* \, H \, \Psi \, d\tau \geq E \tag{4}$$

is greater than or equal to E for the ground electronic state. Here

Ψ^* is the complex conjugate of Ψ. Equation (4) provides an excellent means to seek approximate solutions to the Schrödinger equation for the ground state of molecules, by variation of wave function parameters to minimize $\overline{E}$.

Three decades of computational experience[6] has shown that a trial wave function of determinantal form,

$$\Psi = \frac{1}{\sqrt{n!}} |\phi_a(1)\overline{\Phi}_a(2)\phi_b(3)\overline{\Phi}_b(4)\ldots\overline{\Phi}_{\frac{n}{2}}(n)| \tag{5}$$

is a very effective approximate wave function for closed shell molecules. Here the n electrons of a molecular system are assigned in pairs to the molecular orbitals ϕ_i, with electron spin up or down as denoted by the bar over ϕ_i. The computational studies here employ a determinantal wave function.

The molecular orbitals ϕ_a are linear combinations of basis functions χ_j:

$$\phi_a = \sum_i c_{aj}\ \chi_j \tag{6}$$

In the self consistent-field (SCF) molecular orbital (MO) calculations we report here, the coefficients c_{aj} are varied to minimize $\overline{E}$ for the trial determinantal wave function.

The basis function χ_j may be taken to be fixed linear combinations of primitive basis functions η_i:

$$\chi_j = \sum_i d_{ji}\ \eta_i \tag{7}$$

In the computational studies we have employed Gaussian primitive functions of the form

$$\eta_i(r_A) = N_i X_A^{\ell} Y_A^{m} Z_A^{n} e^{-\alpha_i r_A^2} \tag{8}$$

Here the primitive function is centered on atom (A). The variables X_A, Y_A, Z_A and r_A are electron coordinates from that nuclear center. The normalization constant is denoted by N_i.

Two sets of basis functions denoted by STO-3G and 4-31G are employed in the computational work reported here. The STO-3G basis has one basis function for each atomic orbital. Each basis function is a fit using three Gaussian primitive basis functions, of a Slater orbital for the atom.[7] The 4-31G basis set[8,9] provides a single linear combination of 4 primitive functions for the 1s orbital, but two basis functions for each valence atomic orbital;

an inner basis function built from 3 primitives and an outer basis function built from one primitive. This extended basis set, often referred to as a "double-zeta" basis set, provides important flexibility which permits an adjustment of the size of atoms to reflect their molecular environment.

We have carried out our computations of wave functions using the computer program[10] Gaussian-70 on an IBM 370/168. The computations of electric field gradients for these wave functions were made with the computer program POLYATOM[11] on a Honeywell 6070 computer.

The computation times using the STO-3G and 4-31G basis sets are given in Table I for orthoboric acid and the larger molecule dihydroxyl (dihydroxylboryl) boroxole. Times in parenthesis are estimates for work not yet completed. We also include estimates of the cost of commercial rates for one-day service.

TABLE I

Resources Required for Ab-Initio Calculations

		STO-3G		4-31G	
		sec.	$	sec.	$
Orthoboric acid	SCF	45	6	315	40
$B(OH)_3$	q_B	33	5	58	10
Dihydroxyl(dihydroxyl-boryl)boroxole	SCF	840	100	(5880)	(700)
	q^a	2919	140	(5130)	(246)

[a] At twelve nuclear centers.

The program Gaussian-70 is written for a maximum of 70 basis functions. Thus we were able to treat orthoboric acid with both basis sets. However the larger molecule is just within the capacity of the program with an STO-3G basis: We intend to increase the program dimensions to make a 4-31G basis useable in this case.

The application of the methods used here for silicon bearing glasses would be limited to molecules about half as large, because additional basis functions are required for the 2s-2p core electrons of silicon. Attempts to overcome this limitation are being made by the development of effective potentials[12] which represent the effect of core electrons by an empirical potential. Core electrons are then omitted from explicit inclusion in the wave function.

It is also possible to proceed in a more empirical manner. Some of the integrals, or a group of integrals, which occur in equation (4) may be taken to be empirical parameters. The Huckel molecular orbital method employed by Coulson and Dingle[2,3] is of this type.

CREDIBILITY OF AB-INITIO METHODS

We will apply the computation of determinantal SCF-MO wave functions and energy expectation values to estimate three properties of molecular systems related to borate glass structure. The first application is a computation of heat of reaction for condensations of orthoboric acid with water elimination to form higher polymers. The second application is an exploration of the change in energy accompanying changes of conformation of orthoboric acid and pyroboric acid. The third application is the computation of electric field gradients in orthoboric acid and condensation products for comparison with the results of nuclear quadrupole resonance (NQR) experiments.

It was shown almost a decade[13,14] ago that heats of reaction are given fairly accurately as a difference of SCF-MO total energies ($\overline{E}$) for reactions having closed shell reactants and products. The accuracy of this method of estimating heats reaction is illustrated for various hydrocarbon reactions by the data in Table II, with computations in the STO-3G and 4-31G basis sets published by Pople et al.[15] The accuracy of predicted heats of reaction for the top three reactions in Table I are typical of those for unstrained hydrocarbons of this size, about ± 3 kcal/mole in both the STO-3G and 4-31G basis sets. A large sample of reactions would show the

TABLE II

Thermochemistry of Closed-Shell Molecules

Reaction	ΔH (kcal/mole)		
	STO-3G	4-31G	Exp.
$CH_3CH_3+H_2 = 2CH_4$	-19.0	-22.9	-18.1
$CH_3CH=CH_2+CH_4 = CH_3CH_3+CH_2=CH_2$	4.1	4.0	5.0
$CH_3OCH_3+CH_4 = 2CH_3OH$	2.3	3.3	5.3
$C_3H_6^a+3CH_4 = 3CH_3CH_3$	-48.3	-30.4	-23.5
$C_6H_6^b+6CH_4 = 3CH_3CH_3+3CH_2=CH_2$	72	64.2	61.6

[a] Cyclopropane. [b] Benzene

4-31G basis to give somewhat better predicitions of heats of reaction. This is exhibited by the fourth reaction of Table I, in which the strained molecular cyclopropane reacts with methane to form ethane. The STO-3G basis predicts -48.3 kcal/mole for the reaction. This is 25 kcal/mole too exothermic and corresponds to computed cycloprepane being too unstable. The 4-31G basis also is not completely satisfactory for this reaction, giving a predicted heat of reaction 7 kcal/mole to exothermic. It is known that the addition of d functions to the basis set improves the predicitions of thermochemistry of strained molecules, which are thought to have "bent bonds". The bottom reaction of Table II involves benzene, which has an empirical resonance energy of 36 kcal/mole. The fact that the predicted heat of reaction is only 11 kcal/mole greater than the observed shows that STO-3G basis is capable of predicting 70% of the resonance. The 4-31G basis[16] predicts a heat of reaction for benzene which is 3.1 kcal greater than observed, and thus predicts 91% of the resonance energy of benzene.

Table III compares bond lengths and angles computed for H_2, CH_3, NH_3, and H_2O by minimizing the total energy with respect to the geometry variables. Using the 4-31G basis,[17] one computes bond lengths about 0.01 Å shorter than those observed, and bond angles

TABLE III

Molecular Geometry

Molecule	Bond Length Å			Bond Angle (°)		
	STO-3G	4-31G	Exp	STO-3G	4-31G	Exp
H_2	0.71	0.73	0.74			
CH_3	1.08	1.07	1.08	118	120	120
NH_3	1.03	0.99	1.01	104	116	107
H_2O	0.99	0.95	0.96	100	111	104.5

within about 4° of the observed. The exception is NH_3, for which it is known that d functions must be included with a high quality s-p basis to obtain the observed non-planarity. Bond lengths computed with the STO-3G basis are within 0.02 Å of those observed and bond angles within about 5° of the observed. In larger hydrocarbons, angles between C-C bonds are usually given within 3° by both basis sets.

Electric field gradients q computed for boron bearing molecules[18,19] from wave functions in a "double-zeta" basis set (comparable to 4-31G) are displayed in Table IV. The corresponding values of the ^{11}B quadrupole coupling frequency are about 91% of the observed frequencies. The STO-3G basis is much poorer for the computation of electric field gradients at boron, being only 36% of the observed value for orthoboric acid.

TABLE IV

Boron Nuclear Quadrupolar Coupling

Molecule	q(au)	e^2q Q(Mhz) Theory	Experiments
BF_3	+0.222	1.85	2.64
$B(OH)_3$	+0.260	2.17	2.56
$B(NH_2)_3$	∿		
$B(CH_3)_3$	+0.548	4.57	4.87
BH_3	+0.568	4.74	
BH_3CO	+0.248	2.07	1.55
B_2H_6	-0.307	2.56	2.85
	η=-0.84		$\eta=\pm.69$

The inadequacy of the STO-3G basis for computing the quadrupole coupling of ^{11}B is attributable to the fact that the orbital exponents were chosen for the neutral atom. The boron atom in borates bears of order a unit of positive charge. Under this circumstance the optimum boron valence orbital size is reduced. The STO-3G basis cannot accommodate this change. The result is that the electric field gradient operator, which depends on $\frac{1}{r^3}$, is computed to be too small. Because the oxygen atom of borates has a negative charge, the electric field gradient computed with the STO-3G basis is expected to be too large. The 4-31G basis does not suffer from this defect.[19] There are two basis functions for each atomic orbital, inner and outer components. Variation of the relative coefficients of these two component basis functions during the SCF calculation permits the atom to expand or contract in response to the net charge it bears.

APPLICATIONS TO BORATE GLASS SCIENCE

We have made SCF-MO calculations on compounds chosen to model glass properties; orthoboric acid (I), pyroboric acid (II), metaboric acid (III), and dihydroxy(dihydroxyboryl) boroxole (IV). These structures are illustrated on Figure I. Because the computer program Gaussian-70 is limited to 70 basis functions, we have been able to supply the 4-31G basis only to orthoboric acid. The STO-3G basis has been applied to all molecules. We intend to increase the dimensions of Gaussian 70 to accommodate 120 basis functions, so that the 4-31G basis may be applied to all the borate model compounds of this study.

In all computations, the boron-oxygen bond length is taken to be 1.37 Å.[20] The oxygen-hydrogen bond length is 0.96 Å.[21] The angle between bonds to boron and hydrogen from the same oxygen ($\angle$ BOH) is 120°. The angle between bonds to oxygen from the same boron ($\angle$ OBO) is 120°. The angle between bonds to boron from a common oxygen ($\angle$ BOB) is 120° within boroxole rings and 130° exterior to them.

A result of these computations which is of general interest is the net charge computed for atoms in the model compounds. For a given basis the charge on boron, oxygen, and hydrogen varies little within a compound, or between compounds. The net charges, computed from a Mulliken population analysis, for B, O, and H are +0.60, -0.40, and +0.20 respectively when the STO-3G basis is employed. Using the more flexible 4-31G basis we find the net charges on B, O, and H atoms to be +1.25, -0.84, and +0.43 respectively. It is of interest that net charges computed in the 4-31G basis are almost double those for the STO-3G basis. The lower charges exhibited by the STO-3G basis reflect the fact that the atoms cannot expand or contract to accommodate the flow of electrons. Thus we consider the atomic charges given by the 4-31G basis to be more representative of the actual charges in borate compounds. The characteristic net charges on boron and oxygen atoms of borate glasses may be of use in setting up empirical potentials to model the formation and properties of these glasses.

The formation of a B_2O_3 glass may be thought of as the final step of a series of condensation reactions of orthoboric acid with the elimination of water. Little is known about the thermochemistry of these reactions and the relative stability of the condensation products. For this reason we have employed the total energy computed with STO-3G basis for orthoboric acid, water and three condensation products to estimate the heat of reaction of three condensation-dehydration reactions. The predictions for three reactions are shown in Figure I.

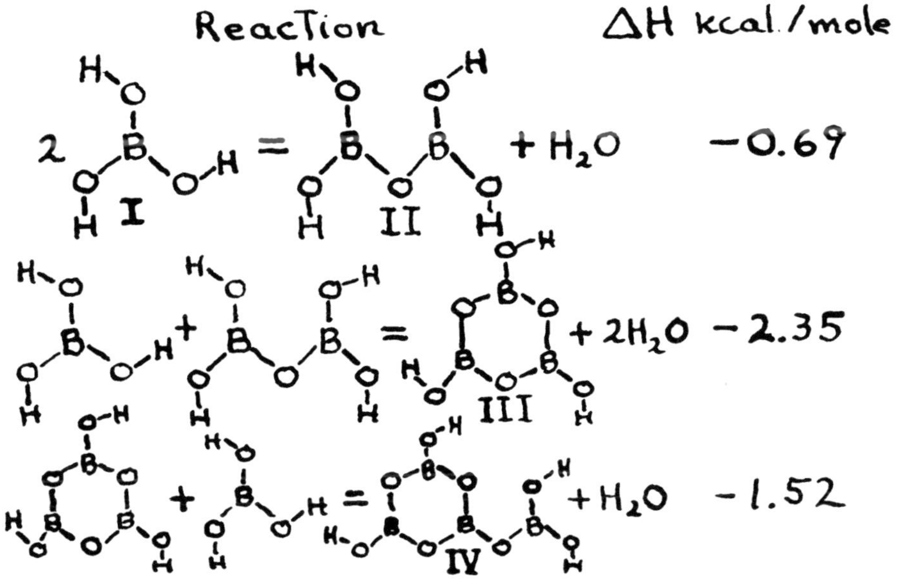

Figure 1. Boric acid thermochemistry.

The formation of pyroboric acid by condensation of two orthoboric acid molecules with the elimination of one water molecule is computed to be exothermic with ΔH = -0.69 kcal/mole. The formation of metaboric acid by the condensation of orthoboric acid with pyroboric acid and the elimination of two water molecules is computed to be exothermic with ΔH = -2.35 kcal/mole. This is 1 kcal/mole more exothermic than would be expected on the basis of the previous reaction. Thus we would estimate the excess resonance energy of the boroxole ring relative to the resonance energy of the reactants is 1 kcal $\pm$ 4 kcal. We find resonance energy of the boroxole ring to be a negligible driving force for the condensation-dehydration reaction. There is little here to expect a large fraction of the BO_3 triangles in B_2O_3 glasses to be incorporated in boroxole rings. The condensation of metaboric acid with orthoboric acid to form dihydroxyl(dihydroxylboryl) boroxole is exothermic with ΔH = -1.52 kcal/mole. Our general result is that the condensation of boric acids with the elimination of water is exothermic by about 1 kcal per mole of water eliminated. We plan to confirm these conclusions with computations in the 4-31G basis. We intend to extend such studies to the thermochemistry of molecules or ions containing tetrahedrally coordinated boron in the borate, diborate, triborate and pentaborate groups. This computational extension has been delayed by uncertainly on how to incorporate the cation.

We have investigated a number of questions about the relative energy of orthoboric acid and pyroboric acid configurations. Knowledge of the relative energy of the configurations of these molecules is a first step toward establishing empirical rules for the relative energy of configurations in B_2O_3 glasses, with the objective of a statistical description of glass structure and properties. Our findings are summarized in Figure II.

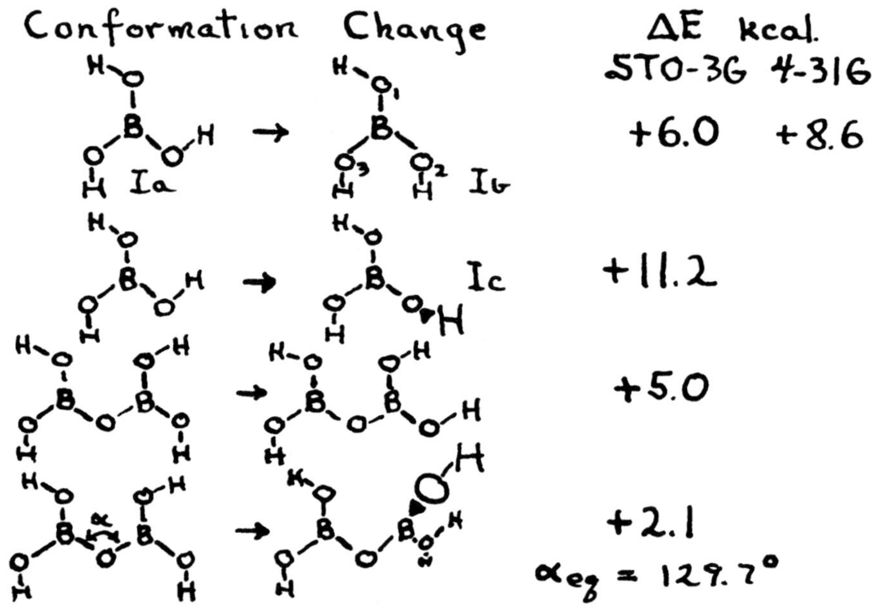

Figure 2. Geometry of boric acids.

Planar orthoboric acid can exist in two forms, the propeller form with a three-fold axis of symmetry (Ia), and the skew form which results from flipping one OH bond over in the molecular plane (Ib). The energy of transition from the propeller to skew form of orthoboric acid is predicted to be +6.0 kcal in the STO-3G basis and 8.6 kcal in the 4-31G basis, qualitatively the same result. The energy of this conformational change is apparently due to the more repulsive interaction of the OH bond dipoles when they are parallel as in the skew form of orthoboric acid.

Another conformational change is the rotation of the OH bond so that is directed away from the molecular plane (Ic). In the STO-3G basis, the required energy is +11.2 kcal. If about +3 kcal of this is due to the interaction of the OH dipoles, as in the previous result, the conformational change appears to be accompanied by an 8.2 kcal reduction of the resonance energy of orthoboric acid. The result suggests that the resonance energy in both orthoboric acid and metaboric acid is significant, even though it does not change in the condensation-dehydration reactions. A related finding by Dill et al.[4] is that the barrier to rotation of the OH bond of hydrooxyborane is 21.7 kcal when computed in the STO-3G basis.

The fourth conformational change of Figure II is an OH bond flip like that first considered for orthoboric acid. The energy required is slightly less in pyroboric acid, amounting to +5 kcal.

The bottom conformational change in Figure II is a rotation by 90° of a $B(OH)_2$ group about its B-O bond to the remainder of pyroboric acid. The required energy is + 2.1 kcal. This is 9.1 kcal less than was found for rotating the OH group of orthoboric acid 90° out of plane. The reduced energy requirement probably reflects a reduction of the repulsion of the net negative charge on oxygen atoms which are adjacent in planar pyroboric acid.

Finally we have varied the pyroboric acid ∠BOB denoted by α in Figure II. We find the total energy to be a minimum for α = 129.7°, close to the value of 130° adopted by Mozzi and Warren[20] in their x-ray diffraction studies of borate glass structure.

The role of nuclear magnetic resonance (NMR), in elucidating the local bonding structure about boron and oxygen in B_2O_3 glasses has been well described by Griscom.[1] The quantitative interpretation for structure of quadrupole couplings in NMR spectra depends on a theory relating molecular geometry to the electric field gradients at nuclei. The simplest widely applied theory is that based on unbalanced p orbital occupancy at the nuclear center. However this theory due to Towns and Dailey[22] is of limited applicability to highly ionic compounds, and those with bent bonds.[19]

The electric field eq_μ at a nucleus μ is an integral of the n-electron wave function:

$$eq_\mu = -e \int \Psi^* \sum_{\ell=1}^{n} \frac{3Z_{\ell\mu}^2 - r_{\ell\mu}^2}{r_{\ell\mu}^5} \Psi d\tau + \sum_m \frac{3Z_{m\mu}^2 - r_{m\mu}^2}{r_{m\mu}^5} \tag{9}$$

where the index m runs over the other nuclei having charge Z_m in units of e, the absolute value of the electron charge. Here eq_μ is the largest (zz) component of the electric field gradient tensor in the principal axis system. The xx and yy components are defined with x and y substituted for z in equation (9). In application to

^{11}B we have taken for the quadrupole moment of the ^{11}B nucleus, $Q_{^{11}B} = 4.0 \times 10^{-2}$ barn (1 barn = 10^{-24} cm^2). The asymmetry parameter η may be computed from

$$\eta_\mu = \frac{q_{xx} - q_{yy}}{q_{zz}} \tag{10}$$

The electric field gradient at boron, q_B, has been computed from STO-3G basis wave functions at all boron centers in model molecules of this study. We have been able to apply the 4-31G basis only to orthoboric acid. A summary of the results is given in Figure III. All field gradients q are given in atomic units, 1au = 32.4140×10^{14} esu-cm^{-3}. For ^{11}B, q = 0.114 corresponds to a quadrupole coupling frequency of 0.95 MHz.

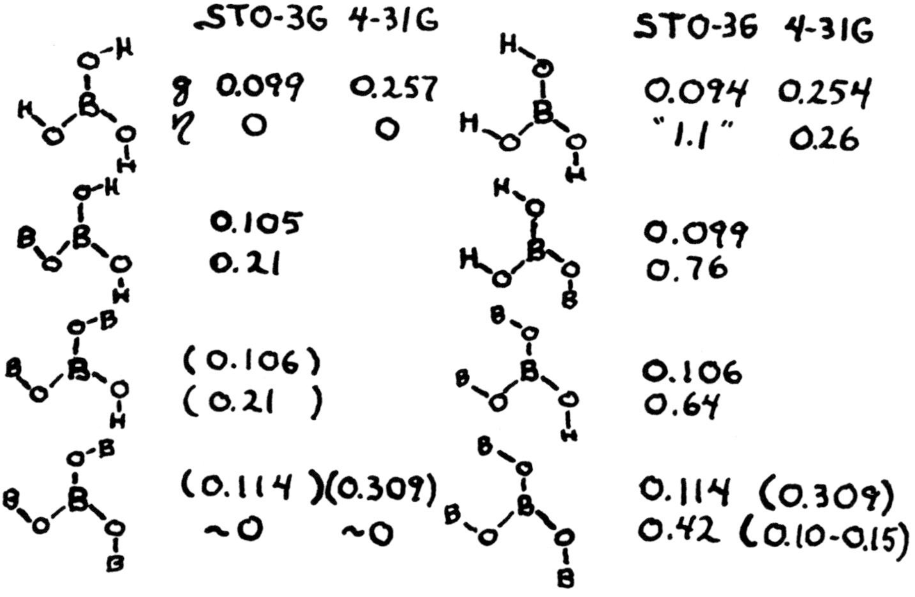

Figure 3. ^{11}B electric field gradient (a.u.)

Boron bound trigonally to oxygen finds itself in a local environment, provided by those oxygen atoms, which has a three-fold axis of symmetry. This environment taken alone would produce an asymmetry parameter $\eta = 0$. For the planar compounds of this study, the next nearest neighbors to boron can occur in the propeller configuration (Ia) or with the skew configuration (Ib), as indicated in Figure II. In the compounds of this study the next-nearest-neighbor atoms of boron are hydrogen or boron. We list in the left column of Figure III computed field gradients and asymmetry parameters for the propellar conformation. The number of boron next-nearest-neighbors to the central boron varies from 0 to 3, from top to bottom. In the right column is a corresponding list for boron with next-nearest-neighbors in the skew conformation. Field gradient parameters in parenthesis were estimated, because they correspond to conformations not present in our model molecules.

To summarize the result of the field gradient calculations in the STO-3G basis, one may note that for all conformations q_B varies only from 0.094 to 0.114 au. the larger values occur with the larger number of next-nearest-neighbor boron atoms. For the same number of next-nearest-neighbor borons, the smaller η corresponds to the propeller conformation. For the propeller conformation η is 30% or less of the value for the skew conformation. We expect that with extension of our computations in the 4-31G basis, that the STO-3G electric field gradients will change in ways analogous to the changes shown in Figure III for orthoboric acid. We expect that the q_B values will increase by a factor of $\sim$ 2.65 over their STO-3G values, due to the increased one center term of smaller atoms. Since the main source of asymmetry is from next-nearest-neighbors, we expect that the η value will be reduced by a factor of order 3. This is the basis for estimated values of q and η. Thus we expect that, in the 4-31G basis, boron in the propeller or skew configuration with three boron next-nearest-neighbors will have $q \approx 0.303$ au or a quadrupole coupling frequency for ^{11}B of about 2.6 MHz. It is also reasonable to expect η to range from 0.10 to 0.15 for the skew configuration. These expected values of q and η are fairly close to the range of values for B_2O_3 glass derived from ^{11}B nuclear magnetic resonance.[23]

Our computations of electric field gradients at oxygen, have been less extensive. For O_2 of configuration Ib of orthoboric acid in Figure II, the STO-3G basis gives $q_o = 2.30$ au, while the 4-31G basis gives $q_o = 1.65$ au, a smaller value as expected for an expanded atom. The asymmetry parameter η similarly decreases from 0.39 to 0.27.

Only the STO-3G basis has been applied to molecules bearing oxygen bonded to a pair of boron atoms. For oxygen in a boroxyl ring we compute $q_o \approx 1.6$ au, with the principal axis in the direction of the vector connecting the boron atoms. For oxygen exterior to a boroxyl ring we find $q_o = 1.76$, where $q_o = 1$ au corresponds to a quadrupole coupling frequency for ^{17}O of 6.1 MHz. For both the above bonding arrangements of oxygen we find $\eta \leq 0.08$, which is in poor agreement with $\eta \approx 0.5$ found in experiments by Jellison.[23] However completion of the computational work in the 4-31G basis may produce a significantly larger theoretical η.

SUMMARY

The quantitative methods developed by quantum chemists have hardly been applied to borate glasses. There appears to be a significant opportunity to augment our knowledge of the structure of borate glasses. It may be possible to provide thermodynamic information on configurations proposed to occur in the glass, in preparation for a statistical description of the networks of covalent bonds of a B_2O_3 glass.

ACKNOWLEDGEMENT

The great contribution of Zelda Wasserman in carrying out the computations reported here is gratefully acknowledged.

REFERENCES

1. D. L. Griscom, Preconference manuscript on Borate Glass Structure.
2. C. A. Coulson, Acta. Crystal 17, 1086 (1964).
3. C. A. Coulson and T. W. Dingle, Acta. Crystal B24, 153 (1968).
4. J. D. Dill, P. V. R. Schleyer and J. A. Pople, J. Am. Chem. Soc. 97, 3402 (1975).
5. E. Schrödinger, Ann. D. Phys. 79, 361 (1926).
6. Modern Theoretical Chemistry, Vol. II, Electronic Structure; Ab-initio Methods, Ed. H. F. Schaefer, Plenum Press (1977).
7. W. J. Hehre, R. F. Stewart and J. A. Pople, J. Chem. Phys. 51, 2657 (1969).
8. R. Ditchfield, W. J. Hehre and J. A. Pople, J. Chem. Phys., 54, 724 (1971).
9. W. J. Hehre and J. A. Pople, J. Chem. Phys., 56, 4233 (1972).
10. W. J. Hehre, W. A. Lathan, R. Ditchfield, M. D. Newton and J. A. Pople, QCPE #236: "Gaussian-70 Ab-initio SCF-MO Calculations on Organic Molecules," Quantum Chemistry Program Exchange, Indiana University, Bloomington, Indiana 47401.
11. D. B. Neumann, H. Basch, R. L. Kornegay, L. C. Snyder, J. W. Moskowitz, C. Hornback and S. P. Liebmann, QCPE #199, "The POLYATOM (version 2) System of Programs for Quantitative Theoretical Chemistry."
12. L. R. Kahn and W. A. Goddard III, J. Chem. Phys. 56, 2685 (1972).
13. L. C. Snyder, J. Chem. Phys. 46, 3602 (1967).
14. L. C. Snyder and H. Basch, J. Am. Che. Soc. 91, 2189 (1969).

15. W. J. Hehre, R. Ditchfield, L. Radom and J. A. Pople, J. Am. Chem. Soc., 92, 4796 (1970).
16. W. J. Hehre and J. A. Pople, J. Am. Chem. Soc., 94, 6901 (1972).
17. W. A. Lathan, W. J. Hehre, L. A. Curtiss and J. A. Pople, J. Am. Chem. Soc., 93, 6377 (1971).
18. L. C. Snyder and H. Basch, "Molecular Wave Functions and Properties" John Wiley (1972).
19. L. C. Snyder, G. E. Peterson and C. R. Kurkjian, J. Chem. Phys. 64, 1569 (1976).
20. R. L. Mozzi and B. E. Warren, J. Appl. Cryst., 3, 251 (1970).
21. T. L. Cottrell, "The Strengths of Chemical Bonds," Buttersworths, London (1954).
22. T. P. Das and E. L. Hahn, "Nuclear Quadrupole Resonance Spectroscopy," Academic, New York (1958).
23. G. H. Jellison, Jr., L. W. Panek, P. J. Bray and G. B. Rouse, Jr., J. Chem. Phys. 66, 802 (1977).

TOPOLOGICAL CONSIDERATIONS OF TRIANGULARLY CONNECTED NETWORKS

Alfred R. Cooper

Case Western Reserve University
Cleveland, Ohio 44106

The importance of topology to glass formation was recognized by Zachariasen. While he never used the word explicitly, his famous rules for glass formation (1932) are all topological.(1) They relate solely to the manner in which the atoms are connected. His first three rules appear to be the necessary and sufficient condition to permit formation of a non-crystalline network from perfectly regular polyhedral coordination shells of oxygen atoms. (2) While the determination of chain length distribution of alkali phosphate glasses (3) was perhaps the first experimental measurement of topologic properties of a glass, interest with topological consideration of amorphous networks remains active, particularly in the area of metallic and chalcogenide glasses. (4)

While fourfold and twofold connected structures are present in the majority of the glass forming systems, there are important oxide, (B_2O_3), sulfide, (As_2S_3) and elemental, (As), glasses that are based on triangularly coördinated networks. Further in many commercial glasses the SiO_4 tetrahedrons are each connected only to approximately three other tetrahedrons.

Closure conditions i.e. the necessity that closed rings and shells be produced, provides a basis for correlation of atomic distances. This has been demonstrated in two dimensions for regular hexagons (5) and is similar to the correlation obtained from the weaker condition of non overlap of atoms which has been treated exactly in 1 dimension (5,6).

This contribution will endeavor to deduce and enumerate properties of triangularly connected networks. Application to atomic structure of glass provides both the motivation and the informal justification for some of the simplifications.

CONNECTIVITY PAIRS

It is convenient to classify networks of metallic oxides according to a scheme where the coordination number about the cation, CN_c, and the average coordination number about oxygen, CN_o, are given as an ordered pair of numbers termed the "connectivity pair". Thus [3,2], [4,2] and [4,8/5] are appropriate for B_2O_3, SiO_2 and P_2O_5 respectively. For elemental materials like arsenic and silicon the connectivity pairs would conventionally be written [3,3] and [4,4] respectively. However we can adopt a somewhat unorthodox but more illuminating viewpoint and consider the inter-atomic bond between nearest neighbors to be equivalent to a cation and the atom to be equivalent to the anion. From this viewpoint the connectivity pairs for arsenic and silicon would be written [2,3] and [2,4] respectively.

The essence of Zachariasen's first three rules is that identical equilateral triangles or identical equilateral tetrahedrons can be arranged with each corner shared without long range order provided the center-vertex-center angles are not fixed. Structures in which the elementary units are equilateral and identical will be termed "regular", and in this paper attention will be confined to regular structures formed from [3,2] and [2,3] connectivity pairs. (For [2,3] pairs "regular" implies that all bonds are of equal length.)

TWO DIMENSIONAL (2D) STRUCTURES

The familiar classic 2D schematic of glass structure (1) shows the cation as a small solid circle and the anion as a larger open circle. Another convenient way to display such a regular [3,2] configuration (Fig. 1a) is to arrange equilateral triangles so that each corner is shared by two triangles. Clearly the vertices represent the anion positions and the triangle centers the cation positions. The regular polygonal interstices formed by the triangle edges are termed rings. By noting that equilateral triangles have internal angles of $\pi/3$, it follows that when two triangles meet at a corner the average angle, $\langle\Theta\rangle$, between the triangles edges must be $(2\pi-2\pi/3)/2 = 2\pi/3$. Thus, if the structure is sufficiently large that the average ring size has reached a limiting value this average size will be 6, a result which could have been deduced using Euler's Theorem. Of course as seen in Fig. 1a some rings may be larger or smaller than average.

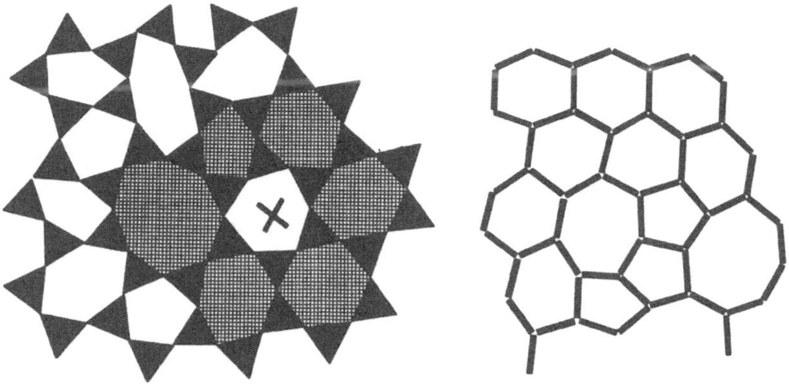

Fig. 1(a) A finite [3,2] configuration of equilateral triangles without obvious crystalline regularity. X identifies an arbitiary hexagonal ring and the shaded rings are the first shell of rings about it. (b) A [2,3] configuration of equal line segments.

The configuration displayed in 1a appears to be amorphous with considerable freedom of angles. If, however, we consider the construction of such an arrangement by making successive shells of hexagonal* rings of triangles about a central hexagonal ring we can calculate the number of angles, n_f, between triangle edges that can be chosen freely and compare it with the total number of angles, n, between edges. By numbering the layers of rings by the index p, and using the convention that for the center ring of triangles p=0, we obtain the following result for the free angle ratio.

$$\frac{n_f}{n} \text{ (2D, [3,2])} = \frac{3 + 6p}{6\left(1 + 6 \sum_{o}^{p} j\right)} \simeq \frac{1 + 2p}{2\left(1 + 6p\,(p+1)\right)} \tag{1}$$

Taking the limit as n (or p) $\rightarrow \infty$ one gets

$$\lim_{n \to \infty} \frac{n_f}{n} \text{ (2D, [3,2])} = \frac{2}{p+1} = 0 \tag{2}$$

---

* A distribution of ring sizes causes only a slight discrepancy in eqn. 1.

This limiting result seems consistent with the observation that for the equilateral polygonal rings three angles are determined per polygon by the closure condition. (e.g. no angles can be chosen in an equilateral traingle, only one can be chosen in a rhombus, and only two in an equilateral pentagon, etc.). The average ring is a hexagon; hence half of the angles are specified by the closure condition. Also half the angles at each vertex are specified because the sum of the vertex angles is 2π. Since these relations are independent, all of the angles are specified and none are free to vary. How can this result be consistent with Fig. 1a? The answer must relate to the fact that Fig. 1a has many external angles which are not affected by the closure condition. In fact, the numerator of eqn. 1 is a very close approximation* equal to half the number of external angles in a layer or rings (e.g. the slightly shaded layer in Fig. 1a). The number of free angles is half the number of external angles because of the vertex conditions at the boundary.

If rather than being unbounded on a planar surface as shown in Fig. 2, the triangles were placed on a closed simple surface such as a spherical surface, then there would be no external angles - and provided the ring structure was specified there would be only one way in which the triangles could be arranged on the surface. Different ring structures might be possible but for a given ring structure there would be no freedom of choice of angles.

It is suggested that to be termed "amorphous" a structure must have geometric freedom. i.e. $(n_f/n) \neq 0$. On this basis the lack of crystalline regularity is a necessary but not a sufficient condition for a configuration to be judged amorphous.

Fig. 1b displays a regular [2,3] configuration. In an identical manner to that used for [3,2] structures, we can calculate the ratio of free angles in a [2,3] structure with uniform bond lengths but variable bond angles with result:

$$\frac{n_f}{n}(2D,\ [2,3]) = \frac{3 + 6p + 6\sum_{o}^{p} j}{6\ (1 + 6\sum_{o}^{p}} = \frac{1/2p + (p+2)}{1/p \ \ +6(p+1)} \qquad (3)$$

$$\lim_{n \to \infty} \frac{n_f}{n}(2D,\ [2,3]) = \frac{1}{6} \qquad (4)$$

*It is exactly so if all rings are hexagonal.

Fig. 2. A boroxyl ring of boroxyl rings of boroxyl rings of $BO_3^{\equiv}$ triangles. Notice the largest ring $B_{27}O_{39}^{\equiv}$ has the same charge (-3) as the $BO_3^{\equiv}$ triangle and analogous to the triangle it is available for bonding only at corners.

It is again easy to understand the limiting value based on the vertex and closure conditions. [2,3] pairs also form configurations whose average ring size is 6 hence half the angles are specified by closure, but a [2,3] configuration has three angles per vertex. Therefore a third of the angles are determined by the vertex condition, half by closure and the remainder, a sixth, can be freely chosen. Thus, provided the bond angles are allowed to vary, regular [2,3] structures can be amorphous, because they have geometric as well as topological freedom even on closed surfaces.

By joining the centers of the triangles we can transform a regular [3,2] configuration into a [2,3] configuration. However, in general it will not be a regular [2,3] array.

Characterization of Topology

The topologic characterization of two dimensional networks is first given by the connectivity pair $[CN_c, CN_o]$. This pair determines the average ring size, $\langle F \rangle$, but leaves the arrangement of rings relative to each other unspecified. Even the ring size distribution does not fully define the topology because it does not give the adjacency conditions. Conditional probabilities e.g. P(5|6) the probability that a ring adjacent to a 6 member ring has five members, and higher order conditional probabilities e.g. P(5|6,6) etc. are necessary to describe the topology fully. Thus just as the pair correlation function only incompletely describes the geometric relations of an amorphous structure, conditional ring pair probabilities do not fully describe its topology. However such incomplete descriptions may be adequate for many purposes.

Boroxyl Ring

The Boroxyl Ring is a planar configuration of much importance in borate glasses. (7) As is well known it consists of three $BO_3\equiv$ triangles joined together in a simple way so as to form a planar 3 membered ring. The fact that it contains two distinct types of oxygen those at the corner, and those along the edge of a ring has recently been noted. (8) The Boroxyl Ring can itself be represented as an equilateral ($B_3O_6\equiv$) triangle. Thus, it has all the topologic characteristics of the elementary $BO_3\equiv$ triangle itself. In simple B_2O_3 glass the boroxyl ring must share all of its corners and only its corners in exactly the same way as an elementary triangle. Likewise 3 boroxyl rings can be joined together to form a boroxyl ring of boroxyl rings etc. See Fig. 2. Thus, the boroxyl ring has ideal properties to permit higher organization structures than the elementary triangle to be used as the building blocks for triangularly coordinated nets.

It is obvious that if the average ring size of [3,2] configurations is 6, introduction of boroxyl rings causes the necessity of some rings of larger size. While Fig. 2 reveals one 12 membered ring it is not representative of an expandable structure because all external angles are π. A 2D structure composed entirely of boroxyl rings can be constructed with half as many 12 membered rings as boroxyl rings. (See Fig. 3)

The number of atoms in a boroxyl ring is triple the number in an elementary triangle and the characteristic linear dimension of the boroxyl ring is double that of the elementary triangle. Therefore, if the boroxyl triangle is directly substituted for the elementary triangle in planar configurations (e.g. consider the triangles in Fig. 1a to be boroxyl rings) the triangle density is reduced to 3/4 of its original value and likewise if they are directly substituted in a three dimensional configuration the density is reduced to 3/8 of its original value.

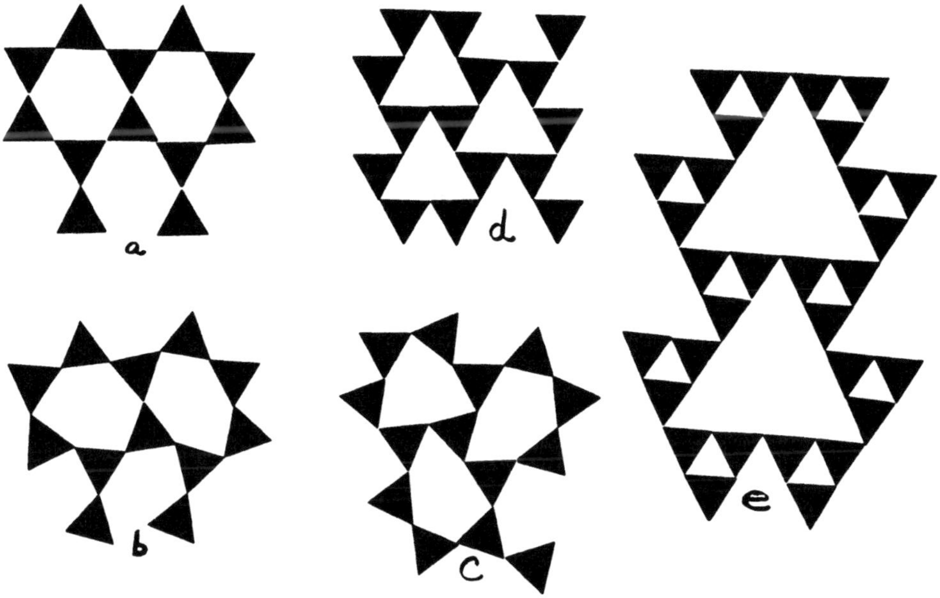

Fig. 3. The four configurations a, b, c, and d have identical topology. Their density increases from upper left (1/4) continuously in a counterclockwise way to (1/3) at upper center. The larger configuration on right (e) has same angular distribution but different topology than the upper center configuration (d).

Configurational Expansion (Geometric and Topologic)

A usual characteristic of glass forming systems is a thermal expansion increase above the glass transition. This is explained by the existence of an expansion associated with a change in configuration with temperature as well as an increase of the nearest neighbor spacing with temperature. Earlier (9) it was suggested that both topologic and geometric change contribute to this expansion. The [3,2] structure provides a good example of a model system that demonstrates such behavior. Shown in Fig. 3 are several schematics of [3,2] configurations. The four configurations on the left are topologically equivalent but have different triangle densities which increase from 1/4 to 1/3 as one proceeds CCW from the upper left to the upper right. The figure displays only two of the continuous distribution of geometries many of them "non crystalline" which lie between these two extreme densities. The denser structures have alternate angles of the rings markedly unequal while the less dense structures have more nearly uniform alternate angles. Thus geometric configurational thermal expansion in this simple model could be associated with a continuous change with temperature from a bimodal distribution of ring angles toward a unimodal distribution of ring angles. An equivalent behavior is observed in [2,3] regular structures with all 6 membered rings. If (to avoid overlap of atoms) the angles, Θ_i, are required to fulfill the condition $\Theta_i > \pi/3$ then: the minimum density, ρ^o, occurs when all angles are equal; a greater density, $(3\sqrt{3}/4)\rho^o$, is obtained from a bimodal distribution of angles such that every third angle is π and the other angles are $\pi/2$; while the highest density, $(3/2)\rho^o$, occurs with a trimodal distribution of angles $\pi/3$, $2\pi/3$ and π each in equal numbers.

It is interesting that an opposite density variation occurs with topoligic changes. As already mentioned in the previous section, the boroxyl ring causes a decrease in density compared with $BO_3\equiv$ triangles. Fig. 3D and e display structures which have the same angular distribution, but 3e displays a bimodal distribution of 3 membered and 12 membered rings which has a density of 1/4, while fig. 3d with only six membered rings yields a density of 1/3. Thus, in this case, at least, the more uniform ring size distribution gives a higher density.

The principle is also followed in [2,3] structures e.g. a 33% decrease in density results when a uniform 6 member ring structure with a bimodal (π, $\pi/2$) distribution of angles has its ring structure modified to 10 membered and 4 membered rings without changing the angular distribution.

The opportunity to have a <u>continuous</u> change in density due to a change in topology is not evident in the small number of elements in the schematics Fig. 3d and 3e. However, when large number of triangles with ring sizes ranging from 3 to 12 are considered, the possibilities for an essentially continuous density distribution become evident.

The wide range of densities available to [3,2] 2D configurations may relate to the large volume increase of B_2O_3 above its glass transition (10). The lack of noticeable configurational expansion at 1400°C and above would seem to indicate tha the most open network structure possible is nearly attained at this temperature.

THREE DIMENSIONAL (3D) CONFIGURATIONS

We may hope that our brief study of 2D configurations can be rewarding in providing insight into 3D [3,2] and [2,3] structures. A major advantage of 2D is that there is no ambiguity about the definition of a ring. Since this clarity is not obvious in 3D and since 2D lends itself to easy representation and clear description, it is reasonable to try to map 3D configurations onto 2D space with a minimum loss or distortion of information. Such an approach may be particularly rewarding for [3,2] and [2,3] structures because in contrast to [4,2] tetrahedral structures [3,2] and [2,3] connectivity pairs do (as is obvious from the previous sections) have configurations in 2D.

A natural mapping process consists of untwisting and unfolding the 3D structure of triangles onto a simple surface. In some cases this may be possible without the necessity of severing a single bond or it may be necessary to sever one or more bonds before this unfolding can be achieved. The ratio of bonds which require severance to the total number of bonds is termed R, the 3D connectivity. It must be noted that for many cases there may not be a unique way of choosing the bonds to be severed. Thus, strictly speaking, R should represent the limiting value of the ratio averaged over a number of independent mappings from 3D to 2D.

The simplest type of 3D configuration to consider is one which can be completely unfolded and unwound so as to produce a non-overlapping layer of triangles without breaking any of the bonds. (i.e. R=0) Prior to unfolding and untwisting, a non-overlapping portion of this presumably puckered layer of triangles can be projected onto a plane. As seen in the schematic, Fig. 4, projection causes triangles to lose their equality and transform from equilateral to scalene. Notice, however, that, as before, the average ring size must be 6 because $\langle\Theta\rangle$ is still $2\pi/3$. The relations between the angles have been reduced. The sum of the opposite ring angles at a vertex is no longer fixed and there are only two angles fixed per ring by the closure condition. Thus the additional freedom provided by the third dimension, which permits the rings to pucker, is such that on the projection

$$\lim_{n\to\infty} \frac{n_f}{n} \;(3D,\ R=0\ [3,2]) = \frac{2}{3} \tag{5}$$

Another way of appreciating this additional freedom is to recognize that the orientation between two equilateral triangles on a surface is described by a single angle while in three dimensions it requires three angles: the center-vertex-center angle, and the two angles necessary to describe the relation of the normal of one triangle to that of the other. These latter two angles can be associated with the twist and the tilt of one triangle with respect to the other.

Fig. 4. A schematic of a regular 3D showing the projection onto a planar surface of a portion of a regular, 3D, R = 0, [3,2] configuration.

The greater freedom present in three dimensional 2,3 structures is revealed in the amorphous arsenic models built by Greaves and Davis (11), where fixing of some of the bond angles did not prevent the freedom to attain amorphous configurations.

It is observed that when R = 0 (see Fig. 4) each triangle edge contributes to one and only one polygonal ring. When R > 0, however, some of the triangle edges contribute to more than one ring. To see the relation between R and the average ring size $\langle F \rangle$, we first recognize that many rings remain intact after the necessary bonds have been severed. These rings can be observed when the severed configuration is unfolded and deformed onto a planar surface as shown in Fig. 5. In this schematic, line segments represent the edges of triangles that are opposite from a severed vertex. Notice that these edges are part of two rings. The triangles themselves in Fig. 5, represent elements without bond severances. Since each severance produces two severed triangles, the ratio of line segments to triangles plus line segments is equal to 3R provided the projection contains a sufficiently large number of units to minimize any boundary effect.

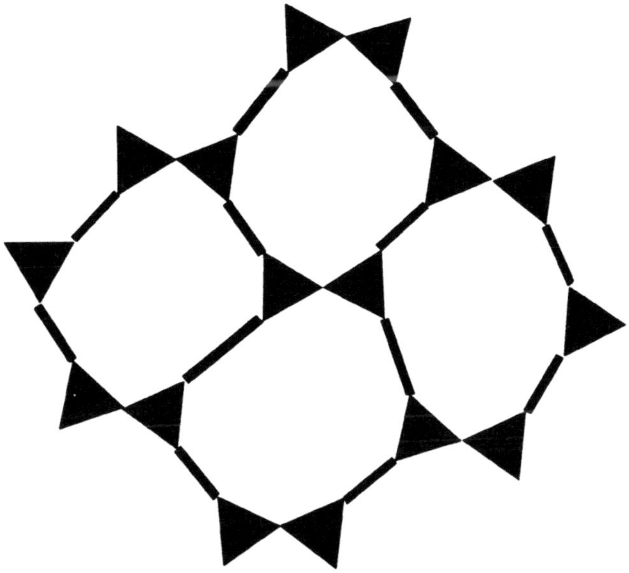

Fig. 5. A schematic of a regular 3D, R = 2/15 [3,2] network after severing, unfolding and deforming onto a surface. The lines are triangle edges that are opposite a severed vertex.

The average value of the ring size depends on R because line-line, line-triangle and triangle-triangle corners produce different average internal ring angles (π, $5\pi/6$, $2\pi/3$, respectively). From this it is easy to show that the average value of the internal ring angle $\langle\Theta\rangle$ relates to R as follows

$$\langle\Theta\rangle = \frac{2\pi}{3}(1 + 3R/2) \tag{6}$$

Equation 6 permits tabulation of the values of R associated with several integral values of the average ring size $\langle F\rangle$ as shown in Table I.

TABLE I

Comparison of Connectivity Parameter, R, and the Average Ring Size $\langle F\rangle$

R	$\langle F\rangle$	R	$\langle F\rangle$	R	$\langle F\rangle$	R	$\langle F\rangle$
0	6	1/12	8	2/15	10	1/6	12
1/21	7	1/9	9	2/13	11	1/3	∞

The entry in Table I--$R = 1/3$, $\langle F \rangle = \infty$ -- relates to a case where bond severing was continued beyond the attainment of a 2D structure to a 1D chain. It is not, therefore, an allowable value of R as we have defined it. Wells (12) has stated that the largest ring size in a uniform 3 connected structure is 12. While larger average ring sizes may exist it seems that $\langle F \rangle = 12$ and $R = 1/6$ represent a good estimate for the upper bound for these quantities for structures derived from [3,2] connectivity pairs.

A pair of triangles which has suffered a single bond severance is topologically equivalent to a tetrahedron, since, if the joint vertex is ignored, we see there are four remaining vertices. The configurations shown in Fig. 5 with $R = 2/15$ can be considered to be composed of 2/3 psuedo tetrahedrons and 1/3 triangles. At $R = 1/6$ the structure would consist solely of psuedo tetrahedrons. Thus [3,2] structures can have all the topological features of [4,2] tetrahedrial structures plus all the structures for which $R < 1/6$ that are not available to [4,2]tetrahedral configurations.

When $R \neq 0$ closure conditions, as always, specify $2/\langle F \rangle$ of the angles on the projection. There are also additional conditions imposed when a triangle or a group of triangles are completely enclosed by line segments. In Fig. 5 for example these relations fix 2/15 of the angles with the result that for this structure the infinite limit of (n_f/n) is 2/3.

Topological Parameters

All of the topological parameters for 2D structures are appropriate to 3D configurations. The primary new parameter introduced by the additional dimension is, R, the 3D connectivity. As seen in Table I, the average ring size $\langle F \rangle$ is a function of R and conversely R is a function of $\langle F \rangle$. Thus if it is preferred $\langle F \rangle$ can be utilized as the new topological parameter. An important ommission which is beyond the scope of this paper is the selection of appropriate topological parameters to characterize to the cages which occur when $R \neq 0$ in 3D [3,2] structures.

Crystalline B_2O_3

The structure for crystalline B_2O_3 has been recently established (13). It consists of three sets of intersecting infinite ribbons of BO_2^-. The ribbons are planar and essentially two triangles wide. As already mentioned it has been noted (8) that boroxyl ring structures have two distinctly different kinds of oxygen. Crystalline B_2O_3 has a similar characteristic. The types in the crystal are: (1) the oxygens that connect triangles within the plane, (ii) the oxygens which lie at the vertices on non coplanar triangles. In each $BO_3{\equiv}$ triangles there are two of the "planar triangle" oxygens and one "non planar".

Ten membered* rings are present in the B_2O_3 crystal structure. They can be seen in a 3D model at the structure** and (although not so clearly) on the schematic of the structure given by Gurr et al (13) and (quite plainly) on a drawing from Wells (12) which focuses on the so called "10-gon net."

According to Table I these ten membered rings of crystalline B_2O_3 correspond to R = 2/15. The schematic on Fig. 5 therefore can be considered to be the result of a mapping of the topology of the B_2O_3 crystal structure onto the plane.

The specific gravity of crystalline B_2O_3 is about 2.56 while glassy B_2O_3 has a specific gravity of only about 1.96. It is clear that substituting boroxyl rings for $BO_3{\equiv}$ triangles into the crystalline B_2O_3 structure will give a specific gravity (3/8 x 2.56 = 0. 96) much lower than the specific gravity for B_2O_3 glass. Since it is hard to imagine configurations much more compact than the crystal it seems unlikely that amorphous B_2O_3 glass can consist primarily of boroxyl rings. It may be that the recent suggestion, based on the NMR studies of B_2O_3 glass by Jellison et al. (8), that boroxyl rings are joined by $BO_3{\equiv}$ triangles may permit actual densities to be achieved with a large number of boroxyl rings in an amorphous network.

Random Network, R. N.

The appropriateness of the term "random network", R. N., used by Zacharaisen as a description for atomic configurations in glass has been attacked, defended and argued over since its introduction. Yet according to the recollection of the present author, it has never been precisely defined, either by Zacharaisen himself his "supporters" or his "foes". The reason surely is that just like "glass", R. N. is impossible to define without some arbitrariness. It seems fitting to attempt a clarification by considering the meaning of the words themselves.

Presumably, by using the word "network", Zacharaisen was indicating that closure and vertex conditions of a network limited the freedom of arrangement of the structure. By calling it "Random" he seemed to be requiring that some freedom be permitted.

---

*Thanks to D. Evans, Corning, for pointing out 10 membered rings

**Thanks to Mahmound Abou el Leil for constructing the model.

with the result that configurations are not fully specified without giving the coordinates of each atom. Practically speaking, therefore, the atomic configuration of a macroscopic random system can only be defined in a statistical way.

This freedom from easy specification can be either geometric or topologic. The geometric freedom can be treated quantitatively in terms of the parameter (n_f/n) considered previously. When n_f/n is sufficiently greater than zero, a macroscopic system has a configuration that requires a statistical description. The difficulty lies in defining how much freedom is necessary for a structure to be considered random. On the one hand we know that factors in addition to closure and vertex relations can diminish the configurational freedom without causing a network structure to be regular. Greaves and Davis, for example, found a broad distribution of ring sizes even when they severely restricted bond angles in a [2,3] structure. Likewise evidence exists for the presence of some boroxyl rings in vitreous B_2O_3. Should such restrictions prevent a network from being termed "random"? Probably it is unreasonable to expect that Zacharaisen had such a narrow view. On the other hand defects like grain boundaries and dislocations in crystalline structures cause (n_f/n) > 0 and cause some variation in ring and cage structures. Calling such defective crystalline structures R. N. would seem to strip the term R. N. of any useful meaning.

Clearly what is sought is a well defined, though perhaps inexact, boundary that will discriminate between randomness and non-randomness of an atomic network. While lack of space and lack of wisdom prevent inclusion of such a quantitative criterion here, it is urged that a semantic agreement be achieved before the R. N. controversy is fruitlessly reignited.

SUMMARIZING REMARKS

The characteristic aspect of glasses is not in their nearest neighbor arrangements. These are frequently identical to crystals. Rather it is the way the network is connected to give no long range spatial correlations. Spectral information describing the near environment about many atoms in glass forming systems has been accumulating. Boric oxide glass has not been overlooked, as an impressive body of spectral data fortunately exists. (14).

In addition to the importance of obtaining more complete spectral data, and perhaps more sophisticated interpretations, there is another task that is worthwhile - namely the effort to try to fit these data into a three dimensional network model for B_2O_3 glass. Such a model needs to be consistent with the known mass density of B_2O_3 glass. Comparison of the predicted and measured entropy difference between crystal and glass and of properties like specific heat, thermal expansion and compressibility, will provide severe tests that may require major model modifications.

Crystallization kinetics of B_2O_3 are perhaps best viewed as being determined by the kinetics of configurational change, probably predominately topologic change. Gaseous permeation through glass relates to the size, shape, and distribution of rings. Detailed interpretation of crystallization kinetics and permeation data may aid in the characterization of the topology and geometry of glassy networks.

It is expected that some of the concepts of the topology triangular networks will likewise be helpful with this task.

ACKNOWLEDGEMENT

Financial support from C.W.R.U. - MRL from NSF grant DMR 74 01878-A02 is gratefully acknowledged as are critical comments by D. Singer.

REFERENCES

1. W. H. Zacharaisen, J. Amer. Chem. Soc. 54 1932, 3841
2. A. R. Cooper, Submitted to Phys. & Chem. Glass. (1977).
3. A. E. Westman, Modern Aspects of Vitreous State J.D. MacKenzie Ed. Butterworths London, 1960 p. 63.
4. See for example chapters by David Turnbull, William Paul and G.A.N. Connell, and D. Wearie in Shashanka S. Mitra Physics of Structurally Disordered Solids Plenum Press, N.Y., 1976.
5. A. R. Cooper and P. F. Aubourg, Physics of Non-Crystalline Solids.
6. J.G. Kirkwood, Theory of Liquids, Gordon and Breach 1968.
7. J. Krogh-Moe Phys. Chem. Glasses 3 1962 1.
8. G. E. Jellison, Jr., et al. J. Chem. Phys. 66 (1977) 802.
9. A. R. Cooper, Phys. Chem. Glasses 17 1976 38.
10. (a) G. S. Parks and M. E. Spaght, J. Phys. Chem. 38 1934 103.
    (b) P. C. Li, et al. J. Amer. Ceram. Soc. 45 1962 89.
    (c) P. B. Maado, This volume p
11. G. N. Greaves and A. E. Davis Phil. Mag. 29 (1974) 1201.
12. A. E. Wells Structural Inorganic Chemistry 4th Ed. 1975 Clarendon Press Oxford 1975. p. 96 and p. 100 ff
13. G. E. Gurr, et al. Acta. Cryst. B26 1970 906.
14. David L. Griscomb, This Volume p. (1-128).

ON THE FLUCTUATION STRUCTURE OF VITREOUS BORON OXIDE AND TWO-COMPONENT ALKALI BORATE GLASSES

E. A. Porai-Koshits, V. V. Golubkov and
A. P. Titov

Institute of Silicate Chemistry of the Academy of
Sciences of the USSR
Leningrad, USSR

By means of the small-angle X-ray scattering (SAXS) technique, studies were made of vitreous B_2O_3 and binary alkali borate glasses containing lithium, sodium, potassium, rubidium and cesium oxides in the range from room temperature to that exceeding T_g by 150°C and sometimes by 300°C.

There was no evidence of submicroheterogeneous structure in vitreous B_2O_3; only thermal density fluctuations were observed.

Alkali borate glasses showed a submicroheterogeneous structure with large and small inhomogeneity regions and a structure of thermal density and concentration fluctuations. The sizes of large inhomogeneity regions were found to be 200-300 Å. The formation of small, inhomogeneity regions of 15 to 20 Å in size was observed in low-alkali glasses. These regions are a manifestation of both grouping of alkali metal ions and the change of the short-range order structure. The mean square difference of electron densities of these inhomogeneity-regions structure increases with temperature over the temperature range studied.

SAXS intensity by thermal density fluctuations of B_2O_3 network increases in strict proportion to temperature as in vitreous B_2O_3 and in alkali borate glasses.

No irreversible change in SAXS intensity that could be related to phase separation processes was observed in the case of sodium, potassium, rubidium and cesium borate glasses containing more than 1% R_2O at temperatures above T_g. Thus, no miscibility gaps are present in these systems.

A miscibility gap is present only in the lithium borate system. The critical composition and temperature are 4.5% Li_2O (mol. %) and 453°C, respectively.

INTRODUCTION

The fluctuation structure is the low limit of inhomogeneity that can be reached by glass. At the same time this is the primary structure in the course of phase separation and its study is important for understanding both the processes that take place at the initial stages of phase separation and the most general process being the basis of glass formation. The structure of phase separated glasses has been studied sufficiently; the same cannot be said about the fluctuation structure.

The study of the fluctuation structure of alkali borate glasses is a logical continuation of our investigations of the fluctuation structure of one-phase glasses[1].

Alkali borate glasses have long ago attracted attention of investigators which was due, to a considerable degree, to the problem of "boron anomaly". When investigating these glasses, in addition to the study of properties, structural methods were used: X-ray scattering at large angles, electron microscopy as well as NMR and IR-spectroscopy. The results were reported in[2, 3].

These investigations permitted one to study the short-range order structure, to establish the changes of the boron coordination state and to relate the "boron anomaly" with the transition of boron from three- to four-coordinated state. Only brief information was obtained about the structure of glasses outside the first coordination spheres. In particular, the problem of the existence of miscibility gaps in alkali borate systems was not cleared up.

The character of the change in liquidus curve of a sodium-borate system allows the possible existence of sub-liquidus phase separation[4]. Electron-microscopic studies of the heated samples of glasses containing from 6 to 26% Na_2O revealed structures which were explained to be due to the phase separation processes[5]. The possible influence of phase separation on the character of changes in electrical conductivity of sodium borate glasses with composition and temperature was pointed out in Reference 6. The values of critical composition reported in References 5 and 6 agree.

On the basis of electron-microscopic investigations of lithium borate glasses, Krogh-Moe realized the possibility of phase separation at a content of 10% and less Li_2O[7].

The most systematic study of alkali borate glasses using a transmitting electron microscope was made by Shaw and Uhlmann[8]. These authors observed a noticeable difference in the structure of samples heated to different temperatures. They found miscibility gaps in all alkali borate systems and their results on the sodium borate system agree well with the data reported in Reference 5. All of this proved that the proper existence of phase separation in alkali borate glasses was genuine.

EXPERIMENTAL

All glasses were obtained by melting chemically pure reactants in a platinum crucible at 1100-1400°C. The resulting specimens were melted in vacuum at 1000°C for several hours until fully degassed. Samples 0.1 to 1mm thick were obtained 1) by polishing, using non-aqueous liquids; 2) by melting a piece of glass placed between two mica plates and 3) in the form of a film melted on a platinum support. In the last case the sample surfaces were most perfect. The samples were practically void of water and did not crystallize even at prolonged isothermal holds. The structure was studied from T_g to a temperature 150°C higher than T_g. In some cases the sample was placed in a cell with mica windows. This enabled structural studies up to 650-700°C.

SAXS intensity was measured on a small-angle apparatus with a high-temperature attachment[9]. This permitted study of structures with such short relaxation times that they could not be quenched.

SAXS technique applied in this study permits obtaining information about the submicroheterogeneous structure as well as about thermal density and concentration fluctuations. The submicroheterogeneous structure is understood as regions of inhomogeneities of sizes from tens to hundreds and more angstroms where the electron density differs from the mean electron density of the sample. The presence in the sample of the submicroheterogeneous structure leads to the appearance of a characteristic portion in the SAXS curve where the intensity decreases with increasing scattering angle. By means of this portion of the SAXS curve, one can determine the sizes of inhomogeneity regions and find the values of mean square differences of electron densities. In case of the fluctuation structure, owing to the absence of sharp phase boundaries and to the unknown character of the distribution of electron density in a separate region, the understanding of size becomes uncertain to some extent. In the present work the values of the size R and mean square difference of electron densities were determined on the assumption that the regions of inhomogeneities represent spherical particles with a constant electron density. In this case the following expressions apply:

$$I(s) = Nn^2 \exp(-S^2R^2/5) = (\bar{\rho}_1-\bar{\rho}_2)^2 \, w_1 v_o V \exp(-s^2R^2/5) \quad (1)$$

$$(\bar{\Delta\rho})^2 = (\bar{\rho}_1-\bar{\rho}_2)^2 \, w_1 w_2 \quad (2)$$

where N is the number of inhomogeneity regions in volume V; n is the number of electrons in one inhomogeneity region; R and v_o, radius and volume of the inhomogeneity region, respectively; $s = 4 \sin \rho/2 / \lambda$; ρ, scattering angle; λ, wave length; $(\bar{\Delta\rho})^2$ is the mean square difference of electron densities; $\bar{\rho}_1$ and $\bar{\rho}_2$, electron densities; w_1 and w_2, the relative volumes of the inhomogeneity regions and the remaining glass volume, respectively.

The intensity of scattering by thermal density and concentration fluctuations is independent of scattering angle. This intensity is usually denoted I(O). In the general case, the intensity I(O) is the sum of scattering intensities by thermal density, $I_\rho(O)$, and concentration, $I_c(O)$, fluctuations. For $I_\rho(O)$ the following equality holds:

$$I_\rho(O) = \bar{\rho}^2 \, k \, T \, \beta_T \, V \quad (3)$$

where $\bar{\rho}$ is the electron density; T is the absolute temperature; β_T is the isothermal compressibility; k is the Boltzmann constant.

RESULTS AND DISCUSSION

Phase Separation in Alkali Borate Systems

SAXS technique is quite convenient for finding and studying phase separation. If one assumes the validity of phase diagrams given in Reference 8, then one can calculate the expected values of the mean square difference between electron densities of phases in a two-phase system. Values of $(\overline{\Delta\bar{\rho}})^2$ in the samples of critical composition heated at 500°C should be 5.5 10^{-3}, 2.5 10^{-3}, 5 10^{-3} $(e/\text{Å}^3)^2$ for sodium, potassium, rubidium and cesium borate systems respectively. Yet, the sensitivity of a small-angle apparatus permits registration of SAXS by a structure with

$$(\overline{\Delta\bar{\rho}})^2 \sim 10^{-7}(e/\text{Å}^3)^2$$

Therefore, even the initial stages of phase separation can easily be detected.

In the present work, studies were made of two-component alkali borate glasses containing from 1 to 20% R_2O. Special attention was paid to sample glasses containing 10% Li_2O, Rb_2O, K_2O, Cs_2O and 16% Na_2O. These samples were held for several hours at 500°C,

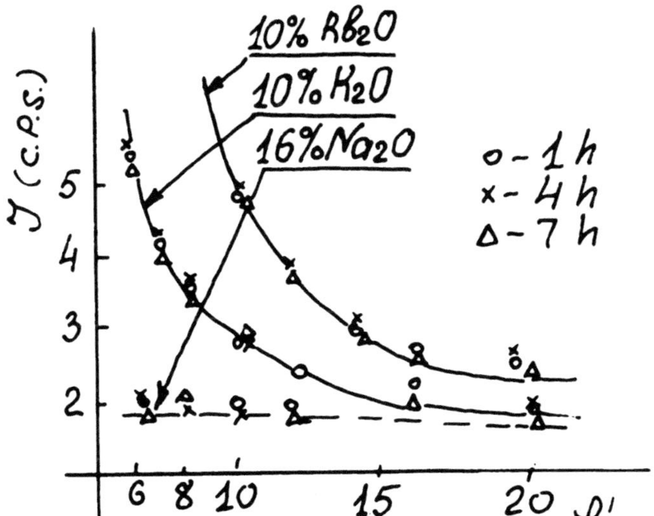

Fig. 1. SAXS intensities of alkali borate glasses at different isothermal treatments.

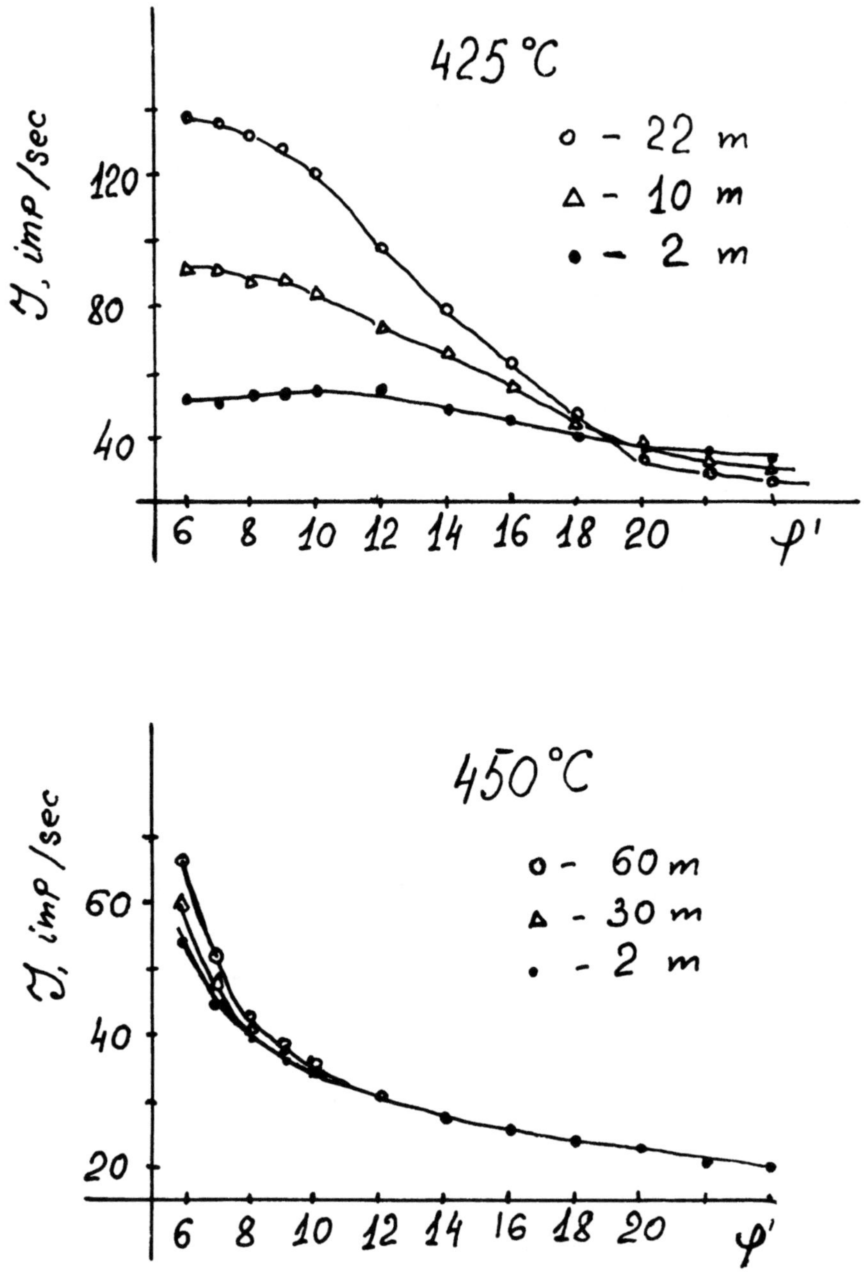

Fig. 2. SAXS intensities of glass containing 3.36% Li_2O heated for different times at several temperatures.

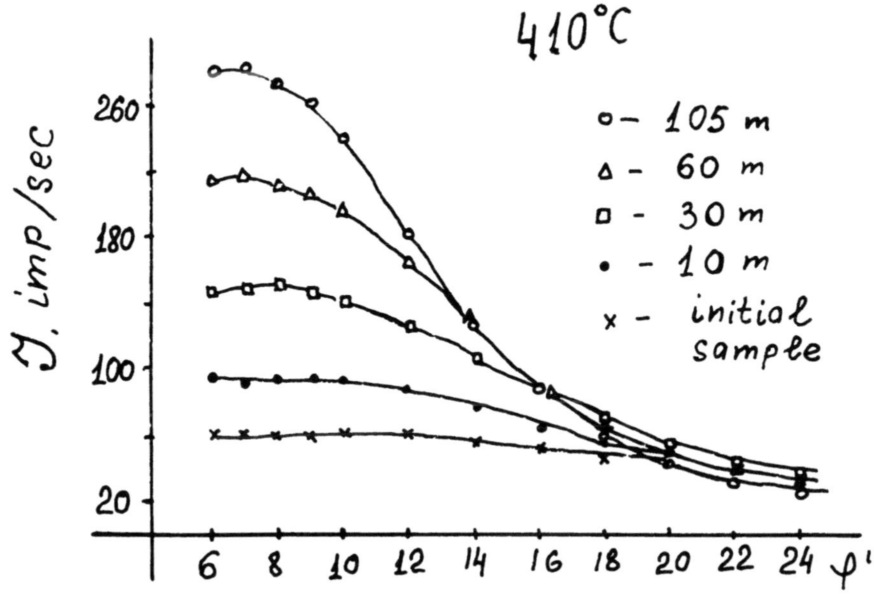
410°C
o – 105 m
Δ – 60 m
□ – 30 m
• – 10 m
x – initial sample
J, imp/sec
260
180
100
20
6 8 10 12 14 16 18 20 22 24 φ'

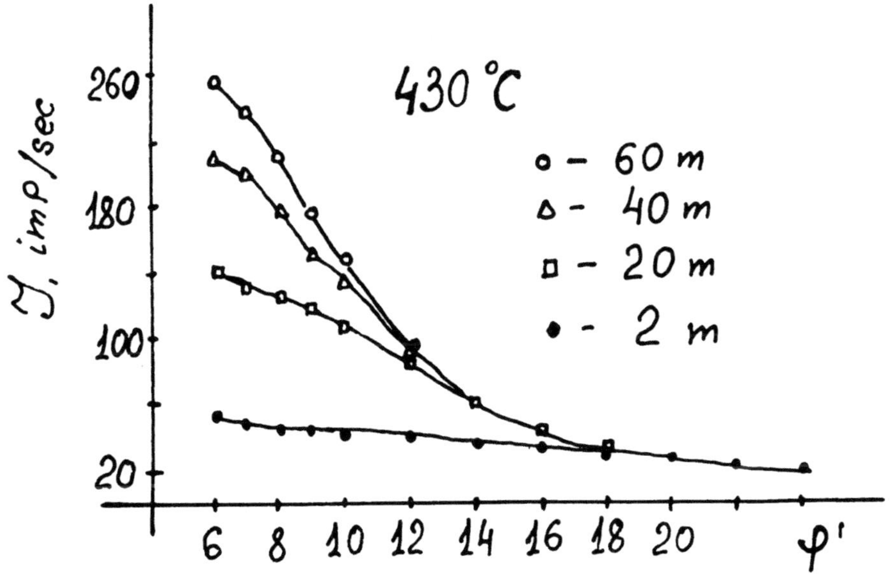
430°C
o – 60 m
Δ – 40 m
□ – 20 m
• – 2 m
J, imp/sec
260
180
100
20
6 8 10 12 14 16 18 20 φ'

and in the case of glass with 16% Na_2O at a temperature to 550°C, the SAXS intensity being measured every hour. The SAXS curves for glasses of these compositions at different heating times are shown in Fig. 1. As seen, SAXS intensity does not change on heating up to 7 hours, i.e. the glass is in the equilibrium state and no structural changes occur. Note that the heating temperature was close to the softening point, and at a temperature 10-20°C higher the sample flowed down the platinum support. Thus the absence of phase separation cannot be related to an insufficiently high sample temperature.

No irreversible change in SAXS intensity that could be related to phase separation processes was observed in the case of sodium, potassium, rubidium and cesium borate glasses of other compositions and heated at lower temperatures. Thus, no miscibility gaps are present in these systems, and structures observed by electron microscopy are not related to the processes of phase separation.

Fig. 2 shows SAXS intensities by the sample of a lithium borate glass containing 3.36% Li_2O heated for different times at 410°C, 426°C, 430°C and 450°C. The sample was preheated at 500°C and then cooled quickly to the temperature indicated. For such a sample, the SAXS intensity does not depend on the time of heating at temperatures above 443°C. Thus, this temperature corresponds to the binodal temperature.

The change in SAXS intensity with increasing duration of isothermal heating is evidence for structural changes occurring in the sample due to the processes of phase separation. It is seen from Fig. 2 that the character of changes in SAXS curves with increasing duration of heating depends on the temperature. At temperatures below 426°C the SAXS curves show a maximum and the point of intersection shifts to lower angles. At temperatures of 430°C and higher the maximum is absent, the curves do not intersect and each of the subsequent curves approaches more and more a certain asymptotic curve. In accordance with Langer's theory [10] the character of changes in SAXS curves at temperatures below 426°C is evidence for spinodal decomposition. The temperature 430°C is somewhat higher than the spinodal temperature. Thus, SAXS data permit the determination of spinodal temperature as between 426 and 430°C.

Table I. Binodal temperatures, T_b, and spinodal temperatures, T_s, for a number of glass compositions of the lithium-borate system.

Content of Li_2O mol. % by analysis	T_s °C	T_b °C
3.36	427	443
4.2	447	450
4.5	453	453
4.8	451	452
5.3	445	450
5.6	---	439

The spinodal and binodal temperatures were determined for a number of compositions of lithium borate glasses. The results are shown in Fig. 3 and Table I. The critical composition and temperature are equal to 4.5% Li_2O and 453°C, respectively.

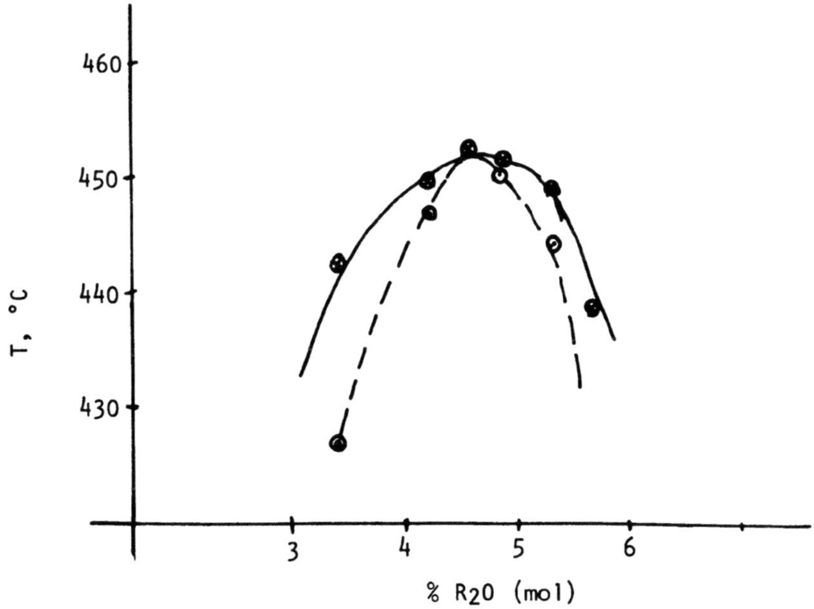

Fig. 3. Phase diagram showing subliquidus miscibility gap in the Li_2O-B_2O_3 system.

Structure of Vitreous B_2O_3

SAXS intensities by the sample of vitreous B_2O_3 at various temperatures are shown in Fig. 4. The SAXS curve referring to the primary quenched sample is due to stresses and submicroscopic cracks which appear on sharp quenching. The rates of temperature changes were then small and the SAXS intensity corresponded to scattering by the true structure of the sample. As follows from Fig. 4a, the SAXS intensity at all temperatures is independent of the scattering angle and increases with increasing temperature. Such a character of intensity is evidence for the complete absence of submicroheterogeneous structure in the sample; SAXS intensity is due solely to thermal density fluctuations.

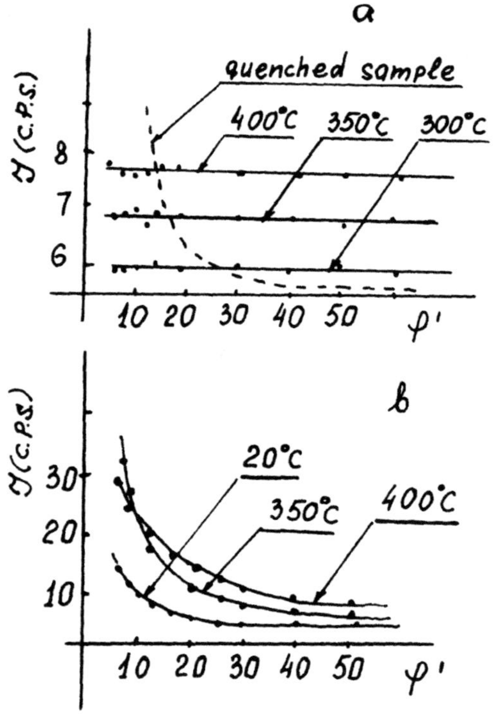

Fig. 4. SAXS intensities by a sample of vitreous B_2O_3 at different temperatures: (a) non-aqueous sample, (b) the same sample after its being in air for several hours.

The dependence of intensity I(0) on temperature is shown in Fig. 5. At temperatures above T_g the SAXS intensity increases strictly proportionally to temperature. At a temperature close to T_g, freezing of the structure occurs, and at lower temperatures I(0) is independent of temperature. The continuation of the plot of I(0) against T does not pass through 0°K which is due to the dependence of β_T on T.

The values of isothermal compressibility calculated using Eq. (3) are given in Fig. 6. They agree well with those reported in [11] where β_T was determined using the data of optical and ultrasonic studies. The value of I(0) by a sample of vitreous B_2O_3 at T_g is equal to 0.8 e.u./Å^3 (e.u. - electron unit). It was obtained using as a standard the sample of quartz glass for which I(0) was assumed to be equal to 25 e.u./V_{SiO_2}.

It is worth noting that SAXS intensity by the same sample increased sharply after aging in air for several hours, Fig. 4b. A submicroheterogeneous structure (large inhomogeneity regions of 250-300 Å in size) appeared in the sample. When heated the SAXS intensity by this structure first increases, then decreases but remains significant even at 400°C. It is possible that this structure was the one observed by electron microscopic studies of vitreous B_2O_3 and alkali borate glasses in Reference 7.

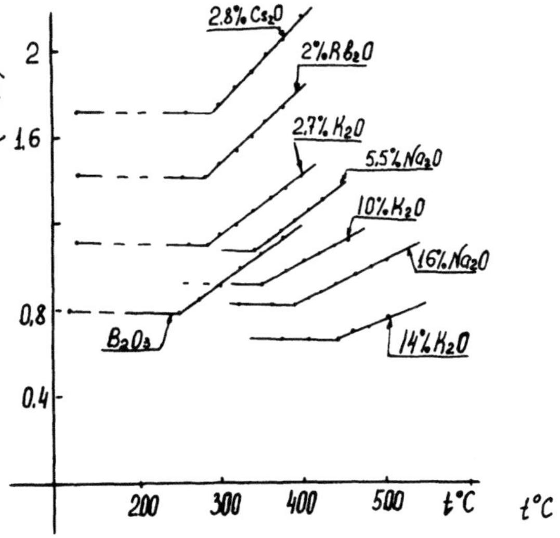

Fig. 5. Temperature dependence of I(0) for several glasses.

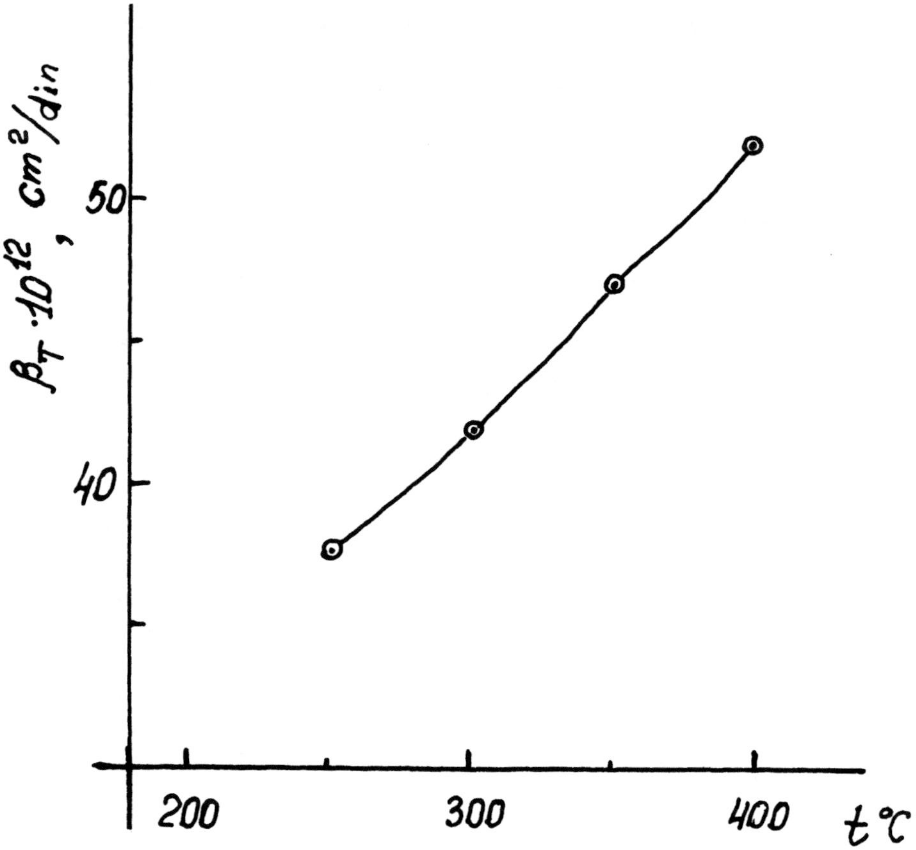

Fig. 6. Temperature dependence of isothermal compressibility for vitreous boron oxide.

It should also be noted that the absence of submicroheterogeneous structure in vitreous B_2O_3 is evidence in favor of the earlier made assumption that only thermal density fluctuations should be present in one-component glasses and that the submicroheterogeneous structure of quartz glasses is due to the technology of their production.

Structure of Alkali Borate Glasses

The character of dependence of SAXS intensity on the temperature and composition of the sample is the same for all systems investigated, therefore this dependence will be considered on potassium borate glasses.

The SAXS curves of potassium borate glass containing 2.7% K_2O are shown in Fig. 7 for different temperatures. Each of the curves can be divided into three parts: 1. at scattering angles from 6' to 20' the SAXS intensity decreases sharply; 2. at angles from 20' to 200' the further decrease of intensity occurs, though not so noticeably; 3. at angles more than 200' the SAXS intensity is independent of scattering angle. Such a character of intensity change implies that large inhomogeneity regions of 250-300 Å (scattering angles of 6'-20'), small inhomogeneity regions of 15 Å (scattering angles of 20'-200') and a structure of thermal density and concentration fluctuations (scattering angles of > 200') are present in the sample.

At temperatures above T_g the SAXS intensity, I(0), increases in strict proportion to temperature (Fig. 5). The SAXS intensity by large inhomogeneity regions, i.e. the SAXS intensity at scattering angles from 6' to 20' (Fig. 6a) after subtraction of SAXS intensity by small inhomogeneity regions, is practically independent of temperature. The intensity at scattering angles of 20'-200' increases with increasing temperature; the slope of the

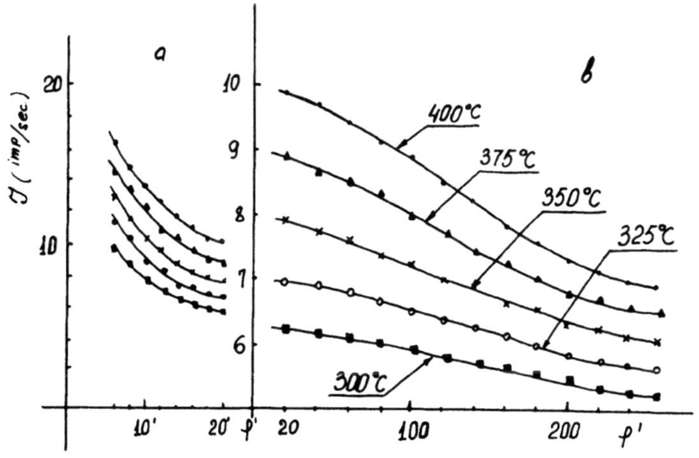

Fig. 7. SAXS curves by a sample containing 2.7% K_2O at different temperatures: (a) φ = 6'-20'; (b) φ 20' (in enlarged scale).

Guinier plot is independent of temperature. Consequently, with increasing temperature the sizes of small inhomogeneity regions do not change but the mean square difference of electron densities increases significantly. With increasing R_2O content the value of $(\overline{\Delta\rho})^2$ decreases for small inhomogeneity regions. The latters are practically inobservable at 10% R_2O.

The common feature of all alkali borate glasses is the presence of submicroheterogeneous structure and thermal density and concentration fluctuations.

The submicroheterogeneous structure is absent in vitreous B_2O_3 and consequently, its appearance is due to the presence of alkali metal ions and to their interaction with the B_2O_3 network.

Regions of inhomogeneity of 15-16 Å are observed only in low-alkali glasses, their sizes being approximately the same for all glasses of the systems studied. Fig. 8 shows the dependencies of $(\overline{\Delta\rho})^2$ on composition for these regions. In potassium borate glasses the value of $(\overline{\Delta\rho})^2$ is minimal. An increase of $(\overline{\Delta\rho})^2$ in rubidium borate glasses can be due to the fact that the contribution of rubidium electrons to the electron density of a separate region is significant for these glasses. It can be concluded that alkali metal ions are grouped in these regions. At the same time the value of $(\overline{\Delta\rho})^2$ is the same for sodium and rubidium borate glasses. The contribution of sodium electrons to the electron density of a separate region is small. Therefore the changes in the short-range order structure caused by sodium ions are the main reason of the appearence of the difference between electron densities; apparently the transition of boron from three- to four-coordinated state shows up here. Thus small inhomogeneity regions are the result of both structural transformation (changes in boron coordination) and clustering effect. In addition to this the changes of the short-range order structure are greatest in sodium borate glasses; but they diminish with increasing order number of the alkali element. Therefore in cesium borate glasses the values of $(\overline{\Delta\rho})^2$ are almost the same as in sodium and rubidium borate glasses despite a considerably greater amount of electrons in a cesium ion.

In lithium borate glasses the same structure was also observed but the SAXS intensity by this structure is much less than that by the structure appearing on phase separation or the SAXS intensity by the supercritical fluctuations structure at a temperature higher

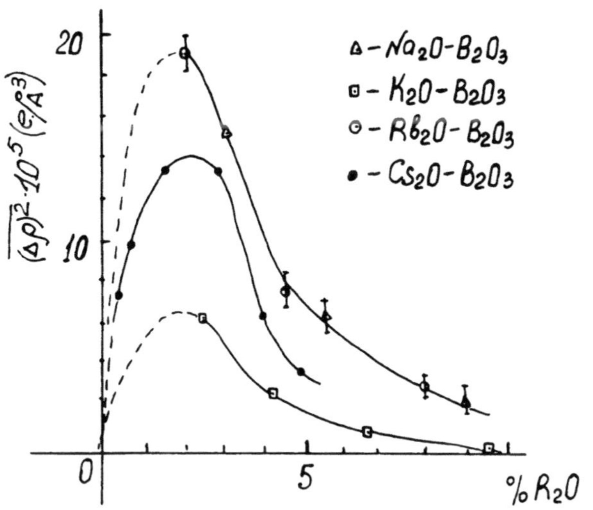

Fig. 8. Dependence of $\overline{(\Delta\rho)^2}$ on glass composition at 400°C.

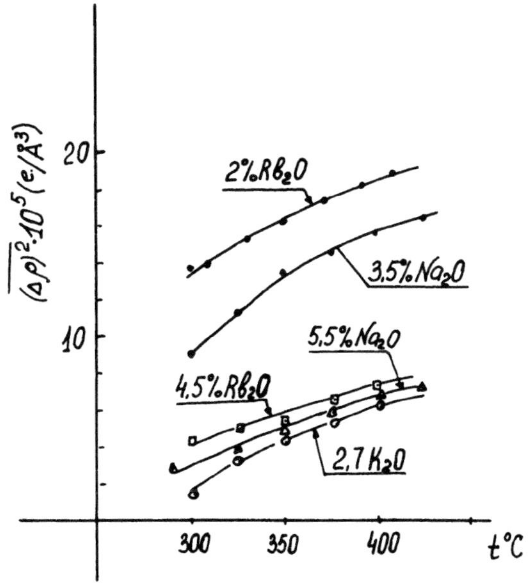

Fig. 9. Temperature dependence of $\overline{(\Delta\rho)^2}$ for the investigated glasses.

than the binodal temperature. It is difficult to distinguish the SAXS intensity by this structure and obtain exact values of $(\bar{\Delta}\bar{\rho})^2$. Therefore the structure of lithium borate glasses is not discussed in the present work.

Fig. 9 shows the dependences of $(\bar{\Delta}\bar{\rho})^2$ on temperature for a number of glasses of the studied systems. A distinguishing feature of alkali borate glasses is an increase of $(\bar{\Delta}\bar{\rho})^2$ with increasing temperature which indicates a strengthening of the processes of the short-range order structure change with increasing temperature. The value of $(\bar{\Delta}\bar{\rho})^2$ reaches a maximum and then does not change up to 650°C.

Large inhomogeneity regions in alkali borate glasses are less pronounced than in silicate glasses. Their sizes are 200-300 Å and $(\bar{\Delta}\bar{\rho})^2 \sim 10^{-6}$ (e./Å^3)^2. The sizes of the regions and the values of $(\bar{\Delta}\bar{\rho})^2$ are independent of temperature in all glasses. These regions are most pronounced in glasses containing ~10% R_2O; they are apparently formed by merging of small regions. With increasing R_2O content this process strengthens and at 10% R_2O small regions are not observed.

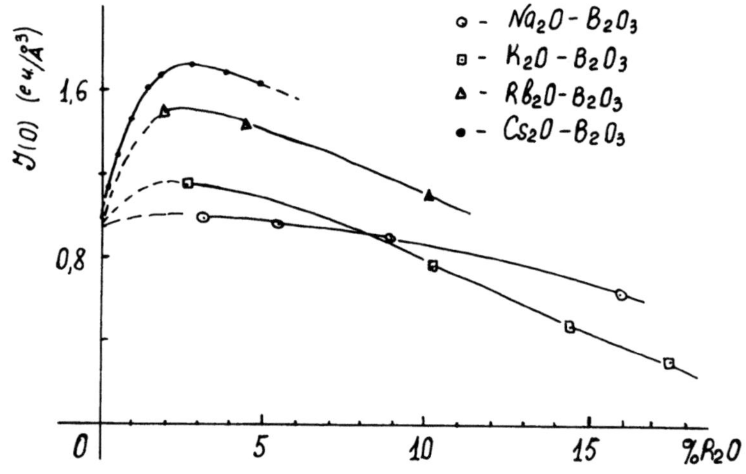

Fig. 10. Dependence of I(0) on glass composition at 300°C.

Fig. 5 shows the temperature dependences of intensity I(0) by several samples of glasses of the studied systems. For all samples at a temperature above T_g the SAXS intensity is proportional to temperature. The plots of I(0) against T in a first approximation forms a fan one ray of which corresponds to vitreous B_2O_3. Such a form of the plots permits a conclusion that the thermal density fluctuations of B_2O_3 network can be observed in all glasses.

The dependence of I(0) on concentration at 300°C is shown in Fig. 10. At small R_2O contents the intensity increases, the greatest increase being observed in cesium borate glasses. The intensity reaches a maximum value at 2-3% R_2O and then decreases. A sharp increase of I(0) is due to concentration fluctuations of alkali metal ions. A decrease of I(0) with increasing R_2O content is caused by increasing volume of the glass in which the change of boron coordination occurs and compressibility decreases accordingly.

REFERENCES

1. E. A. Porai-Koshits, V. V. Golubkov, A. P. Titov, Proc. X Intern. Congr. on Glasses, Kyoto, 12, N 1, 1, 1974.
   V. V. Golubkov, A. P. Titov, E. A. Porai-Koshits, Fiz. Khim. Stekla, 1, 394, 1975.
2. J. Krogh-Moe, J. Non-Crystalline Solids, 6, 269, 1969.
3. J. Krogh-Moe, Phys. Chem. Glasses, 1, 26, 1960.
4. R. Roy, Proc. Symposium on Nucleation and Crystalliz. of Glasses and Melts, Ed. by M. K. Reser, G. Smith, H. Insley, Am. Cer. Soc. Inc., Columbus, Ohio, p. 39, 1962.
5. W. Skatulla, W. Vogel, H. Wessel, Silikattechnik, 9, 51, 1958.
6. F. E. Wagstaff, R. J. Charles, Am. Cer. Soc. Bull., 45, 420, 1966.
7. J. Krogh-Moe, Arkiv Kemi, 14, 1, 1959.
8. R. R. Shaw, D. R. Uhlmann, J. Am. Cer. Soc., 51, 377, 1968.
9. V. V. Golubkov, A. P. Titov, E. A. Porai-Koshits, Prybory Tekhn. Exper., N 1, 215, 1975.
10. J. S. Langer, M. Bar-on, H. P. Miller, Phys. Rev., 11, 1417, 1975.
11. J. A. Bucaro, H. D. Dardy, J. Appl. Phys., 45, 2121, 1974.

DIFFRACTION ANALYSIS OF VITREOUS AND MOLTEN B_2O_3

J. ZARZYCKI

Laboratory of Materials Science and C.N.R.S. Glass

Laboratory. University of Montpellier, France

X-ray diffraction studies of vitreous B_2O_3 provide a good illustration of the progress in structure determination of inorganic glasses in general. An important glass network former with interesting physico-chemical properties, B_2O_3 has been studied almost as extensively as vitreous SiO_2 ; the determination of its structure met, however, with additional difficulties, the main one being for a long time the lack of reliable information on the structure of the crystalline B_2O_3 itself.

The first structural analysis by WARREN, KRUTTER and MORNINGSTAR (1) was followed by those of RICHTER et al (2), ZARZYCKI (3) for both vitreous and molten B_2O_3 up to 1600 °C, DESPUJOLS (4), MILBERG (5) and others. In all these studies the interpretation was based on the approximate method for heteroatomic samples set up by WARREN in his original study. Classically, from corrected scattered intensity $I(k)$ fitted to the (calculated) independent scattering for high values of the scattering vector $k = (4\,\Pi \sin \Theta)/\lambda$ an intensity function $ki(k)$ is obtained which is then Fourier transformed to give :

$$[1] \quad \sum_{u.c.} K_j 4\Pi r^2 \rho_j(r) = 4\Pi r^2 \rho_e \sum_{u.c.} K_j + \frac{2r}{\Pi} \int_0^{\infty} ki(k) \sin rk \, dk$$

In this expression the subscripts u.c. mean unit of composition (B_2O_3) and j the kind of atom : (B or O).
$\rho_j(r)$ is the average within the sample at a displacement $\vec{r}$ from atom j, ρ_e average electronic density and K_j "effective numbers" defined by $K_j = f_j/f_e$ where f_j is the scattering factor of atom j with atomic number Z_j and f_e an average scattering factor

per electron :

$$f_e = (\sum_{u.c.} f_j) / \sum_{u.c.} Z_j$$

The plot of function [1] leads to the so-called radial distribution function (r.d.f.) the successive peaks of which express the probability maxima of various interatomic distances in the sample, the subtended areas being related to average coordination numbers of the disordered structure. Fig. 1 taken from our earlier study (3) shows as example spectra for vitreous and molten B_2O_3 and Fig. 2 the corresponding radial distribution functions.

In practice, convergence factors are used to calculate the integral in Eq. [1] and the integration range is limited at k_m instead of ∞, which introduces a "termination error" and adds a ripple to the r.d.f.

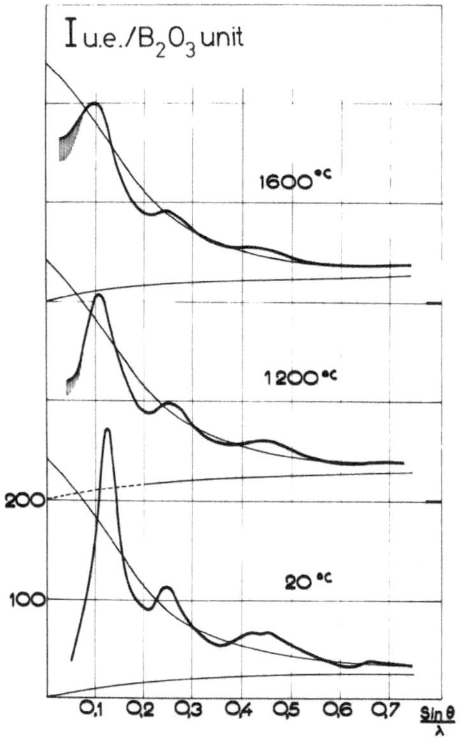

Fig. 1. X-ray diffraction spectra of vitreous B_2O_3 at 20 °C and molten B_2O_3 at 1200 °C and 1600 °C. After (3).

One may safely say that the determination of the structure of vitreous B_2O_3 by this method led to very limited results : first interatomic distances : B-O = 1.37 Å, O-O = 2.40 Å and the coordination number of boron ∿ 3 which indicates that the structure consists of BO_3 triangles. ZARZYCKI (3) used this method to follow also structural changes at high temperature 1200° and 1600° - a very pronounced decrease of the first coordination number was found, amounting to more than 30% at 1600° (Fig. 2).

Attempts to increase resolution by extending the integration range invariably led to ripples or "satellites" and the problem was then to distinguish real structures from artefacts. Fig. 3 shows this effect for WARREN's data on B_2O_3 where the integration range is extended from k_m = 8.0 to k_m = 11.7 and Fig. 4 is relative to a rather extreme case of an r.d.f. determined from data with k_m = 17.3 but with spurious detail possibly from incorrect curve fitting and termination error (2). The improved curve fitting methods have since led to greater confidence in experimental results.

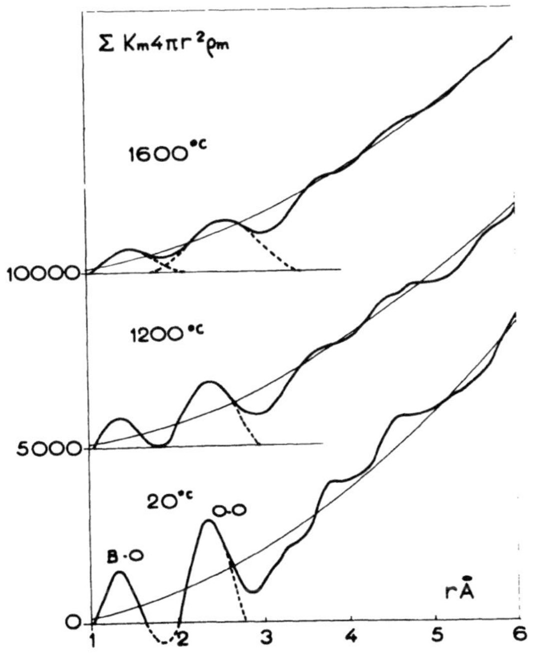

Fig. 2. Radial distribution functions corresponding to the spectra of Fig. 1. After (3).

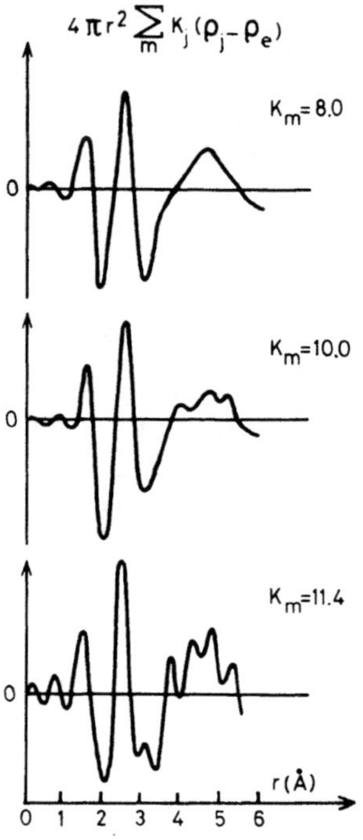

Fig. 3. Radial density distribution curves for vitreous B_2O_3 as calculated from WARREN's data (1) for various integration limits k_m After (6).

Recent advances for B_2O_3 came from the separation of Compton scattering by the method of fluorescence excitation (7) and the introduction of an exact method of interpretation based on the formalism of the Norwegian school of FINBAK and pointed out by WASER and SCHOMAKER (8). The study of MOZZI and WARREN (9) incorporated all these recent advances and led to a substantial improvement in our knowledge of the short range order in vitreous B_2O_3.

The mathematical formulation used in these studies is based on the use of pair functions $P_{ij}(r)$:

$$P_{ij}\,(r) = \int_0^{k_m} \frac{f_i f_j}{g^2(k)}\; e^{-\alpha^2 k^2} \sin kr_{ij} \sin kr \, dk$$

which are used to construct a distribution function :

$$[2] \quad \sum_{u.c.} \sum_{i} \frac{N_{ij}}{r_{ij}} P_{ij}(r) = 2\Pi^2 r\rho_e \sum_{u.c.} Z_j + \int_o^{k_m} ki(k)e^{-\alpha^2k^2} \sin rk\, dk$$

In this expression, N_{ij} is the average of atoms in shell i at a distance j from an atom of type j ; α^2 is a suitable convergence coefficient and $g^2(k)$ a sharpening function which is usually taken to be f_e^2.

The main advantage of this method is that it permits the exact synthesis of the radial distribution function from separate pair functions and automatically takes into account all effects introduced via damping coefficient or termination error, the corresponding ripples being present also in the calculated pair functions P_{ij}. Thus a great part of the dilemma of the original method is removed.

To evaluate the degree of improvement over the old "constant K" method, quantities $(f_i f_j)/f_e^2$ were calculated for B_2O_3 (Fig. 5) ; in the original method they were treated as constants. It can be seen that in the case of B_2O_3 this approximation is poor for all interatomic distances and especially for the B-O pair.

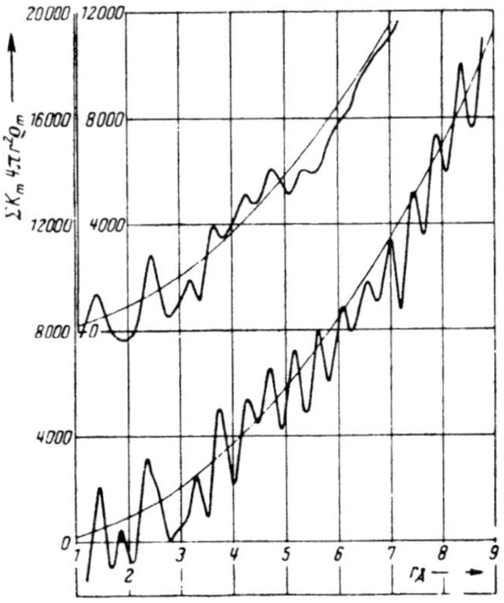

Fig. 4. Radial distribution functions for vitreous B_2O_3 showing spurious peaks ; integration range limited to k_m = 11.9 (upper curve) and k_m = 17.3 (lower curve). After (2).

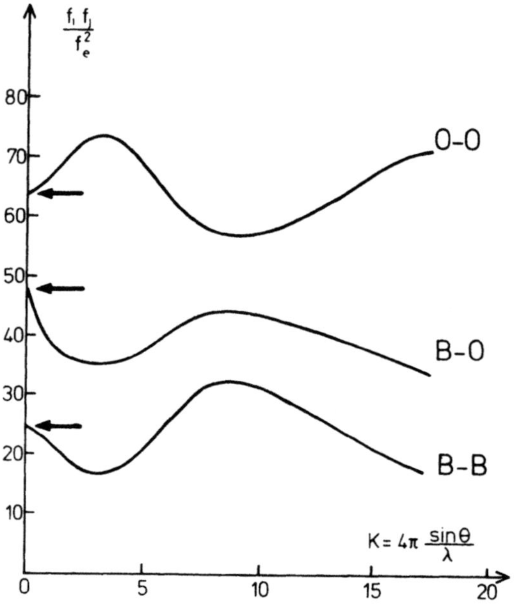

Fig. 5. The ratios $f_i f_j / f_e^2$ for B-B, B-O and O-O pairs, f_e is the average f per electron.

Fig. 6, taken from the recent study of MOZZI and WARREN (9), shows the result of pair function analysis for vitreous B_2O_3.

It seems to give strong support to the "boroxol model", the structure consisting of randomly interlinked (B_3O_6) groups, and indicates that the short-range order in vitreous B_2O_3 is essentially different from that of B_2O_3(I) crystal.

It should be noted, however, that this modern method still contains some approximations. Thus the choice of the damping factor α^2 is not entirely arbitrary as the second term in Eq. [2] does not contain α (10).

More questionable seems to us the method itself of successive peak fitting, used by MOZZI and WARREN who assumed random distribution of elements of structure (e.g. boroxols) around a bond direction. These distributions are certainly not flat - (indeed we are seeking to determine them) - and are not independent but inter-related. The only possible way of obtaining them would be by means of a model, as was done for SiO_2 or chalcogenide glasses (11) but a model is still lacking for the three-coordinated network of B_2O_3.

The apparent absence of detail in the difference curve (C) of Fig. 6 may be due to a lack of resolution of the measured function - still greater resolution is required : it should be attainable by extending the k_m range using modern powerful X-ray sources from linear accelerators such as LINAC (or LURE in France). These data combined with a model could yield additional information about the real distribution of the boroxol groups within the structure.

The further organization or "middle range order" can be studied using small-angle diffraction techniques. Small-angle scattering of X-rays (SAXS) can give valuable information about density fluctuation level in one-component glasses (12). For B_2O_3 glasses no absolute SAXS intensity measurements have been made (most of the studies available dealing with phase-separation in multicomponent systems). However, an estimate of the extrapolated diffused intensity I(O) may be derived using measurements from light diffusion or ultrasonic studies.

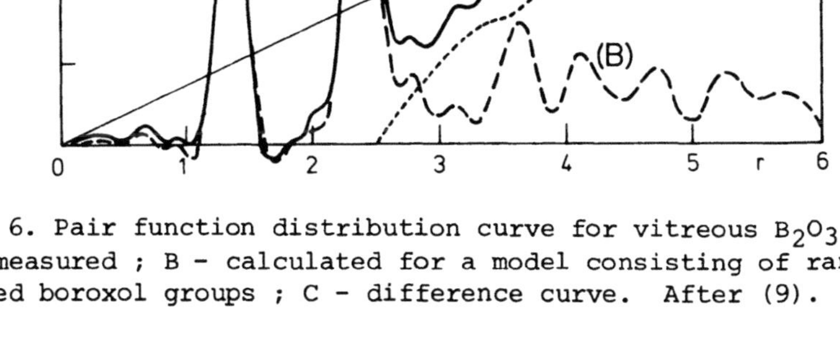

Fig. 6. Pair function distribution curve for vitreous B_2O_3 ; A - measured ; B - calculated for a model consisting of randomly linked boroxol groups ; C - difference curve. After (9).

For a glass, the square deviation of the density fluctuation $< \Delta\rho^2 >$ may be written (13) :

$$[3] \quad <\Delta\rho^2> = \frac{\rho_o^2 \tilde{k}}{V} \left[T_f \left(\beta_{T,o} - \beta_s^{\infty}\right) + T \beta_s^{\infty} \right]$$

where ρ_o is the average density, $\tilde{k}$ the Boltzmann constant, V the volume, T_f the absolute transition temperature, T the absolute temperature, $\beta_{T,o}$ the low frequency isothermal compressibility and β_s^{∞} the high-frequency adiabatic compressibility.

Using the values recently obtained by BUCCARO and DARDY (14), from Rayleigh diffusion studies for a B_2O_3 melt and glass, we have calculated the fluctuation level in vitreous B_2O_3.
With :

$$\beta_{T,o} = 39 \times 10^{-12} \text{ cm}^2 \text{ dyne}^{-1} \quad \text{at } T_f = 548 \text{ °K}$$

and :

$$\beta_{T,\infty} \sim \beta_s^{\infty} = 7.4 \times 10^{-12} \text{ cm}^2 \text{ dyne}^{-1} \quad \text{at } T = 293 \text{ °K}$$

we obtain :

$$\Theta = \frac{V < \Delta\rho^2 >}{\rho_o^2} = 2.7 \times 10^{-24} \quad \text{at 20 °C}$$

To get a feeling of the heterogeneity involved, for a cube of 20 Å edge of a volume $V = 8 \times 10^{-24}$ cm^3, the density fluctuation is : $<\Delta\rho^2>/\rho_o = 0.058$ or roughly 6 per cent.

The expected X-ray diffused intensity I(O) per scattering unit B_2O_3 is (12) :

$$I(O) = \frac{V <\Delta\rho^2{}_e>}{N_V}$$

where N_V is the number of molecules per cm^3 in the material :

$$N_V = \frac{\mathcal{N} \rho_o}{M} \quad \text{(with } \mathcal{N} \text{ Avogadro's number and M molecular weight)}$$

The electronic density fluctuations $\Delta\rho_e$ are related to the density fluctuations $\Delta\rho$ by the relation :

$$\frac{<\Delta\rho^2_e>}{\rho_{oe}^2} = \frac{<\Delta\rho^2>}{\rho_o^2}$$

Whence :

$$[4] \quad I(0) = \frac{V<\Delta\rho^2>}{\rho_o^2} \left(\frac{\sum Z}{M}\right)^2 \mathcal{N} \rho_o$$

Using the previously calculated value of Θ and taking the density ρ_o = 1.84 g cm^{-3}, this leads to an expected value of extrapolated intensity I(O) for vitreous B_2O_3 at 20 °C :

$$I(0)_{20°} = 49.7 \text{ e.u.}/B_2O_3 \text{ unit}$$

which is almost twice the value for vitreous silica (12).

Conventional wide angle X-ray studies of vitreous B_2O_3 provide us with intensity spectra fitted to an absolute scale. Intensity is then expressed precisely in electron units (e.u.) per B_2O_3 unit. Lowest values attained (for k $\sim$ 0.5) are all close to 50 e.u. (1)(2)(3)(4)(5)(9). They are thus in excellent agreement with the above calculations of I(O) limit.

In our initial high-temperature X-ray diffraction measurements on vitreous and molten B_2O_3 (3) we already mentioned the presence of strong diffusion at small angles for melts at 1200 and especially 1600 °C (Fig. 1) ; with the device used the measurements could not, however, be extended to k values lower than 0.5 Å^{-1}.

For these high temperatures the extrapolated intensity I(O) can be easily computed from isothermal compressibility measurements of the melt at temperature T in a way similar to that for a glass :

$$\Theta_T = \frac{V<\Delta\rho^2>}{\rho_T^2} = \tilde{k}\, T\, \beta_{T,o}$$

Using the adiabatic compressibility values β_S measured by BOCKRISS and KOJONEN (15) and setting $\beta_{T,o} \sim \beta_S$, we obtained the following Θ_T and I(O) values (Table 1).

Table 1

t (°C)	ρ_o (g cm^{-3})	β_s (cm^2 dyne^{-1})	Θ_T	I(O)e.u.
1200	1.48	65×10^{-12}	$13{,}21 \times 10^{-24}$	195
1600	1.43	$55{,}8 \times 10^{-12}$	$14{,}2 \times 10^{-24}$	203

This shows the important variation in I(O) which increases fourfold between 20° and 1600 °C.

We have attempted to interpret our calculated I(O) values in terms of a model proposed by BIENENSTOCK and BAGLEY (16) for evaluation of upper SAXS intensity of amorphous materials.

In this model (Fig. 7a) the scattering medium of average electron density $\overline{\rho_e}$ is taken as a random close packing of spheres of radius r_2 ; the central cores of these spheres of a radius r_1 ($r_1 < r_2$) and the interstices between the spheres (r_2) are filled with material of electronic density ρ_M. The spherical shells between spheres (r_2) and (r_1) are filled with the electron density ρ_m. The scattering arises from the difference $\Delta\rho_e = \rho_M - \rho_m$; BIENENSTOCK and BAGLEY suppose $\rho_m = 0$, whence $\Delta\rho_e = \rho_M$. Introducing SCOTT's distribution function of centres of spheres (Fig. 7b) the scattered intensity is easily computed.

We have tentatively assumed that the spherical units of BIENENSTOCK and BAGLEY's model correspond to boroxol groups constituting the network. Taking $r_1 = 2.80$ Å as the radius of a sphere in which the boroxol groups can be inscribed (the linking oxygens are counted for 1/2), such a sphere contains 51 electrons and has an average electronic density $\rho_B = 0.554$. ρ_e values were obtained from macroscopic density data. We than calculated the radii r_2 and the electronic density difference $\Delta\rho_e$ for which the maximum scattered intensity at a given temperature is equal to the previously estimated I(O) value.

Table 2 shows the result of these computations. The calculated scattering curves are represented in Fig. 8 which also shows the tails of wide-angle X-ray spectra taken from our earlier experimental data (3).It can be seen that r_2 is practically constant in the melt and that it undergoes a 5% decrease when the melt is quenched to glass. This is sufficient to bring the fourfold decrease of scattered intensity and gives an indication of variation in the degree of packing of the structural units in the glass and in the melt.

Table 2

t (°C)	ρe (el.Å^{-3})	r_2 (Å)	$\Delta\rho_e$ (el.Å^{-3})
20	0.541	2.99	0.606
1200	0.435	3.14	0.500
1600	0.420	3.15	0.519

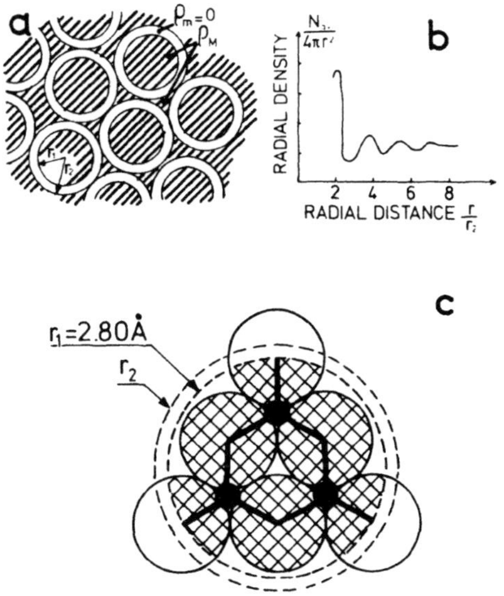

Fig. 7 a) BIENENSTOCK and BAGLEY's model of disordered structure used for SAXS calculation (after 16) ; b) SCOTT's distribution of centres of spheres in a random close packing (after 17) ; c) boroxol unit inscribed in a sphere radius r = 2.80 Å. This sphere contains 3 boron and 6 oxygen atoms (linking oxygens are counted for 1/2).

The compatible electronic density $\Delta\rho_e$ responsible for scattering encompasses the value $\rho_B = 0.554$ for the boroxol group admitted in our analysis.

The presence of boroxol groups is thus compatible with SAXS calculated I(O) intensities both in the glass and in the melt and it can be seen that these agree rather well with the known tails of X-ray spectra. It is clear, however, that complete SAXS intensity distribution measurements would be needed to draw more definite conclusions. A random sphere distribution gives only a poor approximation of the B_2O_3 network and emphasizes the need for a B_2O_3 model.

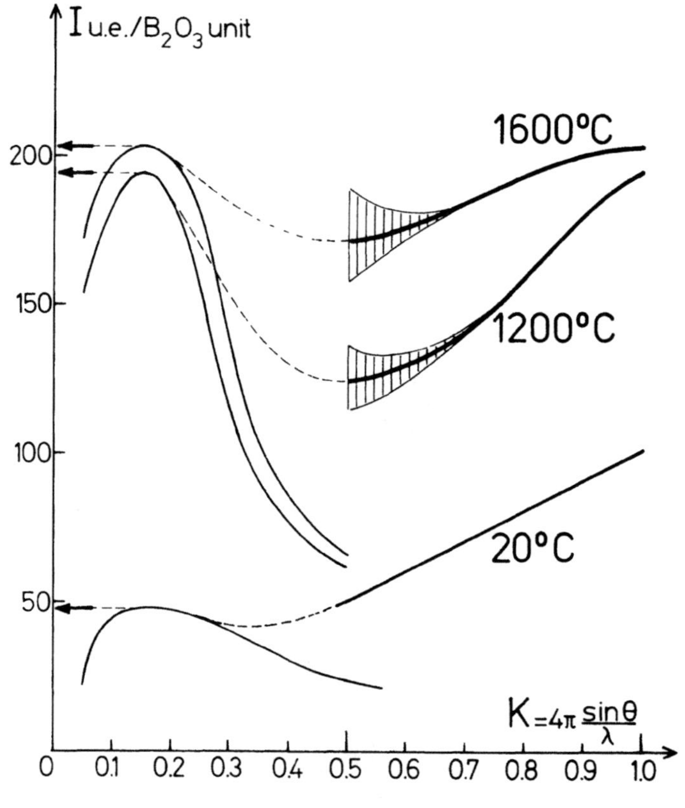

Fig. 8. Calculated SAXS intensity spectra for vitreous B_2O_3 at 20 °C and molten B_2O_3 at 1200 and 1600 °C using BIENENSTOCK and BAGLEY's model with $r_1 = 2.80$ Å. The r_2 radii and $\Delta\rho_e$ values are those given in Table 2. The I(O) values calculated from compressibility data are indicated by arrows. The right portion of the figure (for $k > 0.5$) contains known tails of X-ray scattered intensity taken from ref. (3).

References

(1) B.E. WARREN, H. KRUTTER and O. MORNINGSTAR, J. Amer. Cer. Soc. 19, (1936), 202

(2) H. RICHTER, G. BREITLING, H. HERRE, Z. Naturforsch. 9, (1954), 390

(3) J. ZARZYCKI, Travaux IVe Congrès Intern. du Verre, Paris 1956, VI-4, p. 323

(4) J. DESPUJOLS, J. Phys. Rad. 19, (1958), 612

(5) M.E. MILBERG, F. MELLER, J. Chem. Phys. 32, (1960), 126

(6) K. GRJOTHEIM and J. KROGH-MOE, Glastechn. Tidskr. (1955) N° 2, p. 1

(7) B.E. WARREN and G. MAVEL, Rev. Sci. Instr. 36, (1965), 59

(8) J. WASER, V. SCHOMAKER, Rev. Modern Phys. 25, (1953), 671

(9) R.L. MOZZI and B.E. WARREN, J. Appl. Cryst. 3, (1970), 251

(10) J.R.G. da SILVA, D.G. PINATTI, C.E. ANDERSON, M.L. RUDEE, Phil. Mag. 31, (1975), 713

(11) J. ZARZYCKI, Proc. IV Intern. Conf. The Physics of Non Crystalline Solids Clausthal-Zellerfeld (1976), Frischat Editor, Trans. Tech. Publ. (1977), p. 52

(12) J. ZARZYCKI, Proc. 10 Intern. Congress on Glass, KYOTO, (1974) N° 12, p. 28

(13) N.L. LABERGE, V.V. VASILESCU, C.J. MONTROSE, P.B. MACEDO, J. Amer. Cer. Soc. 56, (1953), 506

(14) J.A. BUCCARO, H.D. DARDY, J. Appl. Phys. 45, (1974), 2121

(15) J.O'M. BOCKRISS, E. KOJONEN, J. Am. Chem. Soc., (1960), 4493

(16) A. BIENENSTOCK, B.G. BAGLEY, J. Appl. Phys. 37, (1966), 4840

(17) G.D. SCOTT, Nature, 194, (1962) 956

LASER-INDUCED FLUORESCENCE LINE NARROWING OF Eu^{3+} IN LITHIUM BORATE GLASS

M. J. Weber*
Lawrence Livermore Laboratory
University of California
Livermore, California 94550

J. Hegarty
Department of Physics
University of Wisconsin
Madison, Wisconsin 53706

D. H. Blackburn
Inorganic Glass Section
National Bureau of Standards
Washington, D. C. 20234

INTRODUCTION

Laser-induced fluorescence line narrowing has recently been used to investigate the local environment and interactions of paramagnetic ions in glass. Since the resulting spectra are sensitive to structural modifications, this technique was applied to study lithium borate glass. In alkali borate glasses, the relative numbers of BO_3 triangles and BO_4 tetrahedra and the existence of subliquidus immiscibility are dependent upon the mole fraction of alkali oxide. Structural changes associated with these properties are, in principle, detectable using the approach of fluorescence line narrowing. Evidence of these effects are presented in this paper.

As a tool to investigate structure in glasses, optical spectroscopy of dilute paramagnetic ions is plagued by two difficulties. First, the paramagnetic ion constitutes a defect. Thus, the presence of the ion perturbs the local field from that of the original "perfect" glass. While this may temper one's enthusiasm for the use of impurities as probes of ideal glass structure,[1] it is precisely the field at the

*This work performed under the auspices of the U. S. Energy Research and Development Administration, Contract W-7405-eng-48.

activator site which governs its luminescence properties. An understanding of these fields and the ability to influence luminescence properties by altering the glass composition are of paramount importance for applications and, therefore, are of significant interest.

Second, paramagnetic ions generally enter a glass as interstitial network-modifying cations, not as substitutional ions. Because of differences in the bonding to nearest neighbors and, in multicomponent glass compositions, in the types and statistical distribution of more distant neighbor ions, the local fields at individual paramagnetic ion sites vary. This results in site-to-site differences in the energy levels and the radiative and nonradiative transition probabilities. Under broadband excitation, these features are evident from the large inhomogeneous optical absorption and emission linewidths and nonexponential excited-state decays. The simultaneous observation of many sites has, in the past, hampered analysis of structure arising from changes in borate glass composition.[2]

This second difficulty can be overcome by the use of a narrowband excitation source. In this case, only those ions resonant to within a homogeneous linewidth are excited and a line-narrowed fluorescence is observed. For resonant excitation and observation, under conditions where no cross relaxation between ions occurs, the fluorescence linewidth is given by $2\Delta\nu_{homo} + \Delta\nu_{instr}$, where $\Delta\nu_{homo}$ is the homogeneous linewidth arising from lifetime broadening of the initial and final states and $\Delta\nu_{instr}$ is the instrumental linewidth caused by the spectral resolution of the source and detection system.[3] Fluorescence line narrowing in general, and for Eu^{3+} in alkali borate glass in particular, is not new. A decade ago, fluorescence line narrowing of Eu^{3+} in a sodium borate glass was demonstrated using narrow emission lines from a mercury lamp for excitation.[4]

The use of a laser as the excitation source for fluorescence line narrowing (FLN) in glass was first demonstrated by Riseberg,[5] who used a fitted frequency laser. Later, with the advent of tunable lasers, it was possible to tune across an inhomogeneous absorption line and thereby probe the entire distribution of paramagnetic ion sites. From the resulting line-narrowed spectra, coordinate models have been developed which correlate distortions of the structure with the relative magnitude and sign of measured crystal-field parameters.[6] Recently, the technique of laser-induced FLN has been applied to several rare-earth ions in different glasses to investigate site-dependent variations in energy levels, radiative and nonradiative transition probabilities, homogeneous line broadening, and energy transfer between ions.[7]

In the investigation of alkali borate glass below, we consider only the energy levels and radiative transition probabilities. Trivalent europium is used as the probe ion since it has a relatively simple energy level scheme with levels in experimentally convenient wavelength regions. In addition, Eu^{3+} has been studied in silicate and phosphate glasses using FLN;[8] thus, the behavior in borate glass can be compared with that observed in other oxide glasses.

Lithium borate glasses were investigated. From NMR measurements,[9] the fraction of BO_4 tetrahedra in the lithium-borate glass system increases with increasing lithia content to a maximum of ≈ 0.45 at 40 mole percent Li_2O. Subliquidus immiscibility has been observed[10] for borate glasses with between 4 and 16 mole percent Li_2O. The maximum of the coexistence curve occurs at 10 mole percent Li_2O. For the present studies, glasses containing 10 and 40 mole percent Li_2O and doped with 1 mole percent Eu_2O_3 were prepared. We were unable to introduce similar quantities of Eu_2O_3 into simple B_2O_3.

GLASS PREPARATION

Glasses were prepared from reagent-grade fused boric acid (B_2O_3) and reagent-grade Li_2CO_3 for Li_2O. High-purity (99.99%) Eu_2O_3 was used for the dopant. The sample compositions were (mol.%): 89 B_2O_3, 10 Li_2O, 1 Eu_2O_3 and 59 B_2O_3, 40 Li_2O, 1 Eu_2O_3. Melts were prepared in platinum crucibles in an electrically heated furnace in an oxidizing atmosphere (air). The melt sizes were 100 grams and were stirred by rotation of the crucible. The melting, pouring, and annealing temperatures were 1000°C, 900-800°C, and 520°C, respectively. For annealing, the samples were placed in a furnace for 30-60 minutes; the furnace was then turned off and allowed to cool to room temperature.

The 40% Li_2O borate glass, although prone to crystallization, could be cast in 6-mm-thick specimens without difficulty. Lithium borate glasses containing 1 mol.% rare-earth oxide with 45 mol.% Li_2O could be made in thin sections by rapid cooling between cold plates; glasses with >50 mol.% Li_2O could not be made. When no Li_2O was present, the rare-earth oxide did not go into solution.

Sample sizes were approximately 5 x 5 x 10 mm and polished for spectroscopic studies. Both glasses were clear, but the 10% Li_2O sample had striae and was of poorer optical quality.

EXPERIMENTAL

The glass samples were mounted in a dewar and maintained at 1.7°K by immersion in liquid helium pumped below its lambda point. Absorption measurements were made with a tungsten filament lamp as the light source. For fluorescence studies, the output of a pulsed nitrogen-laser-pumped dye laser was focused onto the sample. Rhodamine 6G and Coumarin 102 dyes were used for excitation of the $^5D_0$ and $^5D_2$ states of Eu^{3+} respectively. The dye laser was tuned with a collimator and a grating, and operated at a repetition rate of 5-20 Hz. The output pulse was about 7 ns in duration and had a spectral width of less than 1 cm^{-1}.

Fluorescence from the sample was analyzed with a 0.85-m double grating monochromator and detected with a cooled EMI 9658R photomultiplier tube having an extended S-20 response. All spectra were recorded with a resolution varying between 5 and 7 cm^{-1}. Individual photon pulses from the photomultiplier tube were shaped into a standard form by an amplifier/discriminator and subsequently counted by gated digital electronics. Spectral measurements were made with one-channel pulse counter which had a variable gate width and a variable gate delay. Lifetime measurements were made with a multichannel counter which sampled the fluorescence decay at 120 consecutive points after excitation. Data from the counters were outputted to a chart recorder.

RESULTS

To reduce the homogeneous linewidth, all spectra were recorded at 2 K. Figure 1 shows the absorption spectra of the $^7F_0 \rightarrow {}^5D_0$ and

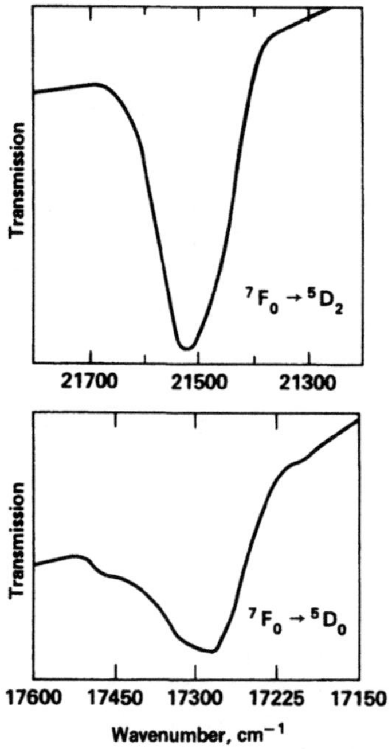

Fig. 1. Absorption bands of Eu^{3+} in 40 Li_2O·60 B_2O_3 glass at 2 K.

To reduce the homogeneous linewidth, all spectra were recorded at 2 K. Figure 1 shows the absorption spectra of the $^7F_0 \rightarrow {}^5D_0$ and $^7F_0 \rightarrow {}^5D_2$ transitions used for excitation. Since the homogeneous width of the $^7F_0 \rightarrow {}^5D_0$ transition is <<1 cm^{-1}, the observed width is a measure of the inhomogeneous broadening. Because of the inhomogeneous broadening, the five-line Stark structure of the $^5D_2$ state is not resolved. The absorption bands of the 10 and 40% Li_2O glasses were very similar in shape, but small differences in the wings of the lines could be undetected because of the small signal-to-noise ratio. The $^7F_0 \rightarrow {}^5D_0$ intensity was less in the 10% Li_2O sample.

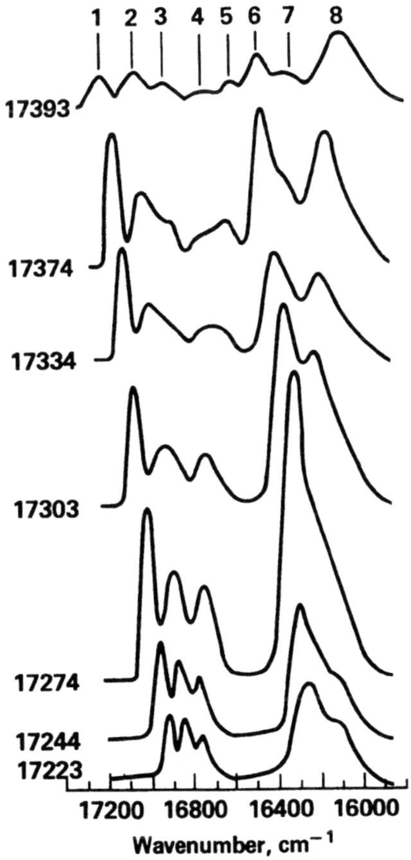

Fig. 2. Fluorescence spectrum of Eu^{3+} in 10 Li_2O • 90 B_2O_3 glass at 2 K as a function of $^7F_0 - {}^5D_0$ excitation energy (cm^{-1}).

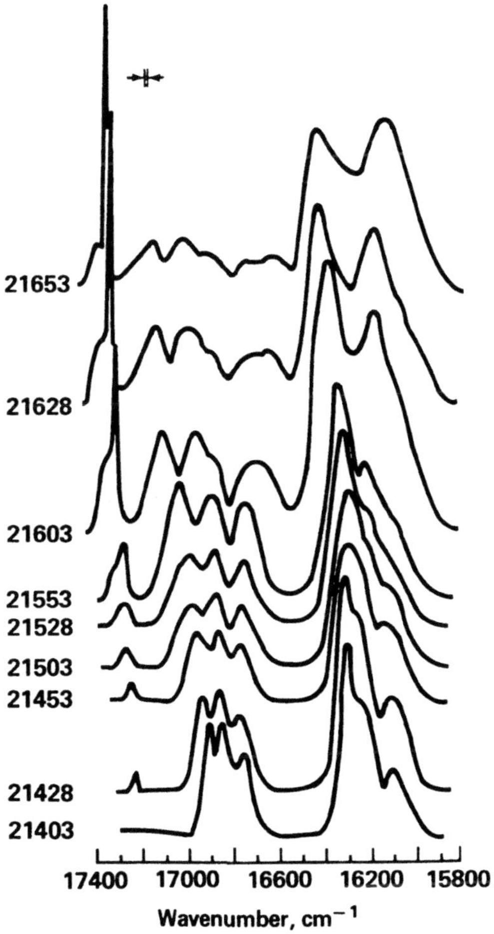

Fig. 3. Fluorescence spectrum of Eu^{3+} in 10 Li_2O • 90 B_2O_3 glass at 2 K as a function of $^7F_0 - {}^5D_2$ excitation energy (cm^{-1}).

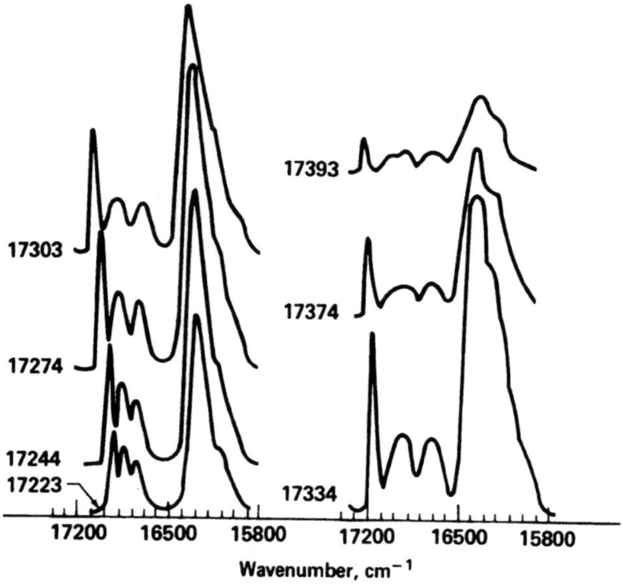

Fig. 4. Fluorescence spectrum of Eu^{3+} in $40\,Li_2O \cdot 60\,B_2O_3$ glass at 2 K as a function of $^7F_0 - {}^5D_0$ excitation energy (cm^{-1}).

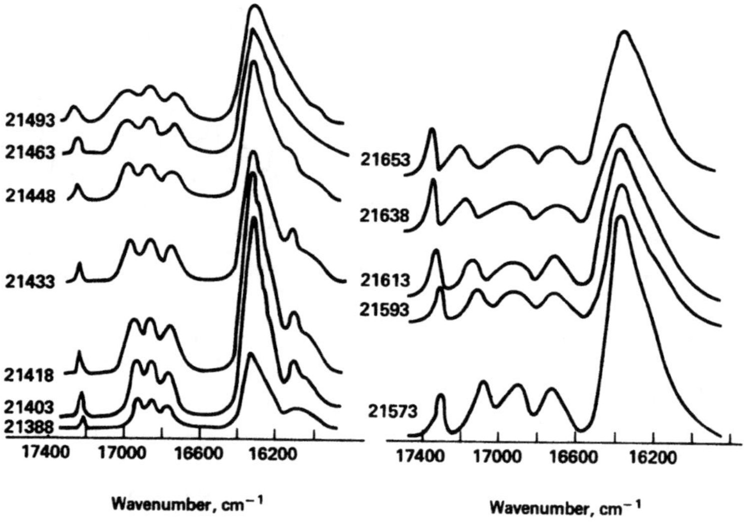

Fig. 5. Fluorescence spectrum of Eu^{3+} in $40\,Li_2O \cdot 60\,B_2O_3$ glass at 2 K as a function of $^7F_0 - {}^5D_2$ excitation energy (cm^{-1}).

Fluorescence was observed from $^5D_0$ to levels of the ^7F multiplet. The intensities to the $^7F_3$, $^7F_5$, and $^7F_6$ states were weak. The most pronounced variations in fluorescence line structure occur for the $^7F_1$ and $^7F_2$ states. These spectra for the 10% and 40% Li_2O samples and $^5D_0$ excitation are shown in Figs 2 and 4. The laser excitation energy (cm^{-1}) is given at the left of each spectrum. The spectra are not normalized to constant excitation. The three lines of the $^5D_0 \rightarrow {^7F_1}$ spectrum occur in the 16600-17200 cm^{-1} region; the five $^5D_0 \rightarrow {^7F_2}$ transitions occur in the 15800-16500 region but are not well resolved.

The resonant $^5D_0 \rightarrow {^7F_0}$ fluorescence was not recorded for $^5D_0$ excitation because of the presence of scattered laser excitation. The $^5D_0 \rightarrow {^7F_0}$ transition in the region 17200-17400 cm^{-1} is observed for nonresonant excitation into $^5D_2$. Ions excited into $^5D_2$ rapidly decay nonradiatively to $^5D_0$ from which fluorescence is observed to levels of the ^7F multiplet. These spectra for the 10% and 40% Li_2O samples are shown in Figs. 3 and 5.

The spectra shown are time-resolved spectra observed with a 100-μs gate immediately following the excitation pulse. Gate widths were varied from 40 to 400 μs without any noticeable change in the spectrum, hence cross-relaxation between ions in different sites on this time scale and at these Eu^{3+} concentrations should be negligible. This is consistent with findings in Ref. 4.

The lifetimes of the $^5D_0$ fluorescence measured as a function of excitation energy are shown in Fig. 6. The fluorescence decays generally exhibited an exponential time dependence over the first two e-folding times (~4 ms) followed by a slower decay tail. As seen in Figs. 2 and 4, for the higher excitation energies, additional lines appear. Referring to the numbered lines on the spectrum at the top of Fig. 2, there are two groups of lines characterized by different lifetimes. The first group consists of lines 1, 2, 5, 6, and 8 and consistently has a shorter lifetime than the second group which consists of lines 3, 4, and 7. These lifetimes are denoted by the circle and triangle data points in Fig. 6. The two main $^5D_0 \rightarrow {^7F_2}$ lines have similar lifetimes.

The fluorescence branching ratios were also examined as a function of excitation energy. The results, corrected for the spectral response of the detection system, are shown in Fig. 7. Fluctuations in the laser intensity made it difficult to achieve a high degree of accuracy for intensity comparison purposes.

DISCUSSION

The laser-excited fluorescence spectra in Figs. 2-5 show some line narrowing, but the widths are still greater than either the

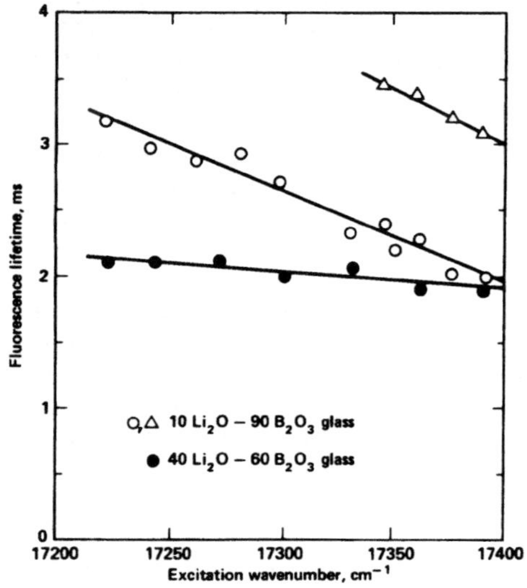

Fig. 6. $^5D_0$ fluorescence lifetime for lithium borate glasses at 2 K as a function of $^7F_0 - {}^5D_0$ excitation energy.

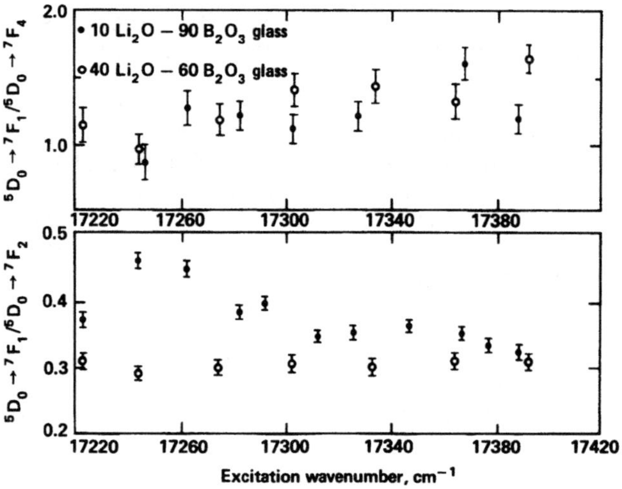

Fig. 7. Fluorescence branching ratios for Eu^{3+} in lithium borate glasses at 2 K as a function of $^7F_0 - {}^5D_0$ excitation energy.

homogeneous or instrumentation widths. This is because the spectra were recorded with nonresonant excitation. In this case, if excitation levels of ions in different sites accidentally coincide, some residual inhomogeneous broadening remains.[3,11] Nevertheless, changes in the spectra with excitation energy are clearly evident. These spectra correspond to the excitation of selected subsets of Eu^{3+} ions.

The large inhomogeneous width of the $^7F_0 \rightarrow {}^5D_0$ transition can arise from site-to-site differences in (1) the electrostatic interaction of the six 4f electrons which determines the separation of the ^5D and ^7F multiplets, (2) the position of the $^7F_0$ level as affected by crystal-field admixing with other ^7F_J states, or (3) by combinations thereof. (1) and (2) are governed by the Racah and the crystal-field parameters, respectively. Studies in oxide crystals[12] and glasses[2] have shown that the position of the $^5D_0 \rightarrow {}^7F_0$ line of Eu^{3+} is sensitive to the host composition and can vary by tens of cm^{-1}.

Since three- and five-line spectra for $^5D_0 \rightarrow {}^7F_1$ and $^5D_0 \rightarrow {}^7F_2$ transitions are resolved in the line-narrowed spectra, the symmetry at the Eu^{3+} site is very low and sufficient to remove completely the (2J+1)-fold degeneracy of the free-ion states. In silicate glasses, trivalent europium has distorted nine-fold oxygen coordination.[6] If the splittings of the $^7F_1$ and $^7F_2$ manifold in the 40% Li_2O glass are compared to those observed in silicate[6] and phosphate[8,13] glasses, one finds that for these three oxide glasses, the pattern of the $^7F_1$ and $^7F_2$ energy-level splittings and their variations are similar, although not identical. The oxygen ligand coordination and bonding

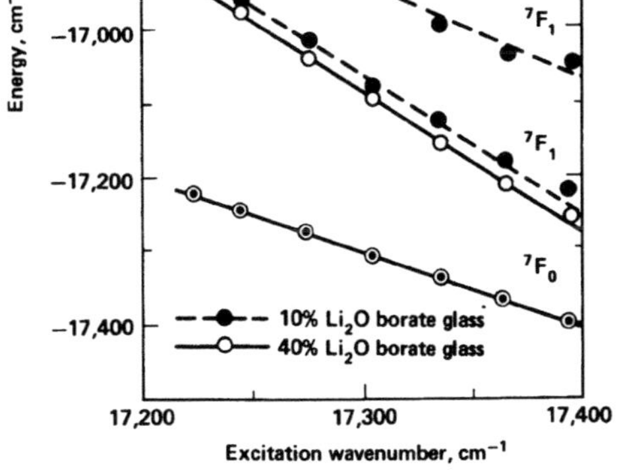

Fig. 8. Positions of the $^5D_0 \rightarrow {}^7F_1$ fluorescence lines as a function of $^7F_0 - {}^5D_0$ excitation energy for Eu^{3+} in two lithium borate glasses at 2 K.

rather than the network-forming or network-modifying cations appear to be predominant in determining the general spectroscopic properties.

The change in the line positions in the FLN spectra with excitation energy arise from differences in the local fields. The energies of the three Stark levels of the $^7F_1$ manifold for the 10% and 40% Li_2O glasses are compared in Fig. 8. The $^5D_0$ level is taken as the zero of energy. As first reported by Motegi and Shionoya,[13] the variation in crystal-field splitting exhibits a monotonic change with $^7F_0 - {}^5D_0$ excitation energy. In the 10% Li_2O sample, Eu^{3+} ions reside in a distribution of sites with fields sufficient to cause splittings ranging from ≈150 to 600 cm^{-1}!

The low-energy excitation portions of the FLN spectra for the 10% and 40% Li_2O glasses are very similar suggesting that they arise from Eu^{3+} ions having approximately equal coordination and bonding. The high-energy excitation portions of the spectra are very different, however. Also, additional lines appear which are attributed to Eu^{3+} ions in sites having a different local environment. These lines are labeled 3, 4, and 7 in Fig. 2 and are also evident in some of the other figures. Their positions are included in Fig. 8.

The fields at Eu^{3+} sites are expected to change as the Li_2O content is increased because boron with three coordinating oxygens is replaced by boron with four coordinating oxygens and by Li^+ network-modifying cations. An increase in the covalency reduces the Racah parameters which, in turn, lowers the $^5D - {}^7F$ energy separation. This is the nephelauxetic effect. Therefore, the low-energy side of the inhomogeneous $^7F_0 - {}^5D_0$ transition corresponds to sites with strong Eu-O interaction and covalency. The converse applies to the high-energy side. This suggests that the high-excitation-energy spectra arise from sites where BO_3 triangles are prevalent.

Since FLN measurements are structure-sensitive, additional lines could also appear if phase separation is present. Metastable immiscibility on a scale of 5-500 nm is known to occur in the lithium-borate system.[10] Lithium-rich and B_2O_3-rich phases are expected to be present. If impurities such as Eu^{3+} enter one or both phases, the excitation energies may be sufficiently different to permit relative probing of local structure in the individual phases. If two phases are present in the 10% Li_2O glass sample, the low-energy FLN spectra suggests that one phase is similar to that of the 40% Li_2O glass. Further FLN studies of phase-separated glasses are warranted to explore this possibility.

Additional spectral lines could also arise from clustering of Eu^{3+} ions or a Eu^{3+}-rich phase. If ion-ion interactions are strong, they can alter the energy levels from those for isolated Eu^{3+} ions. At present, we have no evidence for such clustering. For many rare earths, ion-ion interactions cause concentration-dependent fluorescence quenching. For Eu^{3+}, there are no simple self-quenching processes. At high Eu^{3+} densities, however, nonradiative decay by energy migration to quenching centers is possible.[13,14]

The FLN spectra for $^5D_0$ and $^5D_2$ excitation are similar, but some details and the line broadening are different. As the laser excitation is scanned across the $^5D_2$ manifold, the $^5D_0 \rightarrow {}^7F_0$ line becomes more intense and broader. The increased intensity is associated with increased J-state mixing by which the normally strongly forbidden 0-0 transition becomes allowed. The intensity increase correlates with the increased crystal-field splitting of the $^7F_1$ manifold because both are controlled by the second-order terms in the crystal-field expansion. As the excitation moves to higher energies, higher-lying Stark levels of $^5D_2$ are excited. The larger homogeneous broadening of these levels and increased accidental coincidence account for the increasing $^5D_0 \rightarrow {}^7F_0$ linewidth.

For excitation on the high-energy side of the $^5D_2$ band, a sharp line appears in the region of the $^5D_0 \rightarrow {}^7F_0$ transition. This is clearly seen in the FLN spectra of the 10% Li_2O glass (Fig. 3). The line is resolution-limited and has a width of $< 10\ cm^{-1}$. The lifetime of this component resembles that of the broader background. In addition, the position of the line shifts with the excitation wavelength. If this line is attributed to Eu^{3+}, the narrow width suggests it arises from ions in a more ordered structure.

The radiative quantum efficiency of the Eu^{3+} $^5D_0$ state is unity in most solids, therefore changes in the fluorescence lifetime are attributed to changes in the radiative transition probabilities. Radiative decay from $^5D_0$ involves magnetic-dipole transitions to $^7F_1$ and crystal-field-induced electric-dipole transitions to $^7F_2$, $^7F_4$, and $^7F_6$; transitions to other ^7F states arise from J-state mixing. The spontaneous emission probability for magnetic-dipole transitions is proportional to n^3 and for electric-dipole transitions $n(n^2+2)^2$, where n is the refractive index. If ions occupy glass phases having different refractive indices, the local field correction can cause small differences in the lifetimes.

The fluorescence lifetimes of the 10% and 40% Li_2O glasses decrease with increasing excitation energy. The variation for the 40% Li_2O glass is small; this is qualitatively consistent with the small variations in fluorescence branching ratios in Fig. 7. For all sites examined, the $^5D_0 \rightarrow {}^7F_2 / {}^5D_0 \rightarrow {}^7F_1$ intensity ratio is larger for the glass with the higher Li_2O content, in agreement with the broadband excitation results in Ref. 2. The variation in the $^5D_0$ lifetime with excitation energy for the 10% Li_2O glass is larger; changes in the $^7F_2$ fluorescence branching ratio are also larger. Similar behavior was reported for Eu^{3+} in silicate glass.[6] The lifetime of the 10% Li_2O glass for most excitation energies is much longer than for the 40% Li_2O glass. This indicates a different symmetry and/or strength of the local field. The radiative transition probabilities are larger for sites with larger overall $^7F_1$ splitting, however, the components of the crystal field which affect transition probabilities are not the same as those which affect the energy level splitting. The new lines appearing in the 10% Li_2O glass at high excitation energy have the longest lifetimes and arise from ions in still different local environments.

In conclusion, the present FLN experiments with Eu^{3+} demonstrate that structural variations associated with compositional changes in borate glasses are detectable. The nature of the modifications at the rare-earth sites are not known, however. To develop geometric models for the local coordination and bonding, detailed investigations of laser-excited fluorescence spectra from alkali borate glasses in which both the alkali ion (Li^+, . . . Cs^+) and the alkali ion concentration are varied would be helpful. These results would provide a test of proposed models for the local fields.

REFERENCES

1. D. L. Griscom, "Borate Glass Structure", pp. 11-139, this volume.
2. P. K. Gallagher, C. R. Kurkjian, and P. M. Bridenbaugh, Phys. Chem. Glasses 6, 95 (1965).
3. T. Kushida and E. Takushi, Phys. Rev. B12, 824, (1975).
4. Yu. V. Denisov and V. A. Kizel, Optics and Spectroscopy 23, 251 (1967).
5. L. A. Riseberg, Phys. Rev. Lett. 28, 789 (1972); Phys. Rev. A7, 671 (1973).
6. C. Brecher and L. A. Riseberg, Phys. Rev. B13, 81 (1976).
7. For a review of fluorescence line narrowing studies in glasses and references, see M. J. Weber, Colloque de Lyon, J. Phys. (Paris), in press.
8. C. Brecher, L. A. Riseberg, and M. J. Weber, Proc. 12th Rare Earth Research Conference, Vol. I, 351 (1976).
9. P. J. Bray and J. G. O'Keefe, Phys. Chem. Glasses 4, 37 (1963).
10. R. R. Shaw and D. R. Uhlmann, J. Amer. Ceram. Soc. 51, 377 (1968).
11. M. J. Weber, J. A. Paisner, S. S. Sussmann, W. M. Yen, L. A. Riseberg, and C. Brecher, J. Luminescence 12, 729 (1976).
12. O. J. Sovers, M. Ogawa, and T. Yoshioka, Proc. 12th Rare Earth Research Conference, Vol. II, 728 (1976).
13. N. Motegi and S. Shionoya, J. Luminescence 8, 1 (1973).
14. M. J. Weber, Phys. Rev. B4, 2932 (1971).

MÖSSBAUER INVESTIGATION OF THE INCORPORATION OF TIN AND IRON IN SODIUM BORATE GLASSES

Henning Dannheim and Thomas Frey
Institut für Werkstoffwissenschaften
University Erlangen - Nürnberg, Germany

1. INTRODUCTION

The main intention of this investigation is to study the chemical incorporation of tin and iron as ions in borate glasses and to discuss the results with respect to the structural models of glasses.
In any consideration of glass structure the valency of the incorporated ions is important. If one melts borate glass with tin and iron, all valence states may exist, e.g. Sn^{o}, Sn^{2+}, and Sn^{4+}. In this investigation there had to be limitations on the number of Sn^{2+}, Sn^{4+}, Fe^{2+}, and Fe^{3+} ions, because the metallic species appear only when a highly reducing atmosphere is used.
If different valence states exist, one must also consider the valence ratio. Therefore the furnace atmospheres were chosen, so as to obtain the desired valence ratios, but the influence of furnace atmosphere on the valence ratios was not investigated as such.
A further important point concerns the crystallographic surroundings of the ions, i.e. the co-ordination numbers. In considering the question of the co-ordination numbers of the Sn^{2+} and Sn^{4+} ions and the Fe^{2+} and Fe^{3+} ions in borate glasses, one can look at the field strengths as defined in the Dietzel theory [1].
According to this theory, the oxides of tin and iron are intermediate oxides, i.e. they are able to be network formers or network modifiers, depending on the composition of the glass. If they change their function, there will be a corresponding change of the co-ordination number from 4 (for network formers) to 6 (for network modifiers). The last important point is the influence of the change of co-ordination of the boron itself. The question is whether one can observe the transition of the boron from the co-ordination number 3 to the

co-ordination number 4 with the help of the incorporated ions: For this procedure to be succesful, the ions have to act as a non-participating probe which registers changes in the surrounding network.

2. EXPERIMENTAL

2. 1 Method

The Mössbauer effect is a good method for studying the chemical incorporation of tin and iron in borate glasses. The effect and its application to glass is described in detail in several publications [2],[3],[4],[5],[6]. The single line in the MB spectrum is of Lorentz shape. A computer program fits the experimental results with Lorentz lines and calculates the following parameters: Isomer shift δ, quadrupole splitting Q, and line width. Then the areas under the separated lines are integrated and divided; this simple procedure provides the valence ratio. Fig. 1 shows a generalized Mössbauer spectrum of tin in glasses. All three valence states are represented on this diagram. The metallic tin,

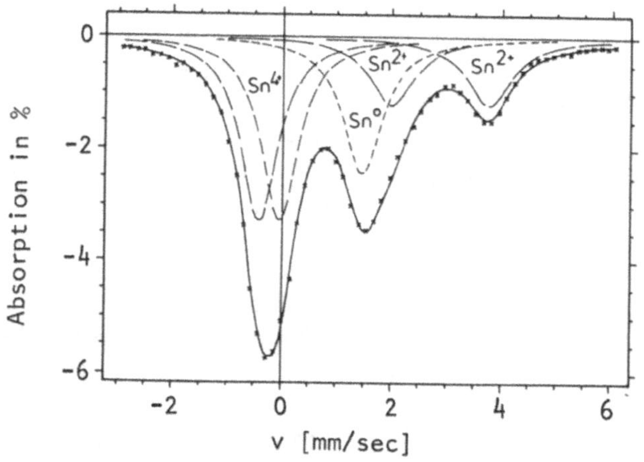

Fig. 1 Generalized Mössbauer spectrum of tin in glasses

having a single line with an isomer shift of about 1 mm/sec, is not of interest in this investigation. The Sn^{4+} has a doublet with a small quadrupole splitting an an isomer shift of about zero. The Sn^{2+} also has a double line with a greater quadrupole splitting and an isomer shift of about 2.5 mm/sec. Both doublets are easy to separate.

Fig. 2 shows the generalized spectrum of iron in borate glasses. In almost all glasses both valence states of +2 and +3 occur; however, the separation is difficult, because two lines of both doublets fall together.

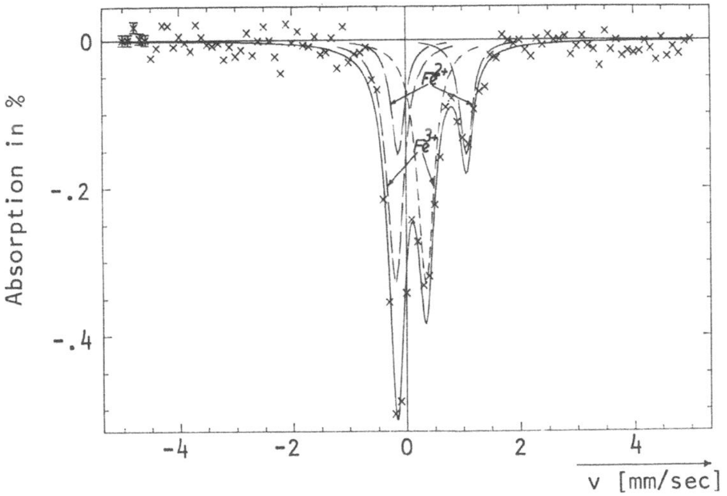

Fig. 2 Mössbauer spectrum of iron in borate glasses

2. 2 Selection and preparation of the glasses

To get the best results, the following three demands must be met:

1. The parameters should be kept to a minimum, i.e. there should be only 3 components in the glass.
2. The whole sample should be vitreous. Even the beginning of crystallisation should be avoided.
3. The variation of the amount of sodium and the variation of the valence ratio should be as great as possible in order to obtain the change of co-ordination of the tin and iron ions as well as the change in co-ordination of the boron itself.

Table 1 shows the selected compositions of tin-borate glasses. In group A, the amount of tin increases. The solubility of tin and iron oxides in B_2O_3 glasses without alkali is very small. Increasing the amount of sodium provides an increasing solubility of the oxides. In series A, with 20 mole % Na_2O, the tin dioxides is up to 7mole% soluble.

The variation from SnO_2 to SnO in group B changes the atmosphere during the melting process and gives, therefore, another valence ratio. Additionaly, the melting atmosphere was altered by using nitrogen instead of air.

Series C with a constant amount of SnO_2 and variable sodium content should be able to show a possible co-ordination change of the boron itself. The regions of glass forming as published by Rawson [7] lie between 100 and 62 mole% and between 33.5 and 28.5 mole% B_2O_3.

Trough the variation of sodium and tin in group D, a change in co-ordination

Table 1
Composition of the tin borate glasses in mole%

	B_2O_3	Na_2O	SnO	SnO_2
A1	79	20		1
A2	77	20		3
A3	75	20		5
A	73	20		7
B1	79	20	1	
B2	77	20	3	
B3	75	20	5	
C1	89	10		1
C2	79	20		1
C3	69	30		1
C4	59	40		1
C5	34	65		1
C6	29	70		1
D1	99	0	1	
D2	93	5	2	
D3	88	10	2	
D4	79	12	9	
D5	79	15	6	
D6	79	19	2	
D7	68	30	2	
D8	58	40	2	

number of the Sn^{2+} and Sn^{4+} ions should be achieved.
Table 2 gives the analogous compositions for the sodium-iron-borate glasses. It should be noted that the concentration of Fe_2O_3 in a glass must be at least 3 mole%, because the natural abundance of the required Mössbauer isotope Fe^{57} is only 2 %. All glasses were melted for an hour in an electric furnace at 1300°C. The melt was poured onto a metal disk to obtain rapid quenching. The glass was then ground, and the powder packed in a plastic form and put into the Mössbauer apparatus. The thickness of the samples was in all cases 0.5 mm.
To test the glassy state of all samples, the following three methods were applied:

a) Polarization microscopy
b) X-ray measurements
c) Appearance of the six-line pattern in the Mössbauer spectrum only for iron glasses

Table 2
Composition of the iron borate glasses in mole %

	B_2O_3	Na_2O	Fe_2O_3
A1	69	27	4
A2	67	27	6
A3	65	27	8
A4	63	27	10
A5	61	27	12
A6	57	27	16
A7	53	27	20
B1	84	10	6
B2	74	20	6
B3	64	30	6
B4	54	40	6
B5	26,5	67,5	6

3. RESULTS

In Fig. 3 the attainable variations in the valence ratio of the tin in borate glasses are shown. On the left side of every individual figure, the ratio of Sn^{2+} to Sn total is noted. The results of this atmospheric-dependent investigation are as follows:
If the borate glasses containing SnO_2 are melted in a normal atmosphere (i.e. air), only the valence state +4 appears (series A and C). The glasses of the series B and D made with SnO contain both valencies, the valence ratio depending on the atmosphere. To get the highest fraction of +2, one needs only to melt in a nitrogen atmosphere. If a mixture of nitrogen and hydrogen is used, the atmosphere becomes too reducing, and metallic tin is separated.

Fig. 4 shows the results for iron in borate glasses in a similar way. It was found that the atmosphere for the iron case must be stronger. To get only the +3 state, the melt had to be performed in an oxygen atmosphere. To achieve only the Fe^{2+}, it was necessary to melt in a very reducing atmosphere which was provided by a mixture of 80% nitrogen and 20 % hydrogen.
The investigated ions belong to the group of intermediate oxides. They can act as network formers or network modofiers, depending on the glass composition.

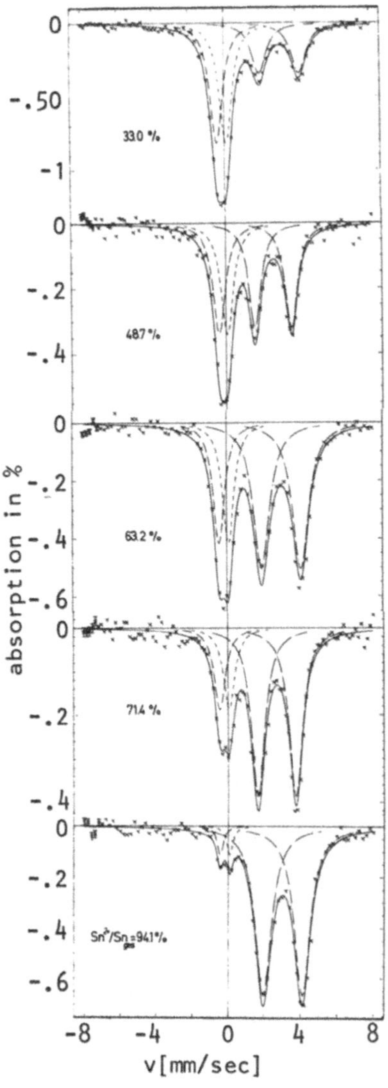

Fig. 3
Tin borate glasses with different valence ratios Sn^{2+}/Sn_{total}

Table 3 shows the co-ordination number and the field strength according to Dietzel for the network formers, modifiers, and the intermediate oxides in general, and for the investigated ions. One can imagine the change of co-ordination in the following way:

In the presence of sufficient modifiers in form of sodium the tin and iron ions occupy the places of the network former. When there is a lack of alkali ions, the iron ions have to take places of the modifier and have a co-ordination number 6, because the ratio between the formes and the modifiers seems to be constant in these glasses.

It was shown by Walker, Wertheim, and Jaccarino [8] that the change of co-ordination can be observed through use of the isomer shift from the Mössbauer investigation. An increase in the co-ordination number is correlated with an increase in the isomer shift. To notice the change, the isomer shift must be observed as a function of the amount of the network modifier, here as a function of sodium.

Based upon the last considerations, the next figures show the isomer shift δ of the Fe^{2+} and Fe^{3+} doublets, as well as the Sn^{2+} doublet, of several glasses as a function of the reciprocal amount of Na^+.

In fact, one can observe an evident change of the isomer shift. The amount of this change for the various ions is as follows: Sn^{2+} about 1 mm/sec, Fe^{2+} about 3 mm/sec, and Fe^{3+} about 1 mm/sec. On the other hand, the Sn^{4+} ions undergoes no such change.

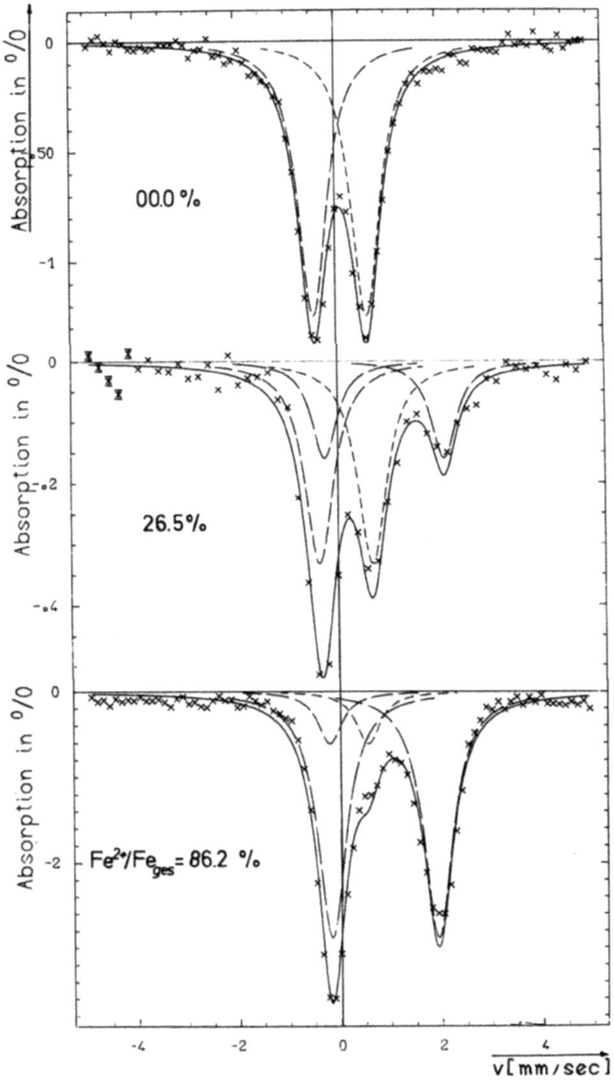

Fig. 4
Iron borate glasses with different valence ratios Fe^{2+}/Fe_{total}

Table 3
Field strength according to Dietzel for the different ions

	Co-ordination number	Field strength
Network formers	3 or 4	1.4 - 2.0
Intermediate oxides	4 or 6	0.5 - 1.0
Network modifiers	6 or 8	0.1 - 0.4
B^{3+}	4 or 3	1.45 or 1.63
Sn^{4+}	4 or 6	1.13 or 1.01
Sn^{2+}	4 or 6	0.46 or 0.41
Fe^{3+}	4 or 6	1.02 or 0.91
Fe^{2+}	4 or 6	0.59 or 0.52
Na^{+}	6	0.19

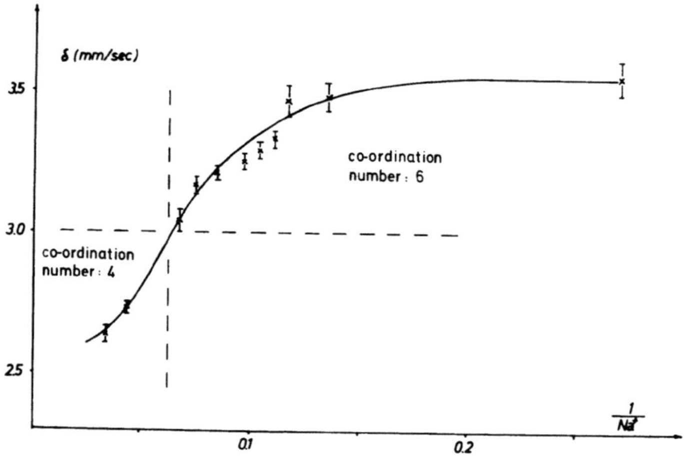

Fig. 5
Isomer shift δ of the Sn^{2+} doublet as a function of the reciprocal Na^{+} content

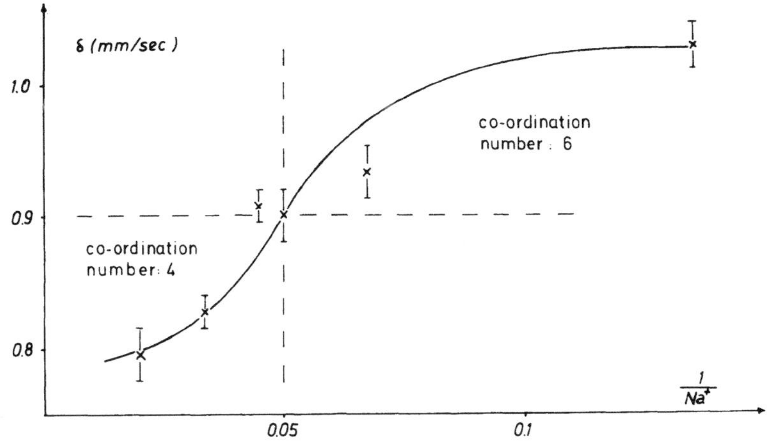

Fig. 6
Isomer shift δ of the Fe^{2+} doublet as a function of the reciprocal Na^+ content

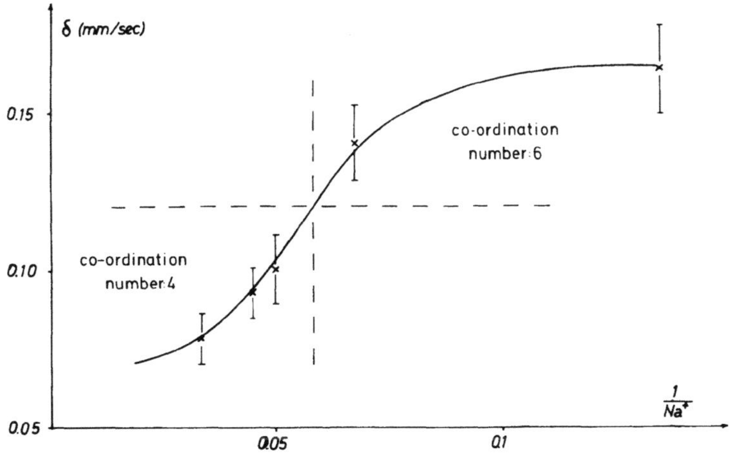

Fig. 7
Isomer shift δ of the Fe^{3+} doublet as a function of the reciprocal Na^+ content

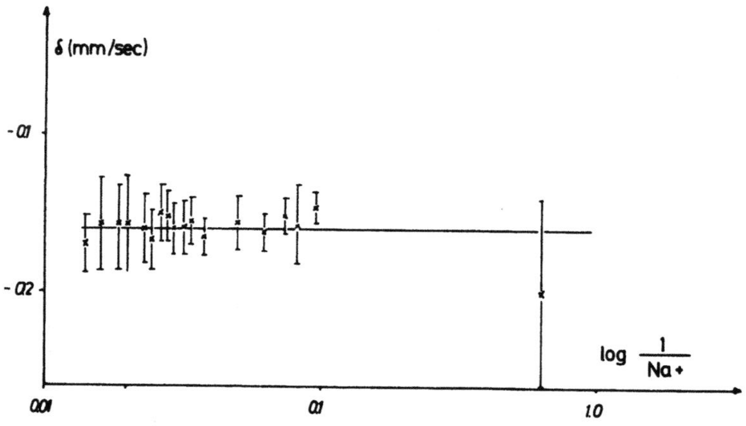

Fig. 8
Isomer shift δ of the Sn^{4+} doublet as a function of the reciprocal Na^+ content

In Fig. 8 one can see the isomer shift of the Sn^{4+} doublet as a function of the logarithmic reciprocal sodium content.

The logarithmic scale was chosen in order to make the diagram easily readable. The isomer shift seems to be constant. If the scale of the y axis is spread one can see a slight variation of about .05 mm/sec, but the curve has another shape, than that seen with the three other ions.

Fig. 9 shows the isomer shift of the Sn^{4+} doublet as a function of the sodium content. This method of presentation allows a comparison to be made with the upper part of the figure, where the co-ordination number of boron is shown as measured by NMR [9]. The isomer shift of the Sn^{4+} varies in about the same way as the boron co-ordination. This variation of the isomer shift can not be a change of the Sn^{4+} itself, because there is no step like that seen with the three other ions, and the value is too small.

4. DISCUSSION AND COMPARISON WITH SILICATE GLASSES

The observation of the co-ordination change of the boron is possible only through the Sn^{4+} ion, because this ion undergoes no co-ordination change itself. The variation of the isomer shift caused by the change in co-ordination of the boron is small compared with the variation of the isomer shift caused by the change in co-ordination of the three MB ions. Only the Sn^{4+} can act as a non-participating probe to study the co-ordination change of the boron.

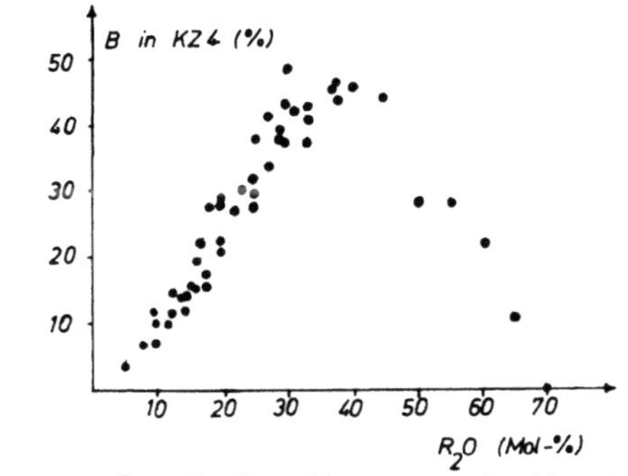

Co-ordination of boron as a function of alkali content measured by NMR

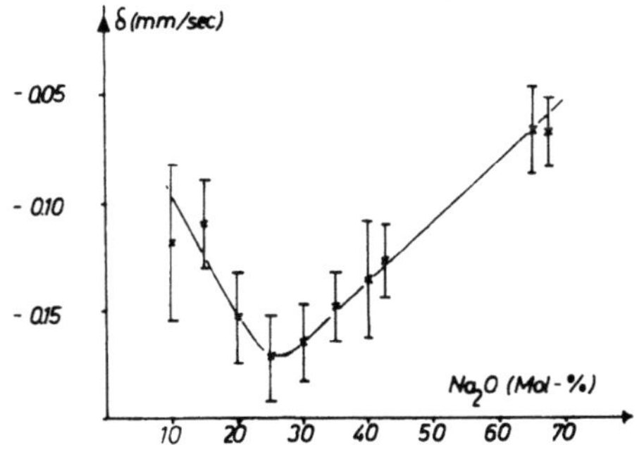

Fig. 9
Isomer shift δ of the Sn^{4+} doublet as a function of the Na_2O content

On the other hand, in silicate glasses the Sn^{4+} ion also makes a co-ordination change from 4 to 6 [6].

Fig. 10 shows the isomer shift of a Sn^{4+} doublet in silicate glasses as a function of reciprocal amounts of modifiers. Notice that in silicate glasses it is not enough to consider only the amount of sodium. All other possible modifiers, here the Sn^{2+}, must be regarded. Because it is a doubly charged modifier, it has the factor 1/2. The variation of the isomer shift is .25 mm/sec. The explanation for the change of co-ordination of the Sn^{4+} ion in silicate but not in borate glasses seems to be connected with the field strength of the network former. The strong Sn^{4+} ion is able to remove boron from tetrahedral co-ordination sites, but not silica, which is a network former with a high field strength. Because the amount of Sn^{4+} in the glass is low, the Sn^{4+} always finds enough BO_4 groups to displace the boron.

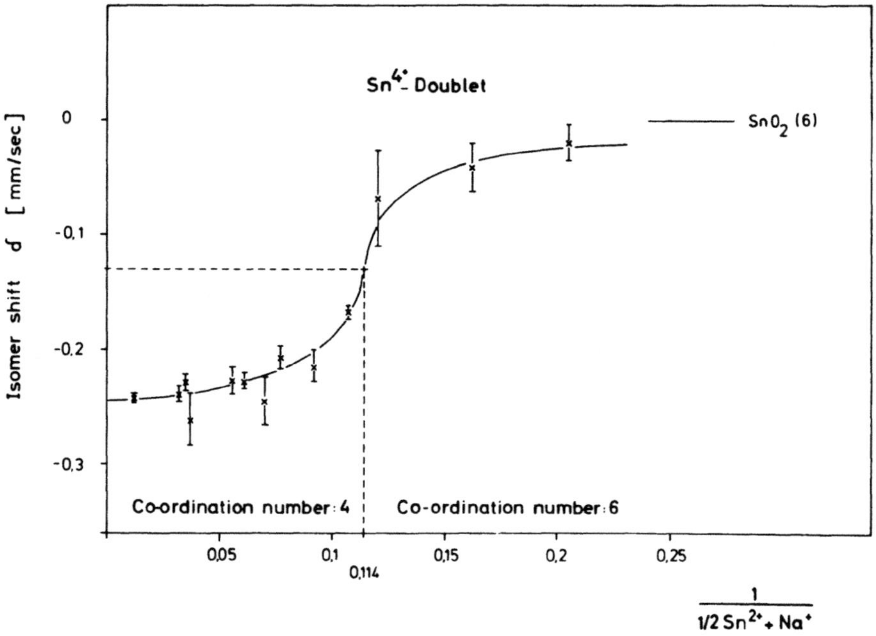

Fig 10
Change of co-ordination of the Sn^{4+} ion in silicate glasses

REFERENCES

[1] A. Dietzel
Glastechn. Ber. 22 (1948/49), 41-50, 81-86, 212-224

[2] G. K. Wertheim
Academic Press: New York 1964

[3] V. I. Goldanskij, R. H. Herber
Academic Press, New York 1968

[4] C.R. Kurkjian
J. Non-Cryst. Sol. 3 (1970), 157-194

[5] G. Tomandl, G. H. Frischat, H. J. Oel
Glastechn. Ber. 40 (1967), 293-298

[6] H. Dannheim, H. J. Oel, G. Tomandl
Glastechn. Ber. 49 (1976), 170-175

[7] Rawson
Inorganic Glass Forming Systems, 98-99

[8] L.R. Walker, G. K. Wertheim. V. Jaccarino
Phys. Rev. Letters 6 (1961), 98-101

[9] P.J. Bray, J. G. O'Keefe
Phys. Chem. Glasses 4 (1963), 37-46

INDUCED SILVER CENTERS IN ALKALI BORATE GLASSES*

A. Bishay, E. Boulos, S. Arafa, F. Assabghy
The American University in Cairo

and

N. J. Kreidl
The University of Missouri at Rolla

INTRODUCTION

Radiation induced color centers in borate glasses have been identified and related to the base glass structure in numerous investigations. Among these centers are those associated with the boron-oxygen network and alkali [1,2,3], with lead [4], cerium [5] and with silver [6,7]. In the present paper, the indentification of silver centers was improved and extended by the use of the single silver isotope ^{107}Ag. While the use of a single isotope is obviously limited, these identifications permit more reliable assignments in borate glasses containing normal silver.

In silver activated oxide glasses, the silver ion is found to play the same role as conventional alkali modifiers in the glass network [8]. On exposure to ionizing radiation, silver may act as a trap for both electrons and holes. The resulting centers are paramagnetic and well suited for ESR and optical studies.

Extensive ESR and optical studies have been made on various glass systems activated by small amounts of Ag. The ESR spectra of centers associated with one Ag^+ trapping an electron have been observed in silicate glasses [9], pyrex [10], phosphate [11,12,13] and borate [9,10] glasses.

* Research sponsored by the U.S. National Science Foundation under Grant GF 3460X.

In a previous publication[14], the authors have examined in detail these atomic electron trap centers associated with one Ag^+, termed Ag^0 and $(Ag^+ + e)$. The ESR spectrum for these centers typically consists of two pairs of hyperfine lines corresponding to the two isotopes of silver, ^{107}Ag and ^{109}Ag, whose nuclear spin is $I = \frac{1}{2}$. The authors have also shown that under suitable conditions these lines coexist with a pair of relatively broader hyperfine lines. The latter were attributed to a center comprising two silver ions having trapped one electron, the so called molecular center Ag_2^+.

The authors have examined the relation between the atomic and molecular centers and their dependence on the glass structure. The formation of these centers is critically dependent on the silver and alkali contents, on the temperature of irradiation and detection, and on subsequent bleaching. Low alkali (< 2 mole %) and low silver (< 1 mole %) glasses favour the formation of the Ag^0 center whereas higher alkali and silver contents favour the creation of the Ag_2^+ center. Low silver, high alkali glasses show a broad resonance different from that of the molecular center which is absent. On the other hand, in the case of high silver, low alkali both resonances may appear together. Heat treatment and bleaching studies have indicated that the atomic centers are less stable then the molecular centers and the former transform to the latter on heat treatment.

The broadness associated with the spectrum of the molecular center complicates the problem of interpreting its features. A convincing attempt at interpreting this spectrum was carried out by Mel'nkov et al[15], who suggested that the broad spectrum is a superposition of three spectra arising from the following possibilities of sharing the paramagnetic electron between two silver isotopes: $(^{107}Ag, {}^{107}Ag)^+$; $(^{107}Ag, {}^{109}Ag)^+$; $(^{109}Ag, {}^{109}Ag)^+$.

As a means of simplifying this problem, in this work, silver is introduced into the glass as the single isotope ^{107}Ag. The resulting ESR spectra associated with atomic and molecular silver in alkali borate glasses are examined in detail and correlated with the glass structure.

EXPERIMENTAL

The sample preparation of all but four glasses, x-irradiation and ESR measurements are described in previous publications[7, 14].

Four glasses were prepared in which Ag_2O was introduced in the form of $^{107}AgNO_3$ precipitated from a solution obtained by dissolving ^{107}Ag in nitric acid in a platinum crucible. The other components were introduced in the usual form (alkali carbonate, boric acid). The batch was then thoroughly mixed, melted and annealed as described elsewhere.

Chemical analysis of the glasses formed was not conducted and the compositions quoted are nominal.

RESULTS AND DISCUSSION

Figure 1 shows the induced ESR spectra for glasses of the molar composition $xAg_2O \cdot (10-x)\ Na_2O \cdot 90\ B_2O_3$ in which both isotopes were present.

Spectrum (a), for base glass ($Na_2O \cdot 9\ B_2O_3$) containing no silver, shows the well known five lines plus shoulder spectrum characteristic of low alkali borate glasses and attributed to boron hole centers [16, 17]. On introducing 0.1 mole % Ag_2O, the ESR lines characteristic of Ag^0 are clearly indicated (spectrum b). The silver spectrum is explained on the basis of the presence of a mixture of two isotopes namely ^{107}Ag and ^{109}Ag, of nearly equal concentration[15]. The resonance associated with each isotope center consists of two hyperfine lines due to the nuclear spin for silver $I=\frac{1}{2}$. On increasing the silver content to 1.5 mole % (spectrum c) an additional resonance to the high field side of the Ag^0 lines overlaps with the one due to ^{109}Ag, and had been attributed to a molecular center Ag_2^+ [15].

In Figure 2, the induced ESR resonance in glasses of the molar composition $xAg_2O \cdot (33-x)\ Na_2O \cdot 67\ B_2O_3$ is shown for x=0.1 and x=1.5. It is observed that the isotopic resolution is less pronounced in spectrum (a) (x=0.1) than that in spectrum (b) of Figure 1 (lower alkali, x=0.1). In addition, the average hyperfine splitting constant A has decreased from 612 gauss in the low alkali glass (Figure 1b) to 584 gauss in the high alkali glass (Figure 2a). On increasing the silver content to 1.5 mole % (Figure 2b) no isotopic resolution could be observed. The broad resonance lines are those associated with the molecular center described by Mel'nkov et al [15] as being the superposition of three underlying resonances. Furthermore, a strong resonance is clearly indicated to the low field side of the complex borate resonance. This strong

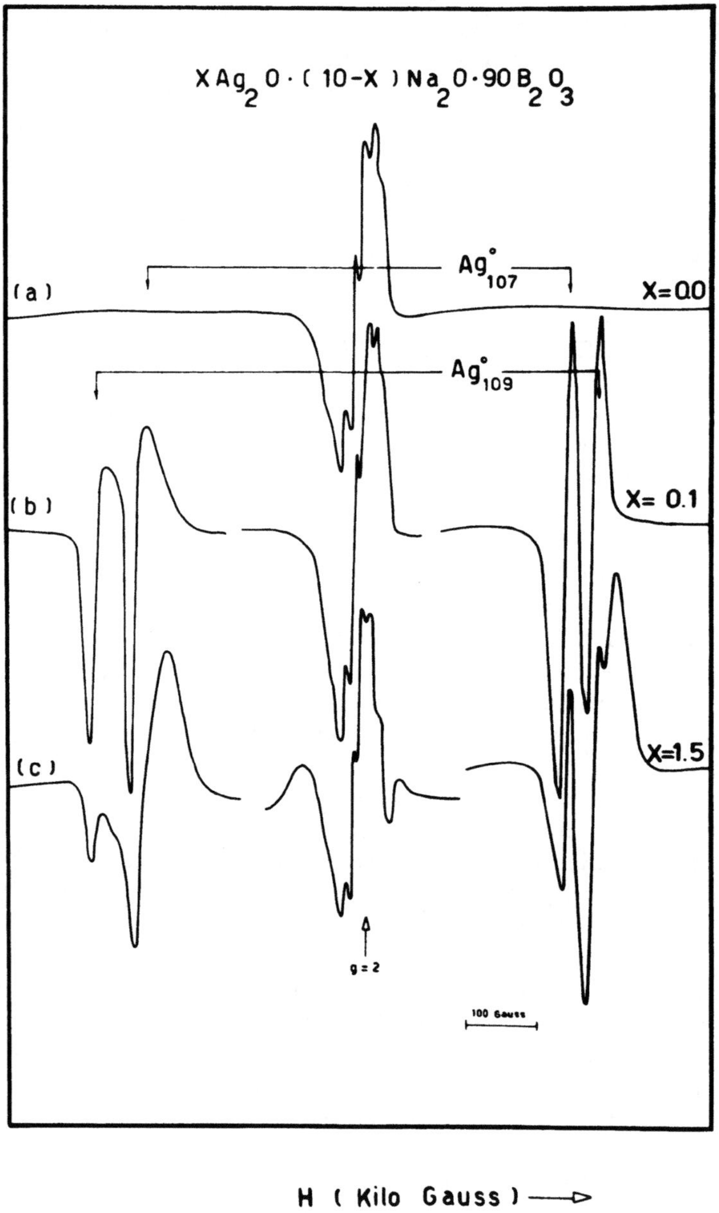

Figure 1. Induced ESR Spectra for x-irradiated glasses of the molar composition: $xAg_2O \cdot (10-x) Na_2O \cdot 90 B_2O_3$ where x = 0.0, 0.1 and 1.5 mole %.

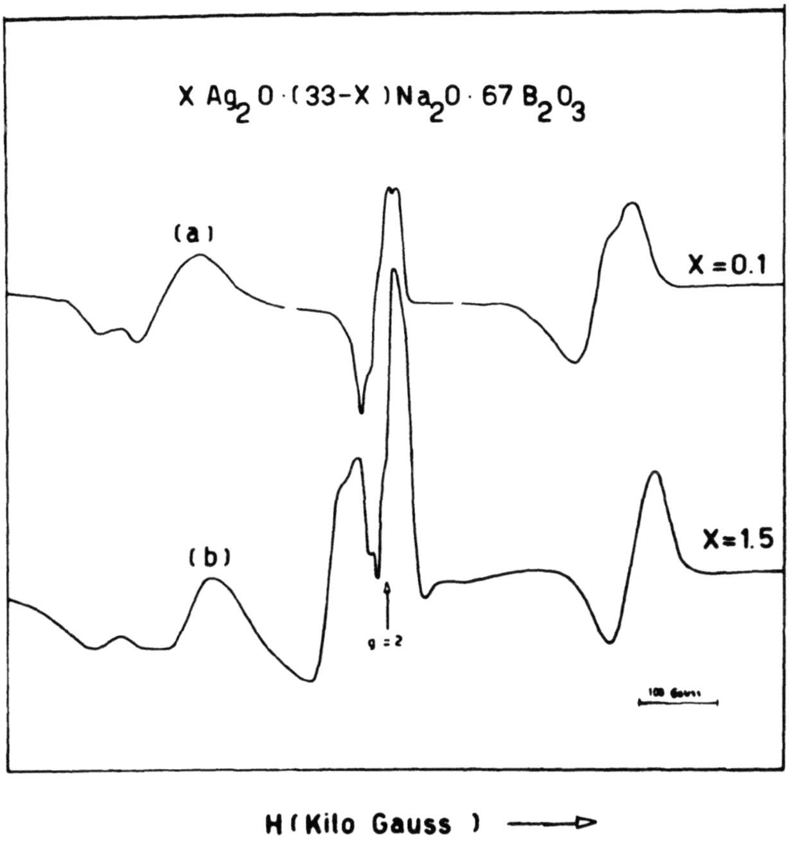

Figure 2. Induced ESR Spectra for x-irradiated glasses of the molar composition: $xAg_2O \cdot (33-x)\ Na_2O \cdot B_2O_3$ where x = 0.1 and 1.5 mole %.

resonance had been attributed earlier [6] to hole centers associated with silver i.e. Ag^{++}. It should be noted that part of the Ag^{++} resonance overlaps with the low field line of the electron trap center. Accordingly, the high field line of the electron trap center will be taken as representative of that and other electron trap centers in the discussion. In general, the high field lines are similar to the low field lines, but are sharper.

Since the observed spectra have been attributed to a mixture of isotopes which leads to complex resonance curves, four potassium borate glasses containing the single isotope ^{107}Ag were prepared. Potassium was used instead of sodium since a better resolution can be obtained in otherwise similar spectra. The following figures present the ESR results of irradiation of these glasses.

Figure 3a shows the induced ESR spectrum for the lower (10%) alkali glass containing 0.1 mole % $^{107}Ag_2O$. The hyperfine lines associated with the atomic $^{107}Ag^0$ center are indicated in the figure. As expected, the two lines that had been associated with ^{109}Ag (compared with Figure 1b) are absent. On increasing the alkali content to 30% (Figure 3b) a broader spectrum is observed in accord with the results reported previously[7, 14]. The weak resonance to the high field side of the $^{107}Ag^0$ will be discussed later.

Figure 4a shows the ESR spectrum for the lower alkali glass containing 1.5% $^{107}Ag_2O$. The hyperfine line associated with ^{109}Ag in Figure 1 is now absent leaving the signal attributed to a molecular, Ag_2^+ center without the interference of an overlap. In the absence of this overlap it becomes apparent that this signal has some structure. Figure 4b shows the ESR spectrum for the corresponding higher alkali glass containing 1.5% $^{107}Ag_2O$. In the absence of the overlap due to ^{109}Ag a strong new resonance is revealed. In order to obtain a better comparison, the induced ESR high field components of these four spectra (for low and high K_2O contents with 0.1 and 1.5% Ag_2O) are shown in Figure 5.

Figure 5a for 10 mole % alkali and 0.1% $^{107}Ag_2O$ shows one strong Ag^0 line associated with ^{107}Ag with a width of 20 gauss. On the high field side of this line now clearly appears a weak, relatively broad line ($\Delta H = 33.3$ Gauss $g = 1.9937$) which had not been observable in glasses containing both isotopes, now labelled Ag_2^+ (R).

Figure 5b shows the induced spectrum resulting from increasing the Ag_2O content from 0.1 to 1.5 mole % (at the expense of K_2O). In addition to the two lines already observed in spectrum 5a and labelled Ag^0 and Ag_2^+ (R) a strong broad line corresponds to the molecular Ag_2^+ center that had been observed easily in equivalent glasses containing both isotopes (Figure1).

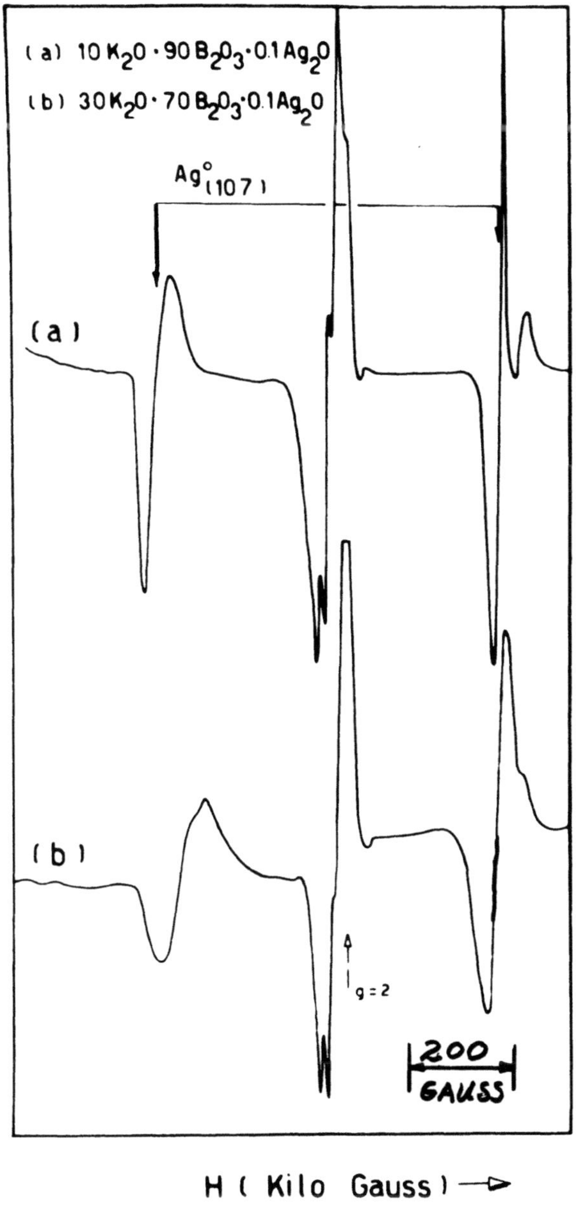

Figure 3. Induced ESR Spectra for x-irradiated potassium borate glasses containing 10 and 30 mole % K_2O, respectively. Both samples contain 0.1 mole % Ag_2O, with silver as ^{107}Ag isotope.

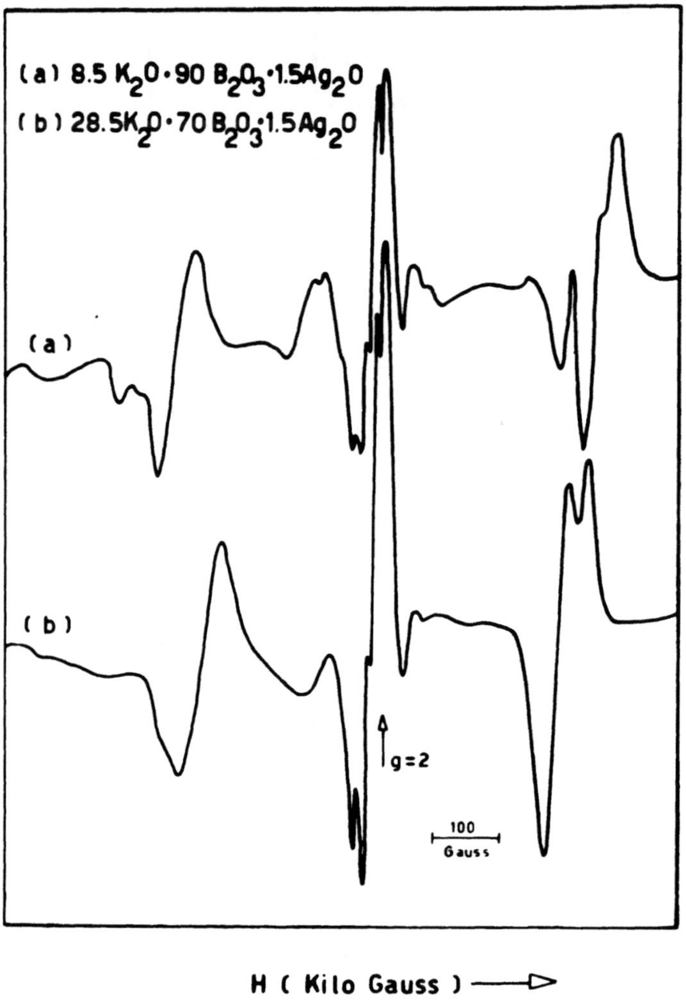

Figure 4. Induced ESR Spectra for x-irradiated potassium borate glasses containing 8.5 and 28.5 mole % K_2O, respectively. Both samples contain 1.5 mole % Ag_2O, with silver as ^{107}Ag isotope.

On increasing the alkali content from 10 to 30 mole % K_2O, the low silver glass (0.1 mole % Ag_2O) gave the induced spectrum shown in Figure 5c. The strong nearly symmetrical line shown in this figure is at a slightly lower field than the Ag^0 lines which appear in low alkali glasses (Figures 5a and 5b). This is associ-

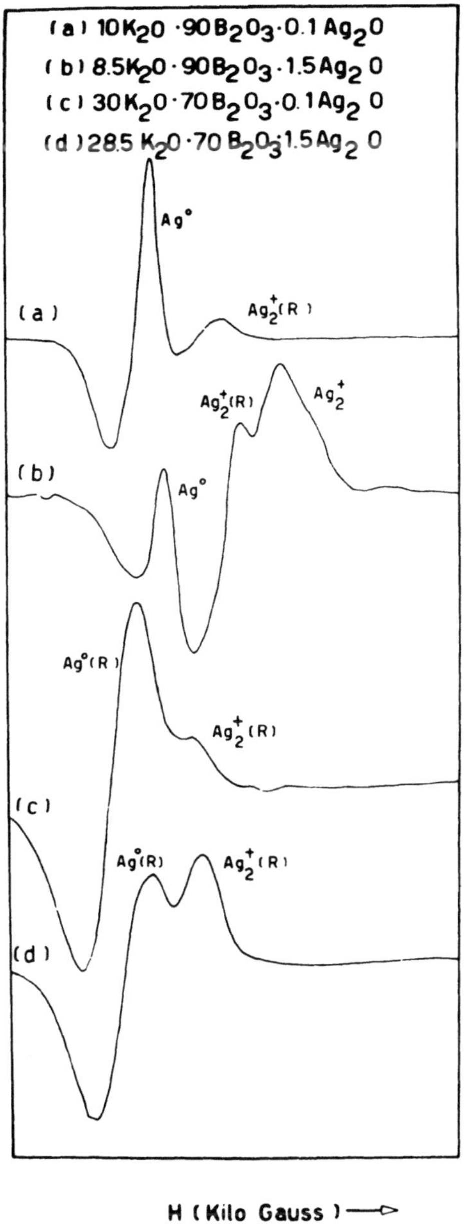

Figure 5. High field induced ESR spectra for x-irradiated low and high potassium borate glasses containing 0.1 and 1.5 mole % Ag_2O, with silver as ^{107}Ag isotope.

ated with a decrease in the hyperfine splitting to about 580 gauss as compared to 604 gauss for the Ag^0 signal. This strong line now labelled Ag^0 (R) is also broader than the Ag^0 line ($\Delta H = 30$, compared to 20 gauss). This behavior represented in (a) and (c) is interpreted as the manifestation of two different atomic electron trap centers, the one Ag^0 which would be favoured in low alkali glasses and the other, Ag^0 (R) favoured in high alkali glasses. The differences in the characteristics of these lines can clearly be interpreted on the basis of some delocalication of the trapped electrons in the center formed in high alkali glasses. Accordingly, atomic centers in low and high alkali glasses are designated as Ag^0 and Ag^0 (R) respectively. The dependence of Ag^0 (R) centers on the amount and kind of alkali is of particular interest.

Similarly, two molecular centers can be distinguished. The relatively weak line on the high field side in Figure 5c corresponds to that observed in (a) and (b) and labelled as $Ag_2{}^+$ (R). Increasing the silver content from 0.1 to 1.5 mole % Ag_2O in this high alkali glass resulted in an increase in the intensity of this $Ag_2{}^+$ (R) line, relative to the Ag^0 (R) line for the same glass (Figure 5d). The difference between the two types of molecular centers namely $Ag_2{}^+$ and $Ag_2{}^+$ (R) is demonstrated by means of bleaching experiments. The results are shown in Figures 6 and 7 for the low and high alkali glasses respectively, both containing 1.5 mole % Ag_2O. The Ag^0 and Ag^0 (R) lines disappear as a result of heating the samples to 100°C. This is in line with our earlier results showing lower stability of the atomic silver center relative to the molecular center. The remaining resonance resulting from heating the low alkali glass (Figure 6b) to 100°C can be attributed to the overlapping of the two molecular centers $Ag_2{}^+$ and $Ag_2{}^+$ (R). On the other hand, heating the high alkali glass to 100°C results in one single resonance corresponding to the $Ag_2{}^+$ (R) center.

The above mentioned assignments may be justified on the basis of the following:

A. Atomic (Ag^0), Ag^0 (R) Centers

The distinction between two types of atomic centers, namely Ag^0 and Ag^0 (R), is dictated by the following considerations:

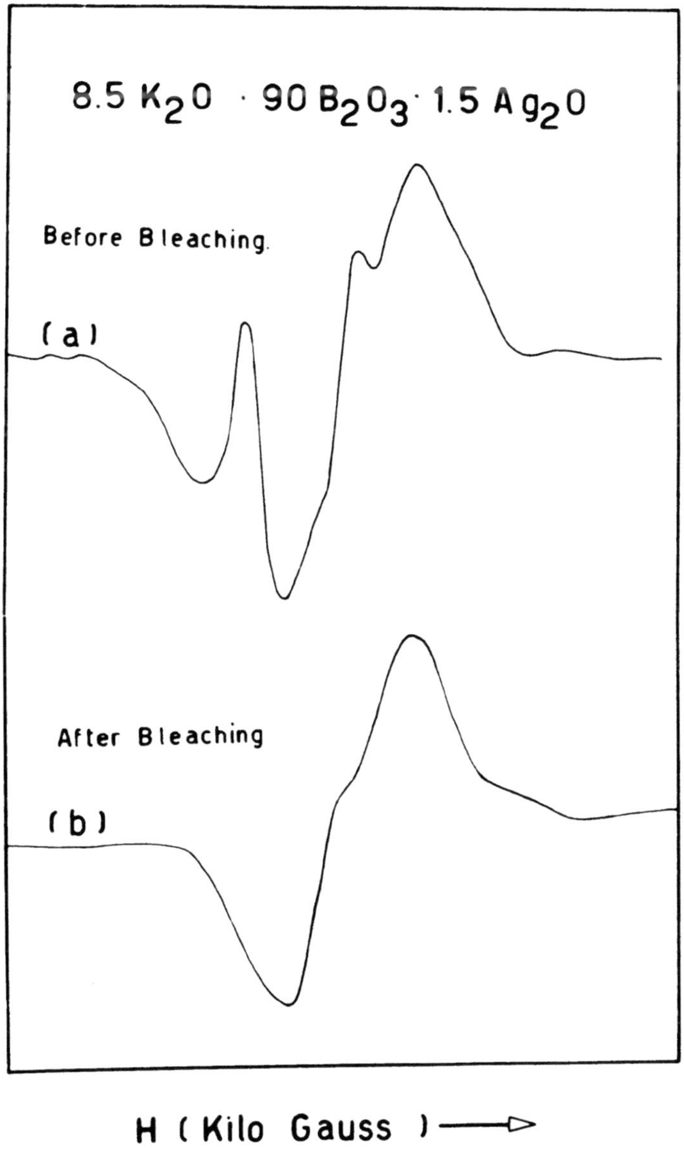

Figure 6. High field induced ESR spectra for x-irradiated $8.5\ K_2O$ - $90\ B_2O_3$ $1.5\ Ag_2O$ glass, before and after bleaching at 100°C (silver as ^{107}Ag isotope).

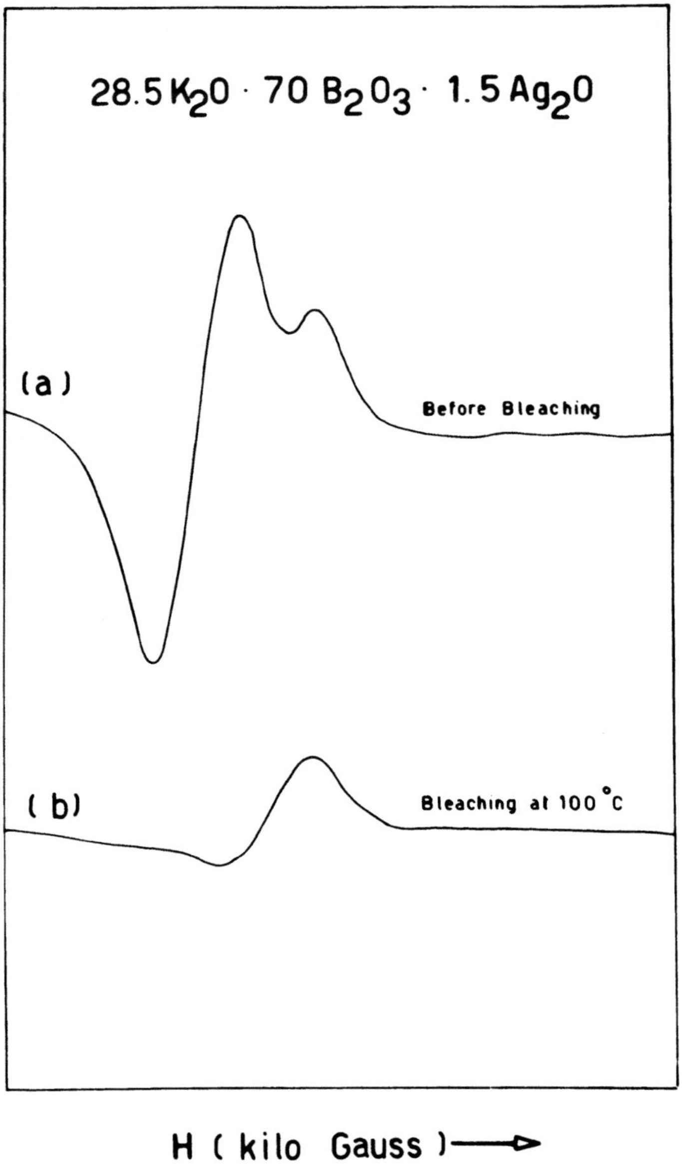

Figure 7. High field induced ESR spectra for x-irradiated $28.5\ K_2O \cdot 70\ B_2O_3 \cdot 1.5\ Ag_2O$ glass, before and after bleaching at 100°C (silver as ^{107}Ag isotope).

i. Unlike Ag^0, the Ag^0 (R) center grows with alkali content and appears to be associated with the vicinity of alkali and non-bridging oxygens and is labelled accordingly. Schematic drawings of models for Ag^0 and Ag^0 (R) are given in Figure 8 (1) and (2).

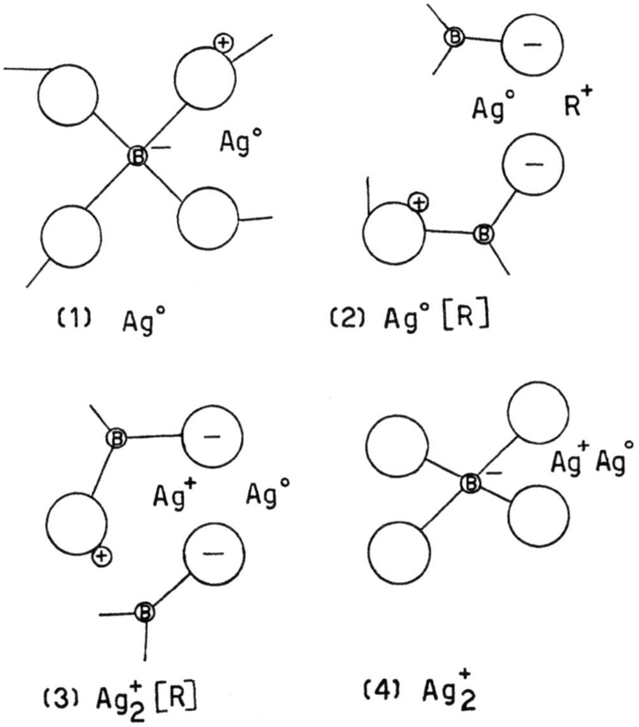

Figure 8. Postulated models for some radiation induced electron-trap silver centers in borate glasses.

ii. This is supported by experiments conducted with high alkali glasses containing both isotopes, which still clearly show the change in hyperfine splitting from Rb to Li as a function of ionic radius (Figure 9, curve b). In contrast, the hyperfine splitting of Ag^0 remains constant in the low alkali glass. (Figure 9, curve a). It should be noted that, the values obtained from Figure 5 are strictly those for a center involving a potassium ion. But no changes in interpretation can be expected for the case of Na.

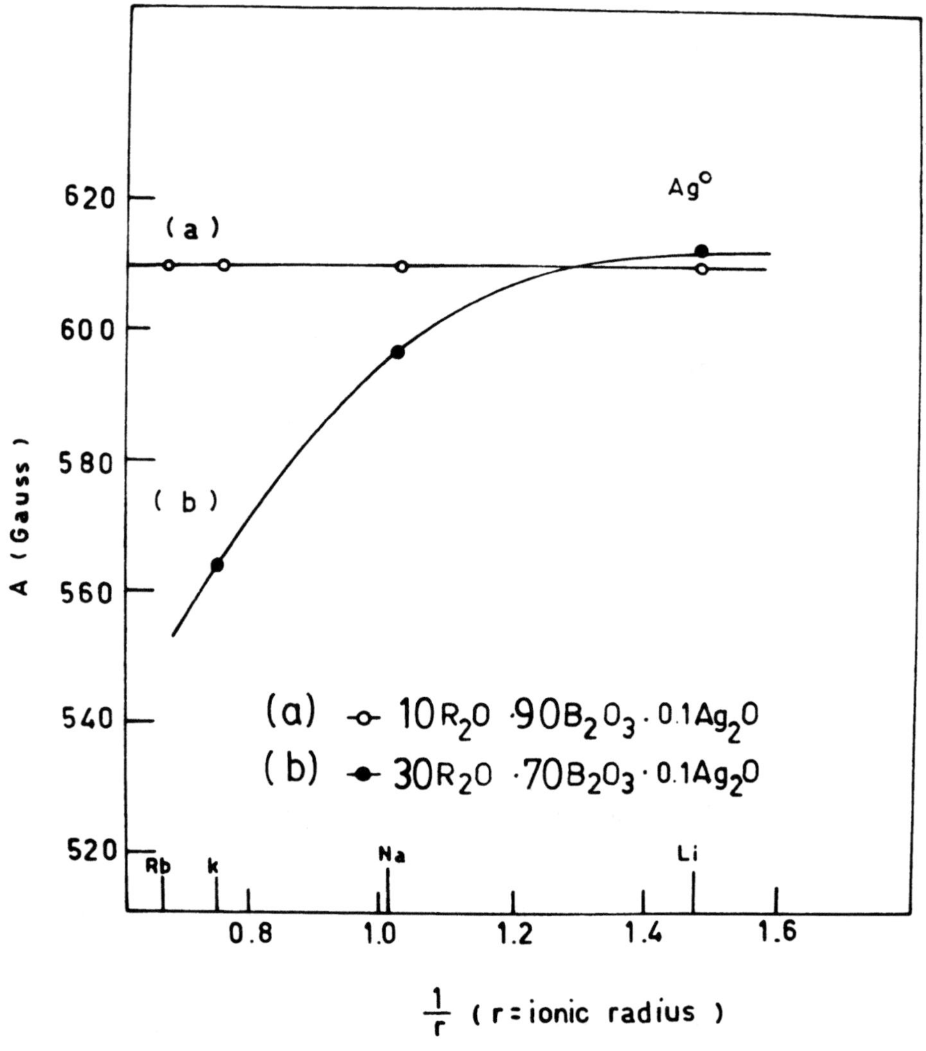

Figure 9. Effect of type of alkali in low and high alkali glasses on Hyperfine Splitting constant (A) for the Ag^0 (107) lines.

iii. The bleaching results show that both centers disappear at the same bleaching temperature, indicating similar stabilities. More accurate bleaching experiments may be needed to show if there is a very slight difference in their bleaching temperatures.

B. Molecular Ag_2^+, Ag_2^+ (R) Centers

The high alkali containing glass when heated to 100°C (Figure 7) indicated the disappearance of the atomic silver resonance Ag^0(R) and a relatively broad line (Δ H=30 G) remained. The high alkali content of this glass, its high silver content, as well as the stability of this resonance strongly suggests that the latter is a molecular type of center in the vicinity of non-bridging oxygens. A schematic drawing of the proposed model for this center is shown in Figure 8 (3), and we will label the center as Ag_2^+ (R) since it is favoured in the high alkali borate glasses.

When a similar experiment was conducted on the low alkali glass containing 1.5 mole % Ag_2O (Figure 6), the atomic silver center bleached out completely and a broad absorption remained. It is clear that the latter resonance can involve more than one line; one which is that attributed to the Ag_2^+ (R) center discussed earlier (Figure 6b). An approximate analysis of this composite resonance remaining after heating the low alkali glass is shown in Figure 10. The analysis suggests that another broader resonance (about 50 gauss) overlaps with the Ag_2^+ (R) resonance (Δ H=30 gauss). The former resonance was labelled Ag_2^+ and corresponds to the molecular center discussed in earlier publications. A schematic drawing of the model proposed for this center is shown in Figure 8. This model is based on the observation that the resonance associated with it could only be identified in glasses containing low alkali (i.e. no or few non-bridging oxygens) and high silver content as well as its high stability.

In conclusion, silver centers have been found more identifiable because of the greater resolution obtained by the use of a single isotope, ^{107}Ag. Obviously, the use of the single isotope is limited. But once the silver centers have been identified in some single isotope glasses, assignments in normal silver glasses become more reliable. It has become possible to differentiate between atomic and molecular centers observed in borate glasses containing two levels of alkali (Ag^0, Ag^0(R), Ag_2^+, Ag_2^+(R).

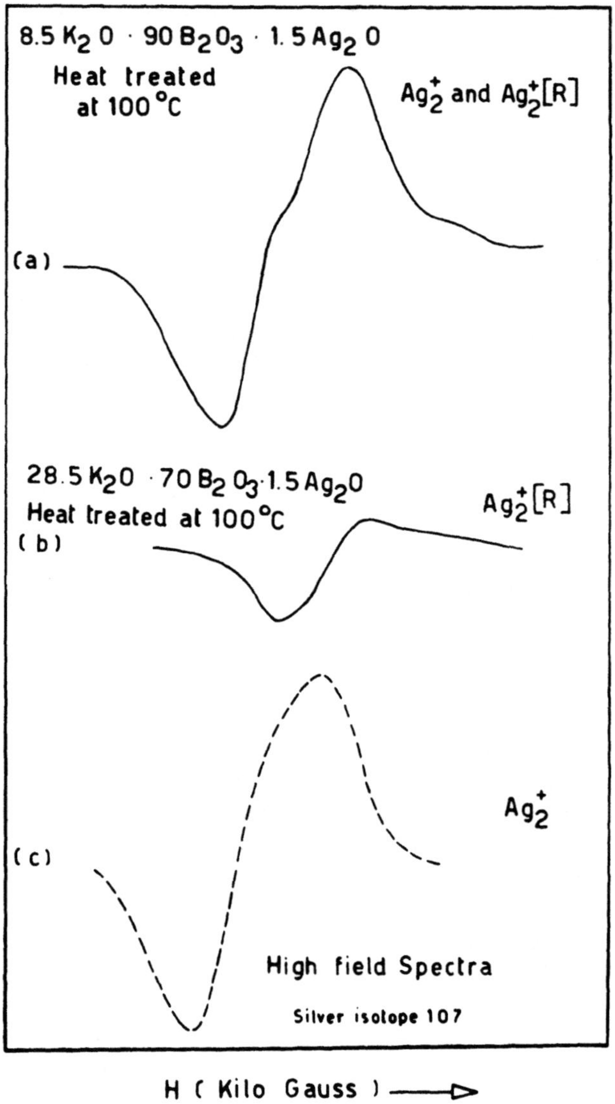

Figure 10. The Ag_2^+ and Ag_2^+ (R) resonance lines.

This differentiation is associated with the observed decrease in hyperfine splitting that clearly starts around 20% alkali content (Figure 11).

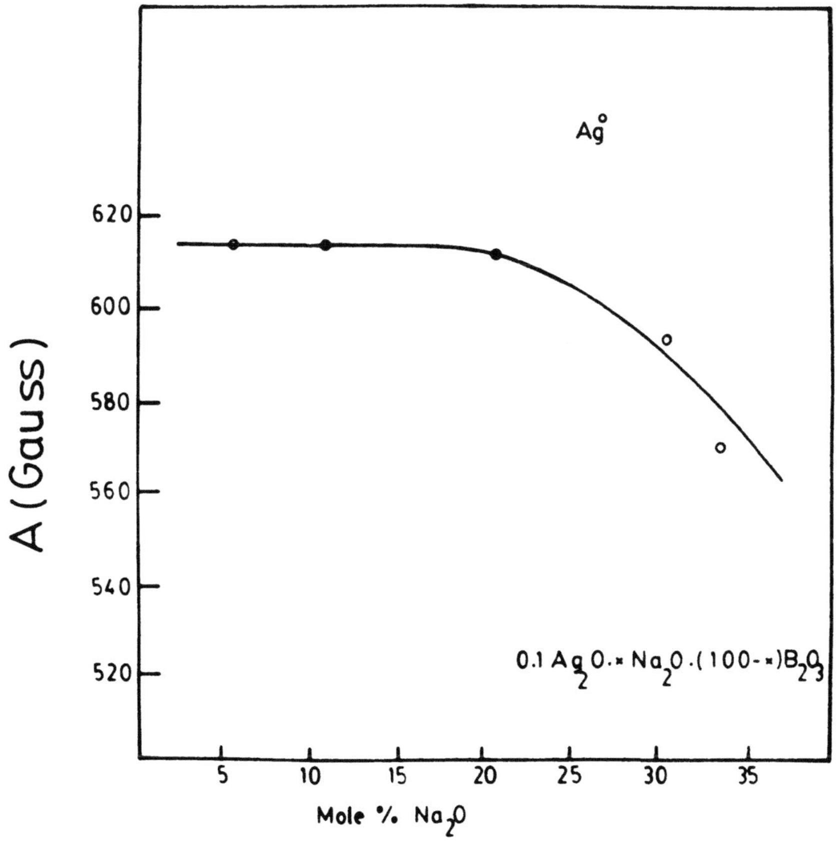

Figure 11. Effect of alkali content on hyperfine splitting constant (A) for the Ag^0 (107) lines.

The models (Figure 8) for the silver centers in high alkali glasses which appear to require some interaction with the alkali shown by their dependence on alkali species, have been related tentatively to the non-bridging oxygen increasingly occurring with increasing alkali content. This assignment is strongly supported by the observation that in silicate glasses a change in hyperfine splitting starts with the first addition of alkali known to be accompanied by the appearance of non-bridging oxygens. On the other hand, in borate glasses the first addition of alkali is exclusively associated with BO_4 group formation and is not accompanied by a change in hyperfine splitting. Thus, the appearance of Ag^0 (R) and Ag_2^+ (R) centers may be taken to indicate the presence of non-bridging oxygens in these glasses. On the other hand, models for centers associated with low alkali glasses were associated with closed structures characteristic of these compositions.

With this improved understanding of induced silver centers achieved as a result of the use of the single isotope ^{107}Ag, a systematic study is currently being made for borate glasses containing various levels and species of alkali, using normal silver. It is hoped to correlate the induced ESR signals in these glasses with models of alkali borate glass structures such as those proposed by Krogh-Moe [18], Bray [19] and Konijnendijk et al [20].

ACKNOWLEDGEMENT

Thanks are due to Uppsala University, Sweden for providing us with the ^{107}Ag isotope.

REFERENCES

1. D. L. Griscom, J. Non-Cryst. Sol. 13, 251 (1974).
2. A. M. Bishay and K. R. Ferguson, Advances of Glass Technology, Plenum Press, New York pp. 133-138 (1962).
3. A. M. Bishay, M. Maklad, I. Gomma and S. Arafa, Interaction of Radiation with Solid, Ed., A. M. Bishay, Plenum Press, New York (1967).
4. A. M. Bishay and M. Maklad, J. Phys. Chem. Glasses, 7 (5), 149-156 (1966).
5. A. M. Bishay, C. Quadros and A. Piccini, Part I, J. Phys. Chem. Glasses, 15 (4), 109-112 (1974). Part II, with R. Weeks, Ibd. Feb. 1977.

6. F. Assabghy, E. Boulos, S. Calamawy, A. M. Bishay and N. J. Kreidl, Recent Advances in Science and Technology of Materials, Ed. A. M. Bishay, Plenum Press, New York (1973).
7. F. Assabghy, S. Arafa, E. Boulos, A. M. Bishay and N. J. Kreidl, Proc. Tenth International Congress on Glass, Koyoto, Japan (1974).
8. E. N. Boulos and N. J. Kreidl, J. Amer. Cer. Soc. 54, 368 (1971).
9. R. A. Zhitnikov and N. I. Mel'nikov, J. Sov. Phys. Solid State, 10, 80 (1968).
10. L. Shields, J. Chem. Phys., 45, 2332 (1966).
11. R. Yokota and H. Imagawu, J. Phys, Soc. Japan, 23, 1038 (1966).
12. E. Lell and N. J. Kreidl, Interaction of Radiation with Solid, Ed., A. M. Bishay, Plenum Press, New York, 199 (1967).
13. R. Yokota and H. Imagawa, J. Phys. Soc. Japan, 20, 1537 (1965).
14. F. Assabghy, S. Arafa, E. Boulos, A. M. Bishay and N. J. Kreidl, J. Non.Cryst. Sol. 23, 81-91 (1977).
15. N. I. Mel'nikov, D. P. Peregood and R. A. Zhitnikov, J. Non. Cryst. Sol. 16, 195-205 (1974).
16. P. Beekenkamp, Ph. D. Thesis, Technische Hochoscule Eindhoven, Netherlands(1965).
17. S. Arafa and A. M. Bishay, J. Amer. Cer. Soc., 53 (7) 390-96, (1970).
18. J. Krogh-Moe, J. Phys, Chem. Glasses, 6 (2), 46, (1965).
19. S. Lee and P. J. Bray, J. Chem. Phys., 39, 2863 (1963), Ibd. 40, 2982 (1964).
20. E. L. Konijnendijk and J. M. Stevels, J. Non-Cryst. Sol. 18, 307-331 (1975).

STRUCTURE OF BORATE AND BOROSILICATE GLASSES BY RAMAN SPECTROSCOPY

W.L. Konijnendijk and J.M. Stevels

Department of Inorganic Chemistry

Eindhoven University of Technology

1. Introduction

This paper summarizes our Raman scattering measurements of borate and borosilicate glasses. These measurements form a part of a more extended investigation into the structure of these glasses. It comprised Raman scattering (1-8 inclusive), infrared absorption (1), spectroscopic calculations (9), viscosity (1, 10), thermal expansion (1, 11), electrical conduction (1, 12), and density and refractive index measurements (1, 13).

The first Raman spectra of vitreous boron oxide and of alkali borate glasses were published long ago. As early as 1936, Kujumzelis (14) described a spectrum of B_2O_3. Bobovich (15) studied a number of sodium borate glasses but the slit width he used in his experimental set up was too large to discover the existence side by side of the peaks at 806 and 770 cm^{-1}, which are really key wavelengths for understanding the structure of these glasses. The advent of laser Raman spectroscopy led various research workers to embark upon a more systematic

study of these glasses.
Several research workers have tried to give a fundamental description of the vibrational spectra of borate glasses. The most recent description has been given by Brawer (16). He showed that the spectra of a glass and a crystal (with the same composition) show considerable agreement if certain conditions are satisfied, the main condition being that the same structural groups should occur both in the glass and in the crystal and that there should be little vibrational coupling between the structural groups.

In our study of the structure of borate and borosilicate glasses we compared the Raman spectra of glasses with those of appropriate compounds whose crystal structure are known.
Together with the conclusions from spectroscopic calculations on the spectra of crystalline borates by Bril (9) it is then possible to obtain a qualitative and sometimes semiquantitative impression of the structure of the glasses. In other words, the Raman spectra of the compounds are used as a fingerprint to reveal the presence of certain structural groups in the corresponding glasses.

From X-ray analysis of many crystalline borates it is known that they are built up of large borate groups (figs. 1 and 2). Already as early as 1962 Krogh Moe (17) postulated that the structure of sodium borate glasses can be described as a random network of large borate groups similar to those present in crystalline sodium borates. In this paper it will be shown that our Raman scattering results confirm this assumption for the structure of alkali borate glasses and that the same types of borate groups are also present in borosilicate glasses. In this way this paper is a tribute to the work of Krogh Moe.

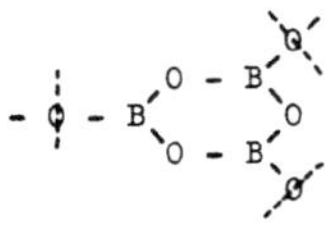

The boroxol ring, observed in vitreous B_2O_3.

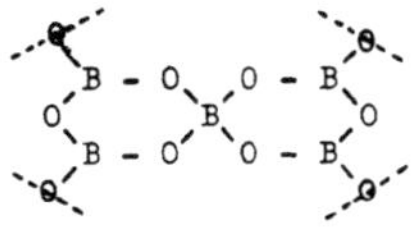

The pentaborate group, observed in the compounds α-$K_2O \cdot 5B_2O_3$ and β-$K_2O \cdot 5B_2O_3$.

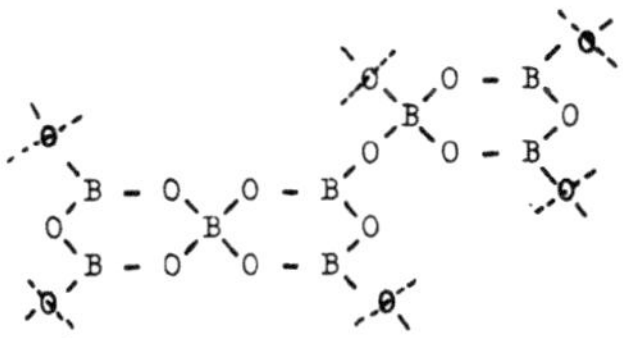

The tetraborate group, observed in the compound $Na_2O \cdot 4B_2O_3$.

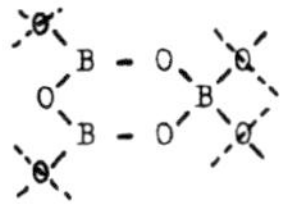

The triborate group, observed in the compound $Cs_2O \cdot 3B_2O_3$.

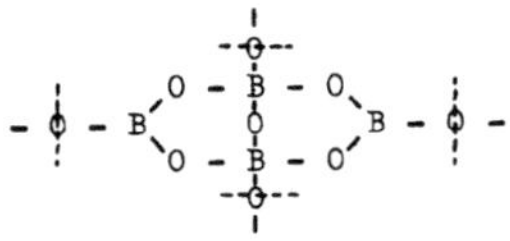

The diborate group, observed in the compound $Li_2O \cdot 2B_2O_3$.

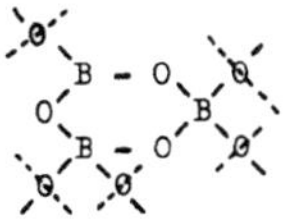

The di-triborate group observed in the compound $K_2O \cdot 2B_2O_3$.

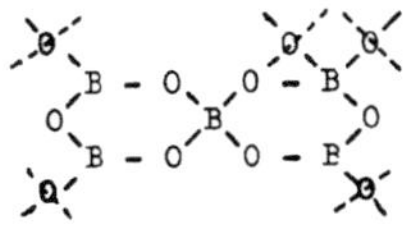

The di-pentaborate group, observed in the compound $Na_2O \cdot 2B_2O_3$.

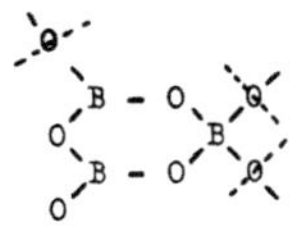

The triborate group with one non-bridging oxygen ion, observed in the compound $Na_2O \cdot 2B_2O_3$.

Fig. 1. Borate groups observed in several borate compounds. Dotted lines through the oxygen ions indicate that they are bridging oxygen ions.

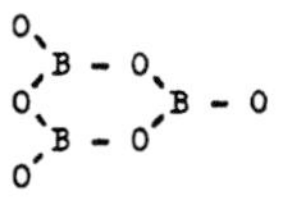

The ring-type metaborate group, observed in the compounds $Na_2O \cdot B_2O_3$ and $K_2O \cdot B_2O_3$

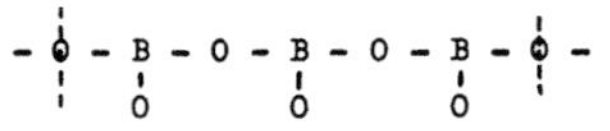

The chain-type metaborate group, observed in the compounds $Li_2O \cdot B_2O_3$ and $CaO \cdot B_2O_3$.

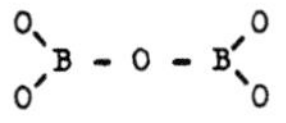

The pyroborate group, observed in the compounds $2MgO \cdot B_2O_3$ and $2CaO \cdot B_2O_3$.

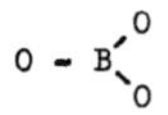

The orthoborate group, observed in the compounds $3MgO \cdot B_2O_3$ and $3CaO \cdot B_2O_3$.

Fig. 2. Borate groups observed in several borate compounds. Dotted lines through the oxygen ions indicate that they are bridging oxygen ions.

2. Experimental results and discussion

2.1 Borate glasses. The Raman spectra of some binary sodium borate glasses are shown in figs. 3 and 4. The Raman spectrum of vitreous boron oxide shows only a strong peak at 806 cm^{-1}.

The most striking feature is the rapid decrease of this peak at about 806 cm^{-1} and the concurrent increase of a peak at about 770 cm^{-1} for increasing sodium oxide content. At 20 mol% sodium oxide the 806 cm^{-1} peak is visible only as a shoulder of the 770 cm^{-1} peak and has completely disappeared at 25 mol% Na_2O. Krogh-Moe (18) and others attribute the 806 cm^{-1} peak to a vibration of the boroxol group whereas Bril (9) showed by calculation, that this vibration is due to the symmetrical breathing vibration of the boroxol ring. The conclusion is that on addition of sodium oxide to boron oxide the boroxol groups are converted into other groups. Based on vibrational spectroscopy analysis Bril assigns the peak that arises at 770 cm^{-1} to a vibration of a six-membered borate ring with one or two BO_4 tetrahedra (cf. fig. 5). This peak at about 770 cm^{-1} is also observed in the spectra of many crystalline borates containing six-membered borate

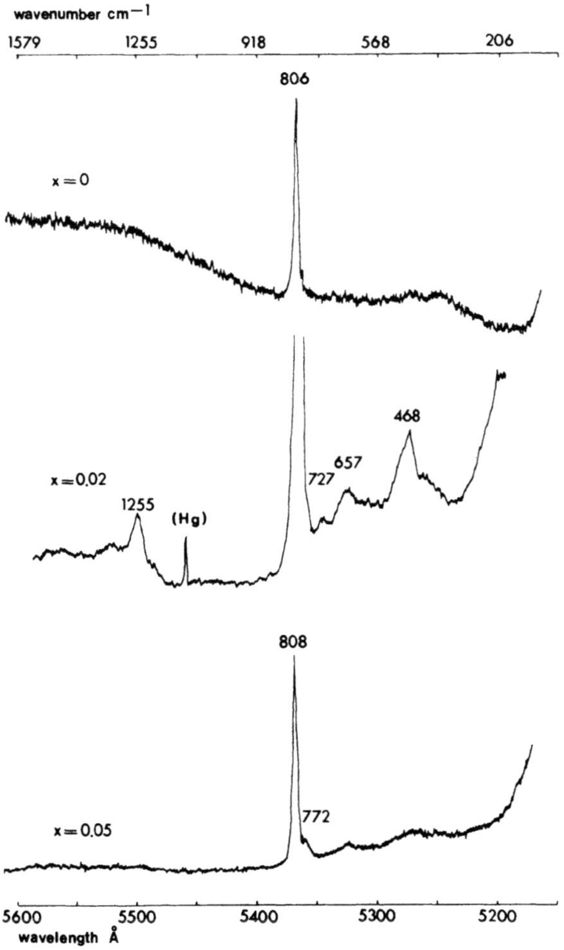

Fig. 3. Raman spectra of glasses of composition $xNa_2O-(1-x)B_2O_3$ (excitation line 514.5 nm Ar^+ ion laser), $0.05 \geq x \geq 0$

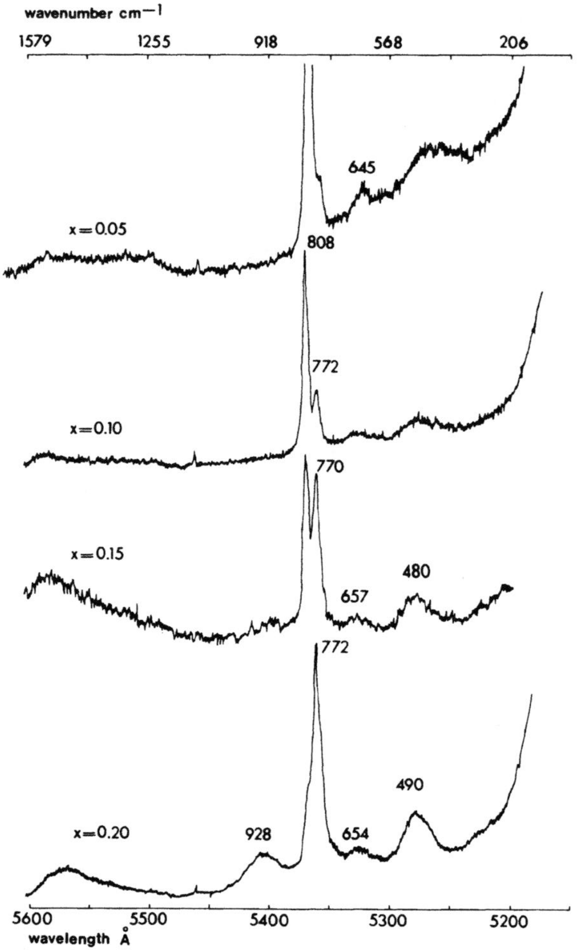

Fig. 4. Raman spectra of glasses of composition $xNa_2O-(1-x)B_2O_3$ (excitation line 514.5 nm Ar^+ ion laser), $0.20 \geq x \geq 0.05$

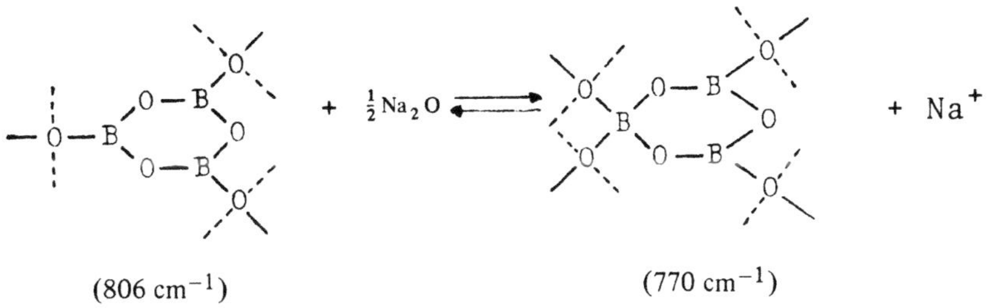

Fig. 5. The equilibrium between the boroxol-ring and the six-membered borate ring with one BO_4 tetrahedron.

rings with one or two BO_4 tetrahedra (1).

From the fact that the 806 cm^{-1} peak is still observable at 20 mol% Na_2O (but not very strongly) it can be concluded that (tetra)borate groups (with an average sodium to boron ratio of 0.25) are formed at this composition because otherwise the boroxol groups would already have been consumed at lower sodium to boron ratios of the glass. The 806 cm^{-1} band would have persisted at higher Na/B ratios of the glass, if a group with a higher Na/B ratio, for instance 0.5 (in other words a diborate group) would have been present. Thus it can be concluded that six-membered borate rings with only one BO_4 tetrahedron are formed in the 20% Na_2O range. Based on the similarity of the spectra of alkali borate glasses with 20 mol% alkali oxide and the spectra of crystalline compounds it is likely that tetraborate groups (that is a couple consisting of one pentaborate and one triborate group) are formed in the concentration range 0-20 mol% alkali oxide.

Based on similar reasonings and on the comparison of the spectra of glasses containing 20-35 mol% alkali oxide with those of crystalline borates in this composition region, it is probable that the tetraborate groups are replaced by diborate groups upon increasing the alkali oxide content (1, 2).

At still higher alkali oxide concentrations the presence of ring type metaborate ($B_3O_6^{3-}$), pyroborate ($B_2O_5^{4-}$) and orthoborate groups (BO_3^{3-}) is observed, with typical bands

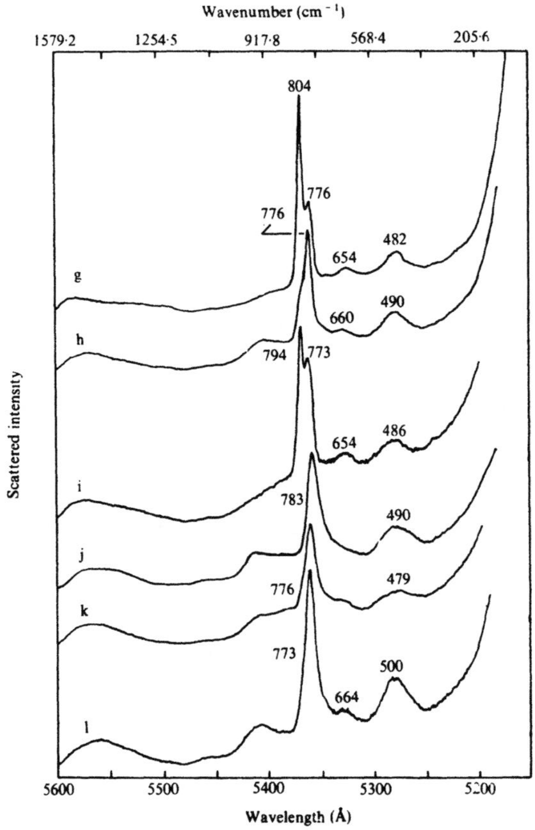

Fig. 6. Raman spectra of $MgO-Na_2O-B_2O_3$ glasses (excitation line 514.5 nm Ar^+ ion laser). g = $0.075MgO-0.075Na_2O-0.85B_2O_3$, h = $0.10MgO-0.15Na_2O-0.75B_2O_3$, i = $0.20MgO-0.05Na_2O-0.75B_2O_3$

at 630 cm^{-1}, 820 cm^{-1} and 940 cm^{-1} respectively. The Raman spectra of calcium-sodium borate glasses are similar to those of the binary sodium borate glasses containing the same amount of B_2O_3(5), indicating a structural similarity between these glasses. However the magnesium-sodium borate glasses show spectra different from those of the calcium-sodium borate glasses (cf. also 5). Figure 6 shows that the peak characteristic of boroxol groups (at about 806 cm^{-1}) has a relatively higher intensity in the spectrum of the glass 0.075MgO-0.075Na_2O-0.85B_2O_3 (curve g of fig. 6) compared to the corresponding calcium containing glass (curve a of fig. 7). For the glasses with 75 mol% B_2O_3, the peak characteristic of boroxol groups is observed and increases when Na_2O is replaced by MgO (curves h and i). They are not observed for the corresponding glasses containing calcium (curves b and c). The recurrence of the boroxol groups in the glasses containing magnesium with 75 and 85 mol% B_2O_3 can be explained by the formation of connected BO_4 tetrahedra in six-membered borate rings, in other words by the presence of diborate groups. For a fixed number of BO_4 tetrahedra, there are more boron ions available for the formation of boroxol groups when BO_4 tetrahedra are present in diborate groups rather than in tetraborate groups.

The preference for the formation of connected BO_4 tetrahedra is also observed for lead borate glasses (8). The conclusion is that Mg and Pb borate glasses behave differently as compared to alkali borate and alkali-Ca borate glasses; The boroxol rings are retained longer because of the formation of diborate groups.

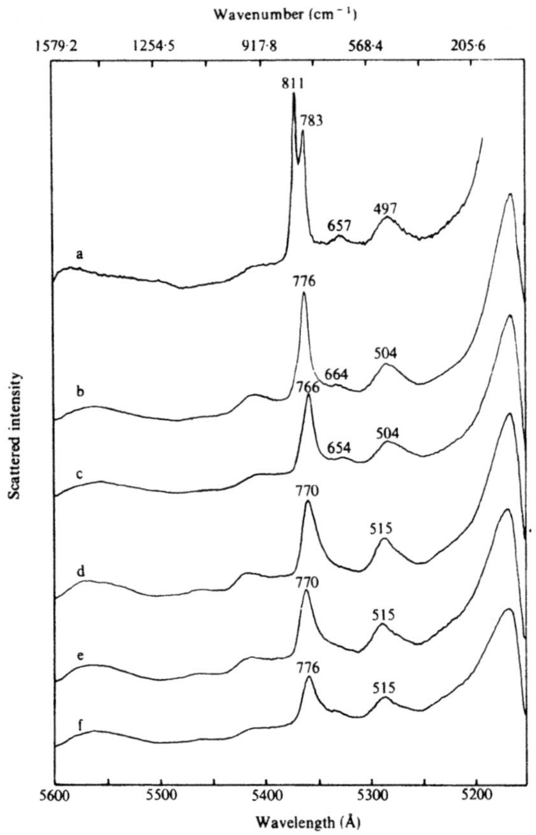

Fig. 7. Raman spectra of $CaO-Na_2O-B_2O_3$ glasses (excitation line 514.5 nm Ar^+ ion laser). a = $0.075CaO-0.075Na_2O-0.85B_2O_3$, b = $0.10CaO-0.15Na_2O-0.75B_2O_3$ c = $0.20CaO-0.05Na_2O-0.75B_2O_3$

2.2 Borosilicate glasses

A comparison of the spectra in series $xK_2O-(1-x)B_2O_3$ with those of series $xK_2O-(0.85-x)B_2O_3-0.15SiO_2$ reveals the same trend in the spectra (figs. 8 and 9). For the glasses in the series $xK_2O-(0.85-x)B_2O_3-0.15SiO_2$, the same decrease in the peak at about 806 cm^{-1} and an increase of the peak at about 770 cm^{-1} is observed.

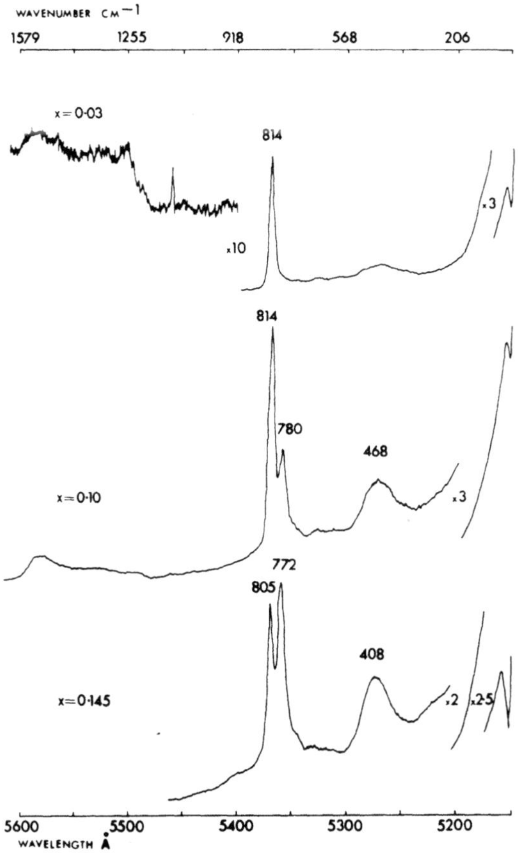

Fig. 8. Raman spectra of glasses in the system xK_2O-$(0.85-x)B_2O_3$-$0.15SiO_2$ (excitation line 514.5 nm Ar^+ ion laser), $0.145 \geq x \geq 0.03$

Therefore, it seems probable that most of the potassium oxide in this composition series is used in forming the borate groups also present in the binary potassium borate glasses. It is observed that at x = 0.20, the 806 cm^{-1} peak has now completely disappeared, hence there are no boroxol groups left in the glass network at this concentration. For values x < 0.20 (just as for binary borate

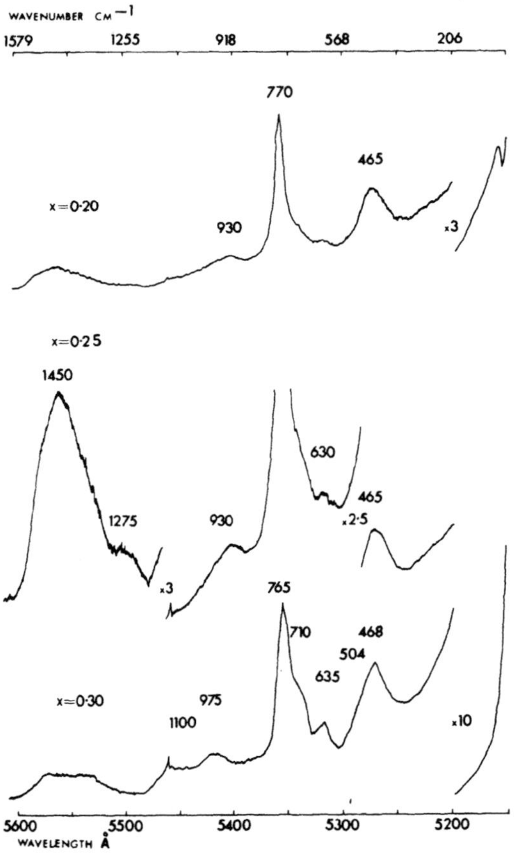

Fig. 9. Raman spectra of glasses in the system xK_2O-$(0.85-x)B_2O_3$-$0.15SiO_2$ (excitation line 514.5 nm Ar^+ ion laser), $0.30 \geq x \geq 0.20$

glass) six-membered rings with one BO_4 tetrahedron are formed (tetraborate groups). However the presence of a small peak at 630 cm^{-1} at x = 0.20 indicates the presence of a small amount of ring-type metaborate groups which are not present in the case of the pure borate glasses. We must also consider the behaviour of the SiO_4 tetrahedra. From the study of pure silicate glasses (3) it is known that SiO_4 with zero, one or two non-bridging oxygen ions show Raman lines at 470 cm^{-1}, 1100 cm^{-1} and 530 cm^{-1} and 970 cm^{-1} and 590 cm^{-1} respectively. For all values of x in the series under consideration the SiO_4 tetraeder with zero non-bridging ions are present, but at x = 0.25 and 0.30 a very small band near 1100 cm^{-1} may be observed, which indicates that a small number of SiO_4 tetrahedra with one non-bridging oxygen ion are formed (1, 3).

We have studied a number of series of borosilicate glasses with 15, 35, 65 and 70% SiO_2. As an example the Raman spectra of borosilicate glasses with 65 mol% SiO_2 are discussed. They behave in a different way, because now the influence of the SiO_2 content is preponderant (cf. fig. 10). In the series of glasses of composition xK_2O-$(0.35-x)B_2O_3$-$0.65\ SiO_2$ it is observed that at low K_2O concentration boroxol groups are present again, as revealed by the presence of the characteristic at about 806 cm^{-1}.The large peak at 470 cm^{-1} indicates that SiO_2 is present in a vitreous silica-like structure (SiO_4 groups with four bonding oxygen ions). In the spectrum of the glass of composition $0.10K_2O$-$0.25B_2O_3$-$0.65SiO_2$ a peak is observed, at 805 cm^{-1} due to boroxol groups, at 765 cm^{-1} due to six-membered ring borate groups with one BO_4 group and at 625 cm^{-1} due to ring-type metaborate groups. For this composition the K/B ratio is 0.4; it is interesting to note that at this ratio (near the diborate composition) there are still boroxol groups present and that hardly any alkali ions are bonded to the silicate network.

Only for x = 0.20 SiO_4 tetrahedra with one non-bridging oxygen ion are present (peaks at 1100 cm^{-1} and

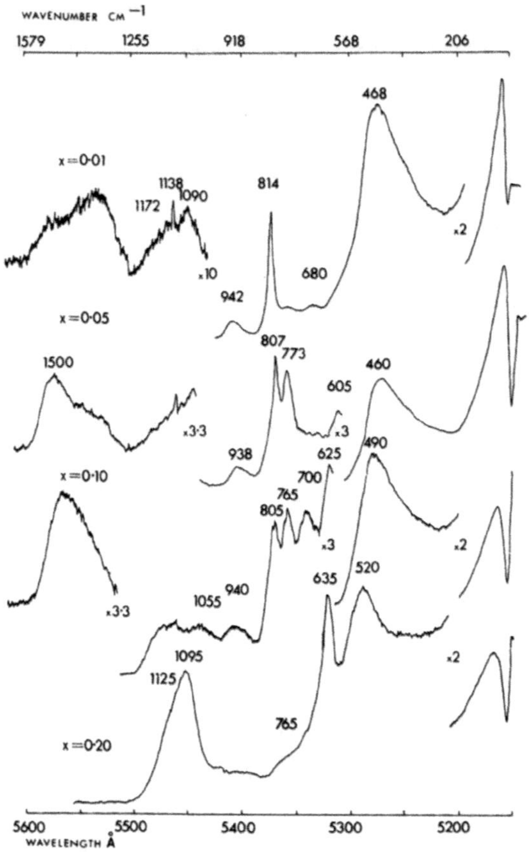

Fig. 10. Raman spectra of glasses in the system $xK_2O-(0.35-x)B_2O_3-0.65SiO_2$ (excitation line 514.5 nm Ar^+ ion laser), $0.20 \geq x \geq 0.01$

530 cm^{-1}). In the spectrum of this glass a peak is observed at 635 cm^{-1} due to ring-type metaborate groups. The band at 765 cm^{-1} reveals the presence of six-membered borate rings with one tetrahedron.

In the composition series $xK_2O-(0.30-x)B_2O_3-0.70SiO_2$ which is very similar to the technical borosilicate glasses of the pyrex type, the same type of spectra are observed as in the series with 65 mol% SiO_2 (1, 7).

We now consider the mixed Na-K borosilicate glasses. The Raman spectrum reveals that in the base glass $0.10Na_2O-0.10K_2O-0.15B_2O_3-0.65SiO_2$, a certain amount (between 10 and 15 mol%) of the alkali oxide is used for the formation of ring-type metaborate groups. The rest of the alkali oxide is used the formation of ring-type six-membered borate groups with one BO_4 tetrahedron and SiO_4 tetrahedra with a non-bridging oxygen ion. Electron microscopy of this glass suggests an irregular phase-separated structure with the composition fluctuations extending over a few hundred ångstroms. The separate phases are a silica like structure and a structure with many metaborate groups. A further analysis showed that the tendency to phase separation extends to a larger region of the system than is usually assumed in the literature (19). An additional conclusion is that boron ions are not incorporated (on a large scale) into the silicate network in a way analoguous to aluminium ions and that the silicon ions are not incorporated in the borate network. This is not unexpected because no crystalline alkali borosilicates could be prepared at atmospheric pressure up to now.

Replacement of Li_2O for sodium and potassium oxide in a glass of composition $0.20R_2O-0.15B_2O_3-0.65SiO_2$ clearly results in different spectra (cf. fig. 11). The peak at about 630 cm^{-1} decreases and ultimately vanishes at $0.20Li_2O$. The ring-type metaborate groups are thus replaced by other borate groups. Crystalline $Li_2O.B_2O_3$ does not contain ring-type metaborate groups but chain-type groups with one non-bridging oxygen ion per BO_3 triangle as is known from X-ray analysis. The Raman spectra of the lithium borosilicate glasses cannot be reconciled with the presence of chain-type metaborate groups, although this

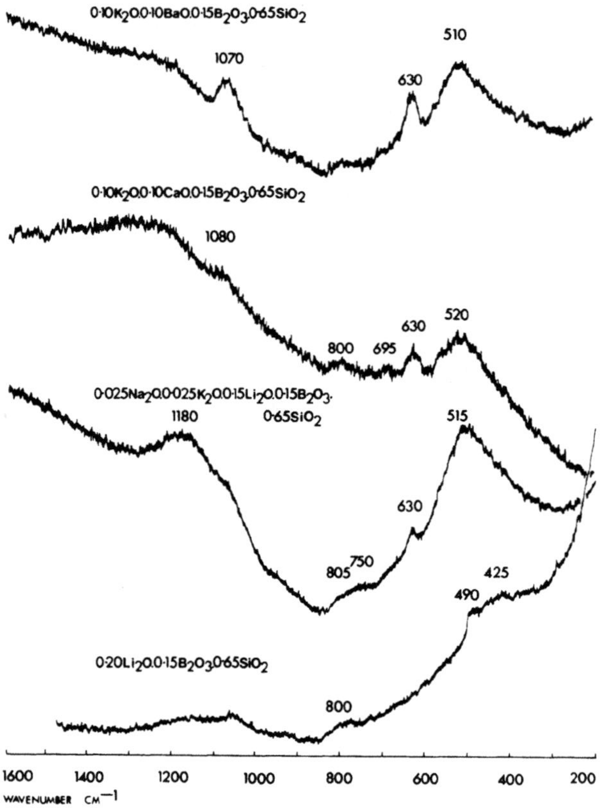

Fig. 11. Raman spectra of some borosilicate glasses (excitation line 632.8 nm He-Ne laser).

possibility cannot be altogether excluded. As a matter of fact so far, a line that must be assigned to chain-type metaborate group has not been discovered.

The substitution of CaO for barium oxide in this series of glasses shows the same tendency as Li_2O in the Raman spectra (1, 4, 7). The introduction of CaO appears to diminish the number of ring-type metaborate groups, as could be expected because crystalline $CaO.B_2O_3$, like $Li_2O.B_2O_3$, is built up of chain type metaborate groups. Here again formation of chain-type metaborate groups in the glass is not indicated by the Raman spectra although this possibility cannot be entirely excluded.

3. Conclusions

The results of the Raman spectroscopy of borate glasses are summarized in table 1. This table gives a qualitative impression of the presence of the borate groups in three composition areas.

Table 1
The presence of borate groups in alkali borate glasses in several composition areas

0-20 mol% R_2O	20-35 mol% R_2O	35-50 mol% R_2O
Boroxol groups	Tetraborate groups	Diborate groups
Tetraborate groups	Diborate groups	Metaborate groups
Loose BO_3 triangles	Loose BO_3 triangles	Pyroborate groups
		Orthoborate units
Loose BO_4 tetrahedra	Loose BO_4 tetrahedra	Loose BO_3 triangles with three bridging oxygen ions
		Loose BO_3 triangles with a non-bridging oxygen ion

Raman spectroscopy of sodium and potassium borate glasses has demonstrated the gradual decrease of the number of boroxol groups with increasing alkali-oxide content in the composition area 0-25 mol% alkali oxide. In this same composition area, tetraborate groups are formed with increasing the alkali-oxide content.

At about 30 mol% alkali oxide, mostly diborate groups occur and a very small number of ring-type metaborate groups. From 40-50 mol% alkali oxide, the formation of metaborate groups, orthoborate groups and pyroborate groups is apparent. The BO_4 tetrahedra are probably mostly present in diborate groups.

Raman spectra of alkaline-earth borate glasses are very similar to those of the corresponding alkali-oxide glasses. At 20 mol% BaO or CaO and 80 mol% B_2O_3, no boroxol groups are detected; probably tetraborate groups are most prevalent in the glasses of this composition. The borate network structures of $CaO-Na_2O-B_2O_3$ and $MgO-Na_2O-B_2O_3$ glasses containing 85 and 75 mol% B_2O_3 are dissimilar. It is indicated that mainly unconnected BO_4 tetrahedra are formed in the glasses containing calcium while in the glasses containing magnesium mainly connected BO_4 tetrahedra are formed. The preference for the formation of connected BO_4 tetrahedra is also observed for lead borate glasses.

The Raman spectra of the borate glasses indicate the presence of boron ions in large borate groups. However, based on the Raman spectra it is also probable that a small part of the boron ions is incorporated in a more random network of loose BO_3 triangles and loose BO_4 tetrahedra not present in a typical borate ring.

In table 2 the results are summarized for the three series of glass with different SiO_2 content.

Table 2

Borate and silicate groups present in alkali borosilicate glasses as indicated by the Raman spectra

	Alkali/boron ratio < 0.5	Alkali/boron ratio > 0.5
15 mol% SiO_2	boroxol groups	diborate groups
	tetraborate groups	SiO_4 (four bridging oxygen ions)
	diborate groups	SiO_4 (one non-bridging oxygen ion)
	SiO_4 (four bridging oxygen ions)	
35 mol% SiO_2	boroxol groups	diborate groups
	tetraborate groups	metaborate groups
	diborate groups	SiO_4 (four bridging oxygen ions)
	SiO_4 (four bridging oxygen ions	SiO_4 (one non-bridging oxygen ion)
65 mol% SiO_2	boroxol groups	metaborate groups
	tetraborate groups	SiO_4 (four bridging oxygen ions)
	diborate groups	
	metaborate groups	SiO_4 (one non bridging oxygen ion)
	SiO_4 (four bridging oxygen ions)	
	SiO_4 (one non-bridging oxygen ion)	

The Raman spectra of the borosilicate glasses indicate that below alkali/boron ratios of 0.5 all, or nearly all, alkali ions are used to form primarily ring-type six-membered borate groups with one or two BO_4 groups, these rings being ordered to form tetraborate or diborate groups. There seems to be some resistance to the formation of six-membered borate rings with two BO_4 tetrahedra; on increasing the alkali/boron ratio above 0.5 non-bridging oxygen ions connected with SiO_4 tetrahedra are formed, whereas below this ratio the SiO_4 tetrahedra carry bridging oxygens only.

At alkali/boron ratios above 0.5, significant amounts of ring-type metaborate pyroborate groups and orthoborate units are also formed, as in the binary alkali borate glasses. At alkali/boron ratios of about one, there seems to be an increase in the formation of ring-type metaborate groups for the sodium and potassium borosilicate glasses upon increasing the SiO_2 content.

Some of the boron ions are present not in typical borate rings, but as loose BO_3 triangles and loose BO_4 tetrahedra, analogous to the binary borate glasses. The Raman spectra of the borosilicate glasses suggest a phase-separated structure extending beyond the usually indicated boundaries of phase separation in these systems. At low SiO_2 content the molecules seem to be agglomerated to a vitreous silica-like structure. At higher SiO_2 and Na_2O or K_2O concentrations, the structure of the continuous phase resembles that of silicate glasses; in these same concentration regions the metaborate groups seem to be agglomerated, too, as suggested by electron microscopy.

References

1. W.L. Konijnendijk, Philips Res. Repts. Suppl. 1975, No 1.
2. W.L. Konijnendijk and J.M. Stevels, J. Non-Crystalline Solids 18 (1975) 307.
3. W.L. Konijnendijk and J.M. Stevels, J. Non-Crystalline Solids 21 (1976) 447.
4. W.L. Konijnendijk and J.M. Stevels, J. Non-Crystalline Solids 20 (1976) 193.
5. W.L. Konijnendijk, Phys. Chem. Glasses 17 (1976) 205.
6. W.L. Konijnendijk, Glastechn. Ber. 48 (1975) 216.
7. W.L. Konijnendijk, Bol. Soc. Esp. Ceram. Vidrio 14 (1975) 517.
8. W.L. Konijnendijk and H. Verweij, J. Am. Ceram. Soc. 59 (1976) 459.
9. T.W. Bril, Philips Res. Repts. Suppl. 1976, No. 2.
10. W.L. Konijnendijk and J. M. Stevels, Verres Réfract. 30 (1976) 821.
11. W.L. Konijnendijk and J.M. Stevels, Verres Réfract. 30 (1976) 371.
12. W.L. Konijnendijk and J.M. Stevels, Verres Réfract. 30 (1976) 519.
13. W.L. Konijnendijk and J.M. Stevels, Verres Réfract. 30 (1976) 223.
14. J. Kujumzelis, Z. Phys. 100 (1936) 221.
15. Ya. S. Bobovich, Op. Spectry 15 (1963) 412.
16. S. Brawer, Phys. Rev. B11 (1975) 3173.
17. J. Krogh-Moe, Phys. Chem. Glasses 3 (1962) 101.
18. J. Krogh-Moe, Phys. Chem. Glasses 6 (1965) 46.
19. W. Haller, D.H. Blackburn, F.E. Wagstaff and R.J. Charles, J. Am. Ceram. Soc. 53 (1970) 34.

STRUCTURE OF BORATE GLASSES BY RAMAN SPECTROSCOPY

William B. White, Steven A. Brawer, Toshiharu Furukawa
and Gregory J. McCarthy

Materials Research Laboratory
The Pennsylvania State University
University Park, Pennsylvania 16802

INTRODUCTION

A large body of literature on the Raman spectra of glasses has now been accumulated. Vibrational spectra of borate glasses have been used to elucidate the structural units present in these glasses as exemplified by the work of Stevels, Konijnendijk, and Bril. Measurements are given in the present paper span for lithium, sodium, and potassium borate glasses. However, we will emphasize the linkages of the structural units and have something to say about phase separation in these glasses.

The spectrum of B_2O_3 glass itself is rather remarkable, and has attracted some attention apart from the binary and ternary glasses [1-3]. It contains a single strong band at 808 cm^{-1} which is very narrow for a vibrational mode of a glass. However, the remainder of the spectrum is very weak and smeared out. This band is considered strong evidence for boroxal rings in B_2O_3 glass [4].

Our interest in the Raman spectra of the alkali borates dates back to 1971 when some spectra of sodium and potassium borate glasses were reported [5]. Comprehensive investigations of alkali borate and borosilicate glasses by Konijnendijk and Stevels [6,7] provide strong evidence for the gradual replacement of the boroxal ring by other structural units as the alkali concentration is increased.

EXPERIMENTAL

All glasses were prepared from reagent grade alkali carbonate and H_3BO_3. Mixtures of appropriate composition were milled together and then heated in ten gram lots in platinum crucibles. The mixes were first fired in air at 950°C for two hours and then heated at 950°C (K_2O-B_2O_3 glasses), 1000°C (Na_2O-B_2O_3 glasses) and 1000°C Li_2O-B_2O_3 glasses) for an additional hour to achieve uniform melting. After this heating procedure, rods approximately 2 mm in diameter were drawn from the melt and rapidly quenched in a cool air stream. Other quenching procedures used in connection with phase separation studies will be described later.

To determine if the glass had crystallized upon cooling, pieces of the bulk glass were powdered and x-ray diffraction patterns were determined. Although the more alkali-rich borate glasses crystallize readily, all Raman data reported here were determined on rods for which there was neither optical nor x-ray evidence for crystalline phases.

As an additional control on composition, a selected group of the sodium borate glasses were chemically analyzed for B_2O_3 and Na_2O. In general the glasses were within one mole percent of the starting composition indicating little change in composition during the melting procedures.

Raman spectra were measured in a 90° scattering configuration from untreated bulk rods. The spectrometer was a Spex Ramalog with an argon ion laser source and cooled RCA 31034 photomultiplier. Best results were obtained from the 488 nm (blue) line using power levels of a few hundred milliwatts.

Griscom [8] lists some of the advantages of Raman spectroscopy as a tool for the investigation of glass structures. We would add another. Raman scattering from bulk rods produces a spectrum in which the line shapes are characteristic of the true spectrum. There is no distortion. This is in marked contrast to infrared spectra measured by conventional techniques of vacuum cold-pressing powdered glass into alkali halide pellets. In this case there are substantial distortions and additional broadening effects on the spectra. We will show that the Raman line shape, particularly the line width, is as important a measurement as the band frequency.

RAMAN SPECTRA OF THE ALKALI BORATE GLASSES

Spectra of the sodium and potassium borates have been published [6,7] and are not repeated here. Spectra of the lithium borate glasses (Fig. 1) illustrate the main features. For purposes of

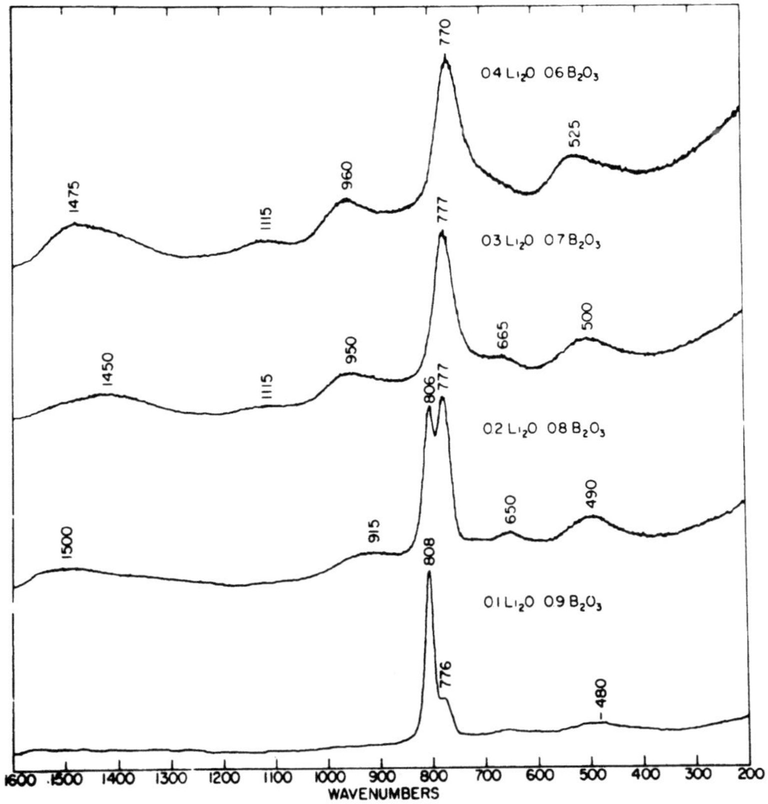

Fig. 1. Raman spectra for $XLi_2O\cdot(1-X)B_2O_3$ glasses.

description and discussion the spectra can be divided into three regions: (i) the high frequency region from 900 to 1500 cm^{-1} where several broad relatively weak bands occur; (ii) the region from 200 to 900 cm^{-1} which is dominated by a pair of sharp intense bands at 806 and 776 cm^{-1}. These bands have also dominated most previous discussion of the alkali borate spectra; (iii) the region below 200 cm^{-1}.

The spectra for Region (i) are shown in Fig. 2. This is the region of the spectrum which for most structures is dominated by bands due to the symmetric stretching modes of the metal-oxygen bonds that make up the structural units. The borate glasses, however, are weak scatterers in this region. Much of the observed scattering may be due to the weakly allowed asymmetric stretching

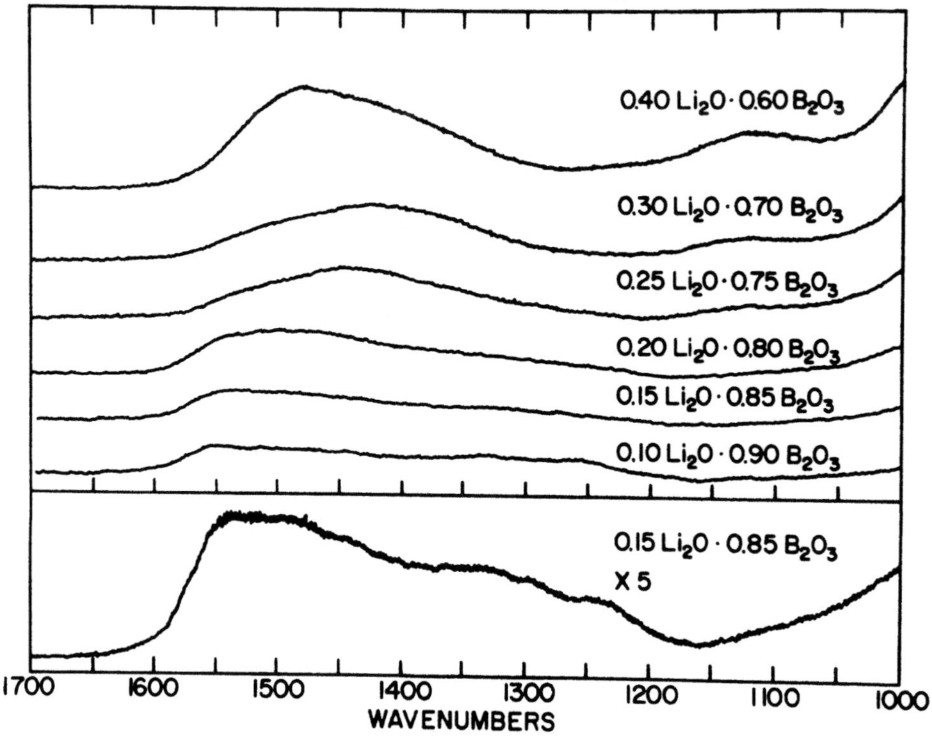

Fig. 2. Raman spectra of lithium borate glasses in Region (i). All measurements were made on pulled rods and intensities are roughly (±20%) comparable.

modes. There is a broad and weak scattering continuum that extends from 1200 to 1600 cm^{-1}. At low alkali content the distribution is weighted towards the 1550 cm^{-1} broad maximum. The continuum changes relatively little until the alkali content reaches 0.25 mole percent where it shifts abruptly to 1450 cm^{-1}. The broad maximum first decreases and then increases in intensity and occurs at higher frequency for the X = 0.4 glass, the most alkali-rich composition examined.

Region (ii) is of considerable interest. Most of the glass compositions examined exhibit two bands in this region (Fig. 3). One near 805 cm^{-1} is very sharp (as glass spectra go) and varies in intensity with alkali content. As the alkali percentage increases the 805 band decreases in intensity and at approximately 25 mole percent alkali it disappears. The second band at 775 cm^{-1} is already visible at 9% alkali and increases in intensity with

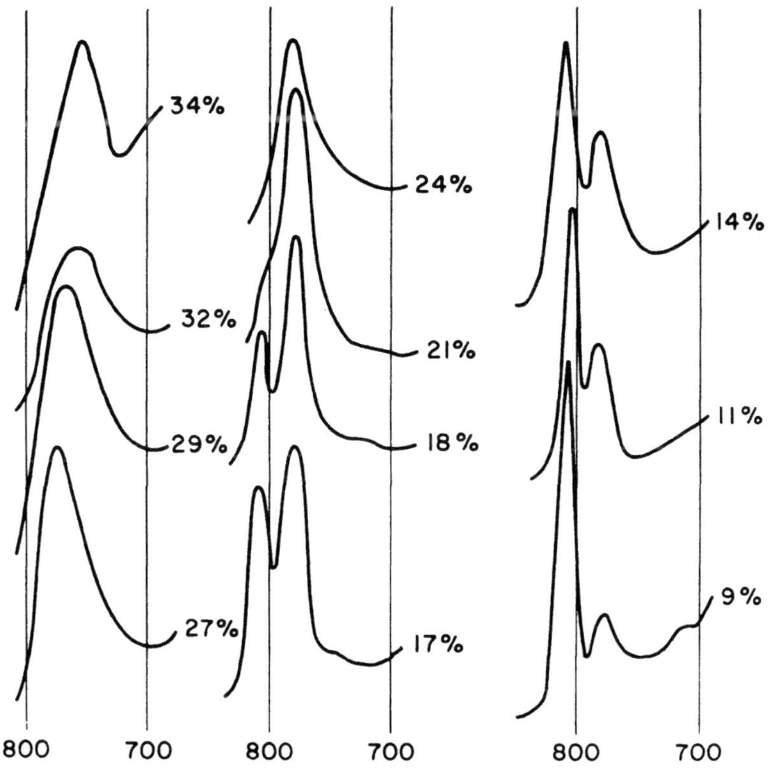

Fig. 3. Variations in the relative intensities of the 805-775 cm^{-1} bands for a series of sodium borate glasses.

increasing alkali content. It also is a sharp band in the low alkali region. The two bands are of comparable intensity at 16 mole percent alkali--the location of the so-called boron anomaly. The 775 cm^{-1} band continues to increase in intensity up to approximately 35 mole percent alkali where it reaches a maximum value. In the range of 25 to 40 mole percent alkali the 775 cm^{-1} band is the only feature in this region of the spectrum. It becomes noticeably broader but remains more or less constant in intensity across this composition range.

The frequency dependence of these two bands is shown in Fig. 4. The high frequency band decreases in frequency very slightly with increasing alkali content. Extrapolating to zero alkali the position of the band agrees exactly with the frequency of B_2O_3 glass

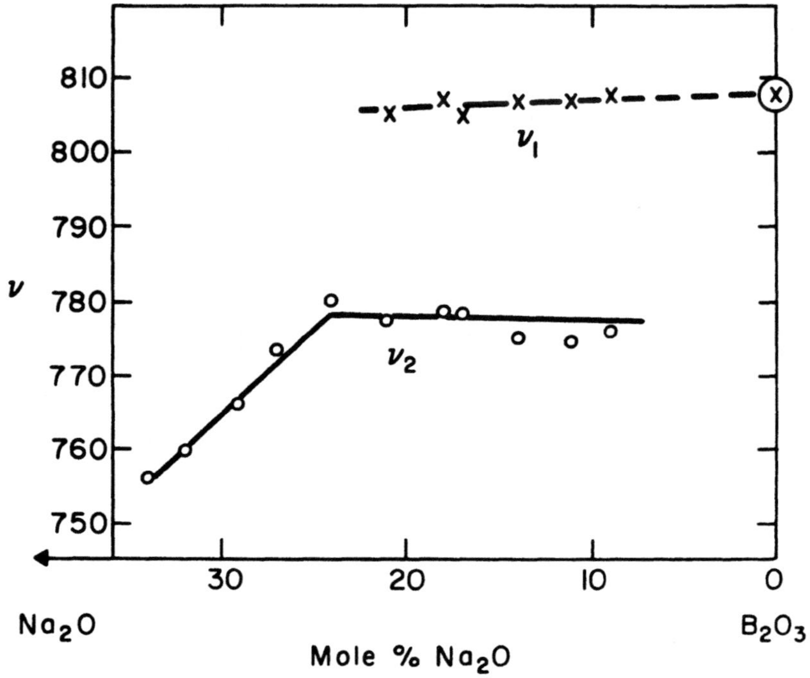

Fig. 4. Variation of band frequency with alkali content for a series of sodium borate glasses.

observed at 808 cm^{-1}. The band at 778 cm^{-1} remains constant in frequency to within experimental error between 9 and 25 mole percent alkali. Between 25 and 35 mole percent alkali the frequency continuously decreases.

The band width of these two sharp peaks also varies with alkali concentration. These widths were obtained by a careful graphical deconvolution of the overlapping bands. The results are shown in Fig. 5 and will be discussed later.

Spectra for Region (iii) are shown in Fig. 6. There is a well defined broad band in this region that is easily resolved from the descending Rayleigh tail. The peak shifts linearly towards higher frequencies as the alkali content of the glass increases. The linear trend in frequency extrapolates to the 24 cm^{-1} value found by Stolen [9] for B_2O_3 glasses. Similar bands have been observed

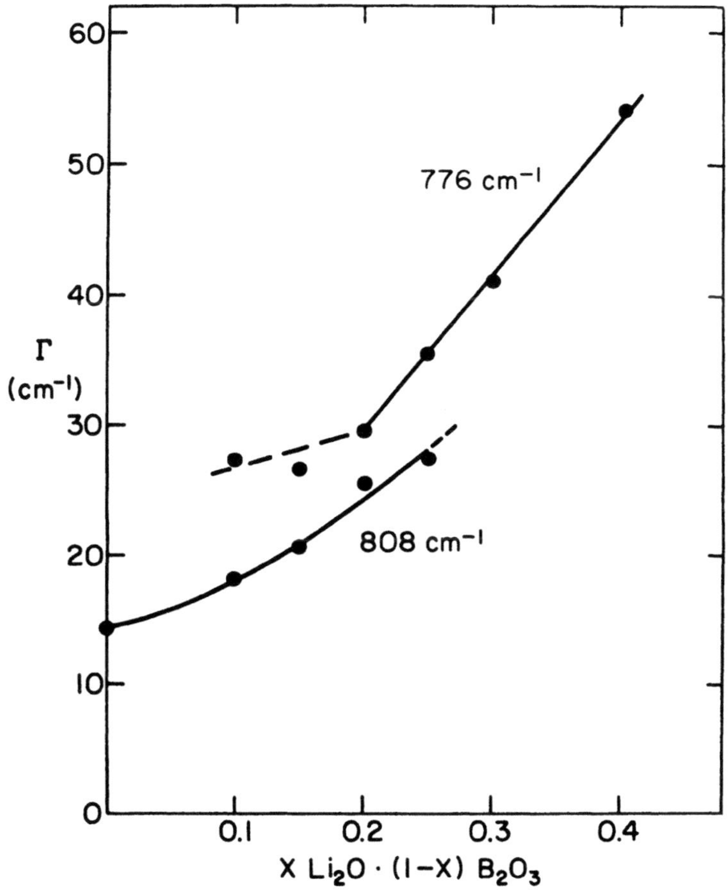

Fig. 5. Variation of full width at half height for a series of lithium borate glasses. The widths were obtained from carefully deconvoluted bands.

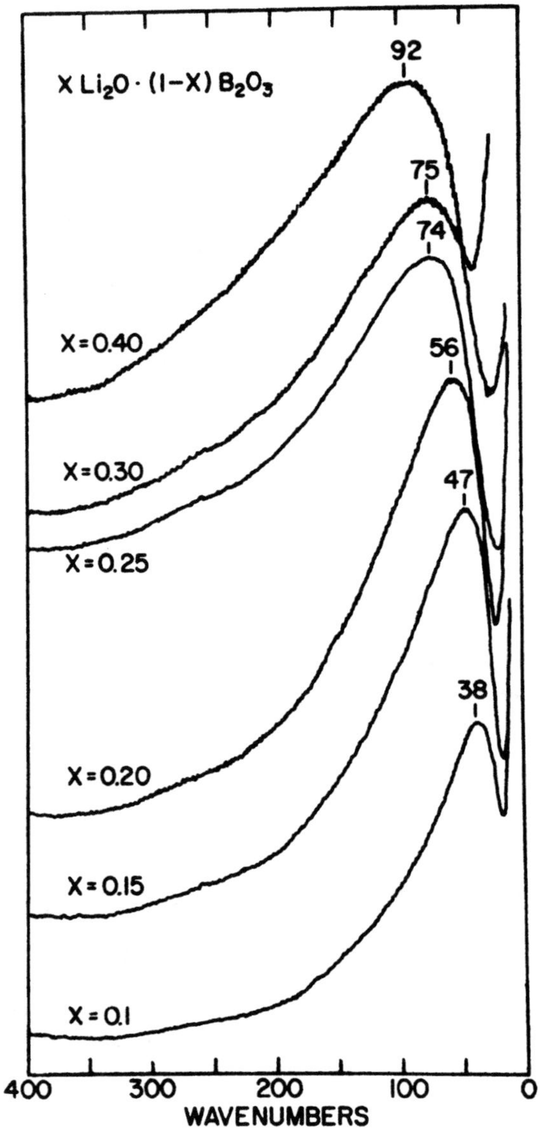

Fig. 6. Raman spectra of Li_2O-B_2O_3 glasses in low frequency region.

for pure B_2O_3, SiO_2 and GeO_2 glasses [9]. It was shown (for a discussion see [10]) that the band is due to a temperature-weighted density of states arising from low-lying optic modes as well as the higher-lying acoustic modes. The shift in band position to higher frequency may be due to the less open structure of the glass with increasing alkali concentration.

PHASE SEPARATION

Alkali borate glasses are known to phase separate on prolonged heating and it is of interest to investigate the effect of phase separation on the Raman spectra. Shaw and Uhlmann [11] using electron microscopic techniques mapped the metastable solvi in the alkali borate systems by first phase separating the glasses at low temperatures and then clearing them by reheating at temperatures above the solvi. Shaw and Uhlmann's solvi for lithium and potassium borate glasses are shown in Fig. 7. Glasses with 0.1 mole fraction Li_2O or K_2O were heat treated at the temperatures shown by the circles in Fig. 7. Three heat treatments were used. First glasses were quenched from the melt at 1000°C by squeezing drops of the liquid between aluminum blocks. Secondly, rods were pulled from the melt in the same manner as the rods used for all other Raman spectra. Thirdly, pulled rods were heat treated at 540°C for 10 hours (potassium borate glass) and 600°C for three hours (lithium borate glass). The main Raman band is shown in Fig. 8 for the potassium borate glass. The lithium borate glass gave an identical result. It can be seen that the line shapes are identical to within experimental error regardless of heat treatment.

The 808 cm^{-1} band and the 775 cm^{-1} band have been interpreted as measures of two quite distinct kinds of ring structures in the borate glass. Phase separation into a boron oxide-rich region and an alkali-rich region would require the formation of different sorts of rings in each of the phase separated glasses which might be expected to produce distinctly different Raman spectra.

DISCUSSION

Before embarking on a detailed analysis of Raman spectra in relation to glass structure, it is perhaps worthwhile to review briefly some of the experimental tools that have been used to investigate glass structures. Many more details are given by Griscom [8]. Each of these tools allows one to "see" in some indirect way the structure of the glass, but their "field of view" is substantially different.

Certain tools, notably NMR, EPR, Mössbauer and optical spectroscopy focus on specific target atoms in the glass. The geometry

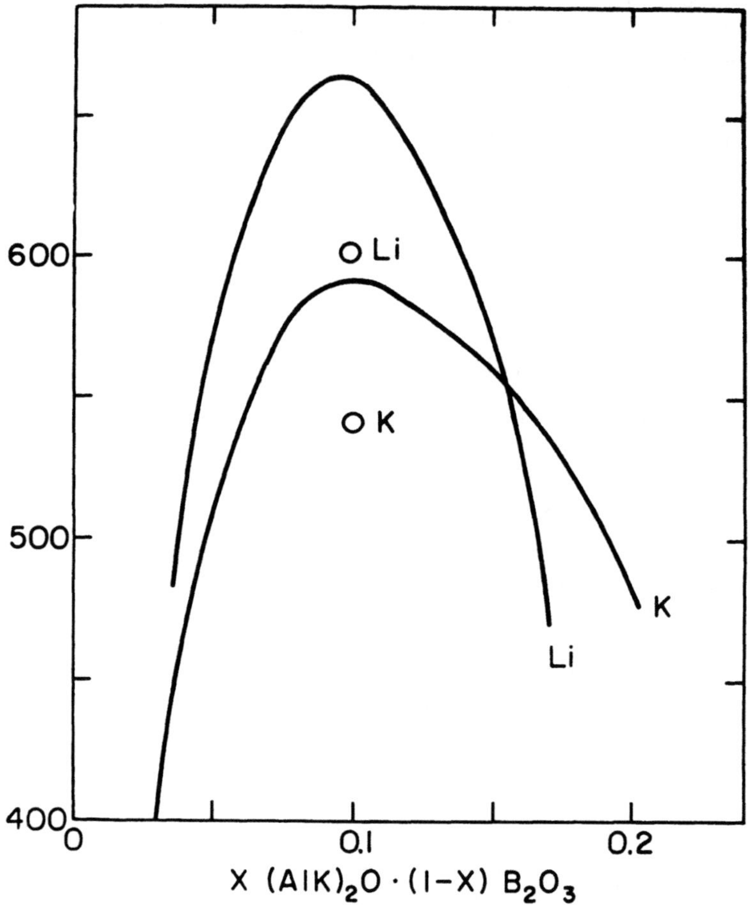

Fig. 7. Metastable solvi for alkali borate glasses from [10]. Circles show temperature-composition of heat treated glasses.

of the nearest neighbor anion coordination polyhedron of the target atoms can be determined with some certainty. The existence of BO_4 tetrahedra and BO_3 triangles as the basic structural building blocks of borate glasses rests on a fairly substantial foundation.

Larger structures in glasses become more equivocal. Infrared spectroscopy should be sensitive to short range order structures and should reveal all primary bonds including those that link the triangles and tetrahedra into larger structural units. Unfortunately, there are few reliable infrared spectra of borate glasses.

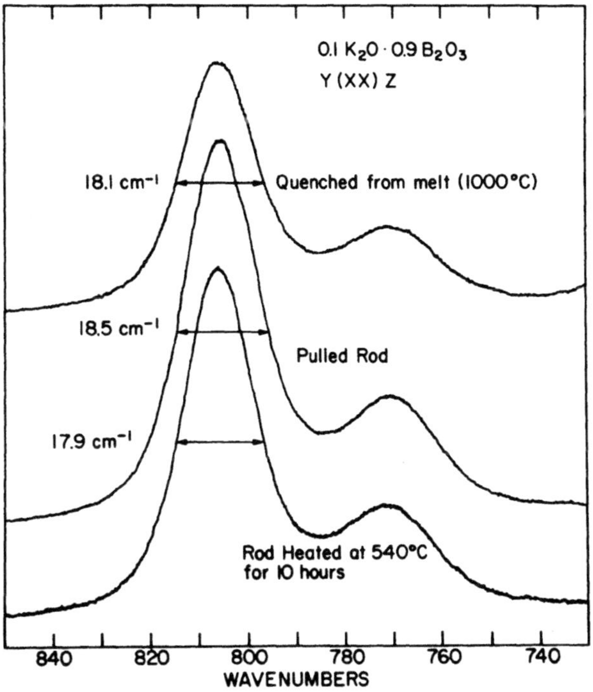

Fig. 8. Raman spectra of potassium borate glass after the heat treatment shown.

Common practice is to grind the glasses to powders, vacuum cold press these into alkali halide pellets, and measure the spectra on the resulting pellets. It is known (for example Ref. 12) that IR spectra of powders have distorted line shapes and shifted peak frequencies. Although the general features of covalent structures such as the borate glasses are likely to be approximately correct, arguments based on small details of band frequency and band shape are not reliable. More measurements using either thin films for direct transmission measurements or measurements by specular reflectance spectroscopy followed by Kramers-Kronig analysis to transform the reflectance spectra into absorption spectra would be helpful.

Raman scattering measurements from bulk glasses give accurate spectral line shapes so that comparisons and analyses of the spectra can be made with some confidence. However, the scale of the structures which will have some influence on the line shapes are less certain. Based on empirical experience with crystalline materials, it appears that Raman spectra are sensitive to structures on a size scale of 2 to 3 nm. Structural detail within volumes of glass of these dimensions can be quite complex and when the Raman spectrum consists of only a few broad bands, there is considerable indeterminacy in the structural interpretation.

Concepts of Glass Structure

Borate glass structures are complex, and there is the possibility of at least four hierarchies of structural organization: (i) the triangles and tetrahedra that act as the basic structural units; (ii) possible closed ring molecular-like structures that make up the building blocks of the Krogh-Moe model [8]; (iii) organizations on a larger scale which might be described as clusters, thermal fluctuations in composition, or incipient nuclei for phase separation, and (iv) distinct phase-separated regions.

These latter two require comment because the term phase-separation is somewhat confused in the glass literature. A two-phase glass must consist of two non-crystalline regions separated by a distinct boundary or interface. The individual phases must have a uniform Gibbs free energy within the bulk, but a discontinuity in the Gibbs free energy across the interface. Phase-separated regions, therefore, are larger structures than the first three and would appear to most of the measuring tools as additive mixtures of two different materials. Clustered glasses are not phase-separated in the phase rule sense but merely consist of regions of differing composition, without any distinct phase boundary between the regions. One could imagine the scale of these regions being rather small, perhaps on the order of 2 to 10 nm, a size range very difficult to observe by any of the available tools.

Interpretation of the Raman Spectra

Returning now to the specific problem of relating observed Raman spectra to borate glass structure, we begin by pointing out three salient features of the spectra: (i) the exceptionally sharp bands at 775 and 808 cm^{-1} that have attracted so much attention; (ii) the absence of any intense or well defined feature in the high-frequency region, and (iii) the overall lack of spectral detail in the low alkali glasses and the gradual transition to more detailed spectra as the alkali content increases. In the glasses of highest alkali content (Fig. 1) all bands are of comparable intensity and width and only these alkali borate glass spectra are comparable to spectra observed in alkali silicate and more complex ternary silicate and even borosilicate glasses [13-17].

The Raman and infrared spectra of borate glasses are generally interpreted as arising from a few well defined ring structures in the glass. This interpretation seems to be most firm for B_2O_3 and alkali borates with low alkali content in which the existence of boroxal rings can be inferred with some confidence.

In a theoretical study [18-19] Krogh-Moe provided strong evidence that such boroxal rings can give rise to a sharp Raman peak at 800 cm^{-1} even if they are joined in a random manner. Moreover, he showed that all other peaks in the spectra should be much broader than the 800 cm^{-1} band. Brawer [14] has shown in a more general study how a similar behavior can arise in silicates. The results of this latter study suggest that for any network glass certain vibrational modes will be relatively insensitive to disorder. These modes will then stand out in the Raman spectra of the glass. The width and frequencies of these bands do respond to a small degree to the general glass structure, but the random nature of the glass is most dramatically apparent in the width of all the other peaks in the spectra.

The interpretation of the Raman spectra of a glass usually reduces to an interpretation of the behavior of the strong Raman peaks. The information obtainable is, generally, the peak frequency and peak width. These two parameters can in no way be used to uniquely deduce the structure of a glass, the random nature of which may require tens of parameters to describe. However, in combination with other physical and chemical evidence some qualitative information may be deduced. It is in this spirit that the Raman spectra are usually interpreted in terms of well defined ring structures [6,7,13-17].

Let us assume, then, that the Raman spectra of alkali borate glasses indicate that, as alkali is added to B_2O_3, the boroxal rings giving rise to the 808 cm^{-1} peak are systematically replaced by some other type of structure (or structures) giving rise to the peak at 776 cm^{-1}. (We will speak here of two structures, although there are possibly more.) The question we would like to ask is the following: Are these different structures intimately joined together or do they occur in different phase-separated regions of the glass? (We will not enter here into a discussion of the nature of these structures, as this has been discussed exhaustively by the Dutch group.)

The experimental evidence bearing on this question consists of: the frequencies of the two peaks at about 805 and 775 cm^{-1} (Fig. 4), the widths of these peaks obtained by careful deconvolution of the spectra (Fig. 5), and the behavior of the spectra for different heat treatments (Fig. 8).

Consider first the peak frequencies. The fact that the peak

frequencies are relatively insensitive to concentration does not mean that the two types of structures giving rise to the peaks are in different phase-separated regions. As discussed above, the strong Raman peaks of some kinds of ring structures can be expected to be relatively insensitive to the way in which the rings are joined together. The constancy of peak frequencies is then taken to indicate that basically the structures are well defined for all compositions.

However, the fact that the peak width of the 805 cm^{-1} band increases monotonically with increasing alkali concentration (Fig.5) does indicate that the environment of the boroxal rings does vary with changing alkali concentration. This in turn suggests that these rings are not in macroscopically phase-separated regions, but rather that the different structures present are linked to each other in some manner.

It does not seem likely, however, that the two different structures are linked together in a completely random manner, but rather that some sort of clustering of like structures occurs. The evidence for this is the behavior of the spectra upon heat treatment (Fig. 8). The fact that the spectra are independent of a heat treatment which, according to Shaw and Uhlmann, causes the glass to phase-separate means that the immediate environment of each type of structure is unchanged by the heat treatment.

Thus we may conclude that the two different ring structures that give rise to the two Raman peaks at 805 and 775 cm^{-1} are clustered in some manner.

We must emphasize that we are not talking about phase separation in the Gibbs phase rule sense. The spectral behavior observed can be accounted for by clusters of some unknown size scale which are strongly linked to each other to form a continuous network. Such glasses would be extremely disordered with a large variability in the B-O-B bridging bond angles. Although the proposed ring structures may be present, certainly there is great disorder in the way in which the rings are linked.

There is some evidence for compositional fluctuations in the thermodynamic behavior of borate liquids. An earlier study [20,21] probed the structure of alkali tetraborate liquids using Ni^{2+} as a probe. Thermodynamic modeling of the liquidus surface suggested that the liquid segregated into boron-rich regions and nickel (and perhaps also alkali) rich regions. Optical spectra of Ni^{2+} in alkali tetraborate glasses provided supporting evidence for the thermodynamic model.

CONCLUSION

Raman spectra of the alkali borate glasses have been presented with emphasis on the implications of detailed frequency shifts and line shapes. The results have been interpreted on a rather complicated three-level hierarchy of glass structures which include boron-oxygen polyhedra, topologically closed rings of polyhedra, and larger scale interconnected clusters of alkali-rich and alkali-poor structures. Both the model and the argument are deliberately vague. The borate glasses appear to be highly disordered materials with a size scale such that no presently available tool gives a really clear view of the structural detail.

Acknowledgment

This work was supported for the most part by the U.S. Energy Research and Development Administration under Contract EY-76-S-02-2754. We thank Joyce McKay for measuring some of the Raman spectra.

References

[1] M.C. Tobin and T. Bååk, J. Opt. Soc. Amer. 60 (1970) 368.

[2] R.S. Krishnan, Indian J. Pure & Appl. Phys. 9 (1971) 916.

[3] A. Bertoluzza, C. Fagnano and P. Monti, Academia Nazionale dei Lincei, Ser. 8, 55 (1973) 506.

[4] J. Krogh-Moe, J. Non-Crystalline Solids 1 (1969) 269.

[5] W.B. White, G.J. McCarthy and J. McKay (abs) Ceram. Bull. 50 (1971) 411.

[6] W.L. Konijnendijk, Philips Res. Reports Suppl. No. 1 (1975).

[7] W.L. Konijnendijk and J.M. Stevels, J. Non-Crystalline Solids 18 (1975) 307.

[8] D.L. Griscom "Borate Glass Structure" pp 1-128, this volume.

[9] R.H. Stolen, Phys. Chem. Glasses 11 (1970) 83.

[10] S.A. Brawer, Phys. Chem. Glasses 16 (1975) 2.

[11] R.R. Shaw and D.R. Uhlmann, J. Amer. Ceram. Soc. 51 (1968) 377.

[12] J.D. Axe and G.D. Pettit, Phys. Rev. 151 (1966) 676.

[13] S.A. Brawer and W.B. White, J. Chem. Phys. 63 (1975) 2421.

[14] S.A. Brawer, Phys. Rev. B11 (1975) 3173.

[15] S.A. Brawer and W.B. White, J. Non-Crystalline Solids 23 (1977) 261.

[16] H. Verweij and W.L. Konijnendijk, J. Amer. Ceram. Soc. 59 (1976) 517.

[17] W.L. Konijnendijk and J.M. Stevels, J. Non-Crystalline Solids 20 (1976) 193.

[18] J. Krogh-Moe, Phys. Chem. Glasses 6 (1965) 46.

[19] L.A. Kristiansen and J. Krogh-Moe, Phys. Chem. Glasses 9 (1968) 96.

[20] J.S. Berkes and W.B. White, J. Crystal Growth 6 (1969) 29.

[21] J.S. Berkes and W.B. White, Phys. Chem. Glasses 7 (1966) 191.

INFRARED SPECTRA AND STRUCTURE OF CVD B_2O_3-SiO_2 GLASSES

Joe Wong

General Electric Corporate Research and Development

Schenectady, New York 12301

INTRODUCTION

Glassy solids are commonly formed by continuous cooling from the liquid state. The procedure consists of quenching a normal liquid into the supercooled liquid range at such a rate so as to bypass its equilibrium crystallization point. Finally, at T_g, the glass transition temperature, the system falls out of its internal equilibrium. The supercooled liquid now transforms to a glass, carrying with it the structural configuration at T_g. However, the essential requirement for glass formation is that the thermal energy of an ensemble of particles, whether molecules, ions, atoms, etc. can be removed at a rate which for kinetic reasons, precludes the organization of these particles into a crystal lattice. For this purpose an initially liquid state of the glassy system is by no means necessary. In fact a more efficient procedure for removing the molecular thermal energy is to deposit the molecules sequentially from a vapor onto a cold substrate (cold with respect to T_g of the deposited material).

Thus, the vapor-phase route provides not only another way of arriving at the glassy state of matter, but often times allows preparation of novel amorphous materials unattainable by conventional techniques of quenching the liquid state. Binary B_2O_3-SiO_2 glasses of composition not readily obtained by melt-quenching can be prepared by chemical vapor deposition (CVD) techniques(1) from one end component to the other.(2) These CVD materials in the form of thin film are ideal for transmission spectroscopy in the frequency region where the materials exhibit high absorption coefficient e.g. 10^4 cm^{-1} in the infrared.(3) Technologically, these binary glass films have wide applications as diffusion source in the planar technology of microelectronic fabrication[2,4] and, more recently, as cladding materials for low-loss SiO_2-core fibers in optical waveguides.[5,6]

In this paper the infrared transmission spectra of binary borosilicate glass films vapor-deposited at low temperatures (350-600C) from N_2-diluted reactive mixtures of B_2H_6, SiH_4 and O_2 are studied in some detail as a function of thermal history and composition. The spectra and structure of CVD B_2O_3 deposited above its T_g (~260C) are contrasted with those of CVD SiO_2 deposited far below its T_g (~1159C). In addition, taking advantage of the accurate control of film thickness during deposition, a differential technique has been used to elucidate some of the complex spectral features in the vibrational spectra of the binary glasses. The analysis and understanding of the infrared spectra of this simple binary borosilicate glass system form a basis of interpretation of the vibration spectra of other complex borosilicate glasses.

EXPERIMENTAL

The deposition of borosilicate glass films on chemically etched high-resistivity single-crystal Si substrates from pre-set mixtures of N_2-diluted B_2H_6-SiH_4-O_2 mixtures (molar ratio of O_2 to total hydride $\geq$ 10) has been described in detail elsewhere.[2,7] Film thickness was monitored in-situ during deposition using the color-thickness chart of Pliskin and Conrad[8] for SiO_2 thermally grown on Si at 1000C. Films containing more than 20 mol percent B_2O_3 are hygroscopic and are prevented from atmospheric moisture degradation during infrared measurement with a covering layer of pure SiO_2 (~400 Å thick) deposited sequentially from a pre-set SiH_4-O_2 mixture. The covering layer of SiO_2 is compensated quantitatively in a Perkin-Elmer double-beam spectrophotometer (Model 457) with an identical SiO_2 layer deposited on a reference Si substrate. Film compositions were determined from the intensity ratio of the bands at ~1300 cm^{-1} and ~1000 cm^{-1} using a composition calibration curve of Tenney.[9]

A differential infrared (DIR) technique applicable to transmission measurement of thin films has been described in some detail elsewhere.[10] For clarity the essential steps will be outlined in the section on DIR data.

Films were heat treated in the deposition reactor to a maximum temperature of 700C for 10 minutes. Under this condition the free-carrier absorption arising from the diffusion of boron into the silicon substrate is negligible so that the infrared spectra below 1000 cm^{-1} are not obscured.

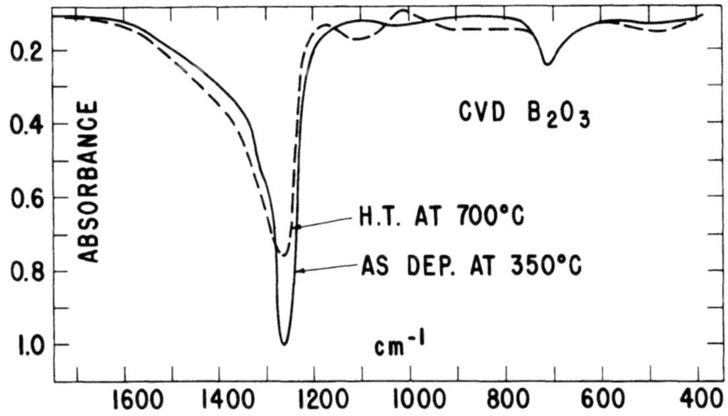

Fig. 1: Room temperature IR transmission spectra of B_2O_3 vapor-deposited on Si. The solid curve denotes that of the as-deposited film deposited at 350°C while the dashed curve is that of the same film heat treated to 700°C. The film was 4000 Å thick and was covered with a 400 Å thin layer of SiO_2 which was compensated in the reference beam with an identical SiO_2 film on a reference Si wafer.

RESULTS AND DISCUSSION

Pure Components

CVD B_2O_3. The room temperature infrared transmission spectrum of pure B_2O_3 deposited at 350C is shown in Fig. 1. Two prominent absorption bands at 1265 cm^{-1} and 715 cm^{-1} are evident. Using the relationship $A = \alpha t$ where A is the absorbance, α the absorption coefficient and t film thickness, the absorption coefficients at these two frequencies are found to be $(2.1 \pm 0.1) \times 10^4$ cm^{-1} and $(0.32 \pm 0.02) \times 10^4$ cm^{-1}. With ^{10}B substitution the 1265 cm^{-1} band exhibits a 21 cm^{-1} upshift(10) and is associated with B-O bond stretching modes in the glassy lattice, while the lower frequency band exhibits only a 6 cm^{-1} upshift and is associated with bond-bending modes.

When deposited above its equilibrium melting point in the range 450-600C (so that B_2O_3 is now deposited as liquid films) the resultant infrared spectra remain invariant and are directly superimposed on to that of the film of the same thickness deposited at

350C shown in Fig. 1. However, when the as-deposited films are subsequently heat treated at 700C for 10 min. in N_2, a large change in the shape of the B-O band at 1265 cm^{-1} is observed (also shown in Fig. 1). The absorbance at band maximum decreases from 0.90 to 0.66 while the bandwidth at half height increases from 65 cm^{-1} to 120 cm^{-1}. The integrated intensity (area under the band) remains constant indicating that the total number of oscillators responsible for this absorption remain unchanged after heat treatment. The increase in bandwidth is indicative of a quenching-in of high temperature liquid configurations which, in general, have a broader distribution.

CVD SiO_2. The above observation in B_2O_3 is contrasted with that of CVD SiO_2 films deposited at temperatures much below its T_g (1159C).[12] When heat-treated to 800C and above the infrared spectrum of SiO_2 deposited at 450C from SiH_4-O_2 mixtures exhibits a progressive bandwidth narrowing accompanied by an increase in both the absorbance and frequency at band maxima.[13] (Fig. 2) The integrated intensity of each band remains invariant as in the case of the B-O absorption. The IR data are correlated with a decrease in the root-mean-squared deviations of atomic distances in the radial distribution function.[13] Together with changes in other physical properties such as density, refractive index, etch rate, etc.[4] the combined vibrational and structural variations are indicative of a densification process brought about in as-deposited SiO_2 as a result of thermal annealing above its deposition temperature.

Binary Glasses

The room temperature IR spectra of as-deposited borosilicate glasses have been studied systematically as a function of composition across the whole binary system.[7] A distinct absorption at 930 cm^{-1} not present in either of the pure component spectra appears in the spectra of the binary glasses, and is associated with a vibration of the B-O-Si linkages in the borosilicate structure. The composition variation of the intensity of this band has been used quantitatively to construct a structural model of random network of BO_3 triangles and SiO_4 tetrahedra bridged by oxygens. Above 1000 cm^{-1}, the absorptions are more complex and will be studied here with a DIR technique that has been proven useful in unraveling the vibrational spectra of CVD P_2O_5-SiO_2 glasses.[10]

In Fig. 3 the curve labeled "BSG" is the normal IR spectrum of an as-deposited borosilicate film containing 24 mol percent B_2O_3 taken versus an identical covering layer of SiO_2 (~400 Å thick) deposited on a bare Si in the reference beam. The curve labeled "SiO_2" corresponds to the normal IR spectrum (compensated by the

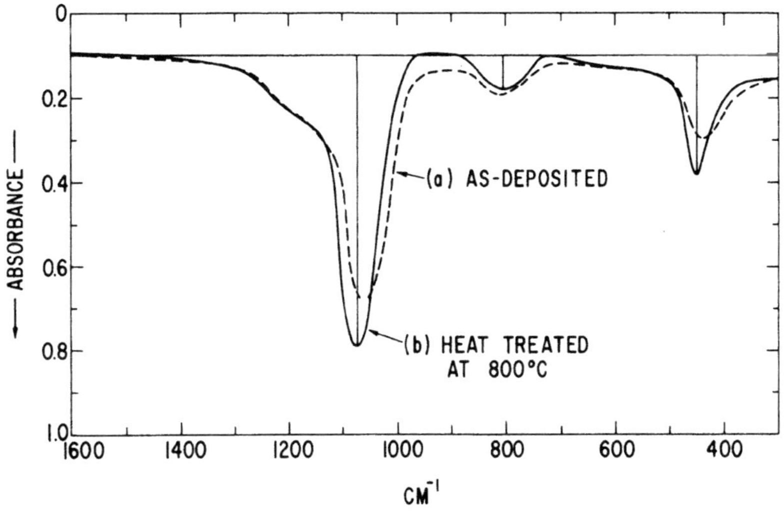

Fig. 2: Room Temperature IR transmission spectra of CVD SiO_2 film, 5200 Å thick: (a) as-deposited at 450°C and (b) the same film heat-treated at 800°C for 15 min.

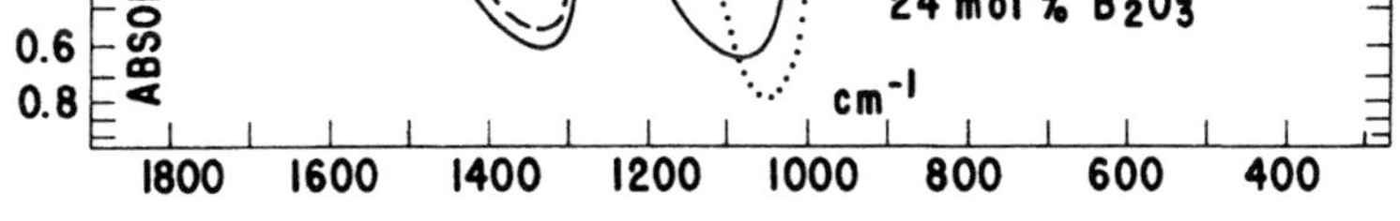

Fig. 3: DIR spectrum of an as-deposited borosilicate film containing 24 mol percent B_2O_3 versus a corresponding SiO_2 film deposited under identical conditions of temperature and flow rates of SiH_4, O_2 and N_2. The curves labeled "BSG" and "SiO_2" are the normal IR spectra taken versus an identical covering layer of 400 Å SiO_2 deposited on a bare Si wafer in the reference beam.

same reference) of a pure SiO_2 film deposited under the same conditions of temperature, time and flow rate of SiH_4, O_2 and N_2 carrier gas as the BSG film. The curve labeled "DIR" was obtained by rescanning the BSG spectrum but now using the SiO_2 film in the reference beam. From the DIR spectrum, a band at 1125 cm^{-1} can clearly be differentiated from the broad absorption at 1100 cm^{-1} in the BSG spectrum. The frequency and halfwidth (110 cm^{-1} $\pm$ 5 cm^{-1}) of this differential band remain constant with composition across the binary system. The intensity of this band (measured as peak absorbance with respect to point X in the DIR spectrum and normalized to a 1-micron-film absorption exhibits a maximum in the vicinity of 40 mol percent B_2O_3 as shown in Fig. 4. This variation is very similar to the behavior of the well-defined Si-O-B vibration at 930 cm^{-1} studied previously.(7) The constancy in frequency and bandwidth suggests that this mode is intrinsic of the borosilicate structure and its intensity variation with composition in analog to that of the 930 cm^{-1} band strongly indicates that it is a bond-stretching vibration of the Si-O-B linkages in the borosilicate lattice. In addition, the broad absorption at 450 cm^{-1} in the BSG spectrum can be seen to consist of overlapping contributions of the Si-O-Si and B-O-B bending modes in the binary mixture.

Similarly DIR spectra have been recorded versus pure B_2O_3. The results for the same 24 mol percent B_2O_3 film discussed above is shown in Fig. 5. The curve labeled "B_2O_3" is that for a pure B_2O_3 film deposited under identical conditions of temperature and time and flow rates of B_2H_6, O_2 and N_2 to those of the BSG film. The DIR spectrum in this case consists simply of a differential band at ~1360 cm^{-1} and is less informative. With addition of SiO_2 to B_2O_3, the frequency of this band increases from the pure B_2O_3 value of 1265 cm^{-1} to 1380 cm^{-1} at 90 mol percent SiO_2. This mode is clearly associated with the stretching vibration of the B-O sublattice and is perturbed by the presence of the less polarizing Si^{4+} cation substituting for an equally polarizing B^{3+} cation as the next-nearest neighbor; thus effectively increasing the force constant of the B-O bond in the mixture.

CONCLUDING REMARKS

The IR transmission spectra of thin borosilicate glass films vapor-deposited on Si from a B_2H_6-SiH_4-O_2 source at low temperatures have been studied systematically across the whole composition of the binary system. The effect of thermal history on the structure of CVD glasses deposited above T_g as in the case of B_2O_3 is contrasted with that of CVD SiO_2 deposited at temperature far below the T_g of the bulk material. In the former case, where configurational relaxation is possible at deposition temperatures, the IR spectrum

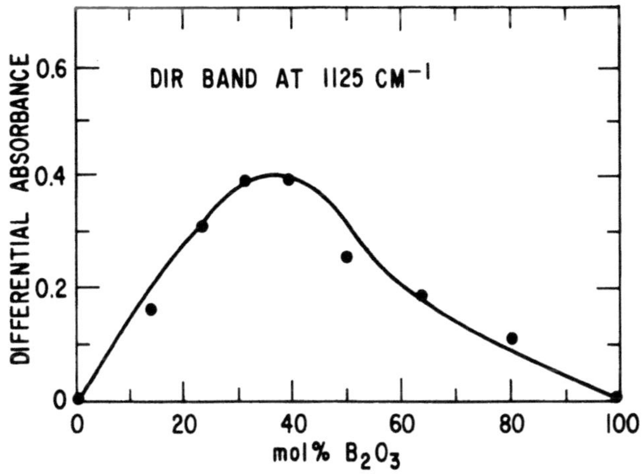

Fig. 4: Compositional variation of the intensity of the DIR band at 1125 cm^{-1} shown in Fig. 3. The plotted differential absorbance is normalized for 1-micron film in all cases.

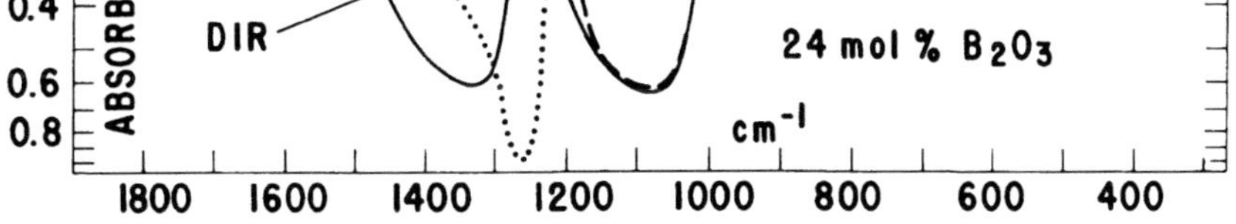

Fig. 5: DIR spectrum of an as-deposited borosilicate film containing 24 mol percent B_2O_3 versus a corresponding pure B_2O_3 film deposited under identical conditions of temperature and flow rates of B_2H_6, O_2 and N_2. The curves labeled "BSG" and "B_2O_3" are the normal IR spectra taken versus an identical layer of 400 Å SiO_2 deposited on a bare Si wafer in the reference beam.

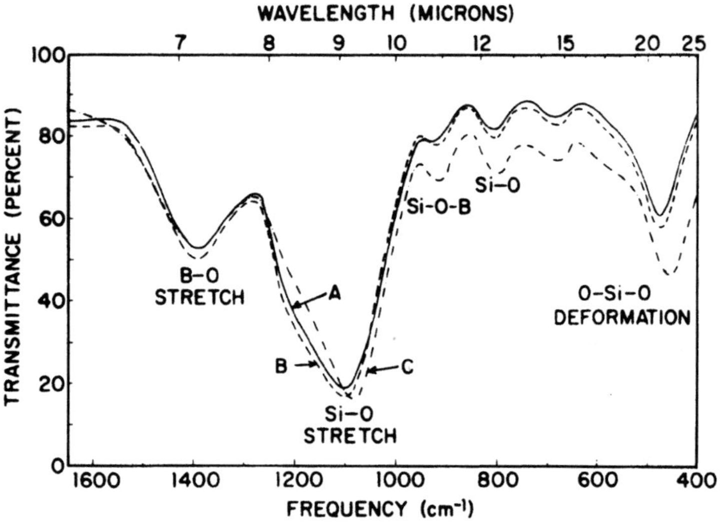

Fig. 6: IR spectra of Corning Pyrex films[15] illustrating the similarity and usefulness of the binary borosilicate spectra in understanding more complex borosilicate system. (A) sedimented powder; (B) annealing (A) at 450°C; (C) fusing (A).

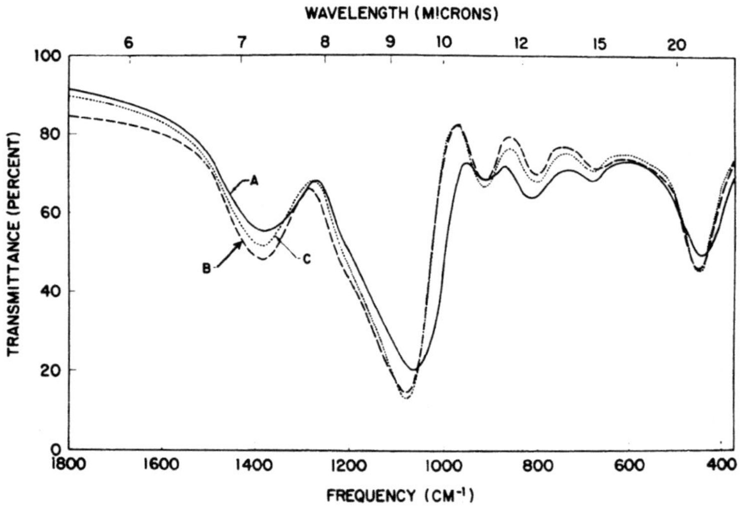

Fig. 7: IR spectra of General Electric GSC-1 glass films (an aluminoborosilicate)[15]: (A) RF sputtered on 100°C Si substrate (0.81 μ); (B) film formed by fusing sedimented powder (0.78 μ); and (C) heating (A) in dry N_2 for 5 min. @ 965°C (78 μ).

and hence the structure of as-deposited B_2O_3 is not dependent on deposition temperature even when deposited as a liquid film at 600C. Heat treating the as-deposited film further to 700C, however, results in an increase in the bandwidth of the B-O band at 1265 cm^{-1}, suggestive of a quenching-in of high temperature liquid configurations which in general exhibit a broader distribution. In the latter case, the IR spectrum of as-deposited CVD SiO_2 exhibits broader absorption bands which progressively narrow with increasing heat treatment temperature and finally approach the bulk values after heat treating at 1000C. The spectral observation is associated with a densification process in glass films deposited far below its T_g and is substantiated by other physical property changes such as increases in density, refractive index, etch rate, etc.

Differential IR data on the binary glasses lead to the resolution of a band at ~1130 cm^{-1} assigned to a bond-stretching vibration of the Si-O-B linkages in the glass structure. The intensity variation of this band with composition lends further support to the structural model proposed earlier for the B_2O_3-SiO_2 glass system in which the boron and silicon are 3-fold and 4-fold coordinated with oxygen respectively such that each oxygen is bridging a pair of B and/or Si atoms to form a 3-dimensional random network.

Finally the analysis of the IR spectra of the binary borosilicate system may serve as a basis of interpreting and understanding the vibrational spectra of more complex borosilicate glasses shown in Fig. 6 for the case of Corning pyrex and in Fig. 7 for the case of General Electric GSC-1 which is an aluminoborosilicate as well as others reported by Plisken.[15]

REFERENCES

1. W. Kern and R.C. Heim, J. Electrochem. Soc. 117, (1970), 568.
2. D.M. Brown and P.R. Kennicott, J. Electrochem, Soc. 118, (1971), 293.
3. J. Wong and C.A. Angell, "Glass: Structure by Spectroscopy", Marcel Dekker Publisher, (1976), Chapt. 8.
4. J. Wong, J. Electronic Materials 5, (1976),113.
5. L.G. Van Uitert, D.A. Pinnow, J.C. Williams, T.C. Rich, R.E. Jaeger and W.H. Grodkienicz, Mater. Res. Bull. 8, (1973),469.
6. W.G. French, A.D. Pearson, G.W. Tasker and J.B. MacChesney, Appl. Phys. Lett. 23 (1973) 338.
7. A.S. Tenney and J. Wong, J. Chem. Phys. 56, (1972), 5516.
8. W.A. Plisken and E.E. Conrad, IBM J. 8, (1964), 43.
9. A.S. Tenney, J. Electrochem. Soc. 118, (1971), 1658.
10. J. Wong, J. Non-Crystalline Solids 20, (1976), 83.
11. T.A. Sodorov and N.N. Sobolev, Opt. Spectrosc. (English Translation) 3, (1957), 360.
12. Fontana and Plummer, Phys. Chem. Glasses 7, (1966), 139.
13. N. Nagasima, J. Appl. Phys. 43, (1972), 3378.
14. P.E. Jellyman and J.P. Proctor, J. Soc. Glass Tech. 39, (1955) 173T.
15. W.A. Pliskin, "Physical Measurement and Analysis of Thin Films", (E.M. Murt and W.G. Guldner, Ed.), Plenum Press (1969), p. 168.

THE VIBRATIONAL ANALYSIS OF BORON IN VITREOUS SILICA

Carlyle F. Smith

Semiconductor Products Dept.
General Electric Company
West Genesee Street
Auburn, N. Y. 13021

I. INTRODUCTION

Glasses containing small amounts of boron in a silicon dioxide matrix are commercially important for their high use temperatures, low thermal expansivity, and resistance to devitrification. This study was undertaken to elucidate the coordination of boron in such glasses by recording and analyzing the infrared (IR) and Raman (R) spectral details that appear as a result of the incorporation of small amounts of boron.

A major problem in the IR recording, namely, the overshadowing of the borate spectrum by that of the silica host, was alleviated by employing a differential infrared (DIR) technique discussed in greater detail elsewhere (1). Briefly stated, the technique depends on the modulation of the reference beam of a double-beam spectrophotometer by a slice of pure material of equivalent optical density and structural composition to that of the host constituent of the doped material under investigation in the sample beam. Thus the sample beam and reference beam are identical in intensity except for spectral details arising only from the interaction of the chemically substituted species with the host network. The nonlinear transmissivity characteristic makes this a very sensitive method for detecting the normally weak dopant species spectra, although it makes thickness control a critical factor in sample preparation.

Differential Raman spectra can be obtained expediently by recording the scattered light intensities with a good machine under identical conditions of illumination; by multiplying the spectral intensities of the pure host material by the mole fraction of the host material in the doped material; and by subtracting these intensities, point by point, from the corresponding intensities obtained from the doped material. Since Raman intensities tend to be monotonically dependent on dopant concentration, this method is not too sensitive to very small quantities of dopant.

The spectral details thus obtained can be treated by normal coordinate analysis (NCA) methods (2-5) to establish the degree of correspondence in spectral position and appearance with predicted vibrational eigenvalues calculated from hypothetical structural models. The task of making the necessary calculations has been greatly simplified by the adaptation of the Wilson GF matrix method to computer programs coded in Fortran IV by Schachtschneider (6). These programs accept input of atomic masses, molecular geometry, vibrational symmetry coordinates, and estimated force constants according to the Urey-Bradley force field (UBFF). The operation is iterative to find the set of adjusted UBFF force constants which bring the calculated spectra into coincidence with that observed. While some judgement in choice of initial force constants and in determining the validity of adjusted force constants is required of the analyst, usually the application of selection rules for IR and Raman activity and comparison of force constant values with those obtained independently for homologous molecules gives sufficient basis for unambiguous confirmation of site group symmetry and structure.

The selection rules indicate that for perfectly symmetric and isolated monoborate site groups of the usual coordination by oxygen, the number and activity of vibrational modes should be as follows:

for BO_4 (T_d): 1 A_1 (R) ; 1 E (R) ; 2 F_2 (IR + R)

for BO_3 (D_{3h}): 1 A_1' (R) ; 1 A_2'' (IR) ; 2 E' (IR + R)

for BO_3 (C_{3v}): 2 A_1 (IR + R) ; 2 E (IR + R)

Thus it is seen that the site groups indicated should produce two, three, or four IR active vibrations, respectively, which should be distict from those of the host network; and either three or four similarly distinct vibrations which are Raman active.

Additional details should appear if the borate group disrupts the silica host network to produce non-equivalent oxygens, thus creating SiO_4 (C_{3v}) defect centers. In this case the splitting of degeneracies by lowering of symmetry leads to the appearance of two new stretching modes and two new bending modes distinct from the host network specta and appearing in both IR and Raman (7).

Finally, if the borate group is not isolated, but rather joined uniformly through all its oxygens to the host network through Si-O bonds, other vibrations may occur under certain conditions. If boron merely substitutes for silicon in tetrahedral coordination, it would not be expected to give rise to modes other than isotope-like shifts of the host network vibrations. The presence of boron as BO_3 groups attached to the network and replacing an SiO_4 group should result in one stretching vibration and one torsional vibration of the BO_3 group as a whole against the host network, over and above the modes appearing due to lowering of symmetry of the host network in the locality. Both these vibrations should be IR and Raman active.

II. EXPERIMENTAL

The glasses employed in this investigation were those listed in Table 1, together with the concentrations of the chemically substituted species as determined by chemical or spectroscopic analysis. The hot pressed glass containing boron was made by mixing one half by weight of Baker Reagent Grade silicic acid with one half by weight of DuPont No. 1 colloidal silica molding powder; making a paste of 5 grams of this mixture with 10 ml of an alcoholic solution of boric acid (4.2 gm H_3BO_3 per 100 ml methanol); drying and crumbling the resultant paste; and then hot pressing the resultant coarse powder in an inductively heated graphite mold for 15 minutes at 1125 ^oC and 3,000 psi pressure. The blank thus fabricated was clear, glassy, free of cristobalite or pore inclusions, and con-

Table 1. Characterization of Materials

Source	Code and Trade Name	Hydroxyl Content (ppm)	Major Dopant Concentration (atom/atom Si)
Amersil	Infrasil[R]	13	-
Amersil	Optosil[R]	179	-
General Electric	Type 124	5	-
Quartz Scientific	TD	20	-
Corning	7943	5	-
Hot Pressed	HP	10	.025 B
Corning	7913 Vycor[R]	1060	.041 B, .005 Al
Corning	7740 Pyrex[R]	450	.362 B, .030 Al .105 (Na,K)

tained 1.4 % by weight of boric oxide, reported as B_2O_3 in the chemical analysis. Samples for the IR investigation were prepared as described by Smith, et. al. (1). The Raman samples were rectangular parallelopipeds of approximately 0.5 x 0.5 x 1.5 cm cut from blocks of the various materials, and polished on all sides to minimize diffuse scattering of light.

Infrared spectra were obtained in the range 4000 to 200 cm^{-1} with a Perkin-Elmer Model 621 double beam diffraction grating spectrophotometer. The wavenumber accuracy of all sharp bands was $\pm$ 2 cm^{-1}. The machine was purged of CO_2 and H_2O using a Puregas heatless gas dryer. Raman spectra were recorded with a SPEX Industries, Inc. Model 1401 double spectrophotometer, with gratings ruled 1200 per mm and blazed for 500 nm. Slit widths were set at at 250 μm because of the broadness and low intensity of the spectral details. Illumination was provided by a Coherent Radiation Laboratories Model 54 argon ion laser operating at the 480 nm blue line with intensity of about 300 milliwatts after exiting the narrow-band-pass interference filter and focussing lens. Spectral intensity was sensed with an ITT Model FW-130 photomultiplier operating in the photon counting mode, and cooled by a Products for Research Model TE-104 thermoelectric refrigerator. Output from the photomultiplier was amplified, and recorded with a Yokogawa Electric Works Ltd. time-base recorder.

Examples of the IR, DIR, and Raman spectra are reproduced in Figures 1 through 4, where the spectra above 2000 cm^{-1} in the IR and above 1300 cm^{-1} in the Raman have been omitted for lack of pertinent detail. The spectra recorded for the various pure vitreous silica samples were in all cases coincident in structure and spectral intensity except for thickness-related absorption in the infrared, and for variations in the OH band at 3675 cm^{-1}.

The DIR spectra indicate the presence of borate-induced bands at 1390, 910, 770, 705, and 678 cm^{-1}. The normal IR spectrum for the HP glass also indicates a band at 475 cm^{-1} in a region of high absorption and low machine sensitivity. The DIR band in Fig. 1 at 1300-1100 cm^{-1} is the major silica network band appearing due to thickness mismatch between the samples, and illustrates the criticality of thickness control. The band at 275 cm^{-1} in Fig. 2 is located in a region of wavelength approaching the sample thickness, and hence is suspect as an interference-related effect. Otherwise, the DIR spectral details for the commercial low-boron glass and for the hot pressed glass are virtually identical in structure and band center location.

The Raman spectra of Fig. 3 are not easy to interpret without further treatment. The graphical reduction of spectral differences, as shown in Fig. 4, shows consistently appearing borate-related

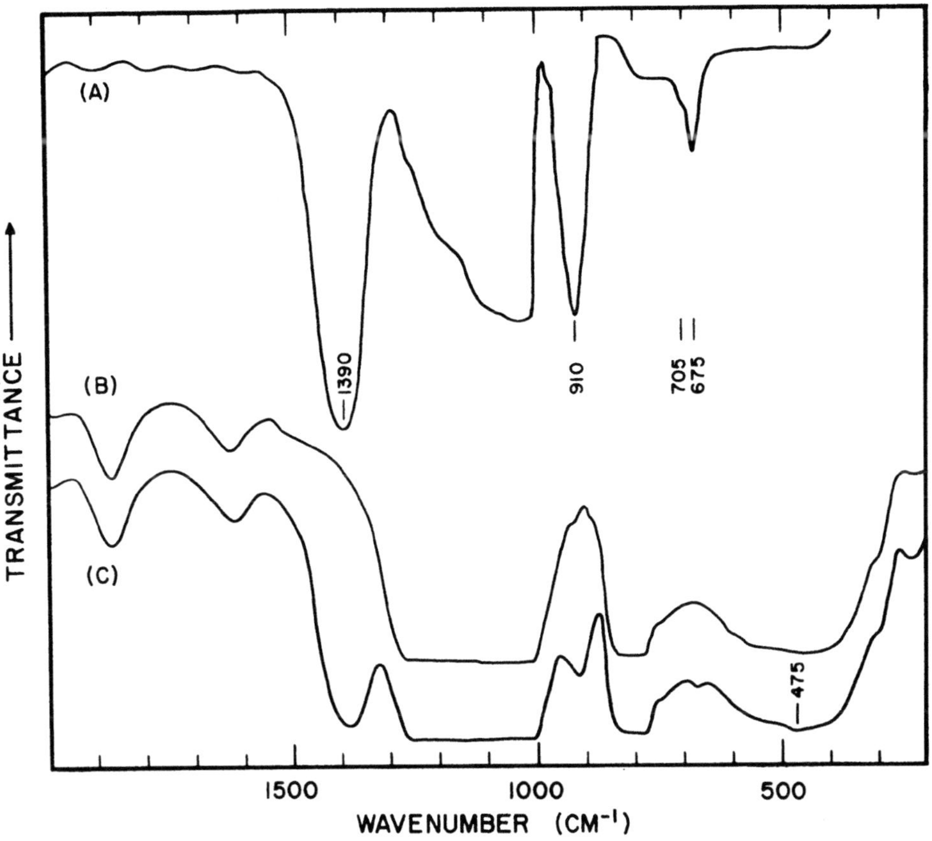

Figure 1. Infrared Spectra for Vitreous Silica and Boron-Containing Glass

(A) DIR spectrum for HP glass vs. Infrasil
(B) Normal IR spectrum for Infrasil, 28.4 μm thick
(C) Normal IR spectrum for HP glass, 34.2 μm thick

peaks appearing at 1132, circa 945, 770, 712, 475, 443, and 38 cm^{-1}. It is seen from Fig. 3 that as the boron content is increased, the host network peaks at 495 and 612 cm^{-1} gradually disappear, while the structure at 67 cm^{-1} gradually grows in intensity and shifts to lower wavenumbers. Also, the shoulder at ca. 830 cm^{-1} gradually disappears. These phenomena appear in the differential Raman spectrum as the growth of underlying broad intensities which gradually submerge, obscure, and shift the fundamental host network intensities. The absolute values of spectral intensities with reference to recorder scale zero were actually the reverse of that depicted

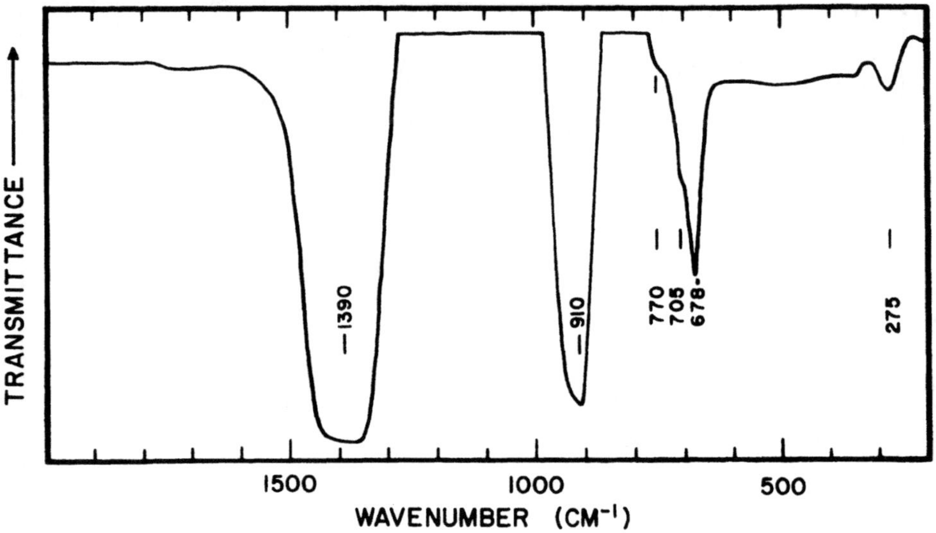

Figure 2. The DIR Spectrum for Vycor, 21.4 um Thick

in Fig. 3, where the intensity is relative to the particular spectrum displayed, only.

III. DISCUSSION AND CONCLUSIONS

In the IR and Raman investigations of sodium and titanium silicate glasses by Smith (7), the IR band circa 910-930 cm^{-1} was clearly shown to be due to the A_1 stretching mode of the vibrations of the basal oxygens of an SiO_4 group containing one symmetrically non-equivalent Si-O bond; and that the Raman peak at 1132 cm^{-1} corresponds to the stretching vibration of the non-equivalent oxygen of such a group. Placing this assignment on the details observed here, this leaves the IR and Raman details at 1390, ca. 935, 770, 712, 705, 678, 475, 443, and 38 cm^{-1} to be analyzed, as well as the broad, weak Raman intensity extending from 300 to 450 cm^{-1}.

In analyzing the spectra, one would first like to estimate whether the boron is present in trigonal or tetrahedral coordination by oxygen, other possibilities being unlikely. One may tentatively exclude the BO_3 (C_{3v}) group from consideration, since this symmetry requires four IR-active bands, and only three bands are apparent which are not attributable to isotopic shift or interaction with the host network. To distinguish between the tetrahedral and trigonal planar possibilities, the examination of the 675 cm^{-1} band is critical because of its intensity and sharpness,

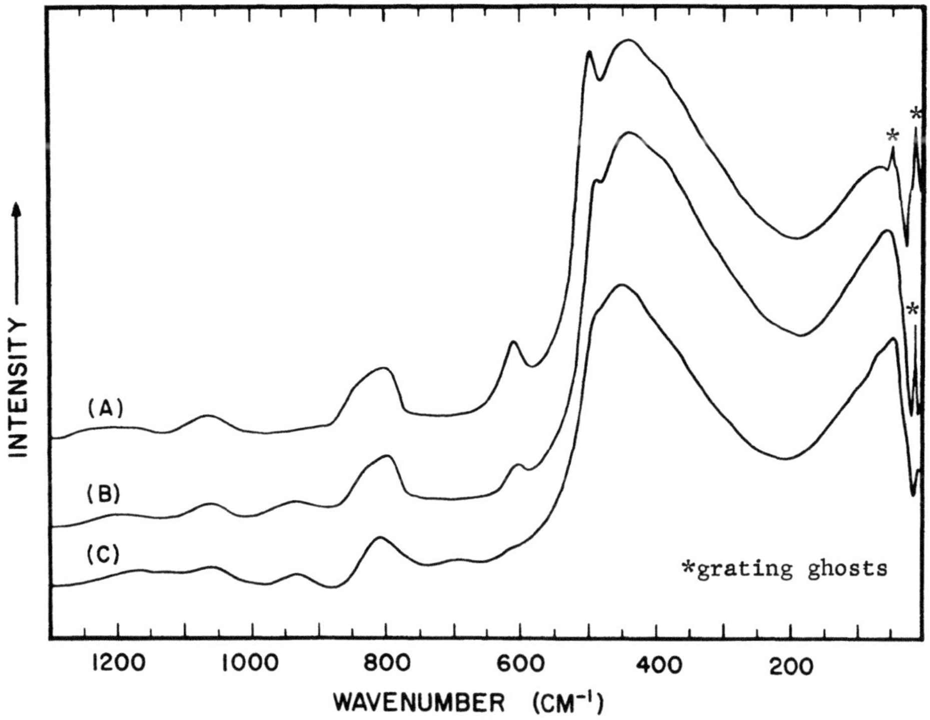

Figure 3. The Raman Spectra for Vitreous Silica and Boron-Containing Glasses

(A) Optosil (B) Vycor (C) Pyrex

and because the shoulder at 705 cm^{-1} suggests the appearance of the isotopic shift effect. The natural isotopic abundance of B^{10} of 19.6 % should give rise to shifted peaks for vibrations in which the boron mass plays a governing role. The effect will not be seen in the symmetric stretching vibration of the tetrahedral or trigonal site group, inasmuch as the vibrational wavenumber is dependent only on the oxygen mass and bond force constants. There would be no effect on the tetrahedral E species bending mode for the same reason. The wavenumber is too low for trigonal or tetragonal asymmetric stretching modes and quite high for asymmetric bending modes for the atomic masses and type of bond involved. Analyzing the band under the assumption that it is an F_2 species bending mode of a BO_4 tetrahedral site group, and making appropriate changes in the G matrix of the analysis, the calculation shows that the maximum isotopic shift could only be 4 to 5 cm^{-1}, rather than the 25 to 30 cm^{-1} actually observed. Furthermore, the analysis

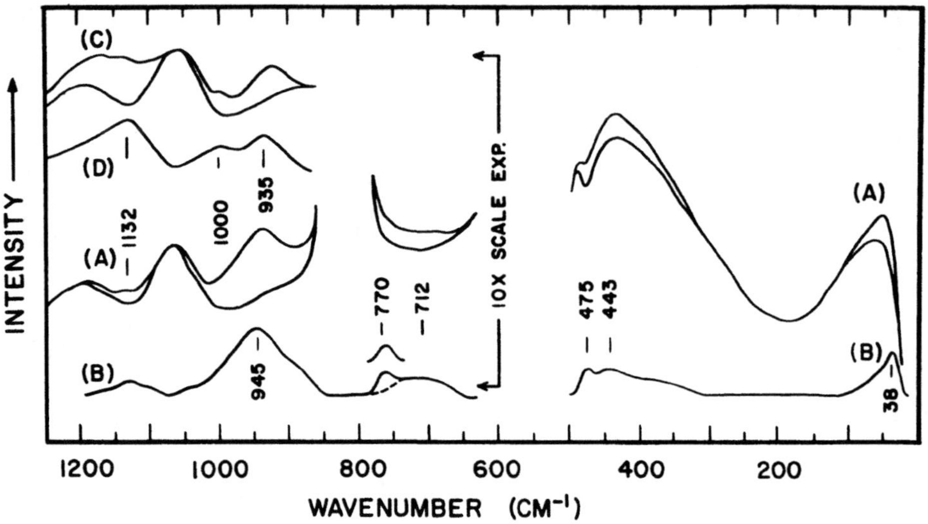

Figure 4. Differential Raman Spectra for Boron-Containing Glasses by Graphical Reduction of Intensities

(A) Spectra of Vycor (above) and Code 7943 (below)
(B) Difference spectrum obtained from (A)
(C) Spectra of Pyrex (above) and Code 7943 (below)
(D) Difference spectrum obtained from (C)

conducted using reasonable values of force constants shows that the vibrational wavenumbers for the tetrahedral E and F_2 species bending modes must lie quite close together and appear some 300-400 cm^{-1} below the 675 cm^{-1} band. The very sharp, intense characteristic of the band structure suggests that a large dipole moment change is occurring, and that the structural groups participating in the vibration are not too distorted by the surroundings to cause band broadening by network-induced distortion. For these reasons, the tetrahedral configuration hypothesis is considered to be quite unlikely.

On the other hand, considering the vibration to be the A_2'' out-of-plane bending mode for a BO_3 trigonal planar site group, a preliminary test calculation can be made to approximate the isotopic shift by using Herzberg's (8) equation (II,210):

$$\lambda_{A_2} = 3 \mu H'$$

where H' is the out-of-plane bending force constant per bond, μ is the reduced mass of the boron-oxygen complex, and $\lambda = (2\pi c\nu)^2$. The result is that the predicted wavenumber for $B^{10}O_3$ is 702-709 cm^{-1}.

The observed shoulder corresponds exactly, and the type of vibration fits the requirements for sharpness and intensity of the main band. The appearance of this band strongly in the IR but not in the Raman also agrees with the selection rules for D_{3h} site symmetry. Therefore one is encouraged to proceed with the NCA based on that assumption.

The machine-assisted computations require a fairly correct initial assignment of the observed wavenumbers to the corresponding species of vibration for iterative convergence. Therefore it is useful to make some further preliminary tests. The selection rules require that the A_1' species symmetric stretching vibration should appear in the Raman but not in the IR spectrum. One can make a fairly close prediction of this mode wavenumber by multiplying the observed value for the BF_3 molecule (9) by the square root of the fluorine/oxygen mass ratio. The result of 960 cm^{-1} is quite close to the 945 cm^{-1} peak appearing only in the Raman spectrum of the Vycor glass. The E species asymmetric stretch must lie above this, and should appear in both IR and Raman spectra. The 1390 cm^{-1} IR band fits this requirement, although it does not appear clearly in the Raman because of the inherent weakness of intensity of the high wavenumber stretching modes and low boron concentration. The E species bending mode vibration should appear below 675 cm^{-1}, and should likewise appear in both IR and Raman, as does the 475 cm^{-1} vibration. The NCA conducted with this set of assignments gives the UBFF force constants K, H, H', F, and F' as 5.384, 0.314, 0.802, 0.951, and -0.027 millidynes per angstrom, respectively. This is in good agreement with other work, as shown in Table 2, where the results for homologous molecular groups are listed for comparison.

The NCA treatment and observed agreement with the selection rules for spectral activity confirms the presence of the trigonal planar BO_3 group as the major molecular vibrating borate in the glasses investigated.

Having isolated and assigned the spectral details related to the BO_3 site group and to the vibrations arising from the creation of non-equivalent Si-O bonds, the interpretation and assignment of the remaining bands at 770, ca. 712, 443, and 38 cm^{-1} must be performed. If the BO_3 group were structurally isolated from the network, it would not be expected to see any additional bands other than perhaps a broad, intense, temperature dependent, low wavenumber scattering due to hindered rotation of the isolated group. on the other hand, if the BO_3 group is uniformly attached to the host network through three equivalent B-O-Si bonds, then the site symmetry of the BO_3Si_3 group thus formed would be C_{3v} as depicted in Fig. 5. Two new vibrations should appear: the first being the vibration of the BO_3 group as a whole against the network; and the second being a torsional rotation of the BO_3 group with respect to

Table 2. Force Constant Comparisons

Site Group	Force Constant (md/Å) K	H	F	Reference
BF_3	6.044	0.212	0.926	9
$B(OH)_3$	5.629	0.401	0.708	9
$(BO_3)^{-3}$	4.19	0.258	1.372	9
$(CO_3)^{-2}$	5.439	0.335	1.736	9
$(NO_3)^{-1}$	5.616	0.541	1.590	9
$(BO_3)^{-3}$	5.384	0.314	0.951	this work

the network. These vibrational modes should appear in both IR and Raman spectra.

For the model situation in which the O-Si and B-O bonds of the terminal oxygens are nearly perpendicular, the effective mass will be one oxygen mass plus one-third of a boron mass for each Si-O bond. A good approximation to the bond stretching force constant can be determined from the general form of the F matrix elements for the A_1 stretching mode of the non-equivalent oxygen of the $O\text{-}SiO_3$ defect group. Using Smith's values (7), a calculated wavenumber of 750 cm^{-1} is obtained, in good agreement with the 770 cm^{-1} value observed. Therefore, this vibrational wavenumber is assigned to the BO_3 group stretching mode against the network (calculation assumes the terminal Si to be of infinite mass).

The attachment of the BO_3 group to the network should also result in an upward shifting of the BO_3 E species bending mode vibrational wavenumber, as well as causing perturbations to the bending mode vibrations of the terminal SiO_4 groups. Therefore, the tentative assignments of the 443 cm^{-1} broad, weak Raman peak to these perturbed SiO_4 vibrations is made.

The torsional mode of the attached BO_3 group should occur at very low wavenumbers, and hence the peak at 38 cm^{-1} is assigned to this vibrational mode.

The origin and assignment of the broad, weak detail of the Raman spectrum centered at ca. 712 cm^{-1} is not known at the time of this writing. The assignment of the peak at 1000 cm^{-1} in the Raman spectrum of the Pyrex glass is not clear, but might be attri-

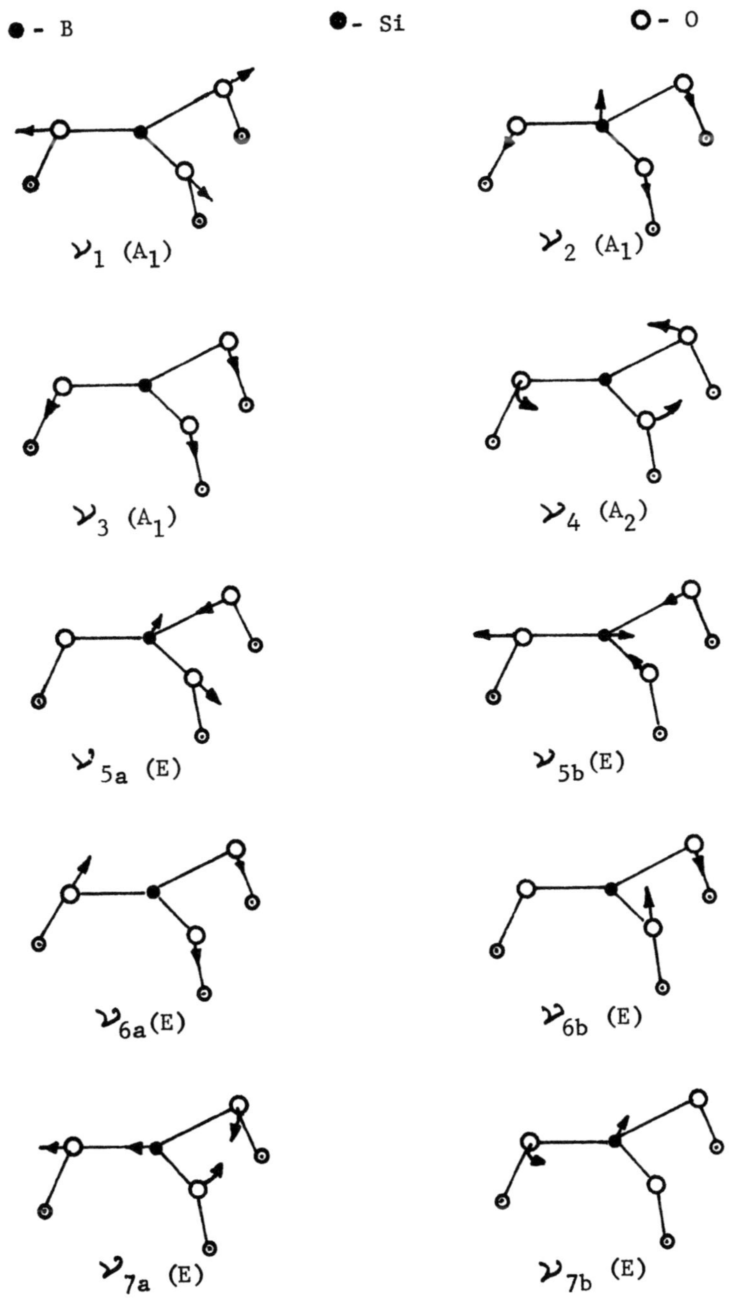

Figure 5. Normal Modes of Vibration for the BO_3Si_3 Site Group

Table 3. Vibrational Band Assignments for the Spectra of the Trigonal BO_3 Group in Vitreous Silica

Wave-number (cm^{-1})	Normal Mode Species	Type of Vibration
945	ν_1 (A_1)	B-O symmetric stretching
678	ν_2 (A_1)	O-B-O out-of-plane bending
770	ν_3 (A_1)	(BO_3)-Si_3 symmetric stretching
38	ν_4 (A_2)	(BO_3)-Si_3 torsional
1390	ν_5 (E)	B-O asymmetric stretching
1132	ν_6 (E)	(BO_3)-Si_3 rocking
475	ν_7 (E)	O-B-O in-plane bending

buted to a stretching vibration of AlO_4 tetrahedra isomorphously substituting for SiO_4 groups in the network.

The assignments for the vibrational bands are listed in Table 2 above, and the corresponding vibrational displacements are depicted in Fig. 5.

For the glasses investigated, the analysis strongly indicates that the boron enters the silicate structure as trigonal planar BO_3 groups uniformly attached to the network through bonds to the network silicon atoms. The force constant magnitudes suggest that the bonding is largely covalent in type.

While analyses of infrared and Raman spectra have long been important to the determination of structural relationships and determination of coordination in glasses, very specialized new techniques in recording, analyzing, and interpreting the spectra can and should play an important role in determining the structural environment of the constituents. The employment of differential methods such as used here, coupled with machine-assisted normal coordinate analyses can help to distinguish between alternative structural hypotheses, as well as to assist in the interpretation of more complex spectra.

*Grateful acknowlegement is made to W. Votava for sample preparation and to G. Cartledge for chemical analyses. The work was performed under a HEW Title IV Fellowship for doctoral studies.

REFERENCES

1. C. F. Smith, R. A. Condrate, Sr., and W. E. Votava, Appl. Spect. 29 (1975) 79

2. E. B. Wilson, Jr., J. Chem. Phys. 7 (1939) 1047

3. E. B. Wilson, J. C. Decius, and P. C. Cross, Molecular Vibrations , McGraw-Hill, New York (1955)

4. F. A. Cotton, Chemical Applications of Group Theory, Interscience, New York (1955)

5. T. Shimanouchi, M. Tsuboi, and T. Miyazawa, J. Chem. Phys. 35 (1961) 1597

6. J. H. Schachtschneider and R. G. Snyder, Spect. Acta 19 (1963) 117

7. C. F. Smith, Jr., Thesis, Alfred University (1973) 83, 92, 118, 111

8. G. Herzberg, Molecular Spectra and Molecular Structure, II. Infrared and Raman Spectra of Polyatomic Molecules, Van Nostrand, Princeton, N. J. (1959) 178

9. C. J. Peacock, A. Muller, and R. Kebabcioglu, J. Mol. Struct. 2 (1968) 163

NMR STUDIES OF BORATES*

P.J. Bray

Department of Physics, Brown University

Providence, Rhode Island U.S.A. 02912

INTRODUCTION

Nuclear magnetic resonance (NMR) studies of the structure of glasses have been carried out at Brown University for some twenty years[1-24]. Much of the work has involved the use of B^{11} NMR spectra to identify BO_3 and BO_4 units in binary borate glasses and to determine quantitatively the fraction N_4 of four-coordinated boron in the glasses. Recent work has extended the systems to silicate glasses, ternary systems, oxygen bonding in glasses, and the identification in borate glasses of the large structural groupings found in the crystalline borate compounds. The studies of borate glasses lend strong support to the model of borate glasses proposed by Jan Krogh-Moe[25]. These recent developments in NMR studies of borate glasses are summarized in this paper after presentation of some basic NMR theory.

BASIC NMR THEORY

NMR studies are made possible by the interaction of nuclear magnetic dipole moments with an applied magnetic field. The nuclei of most atoms in nature do possess a magnetic dipole moment, a vector quantity that can be expressed as

$$\vec{\mu} = g\mu_0 \vec{I} \quad (1)$$

Here $\vec{\mu}$ is the magnetic dipole moment, $\vec{I}$ is the dimensionless spin vector of magnitude $[I(I + 1)]^{\frac{1}{2}}$ which gives the intrinsic spin angular momentum of the nucleus in units of Planck's constant divided by 2π (i.e., $h/2\pi$), the scalar I is an integer or half-integer and is called the spin of the nucleus, μ_0 is the Bohr nuclear magneton, and g is the nuclear g-factor. The interaction between the dipole moment and an applied magnetic field $\vec{H}$ creates a set of energy levels given by

$$E = -\vec{\mu} \cdot \vec{H} = -g\mu_0 \vec{I} \cdot \vec{H} = -g\mu_0 Hm \quad (2)$$

where the quantity m, called the magnetic quantum number, has (2I + 1) values ranging from -I to +I in integer steps. In the absence of other interactions, the (2I + 1) energy levels are equally spaced as shown in Figure 1a. (The levels shown are appropriate for a nucleus of spin I = 3/2, such as B^{11}.)

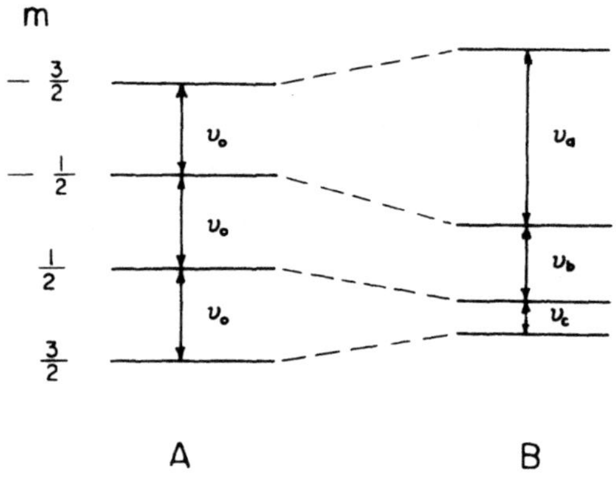

Figure 1. Energy levels arising from the interaction of the nuclear magnetic dipole moment with a magnetic field: a) no electrical quadrupole interaction; b) small quadrupole interaction present.

Transitions between adjacent energy levels in Figure 1a can be induced by bathing the system in electromagnetic radiation whose frequency satisfies the Bohr condition

$$h\nu_o = \Delta E = g\mu_o H \qquad (3)$$

where ΔE is the energy difference between adjacent levels. Since the energy levels are equally spaced, there is only one frequency in this case. If the radiofrequency ν_o or the magnetic field H is slowly swept through the resonance condition, the absorption lineshape for the resonance can be obtained.

Interactions other than the Zeeman interaction of Eq. (2) can split the Zeeman energy levels so that the absorption lineshape has structure or is broadened. In particular, the local magnetic fields at resonating nuclei due to each other or nuclei of other elements always introduce broadening and smoothing of the lineshape which would be a delta function if only the Zeeman interaction of Eq. (2) were operative. A recording of a typical resonance is shown in Figure 2. This is the response for B^{11} nuclei in BO_4

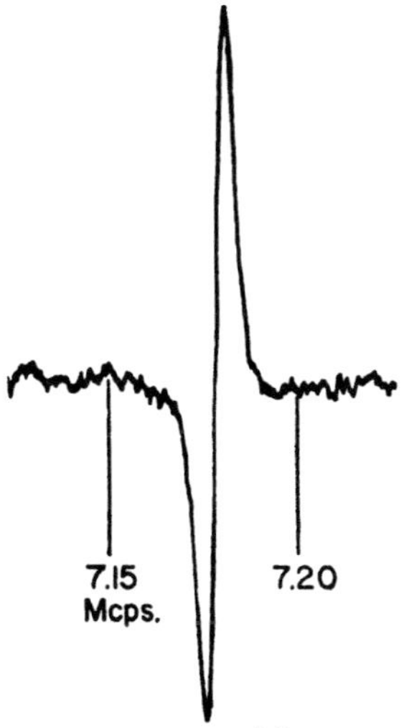

Figure 2. NMR spectrum for B^{11} units in BO_4 units in a glass of composition $Na_2O \cdot 2B_2O_3$. (First derivative of the absorption curve.)

units in a sodium borate glass. It was obtained at ν_o= 7.177 MHz in a magnetic field of 5250 Gauss. (The spectrometer produces the first derivative of the absorption lineshape as a function of the applied field or the spectrometer frequency. Though the result resembles a dispersion curve, it is in fact the derivative of the absorption lineshape.)

O^{17} NMR STUDIES OF OXIDE GLASSES[24]

Most NMR studies of structure and chemical bonding in glasses involve oxide systems for which determination of the oxygen sites and metal-oxygen bonding characteristics are of essential importance; yet, oxygen NMR data do not exist for these glasses. This striking omission is surprising until it is realized that the only oxygen isotope (O^{17}) having a nuclear magnetic moment is only 0.037% abundant in nature. However, recent decreases in the cost of water enriched to as much as 37% in O^{17} have permitted fabrication of glasses with this level of O^{17} content.

Interpretation of the NMR spectra for O^{17} requires a knowledge of quadrupole interactions since the O^{17} nucleus has a spin greater than ½ (in fact I = 5/2 for this nucleus) and thus possesses an electrical quadrupole moment. When a quadrupole interaction is present, the Zeeman energy levels of Figure 1a are shifted, generally by unequal amounts. (This is depicted in Fig. 1b for the case of a nuclear spin I = 3/2.) For cases in which the quadrupole interaction is sufficiently small with respect to the Zeeman interaction, the quadrupole interaction can be treated as a perturbation, yielding an NMR transition frequency given by

$$\nu = \nu_o + \nu_Q(m - \tfrac{1}{2})\, A + \frac{\nu_Q^2}{\nu_o}\, C + \frac{\nu_Q^3}{\nu_o^2}(m - \tfrac{1}{2})E \quad . \qquad (4)$$

Equation (4) is calculated to third-order employing perturbation theory. Here ν_o is the resonance frequency when only the Zeeman interaction occurs (Eq. (3));

$$\nu_Q = 3Q_{cc}/2I(2I - 1)$$

$$Q_{cc} = e^2qQ/h$$

$$\eta = V_{xx} - V_{yy}/V_{zz}$$

where eQ is the electrical quadrupole moment of the nucleus, η is the asymmetry parameter, and V_{xx}, V_{yy}, and V_{zz} are the components of the electric field gradient (EFG) tensor defined so that $|V_{zz}| \geq |V_{yy}| \geq |V_{xx}|$ and eq $\equiv V_{zz}$. The terms A,C, and E in Eq. (4) are complicated functions of m,I, cos θ, cos 2ϕ and η. The quantity m denotes the transition from the energy level labeled by the magnetic quantum number m to the level labeled by m - 1, and θ and ϕ are the Euler angles of the magnetic field H_o with respect to the principal axis system of the EFG tensor.

If the sample is polycrystalline or vitreous, than all of the Euler angles θ and ϕ occur randomly. Therefore, the resonance condition (Eq. (4) must be averaged over all the possible values of the Euler angles. For the case $\eta = 0$ and a quadrupole interaction frequency Q_{cc} that is small with respect to the Larmor frequency ν_o of Eq. (3), the pattern is as shown in Figure 3 for the case $I = 3/2$. (The powder pattern itself is depicted by the dashed line. Dipolar broadening smoothes the curve into the shape given by the solid line.) An example of this case observed in a resonance measurement is depicted in Figure 4 which is the B^{11} resonance for polycrystalline boron phosphate. (The recorded trace is the first derivative of the absorption curve rather than the absorption curve itself.) Here Q_{cc} is 50.4 kHz while the Larmor frequency ν_o of Eq. (3) is 7.177 MHz. The sharp narrow line in the center of Figure 4 is characteristically observed for boron atoms in BO_4 units because Q_{cc} is less than 800 kHz for boron atoms in tetrahedral symmetry; the "wings" on each side of the narrow line in Figure 4 are usually not detected in B^{11} NMR spectra for glasses. (Figure 2 is the B^{11} NMR spectrum for boron atoms in BO_4 units in a sodium

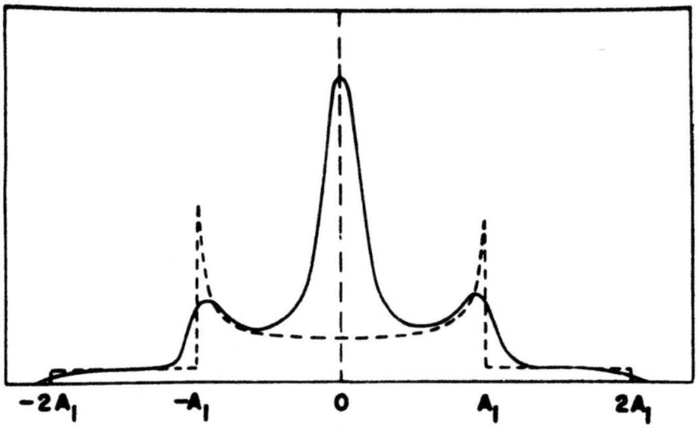

Figure 3. Resonance lineshape (solid curve) for a nucleus of spin I=3/2, such as B^{11}, with a small quadrupole interaction in a glass or polycrystalline powder. $A_1 = 1/4\,Q_{cc}$.

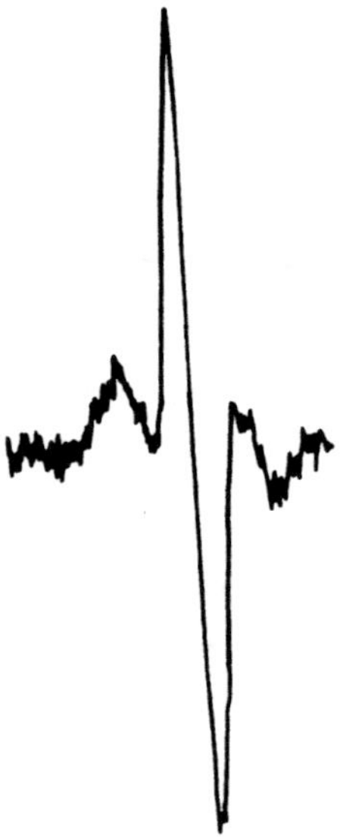

Figure 4. NMR spectrum of B^{11} in polycrystalline boron phosphate (BPO_4). $\nu_o = 7.17$MHz. The first derivative of the absorption curve is presented.

diborate glass.)

When Q_{cc} [and thus ν_Q of Eq. (4)] is not negligible with respect to ν_o, the observable powder pattern is usually confined to the m = ½ ↔ m = - ½ transition and has the shape depicted in Figure 5 (for η = 0). (Again, the powder pattern itself is depicted by the dashed lines. Dipolar broadening smoothes the curve into the shape given by the solid line.) An example of this case observed in a resonance measurement is depicted in Figure 6 which is the B^{11} resonance for vitreous B_2O_3. (Here the spectrometer frequency was held constant at 16 MHz and the magnetic field was scanned through the resonance. The Varian Wide-Line NMR spectrometer was run in the dispersion mode which yields a recorder tracing directly proportional to the absorption curve rather than its derivative.) Here η is not strictly zero (in fact η = 0.12) but the pattern is essentially that for η = 0. The quadrupole interaction is relatively large for this case of borons in the planar triangular BO_3 configuration: Q_{cc} = 2.76 MHz.

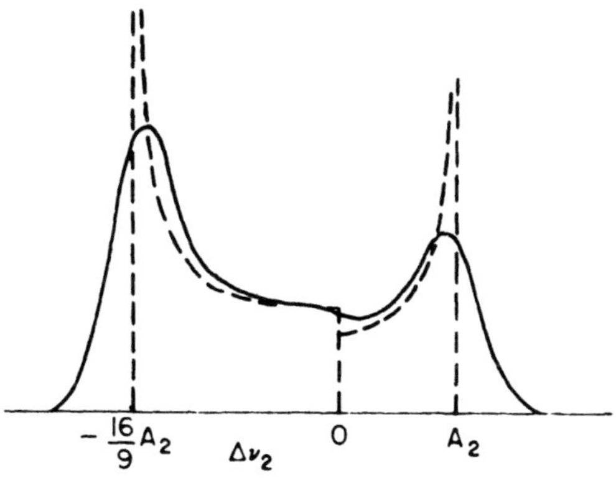

Figure 5. Resonance lineshape (solid curve) for a nucleus with a large quadrupole interaction Q_{cc} and η=0 in a glass or polycrystalline powder. Only the m=½ ⇄ m=-½ transition is presented. For I = 3/2, $A_2 = \frac{3}{64}\frac{(Q_{cc})^2}{\nu_o}$.

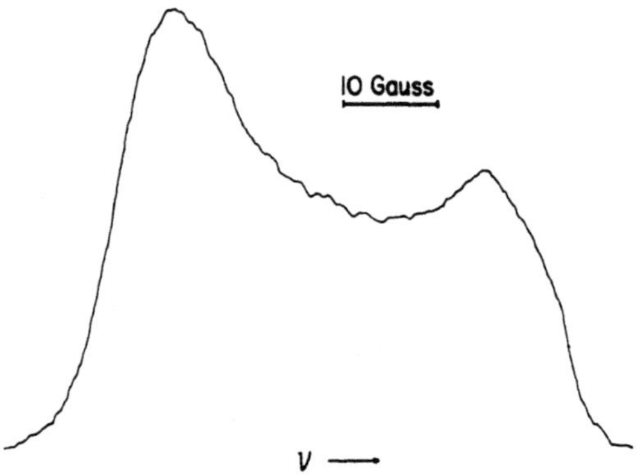

Figure 6. The B^{11} NMR spectrum for vitreous B_2O_3 at 16MHz (dispersion mode).

When η departs substantially from zero, the powder pattern is more complex as shown in Figure 7. A powder pattern for particular values of Q_{cc} and η is also shown in Figure 8a. When the smoothing effects of dipolar broadening are taken into account, and the derivative is taken for comparison with the usual spectrometer tracing, a pattern such as that depicted in Figure 8b is obtained. This simulated spectrum can be compared with the experimental trace for B^{11} given in Figure 8c for polycrystalline calcium metaborate. This material contains chains of the form $(BO_2^-)_n$ as shown in Figure 9. Each boron has two bridging and one nonbridging oxygen. This destruction of 3-fold symmetry about the axis perpendicular to the BO_3 plane produces a large asymmetry ($\eta = 0.54$) in the EFG tensor and, consequently, a fairly complicated NMR spectrum (Fig. 8). However, the coupling constant Q_{cc} of 2.56 MHz is changed relatively little from the value (2.76 MHz) for B_2O_3 in which η is small.

The experimental spectrum for O^{17} in vitreous B_2O_3 (Fig. 10) is

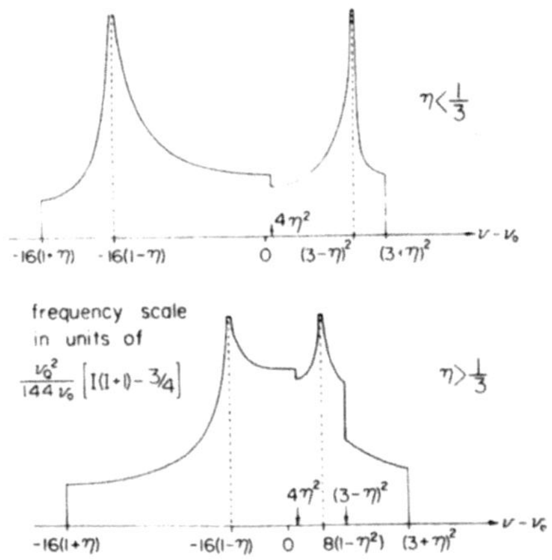

Figure 7. Powder pattern for the central transition ($m = \frac{1}{2} \rightleftarrows m = -\frac{1}{2}$) of the NMR spectrum for a nucleus of half-integral spin.

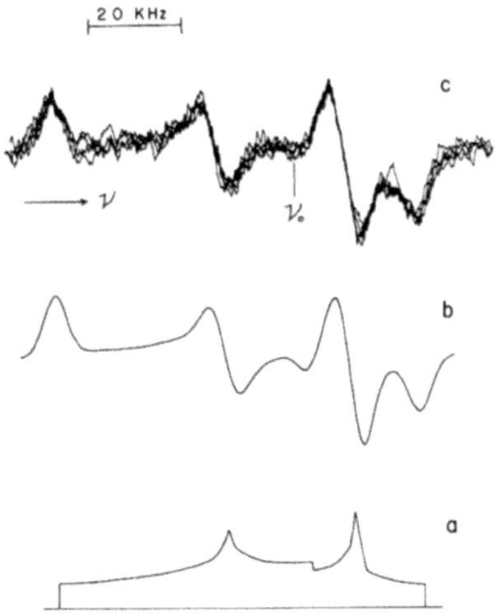

Figure 8. a) Theoretical powder pattern for the $m=\frac{1}{2} \rightleftarrows m=-\frac{1}{2}$ transition with $I = 3/2$, $\nu_0 = 16$MHz, $Q_{cc} = 2.56$MHz, and $\eta = 0.54$.
b) First derivative of (a) after dipolar broadening.
c) Superposition of four experimental traces for B^{11} in polycrystalline calcium metaborate at 16 MHz.

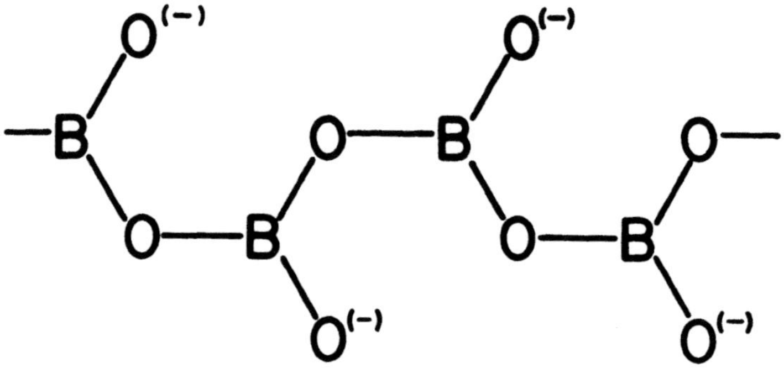

Figure 9. A part of the $(BO_2^-)_n$ chain in calcium metaborate.

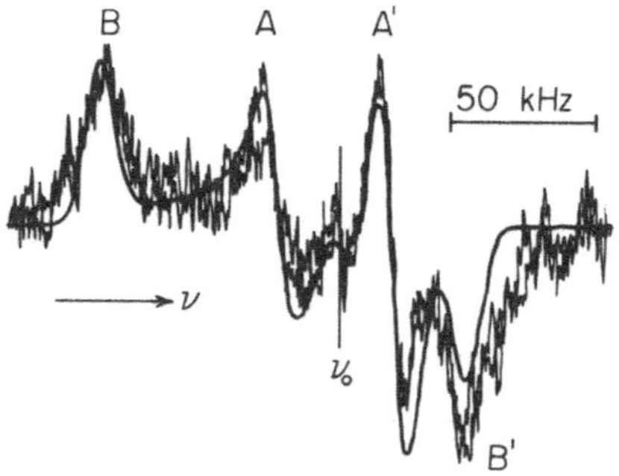

Figure 10. Two superimposed experimental O^{17} NMR derivative spectra, with a one-site computer stimulation (smooth curve).

of the same type as that found for calcium metaborate (Fig. 8c). (The glass is enriched to 37% O^{17}.) The computer-simulated spectrum (smooth line in Fig. 10) is for one oxygen site having Q_{cc} = 4.69 MHz and η = 0.58 with no distributions in either quantity. The sharpness of the central features (A,A') of the pattern confirms the presence of very small distributions in Q_{cc} and η. This is consistent with the model for this glass put forward by Krogh-Moe[26] in which the glass consists of randomly oriented six-membered boroxol rings (Fig. 11). The boroxol rings themselves are relatively stable structures with little distortion in the rings, and, consequently, only small distributions in the parameters Q_{cc} and η for the oxygens in the boroxol rings (denoted as O(R) in Fig. 11).

It is clear that one should expect a second O^{17} spectrum arising from the oxygens that connect boroxol rings to each other or to BO_3 units that are not parts of boroxol rings (denoted as O(C) in Fig. 11). In accord with this expectation, a second oxygen site has been invoked for the simulated spectrum shown in Figure 12. Here Q_{cc} = 5.75 MHz and η = 0.40. The spectrum from this site removes the disagreement [(seen in Fig. 10 in the outer features (B,B')] between the simulated and experimental spectra. Since the

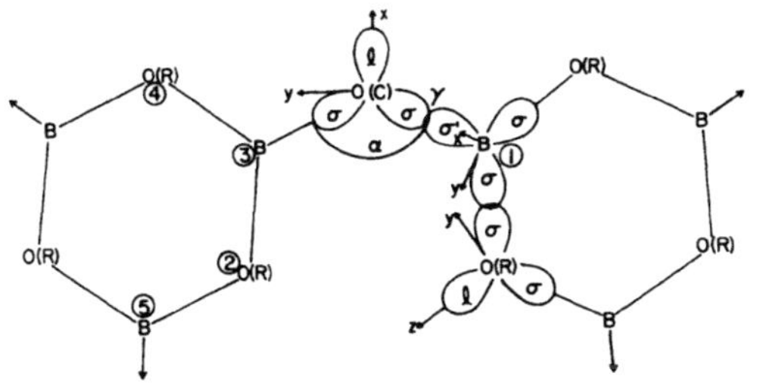

Figure 11. Boroxol-ring model for B_2O_3 glass. O(R) denotes ring oxygens; O(C) denotes connecting oxygens.

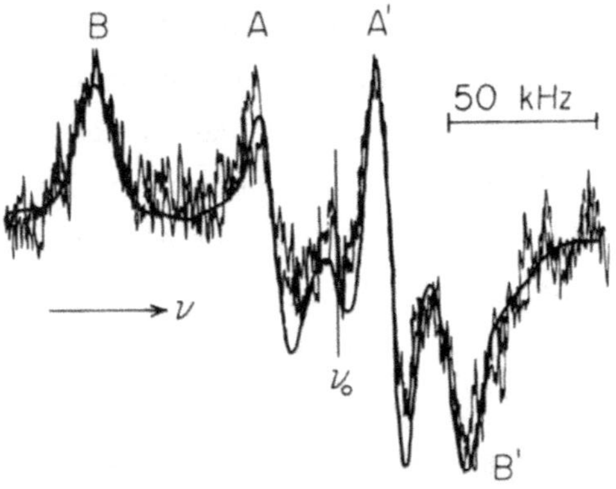

Figure 12. A two-site fit (with distributions) of the superimposed O^{17} NMR derivative spectra displayed in Figure 10.

connecting oxygens O(C) are expected to exhibit a distribution in the B-O-B bonding angle α of Figure 11, it is satisfying to find that the best fit of the simulated and experimental spectra are obtained with a broad distribution (σ_η = 0.20) in the values of η. This spread, when employed in a Townes and Dailey[27] type of analysis, yields a small distribution in α about either 134.6° or 128.1°. Either result is consistent with the x-ray work of Mozzi and Warren[28]. It is clear that the NMR data are in good agreement with the boroxol-ring model put forward by Krogh-Moe and supported by the work of Mozzi and Warren.

The O^{17} work is being carried forward in other materials. In particular, the characteristic NMR spectrum for a nonbridging oxygen (NBO) has been sought experimentally and obtained[29] in polycrystalline lithium metaborate ($Li_2 \cdot B_2O_3$), but extraction of the quadrupole parameters by comparison of experimental and computed spectra is still in progress. It should be noted that comprehensive programs are available[30-32] for computation of simulated spectra, even in cases for which both quadrupole effects and chemical shift anisotropy are present[33]. It will also be of interest to determine the characteristic O^{17} NMR spectrum for the three-coordinated oxygens found, for example, in $PbO \cdot 2B_2O_3$[5] and $SrO \cdot 2B_2O_3$[17].

NMR STUDIES OF B^{10} IN GLASSES

Previous NMR investigations[1-24] of borate glasses have concentrated primarily on the B^{11} nucleus for which I = 3/2. In particular, attention has focussed on the $m = \frac{1}{2} \leftrightarrow m = -\frac{1}{2}$ transition for which second-order quadrupolar effects are dominant. All other transitions for B^{11} (and other half-integer spin nuclei) are affected in first-order and either lie under the less affected $m = \frac{1}{2} \leftrightarrow -\frac{1}{2}$ transition or are so widely distributed as a function of spectrometer frequency or magnetic field that special efforts must be made to detect the very weak responses.

Recently, computations have been made[34] of the NMR spectra for integer spin nuclei, and it has been found that the $m = 0 \leftrightarrow m = -1$ and $m = 1 \leftrightarrow m = 0$ transitions for B^{10} (I = 3) are observable and far more sensitive tools than the B^{11} spectra for investigating structure and bonding. The dipole moment of B^{10} is about 67% of the dipole moment for B^{11}, so that dipolar broadening is reduced. Further, the B^{10} powder pattern for the $0 \leftrightarrow -1$ and $1 \leftrightarrow 0$ transitions extends over a range that is approximately 20 times greater than that for the B^{11} central ($\frac{1}{2} \leftrightarrow -\frac{1}{2}$) transition, and that central transition is not present to obscure the $0 \leftrightarrow -1$ and $1 \leftrightarrow 0$ transitions. These factors decrease the effectiveness of the dipolar interaction as a broadening mechanism by about a factor of 30 for B^{10} NMR as compared to B^{11} NMR. Consequently, more accurate values of the quadrupole parameters can be secured, and the distributions in their values are more observable and subject to measurement.

The powder pattern for the $0 \leftrightarrow -1$ and $1 \leftrightarrow 0$ transitions of B^{10} is displayed in Figure 13a. The location of each of the features a through f has been calculated to third-order in the perturbing quadrupole interaction. Addition of dipolar broadening to Figure 13a and construction of the derivative provides a simulated curve that - with the proper choice of dipolar broadening, Q_{cc}, η, and distributions in the latter two parameters (expected in a glass) - can be fitted to the experimental cases (e.g., the B^{10} spectrum

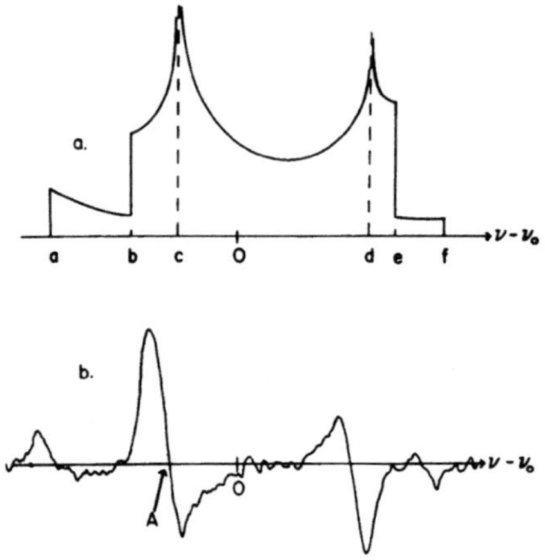

Figure 13. a) B^{10} NMR powder pattern for the $0 \rightleftarrows -1$ and $1 \rightleftarrows 0$ transitions.
b) Experimental B^{10} NMR spectrum for B_2O_3 glass (derivative of the absorption curve).

for vitreous B_2O_3 glass in Figure 13b). Various features in the complex spectrum of Figure 13 are particularly sensitive to one of the quadrupole parameters (Q_{cc}, η) to the exclusion (or almost to the exclusion) of the other. This is another reason why more accurate values of these parameters can be obtained from the B^{10} spectra as compared to values from the B^{11} spectra.

The sensitivity is demonstrated in Figure 14 in which the central feature (A) in Figure 13b has been obtained on an expanded scale and conditions which leave it free of some distortion which influenced it in Figure 13b. It is clear from Figure 14 that this feature is extremely sensitive to the most probable value (η_o) and distribution (σ_η) for the asymmetry parameter. Relatively minor variations in the choice of either η_o or σ_η for the simulated spectrum produce obvious disagreement with the experimental result.

The high sensitivity of the B^{10} NMR structure in studying bonding and structure has been employed in a study[35] of glasses in the system $Na_2O-B_2O_3$. The B^{10} NMR spectra for a series of eight

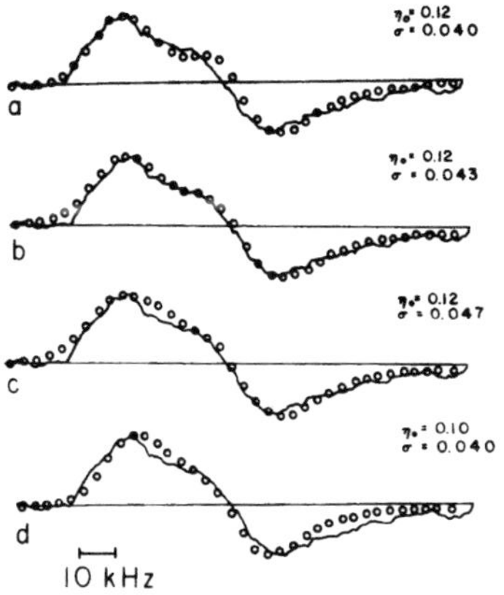

Figure 14. Experimental B^{10} derivative of the main feature (solid line) and simulated spectra (circles) for various values of η_o and σ_η (where Q_{cc}=5.51MHz and $\sigma_{Q_{cc}}$ = 0.21MHz).

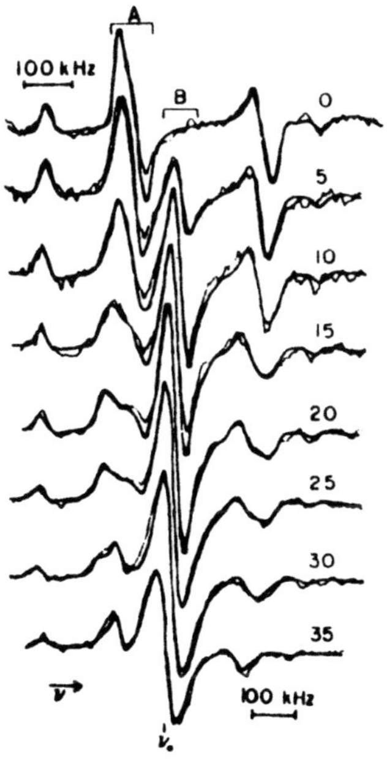

Figure 15. B^{10} NMR derivative spectra for eight Na_2O-B_2O_3 glasses; simulated spectra are superimposed as smooth lines.

glasses are shown in Figure 15. (The glasses were enriched to 92% B^{10} rather than the natural abundance of 18.83%.) A Nicolet signal averager was used to accumulate the signals for several days for each glass. The agreement of the simulated and experimental spectra in Figure 15 is good despite the presence of some distortion in narrow features A and B due to the effects of time constants and relatively large magnetic field modulation. Feature A arises from the 3-coordinated boron atoms and is shown in Figure 16 where the frequency scale is expanded and the experimental conditions are adjusted to assure the attainment of the actual lineshape without distortion. (The procedure for generating the simulated spectra in Figure 16 will be discussed at length below.) Feature B in Figure 15 arises from 4-coordinated borons and will not be discussed in detail here.

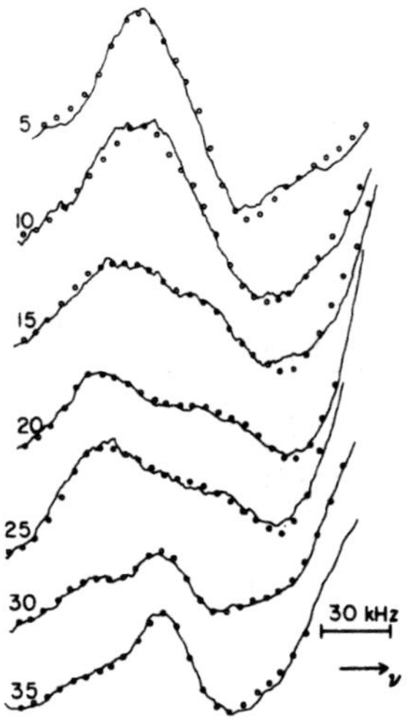

Figure 16. Experimental B^{10} spectra displaying the main feature (feature A of Fig. 15) on an expanded scale for 3-coordinated boron atoms in seven Na_2O-B_2O_3 glasses (solid lines). Simulated spectra are shown by open circles.

In setting up a model with which to compare the experimental results, use was made of the proposal by Krogh-Moe(25) that alkali borate glasses with less than 33 1/3 molar% alkali oxide contain mixtures of four crystalline structural groups: boroxol, pentaborate, triborate, and diborate. (See Fig. 17.) This proposition was advanced on the basis of a comparison of the infrared spectra of the crystalline compounds and the glasses, which show that the spectra of the glasses bear resemblences to the spectra of the crystalline compounds. Recent Raman studies by Konijnendijk(36,37) are consistent with the Krogh-Moe structural model.

Consideration of the NMR Raman data indicated that the following assumptions can be made.

1. "Loose" BO_3 and BO_4 units (i.e., units not included in the structural groupings of Fig. 17) can be neglected. The Raman data(36,37) indicate that minor fractions could be present, but they are ignored here.

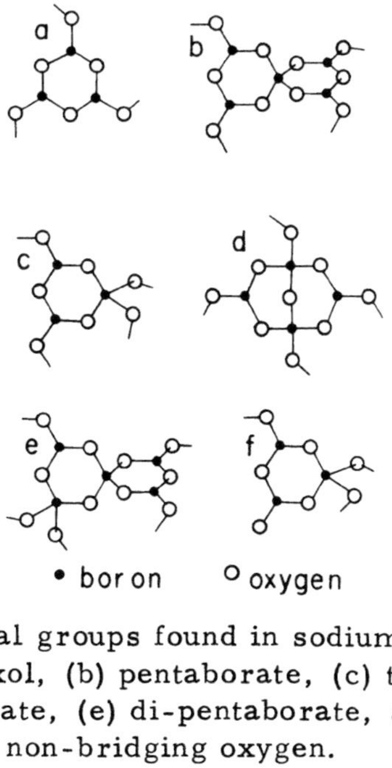

Figure 17. Structural groups found in sodium borate compounds: (a) boroxol, (b) pentaborate, (c) triborate, (d) diborate, (e) di-pentaborate, and (f) triborate with one non-bridging oxygen.

2. Other units that occur in crystalline compounds and might possibly appear in minor amounts in the glasses (such as di-pentaborate units or triborate units with one nonbridging oxygen, see Fig. 17e and f) are ignored.

3. Pentaborate and triborate units (see Fig. 17b and c) are assumed to occur in pairs forming tetraborate groups. The Raman studies[36,37] indicate the absence of lone pentaborate units in the glasses.

4. Below 25 molar% Na_2O, the fraction N_4 of boron atoms in 4-coordination is given by $N_4 = \frac{x}{1-x}$ where x is the molar fraction of Na_2O. This has been confirmed by NMR determinations of N_4[3].

5. Above x = 0.25 the number of boroxol groups is negligible or zero. This is confirmed by the Raman studies[25]. With these assumptions it is possible to fit the experimental NMR spectra for the glasses with weighted combinations of the spectra for the boroxol, tetraborate, and diborate units by adjusting only one parameter designated δ and defined in the following manner: let B,T,D denote boroxol, tetraborate, and diborate units respectively. A superscript 3 or 4 denotes the 4-coordinated borons in each unit, and a subscript 0 labels the amount of each unit (B,T,D) expected by application of the simple lever rule between the two compounds above and below the glass composition. The symbol B,T or D without the subscript denotes the actual amount of each unit present. Then δ is defined by

$$D^3 = D^4 = D_0^3 + \delta \qquad (5)$$

(Note: In the diborate unit there are two 3-coordinated and two 4-coordinated borons; hence, $D^3 = D^4$ and $D_0^3 = D_0^4$. Similarly, $T^4 = \frac{T^3}{3}$ and $T_0^4 = \frac{T_0^3}{3}$.) From this definition of δ and the assumptions noted above one can show[35] that

$$B^3 = B_0^3 + 2\delta$$

$$T^3 = T_0^3 - 3\delta$$

and

$$D^3 = D_0^3 + \delta$$
$$T^4 = T_0^4 - \delta$$
$$D^4 = D_0^4 + \delta$$

The single parameter δ is then varied to obtain the best agreement between the observed and simulated spectra. The process is indicated in Figure 18 which shows the experimental spectrum of the $x = 0.20$ glass superimposed upon three computer-simulated traces for different values of δ(δ = 0.06, 0.08, 0.10 respectively for a, b, and c). It is clear that the degree of agreement is a sensitive function of δ.

It is the "best fit" simulated spectra that appear in Figure 16. Since each fit determines a particular value of δ and, hence, values for the amounts of each structural unit present (see Eqs. (6) above), it is possible to depict graphically the amount of each unit present in each glass. This is done in Figure 19 where the solid lines give the fractional amount of each unit that should be present if the simple lever rule between compounds is obeyed. It is clear that boroxol units persist above their anticipated disappearance at the tetraborate composition ($x = 0.20$, $R = \frac{x}{1-x} = 0.25$); that there is a depletion of tetraborate units around the tetraborate composition because both boroxol and diborate units are present; and diborate units appear at compositions as low as 10% Na_2O.

Further information can be extracted from the data[35], but the point should be clear: B^{10} NMR is sufficiently sensitive to identify quantitatively in glasses the type and amount of each structural grouping present in any significant amount. Certainly the detailed agreement between the experimental spectra and simulated spectra (Fig. 16) is strong support for the model of borate glasses proposed by Krogh-Moe[25].

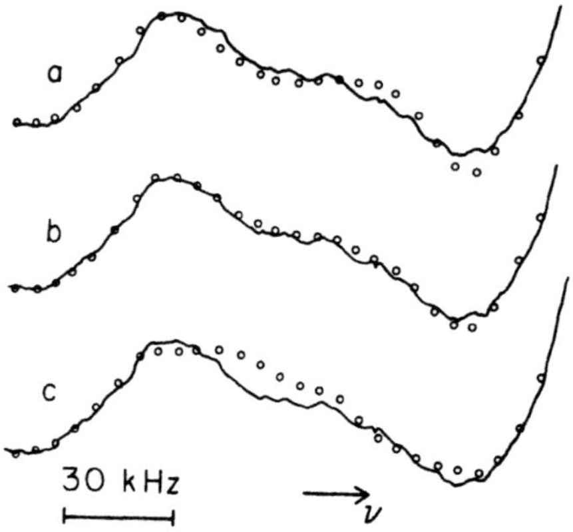

Figure 18. Experimental B^{10} derivative of the main feature (A of Fig. 15) for 3-coordinated boron atoms in a 20 molar % Na_2O glass (solid line). Simulated spectra are shown by open circles for $\delta = 0.06$, 0.08, and 0.10. (See text for definition of δ.)

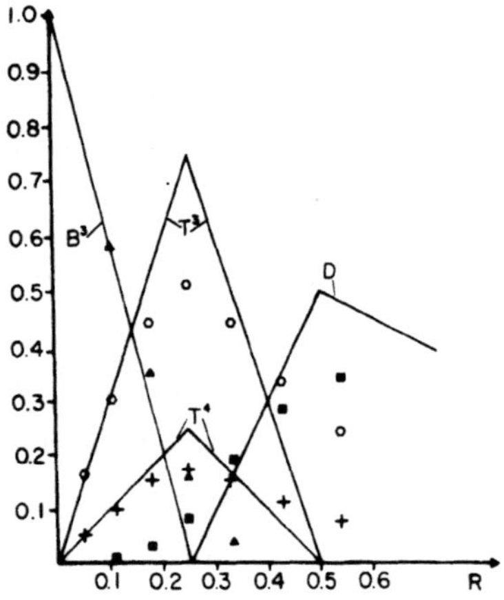

Figure 19. Fractions of B^3 (▲), T^3 (⬡), D^3 and D^4 (■) plotted as a function of $R = x/1-x$, x = molar fraction Na_2O. The straight line segments represent the case when the lever rule is obeyed (see text).

NMR STUDY OF GLASSES IN TERNARY SYSTEMS

Earlier B^{11} NMR studies of binary borate glasses have now been extended to several ternary systems[18,22,38-46]. Recent NMR studies[42-46] have focused on the borosilicate (Na_2O-B_2O_3-SiO_2) glass system, which is of major practical importance, in order to analyze the fraction (N_4) of boron atoms in BO_4 units as a function of composition. Milberg[43], Muller-Warmuth[44], and Zhadnov[42,46] chose to plot N_4 as a function of $R \equiv$ molar% Na_2O/molar% B_2O_3. They have observed that for $R < 0.5$ the sodium oxide is employed entirely to convert BO_3 units to BO_4 units; the glasses behave as if they were binary sodium borate glasses diluted by silica in that $N_4 = R$ as in the binary systems. However, above $R = 0.5$ the data points appear to scatter widely between $N_4 = 1$ and the values of N_4 for the binary system. There has, consequently, been no clear model or quantitative analysis for the ternary system with $R > 0.5$. A model and analysis[47] can now be reported.

Glasses were made in families of constant K where $K \equiv$ molar% SiO_2/molar% B_2O_3; this follows a procedure used by Müller-Warmuth.[44] Typical B^{11} NMR spectra are displayed in Figure 20. The narrow line (indicated by a in Fig. 20) in each case arises from boron atoms in BO_4 units. The single broad line (indicated by c) in Figure 20-(1) arises from borons in BO_3 units having all bridging oxygens (denoted as symmetric BO_3 units). This glass has the composition 0.11 $Na_2O \cdot B_2O_3 \cdot SiO_2$ and displays a spectrum characteristic of all glasses with $R < 0.5$. For glasses with $R > 0.5$, the B^{11} resonance exhibits an additional line indicated by b in Figure 20-(2) and arising from BO_3 units with one or two nonbridging oxygens (denoted as asymmetric BO_3 units). (This glass has the composition 1.3 $Na_2O \cdot B_2O_3 \cdot SiO_2$ and displays a spectrum characteristic of all glasses with $R > 0.5$.) The fraction N_4 of 4-coordinated borons, N_{3S} the fraction of 3-coordinated borons in symmetric BO_3 units, and N_{3A} the fraction of borons in asymmetric BO_3 units can be determined[47] from the experimental spectra.

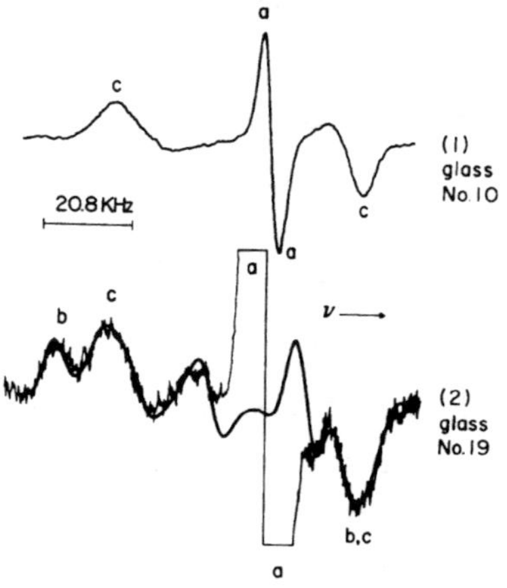

Figure 20. B^{11} NMR derivative spectra at 16MHz for glasses of the following compositions: No. 10, 0.11 $Na_2O \cdot B_2O_3 \cdot SiO_2$; No. 19, 1.3 $Na_2O \cdot B_2O_3 \cdot SiO_2$.

A plot of N_4 versus R is presented in Figure 21. For $R < 0.5$ the values of N_4 are the same for the binary sodium borate and ternary sodium borosilicate glasses, $N_4 = R = \frac{x}{1-x}$; this is in accord with the findings of Milberg[43], Müller-Warmuth[44] and Zhdanov[42, 46]. But for $R > 0.5$, N_4 continues to rise along the $N_4 = R$ line until it reaches a maximum $N_{4_{max}}$ at a value R_{max} that depends on K, and then N_4 decreases linearly with R at a rate that also depends on K; the glasses are clearly separated in this region into families that are governed by K. N_{3A} also depends on K as shown in Figure 22 where N_{3A} is zero below R_{max} and increases linearly with R above that value. The slopes $S_4(K)$ of the N_4 curves versus R above R_{max}, and the slopes $S_{3A}(K)$ of the N_{3A} curves versus R above R_{max}, are shown in Figure 23 and compared with theoretical predictions that will be developed in the remainder of the paper.

A model for the behavior of the binary glasses (K = 0) can be

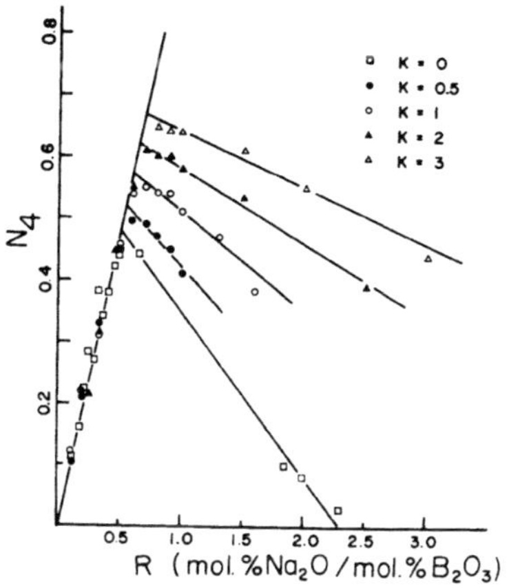

Figure 21. The fraction N_4 of boron atoms in BO_4 units versus R.

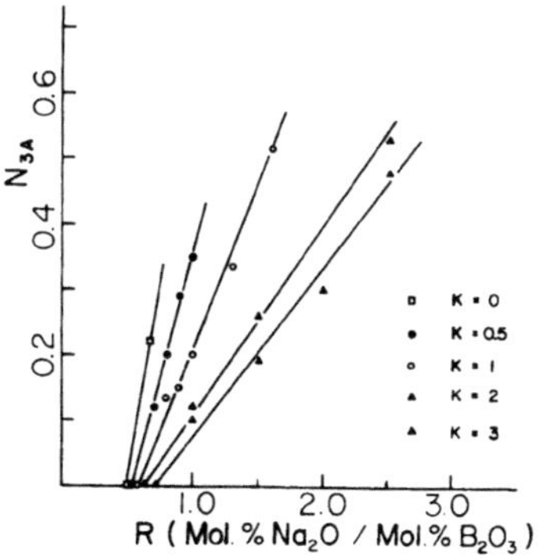

Figure 22. The fraction N_{3A} of boron atoms in asymmetric BO_3 units versus R.

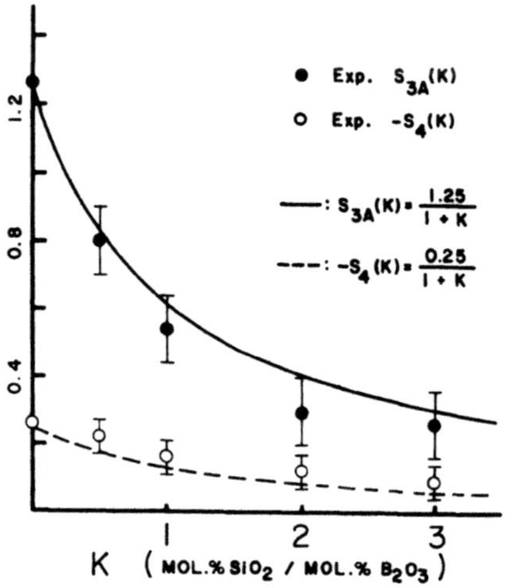

Figure 23. S_4 and S_{3A} versus K.

based[47] on the infrared work of Krogh-Moe[25], the Raman studies by Konijnendijk[36,37], and NMR studies by Rhee[48] which indicate that binary glasses in the region from 33 1/3 to 50 molar% Na_2O consist of diborate units, ring-type metaborate units, and "loose" BO_4 units as shown in Figure 24. If it is assumed that each "molecule" of Na_2O added above 33 1/3% is used entirely to convert diborate units to metaborate units and "loose" BO_4 units, then it is predicted that

$$N_4 = N_{4_{max}} - 0.25(R - R_{max})$$

$$N_{3A} = 1.25(R - R_{max}) \quad .$$

The experimental data are in excellent agreement, with slopes $S_4(0) = - 0.26$ and $S_{3A}(0) = 1.26$ as compared to $- 0.25$ and 1.25, respectively.

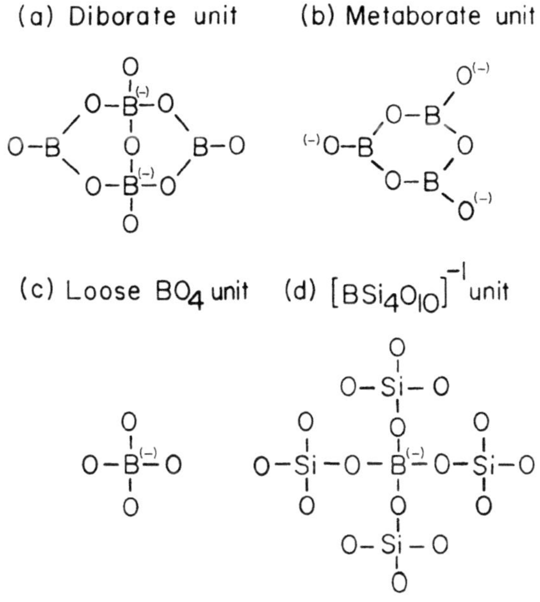

Figure 24. Some structural units found in alkali borate and alkali boro-silicate compounds.

For the ternary system, it is clear that all of the Na_2O is incorporated into the borate network for $R < 0.5$ and the SiO_2 simply dilutes the system. If it is postulated that for $0.5 < R < R_{max}$ the added Na_2O is also associated only with the borate units, and used entirely to convert diborate units into the structural unit $[B\ Si_4O_{10}]^{-1}$ found[49] in the mineral Reedmergnerite ($Na_2O \cdot B_2O_3 \cdot 6SiO_2$) and shown in Figure 24, then agreement of the model and data can be obtained. The formation of the Reedmergnerite unit continues until all of the SiO_2 is employed in the $[B\ Si_4O_{10}]^{-1}$ units. At this point $N_4 = N_{4max}$ at $R = R_{max}$ and the dependence of R_{max} on K can be shown[47] to be

$$R_{max} = \frac{1}{16} K + 0.50$$

which is in agreement with experiment as shown in Figure 25.

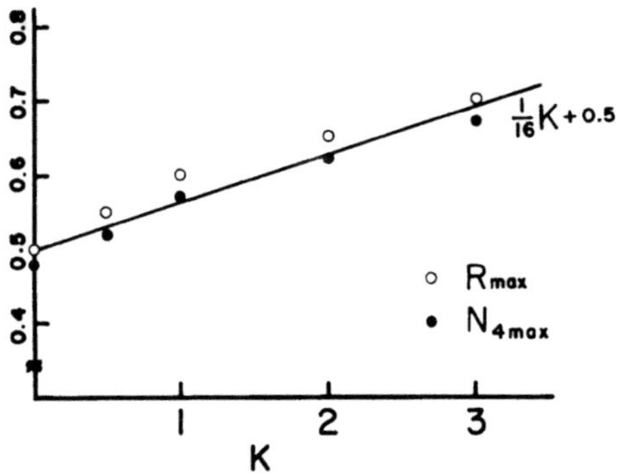

Figure 25. R_{max} and N_{4max} versus K.

The Na_2O added above $R = R_{max}$ is assumed to divide between the borate and silicate units in proportion to the amounts of B_2O_3 and SiO_2 present in the glass. (This proportional division of the added oxide was found in an earlier study[22] for PbO in the ternary lead borosilicate glasses.) With this assumption, the magnitudes of the slopes $S_4(K)$ for N_4 above R_{max}, and $S_3(K)$ for N_{3A} (which are found for $R > R_{max}$) should be reduced from those found in the binary glasses by a factor of (mol.% B_2O_3/mol.% B_2O_3 + mol.% SiO_2)= $\frac{1}{1+K}$. Then

$$S_4(K) = \frac{S_4(0)}{1+K} = -\frac{0.25}{1+K}$$

$$S_3(K) = \frac{S_{3A}(0)}{1+K} = \frac{1.25}{1+K} \quad .$$

These are the solid curves shown in Figure 23 which are in reasonable agreement with the experimental values.

The successful model can be summarized, then, in this fashion. For $R < 0.5$, N_4 is the same for the binary and ternary glasses, and the SiO_2 simply dilutes the sodium borate network. For the region $0.5 < R < R_{max}$ in the ternary glasses, the added Na_2O converts diborate units to Reedmergnerite $[B\ SiO_4O_{10}]^{-1}$ units until all of the

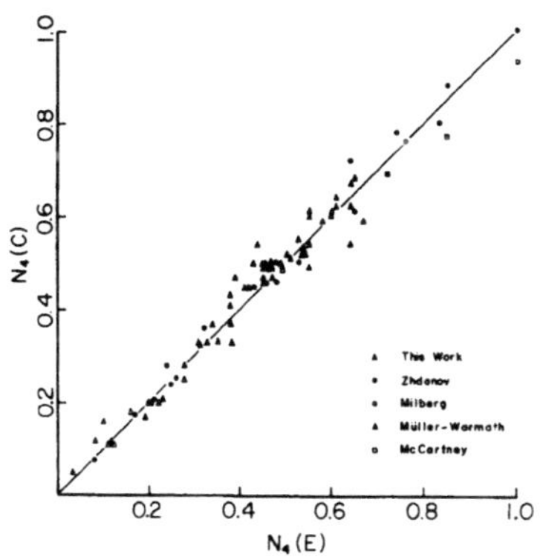

Figure 26. Experimental values of $N_4(E)$ versus values $N_4(C)$ calculated according to the model. A solid line indicates $N_4(E) = N_4(C)$.

SiO_2 is bound into these units, at which point $R = R_{max} = \frac{1}{16} K + 0.50$. Above $R = R_{max}$, added Na_2O is shared between the borate and silicate units in proportion to the amounts of B_2O_3 and SiO_2 in the glasses.

Figure 26 displays a plot of values of N_4 [labeled $N_4(C)$] calculated from the model summarized above versus values obtained from experiment [labeled $N_4(E)$] by means of NMR studies. The solid line is the line $N_4(C) = N_4(E)$. It is clear that the model is a useful one for predicting the numbers of 4-coordinated boron in the ternary sodium borosilicate glasses.

Finally, it should be noted that the relationship $N_{4_{max}} = R_{max} = \frac{1}{16} K + 0.50$ will yield $N_{4_{max}} = 1$ for $K = 8$. The available[43] NMR data for glasses with $K \geq 8$ and $R \geq 1$ suggest that all such glasses have $N_4 = 1$. Since at $K = 8$ and $R = 1$ the glasses consist entirely of Reedmergnerite $[B\ Si_4O_{10}]^{-1}$ units and silica, it is assumed that increases of SiO_2 simply produce glasses consisting of the Reedmergnerite units embedded in a silicate network. The Na_2O excess above that needed to achieve R = 1 for any glass will presumably be used to create nonbridging oxygens on SiO_2 units.

REFERENCES

*Research supported by the National Science Foundation through Grant No. 73-07615 A02, and the Materials Science Program of the National Science Foundation at Brown University (DMR 72-03023-A06).

1. Silver, A.H. and Bray, P.J., J. Chem. Phys. 29, 984 (1958).
2. Bray, P.J. and Silver, A.H.: "Modern Aspects of the Vitreous State," Vol. 1, Chap. 5, Butterworth, London, 1960, pp. 92-119.
3. Bray, P.J. and O'Keefe, J.G., Physics Chem. Glasses, 4, 37 (1963).
4. Bray, P.J., Leventhal, M. and Hooper, H.O., Physics Chem. Glasses 4, 47 (1963).
5. Leventhal, M. and Bray, P.J., Physics Chem. Glasses, 6, 113 (1965).
6. Bray, P.J., Proceedings of the Fourth All-Union Conference on the Vitreous State, Leningrad, USSR, 1964. Science Press, Leningrad, 1965; pp. 237-251.
7. Bray, P.J., Proceedings of the Seventh International Glass Congress, Brussels, Belgium, 1965, pp. 40.1-40.9.
8. Bray, P.J., Kline, D. and Poch, W., Glastechn. Ber. 39, 175 (1966).
9. Bray, P.J., "Interaction of Radiation with Solids," edited by Adli Bishay. Plenum Press, New York, 1967, pp. 25-54.
10. Bray, P.J., Silikattechnik 19, pp. 307-312 and 350-356 (1968).
11. Baugher, J.F. and Bray, P.J., Phys. Chem. Glasses 10, 77 (1969).
12. Bray, P.J., "Magnetic Resonance," Plenum Press, 1970, pp. 11-40. (Proceedings of the International Symposium of

Electron and Nuclear Magnetic Resonance, Melbourne, Australia, 1969.)

13. Landsberger, F.R. and Bray, P.J., J. Chem. Phys. 53, 2757 (1970).
14. Bray, P.J., Proceedings of the Fifth All-Union Conference on the Vitreous State, Leningrad, USSR, 1969. Science Press, Leningrad, 1971, pp. 191-193.
15. Kriz, H.M. and Bray, P.J., J. Mag. Res. 4, pp. 69 and 76 (1971).
16. Kriz, H.M., Park, M.J. and Bray, P.J., Phys. Chem. Glasses 12, 45 (1971).
17. Park, M.J. and Bray, P.J., Phys. Chem. Glasses, 13, 50 (1972).
18. Kim, K.S. and Bray, P.J., Phys. Chem. Glasses 15, 47 (1974).
19. Hendrickson, J.R. and Bray, P.J., J. Chem. Phys. 61, 2754 (1974).
20. Bray, P.J., Proceedings of the Tenth International Glass Congress, Kyoto, Japan, 1974, No. 13, "Glass Structure," pp. 13-1 to 13-20.
21. Bray, P.J., Fizika i Khimiya Stekla (Leningrad) 1, 490 (1975).
22. Kim, K.S., Merrin, S. and Bray, P.J., J. Chem. Phys. 64, 4459 (1976).
23. Bray, P.J., "The Physics of Non-Crystalline Solids," edited by G.H. Frischat, Trans. Tech. Publications, 1977, pp. 65-80. (Proceedings of the Fourth International Conference on the Physics of Non-Crystalline Solids, Clausthal-Zellerfeld, Federal Republic of Germany, 1976.)
24. Jellison, G.E. Jr., Panek, L.W., Bray, P.J. and Rouse, G.B. Jr., J. Chem. Phys. 66, 802 (1977).
25. Krogh-Moe, J., J. Phys. Chem. Glasses, 3, 101 (1962).
26. Krogh-Moe, J., J. Phys. Chem. Glasses, 6, 46 (1965).
27. Townes, C.A. and Dailey, D.P., J. Chem. Phys. 17, 782 (1948).

28. Mozzi, R.L. and Warren, B.E., J. Appl. Crystallogr. 3, 251 (1970).

29. Feller, S.A. and Bray, P.J., unpublished data.

30. Taylor, P.C. and Bray, P.J., J. Mag. Res. 2, 305 (1970).

31. Baugher, J.F., Kriz, H.M., Taylor, P.C. and Bray, P.J., J. Mag. Res. 3, 415 (1970).

32. Taylor, P.C. and Bray, P.J., Bull. Amer. Ceram. Soc. 51, 234 (1972).

33. Baugher, J.F., Taylor, P.C., Oja, T. and Bray, P.J., J. Chem. Phys. 50, 4914 (1969).

34. Jellison, G.E. Jr., Feller, S.A. and Bray, P.J., "NMR Powder Patterns for Integer-Spin Nuclei in the Presence of Asymmetric Quadrupole Effects," to be published in the Journal of Magnetic Resonance.

35. Jellison, G.E. Jr. and Bray, P.J., "Structural Determinations for Sodium Borate Glasses Using B^{10} and B^{11} NMR," manuscript in preparation.

36. Konijnendijk, W.L., "The Structure of Borosilicate Glasses," Thesis, Philips Res. Repts. Suppl. 1975, No. 1.

37. Konijnendijk, W.L. and Stevels, J.M., J. Non-Crystalline Solids 18, 307 (1975).

38. Bishop, S.G. and Bray, P.J., Phys. Chem. Glasses, 7, 73 (1966).

39. Beekenkamp, P. and Hardeman, G.E.G., Verres et Refr. 20, 419 (1966).

40. Kline, D. and Bray, P.J., Phys. Chem. Glasses, 15, 41 (1966).

41. Beekenkamp, P., Phys. Chem. Glasses, 9, 14 (1968).

42. Zhdanov, S.P., Kerger, I. and Koromal'di, E.V., Dok. Akad. Nauk SSSR 204, 622 (1972).

43. Milberg, M.E., O'Keefe, J.G., Verhelst, R.A. and Hooper, H.O., Phys. Chem. Glasses, 13, 79 (1972).

44. Scheerer, J., Müller-Warmuth, W. and Dutz, H., Glastech. Ber. 46, 109 (1973).

45. Brungs, M.P. and McCartney, E.R., Phys. Chem. Glasses, 16, 48 (1975).

46. Zhdanov, S.P., Proceedings of the Tenth International Congress on Glass, Kyoto, Japan. Ceramic Society of Japan, 1974, pp. 13-58.

47. Yun, Y.H. and Bray, P.J., "Nuclear Magnetic Resonance Studies of the Glasses in the System Na_2O-B_2O_3-SiO_2," manuscript in preparation.

48. Rhee, C., J. of Korean Phys. Soc. 4, 51 (1971).

49. Appleman, D.E. and Clark, J.R., Am. Mineral, 50, 1827 (1965).

STRUCTURAL DETERMINATIONS FOR SODIUM BORATE GLASSES USING B^{10} and B^{11} NMR *

G. E. Jellison, Jr.†

U. S. Naval Research Laboratory, Washington, D. C. 20375

P. J. Bray

Brown University, Providence, R. I.

INTRODUCTION

In earlier NMR studies of structure and bonding in borate glasses (1-7), the central transition ($m=\frac{1}{2} \leftrightarrow -\frac{1}{2}$) of the B-11 isotope has been employed almost exclusively. Aside from its usefulness in the determination of N_4, the fraction of four coordinated boron atoms, B-11 NMR has been of limited value in determining the coupling constant (Q_{cc}) and the asymmetry parameter (η) found in the glass. This limitation is due primarily to the importance of dipolar broadening and the existence of distributions in Q_{cc} and η. B-10 NMR, on the other hand, affords 3 advantages over the use of the central transition of B-11 NMR: 1) dipolar broadening is much less important (8); 2) the distribution of Q_{cc} is directly obtainable from a part of the B-10 NMR spectrum (8); and 3) there exists another part of the B-10 NMR experimental spectrum that is sensitive to the distributions of η, but insensitive to distributions of Q_{cc} (9). These features of the B-10 NMR spectrum are employed in this work for the study of glasses in the sodium borate system.

The theory of B-10 NMR powder patterns is presented in ref. 10 and will therefore be only reviewed here. Figure 1 shows a powder pattern for B-10 NMR. It is assumed that typical values of Q_{cc}

*Research supported by National Science Foundation, Materials Science Program (DMR 7203023-A06).
†NRL-NRC Research Associate

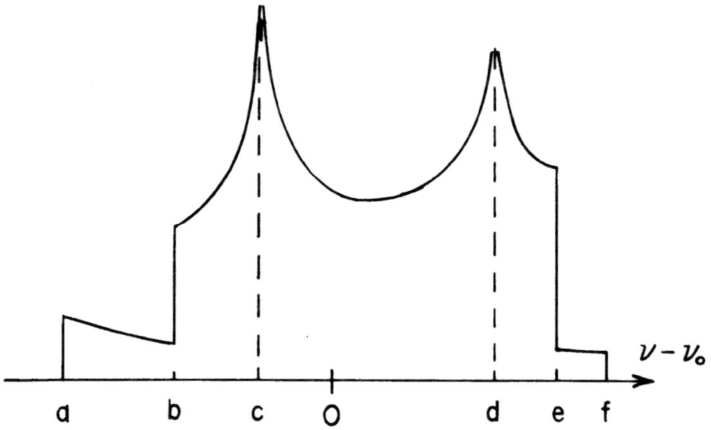

Figure 1: B-10 NMR powder pattern for 2 or the 6 transitions at ν_o = 7 MHz. The shoulders b and f and the divergence c correspond to the 0 ↔ -1 transition and the shoulders a and e and the divergence d correspond to the 1 ↔ 0 transition.

(=5.5 MHz) and ν_o (=7 MHz, ν_o being the spectrometer frequency) are employed. For higher values of β ($\beta=\nu_Q/\nu_o$, $\nu_Q=3Q_{cc}/2I(2I-1)$), additional shoulders and divergences may appear (see reference 10 for details). The positions of the shoulders and divergences of the powder pattern may be calculated exactly (11); the positions of the 6 features a-f have been calculated and are given in references 8-10.

RESULTS

Figure 2 depicts the B-10 NMR derivative spectra for several sodium borate glasses (narrow lines) and computer simulated spectra (wide lines, to be discussed below). Feature A (to the low frequency side of ν_o) is due to boron sites with a large coupling constant (i.e. 3-coordinated borons); feature B, centered at ν_o, is due to boron sites with a small coupling constant (i.e. 4-coordinated borons). Portions of both features A and B were distorted somewhat from the experimental conditions (magnetic field modulation and integration time constant) required to see the entire spectrum. However, scans of feature B, where experimental conditions did not cause distortions, indicated that feature B was symmetric between the highest and lowest points of the derivative (the dot (•) of fig. 2 shows the lowest point of the derivative of feature B, assuming that the highest and lowest points of feature B are equidistant from the baseline (using the highest point as the standard)).

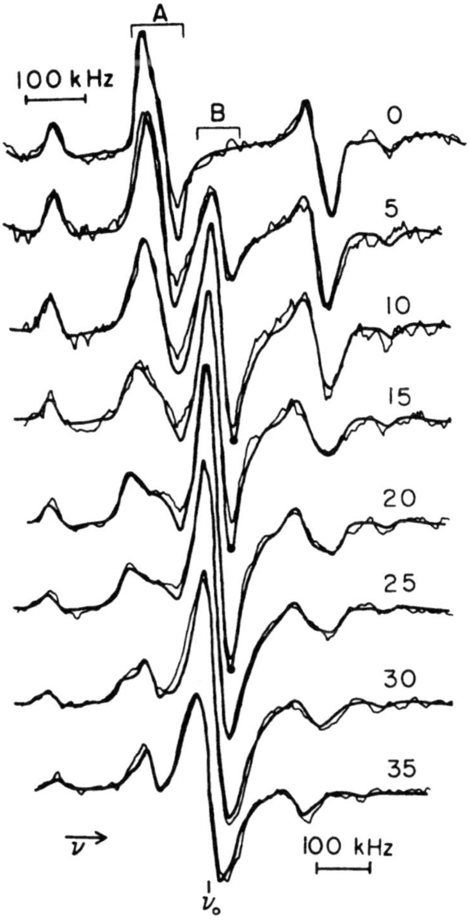

Figure 2: B-10 NMR derivative spectra for eight sodium borate glasses. The computer-simulated spectra (the darker lines in each case) were computed using the five sites characterized in Table I, with the appropriate weighting factors given in Table III. The number to the right of each trace indicates the molar % sodium oxide. The calibration given in the upper left-hand corner is for the glasses of composition 0, 5, 10, 15, 20, 25, and 30 molar % sodium oxide. The calibration given in the lower right-hand corner is for the 35 molar % sodium oxide glass. The features A and B are the features of maximum distortion due to magnetic field modulation and integration time constant for all glasses. The Dot (•) represents the lowest position of feature B if these effects were not present.

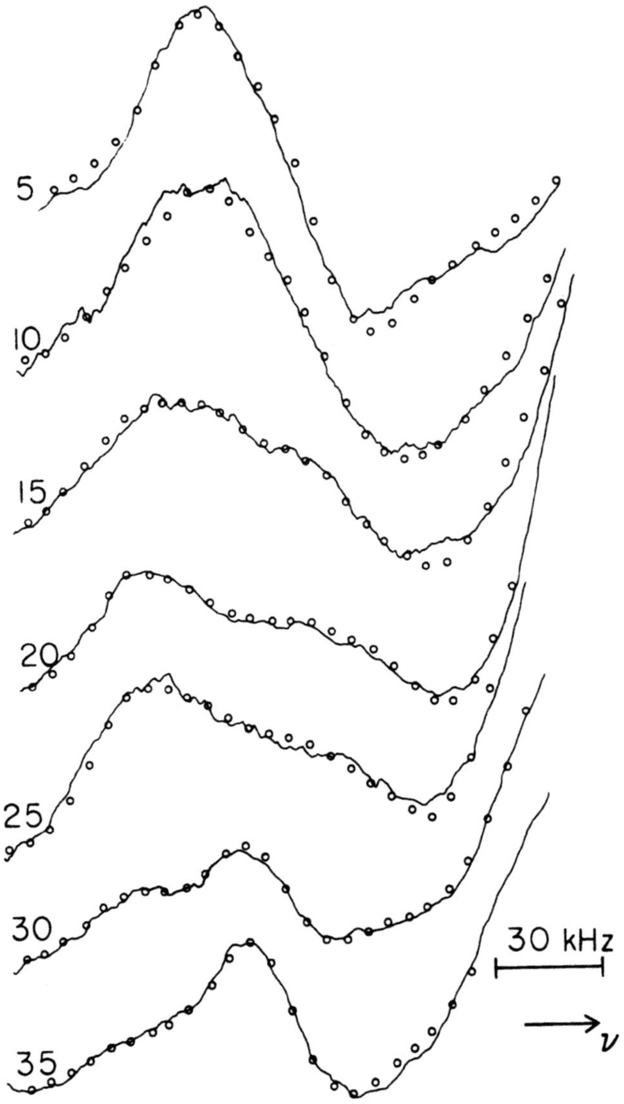

Figure 3. Experimental B-10 derivative of the main feature (feature A of Fig. 2) for 3-coordinated boron atoms for seven sodium borate glasses (solid line). The simulated spectra (circles) were calculated using the five sites characterized in Table I, and the appropriate weighting factors given in Table III.

Likewise scans of feature A were taken such that experimental conditions did not cause distortions; these scans are shown in fig. 3, with computer simulated spectra (circles) superimposed.

The actual lineshape for a glass involving distributions of both ν_Q and η is given by (8)

$$K(\nu') = \int_o^\infty d\nu_Q \int_o^1 d\eta \; p(\nu_Q,\eta) \; R(\nu', \nu_Q,\eta) \qquad (1)$$

where $p(\nu_Q,\eta)$ is the distribution function for ν_Q and η, and $R(\nu',\nu_Q,\eta)$ is the dipolar-broadened shape function. In computing the computer-simulated spectra of figs. 2 and 3, it was assumed that the distribution function consisted of a finite sum of component parts, each component part representing one site in the glass. Furthermore, it is assumed that the variables ν_Q and η are uncorrelated (this assumption is not entirely correct as η is a function of ν_Q. The exact correlation, however, is model-dependent and cannot be determined experimentally), and that the distributions are Gaussian; i.e.

$$p(\nu_Q,\eta) = \sum_i p_i(\nu_Q) \; p_i(\eta). \qquad (2)$$

The calculated spectra, shown in figs. 2 and 3, required 5 sites to fit the experimental spectra of the glasses in the sodium borate system: 2 sites with a low value of Q_{cc} (corresponding to 4-coordinated boron atoms) and 3 sites with a higher value of Q_{cc}, and a low value of η (corresponding to 3-coordinated boron atoms). (The reason 5 sites were chosen will be discussed below; see references 12 and 13 for a description of the simulating procedure.) Each site is characterized by a most probable value and a Gaussian width of ν_Q and η. The parameters of the five sites is given in Tables I and II, while the weighting fractions used for the computer simulations are given in Table III.

ASSIGNMENT OF SITES TO STRUCTURAL GROUPS

The vastly different Hamiltonian parameters listed in Table I for the five sites can be explained by the difference in the electronic distribution of the 3- and 4-coordinated boron atoms. One effect of this difference, which has been exploited extensively (1-7), is that the distribution of electrons about 4-coordinated boron atoms is close to being tetrahedrally symmetric, which produces a weak quadrupole interaction (as measured by Q_{cc}); on the other hand, the distribution of electrons about 3-coordinated boron atoms in borate glasses, for the case when all oxygen atoms connect two boron atoms, is close to being axially symmetric, which produces a larger Q_{cc}, but a small η. Therefore, sites 1 and 2 can be assigned to BO_4 units, while sites 3, 4, and 5 can be assigned to

TABLE I

Characterization of the distributions of Hamiltonian parameters for the 5 sites observed in sodium borate glasses for B-10 NMR

Site	Q_{cc} (MHz)	$\sigma_{Q_{cc}}$ (MHz)	η	σ_η
1	0.8	0.2	0.5	0.5
2	1.5	1.0	0.0	0.5
3	a	a	0.12	0.05
4	a	a	0.26	0.05
5	a	a	0.08	0.05

[a] The values of Q_{cc} and $\sigma_{Q_{cc}}$ for sites 3, 4, and 5 vary from glass to glass and are given in Table II.

TABLE II

Experimental values of Q_{cc} and $\sigma_{Q_{cc}}$ for sites 3, 4, and 5 in sodium borate glasses (from reference 14). Values of η and σ_η for these sites are given in Table I.

Molar % sodium oxide	Q_{cc}	$\sigma_{Q_{cc}}$
0	5.51	0.21
5	5.48	0.22
10	5.48	0.21
15	5.45	0.21
20	5.41	0.21
25	5.41	0.21
30	5.37	0.23
35	5.39	0.23

BO_3 units. Further differences in the electronic configurations can be seen by considering the type of <u>unit</u> a BO_3 or BO_4 unit is bonded. Retrospectively, the reason 5 sites were chosen can now be seen. There are at least 2 possible BO_4 sites: BO_4 units connected to all BO_3 units and BO_4 units connected to a mixture of BO_3 and BO_4 units (the possibility of BO_4 units being bonded to all BO_4 units is neglected, since this possibility has never been observed in crystalline borates). There are 3 possible BO_3 sites: BO_3 units connected to 1) all BO_3 units, 2) a mixture of BO_3 and BO_4 units, and 3) all BO_4 units. Since a boron atom in a BO_3 unit will be in a sp^2 hybridized state, while one in a BO_4 unit will be in a sp^3 hybridized state, the B-O bonds of a BO_3 unit could possibly have some π character, whereas the B-O bonds of a BO_4 unit

TABLE III

The fractions of sites 1, 2, 3, 4, and 5 determined by the computer fitting procedure given in reference 12, for a series of sodium borate glasses. The notation T^4, D^4, B^3, T^3, and D^3 is explained in Table IV. All fractions are rounded to two significant figures.

glass	Site 1 (T^4)	Site 2 (D^4)	Site 3 (B^3)	Site 4 (T^3)	Site 5 (D^3)
5	.05	---	.79	.16	---
10	.10	.01	.58	.30	.01
15	.15	.03	.38	.44	.03
20	.17	.08	.16	.51	.08
25	.14	.19	.04	.44	.19
30+	.11	.28	---	.33	.28
35+	.08	.34	---	.24	.34

+Borons bonded to non-bridging oxygens have been ignored in these glasses.

TABLE IV

Symbols used to denote the fractions of borons in each structural group

Symbol	Unit	Coordination	Site
T^4	tetraborate	4	1
D^4	diborate	4	2
B^3	boroxol	3	3
T^3	tetraborate	3	4
D^3	diborate	3	5

cannot. Therefore the electronic configuration of an oxygen will depend considerably on whether it bonds two BO_3 units, two BO_4 units, or one BO_3 and one BO_4 unit. Therefore, a BO_3 unit that is connected to both BO_3 and BO_4 units will have a higher value of η (indicating a larger deviation from axial symmetry) than a BO_3 unit connected to all BO_3 or all BO_4 units; site 4 belongs to the former class, while sites 3 and 5 belong to the latter. A similar conclusion can be drawn concerning the BO_4 units: a BO_4 unit connected to all BO_3 units will be closer to tetrahedral symmetry than a BO_4 unit connected to both types of units. Therefore site 1 should be due to BO_4 units connected to all BO_3 units, while site 2 should be due to BO_4 units connected to both types of units.

Krogh-Moe's interpretation (14) of infrared data of sodium borate glasses states that sodium borate glasses consist of 4 structural groups: boroxol, pentaborate, triborate, and diborate (see Figure 4 a-d). For glasses below 20 molar % sodium oxide, the addition of one "molecule" of sodium oxide to boron oxide results in the formation of one triborate and one pentaborate group. These groups are connected, so it is convenient to refer to this pair as a tetraborate group. Above 20 molar % sodium oxide and below 33 1/3 molar % sodium oxide, diborate groups will increase at the expense of tetraborate groups. The ideal case of this model, where the lever rule determines the amount of each group, and where the glass is assumed to contain all boroxol groups at 0 molar % sodium oxide, all tetraborate groups at 20 molar % sodium oxide, and all diborate groups at 33 1/3 molar % sodium oxide, is depicted in Figure 5 by the straight lines.

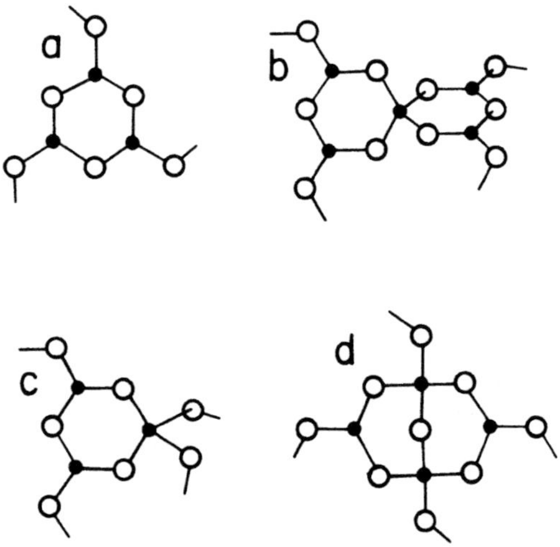

Figure 4: The four crystalline structural groups found in sodium borate glasses (a. boroxol, b. pentaborate, c. triborate, and d. diborate). The tetraborate group is formed by connecting one oxygen atom of the BO_4 unit in the triborate group to a BO_3 unit of the pentaborate group. • boron; ○ oxygen.

The 4-coordinated boron sites (1 and 2) must come from the tetraborate groups and the diborate groups. Site 1 could only come from tetraborate groups, so this assignment is trivial. Site 2, however, could come from tetraborate BO_4 units, as well as diborate BO_4 units, if two tetraborate groups were connected by BO_4

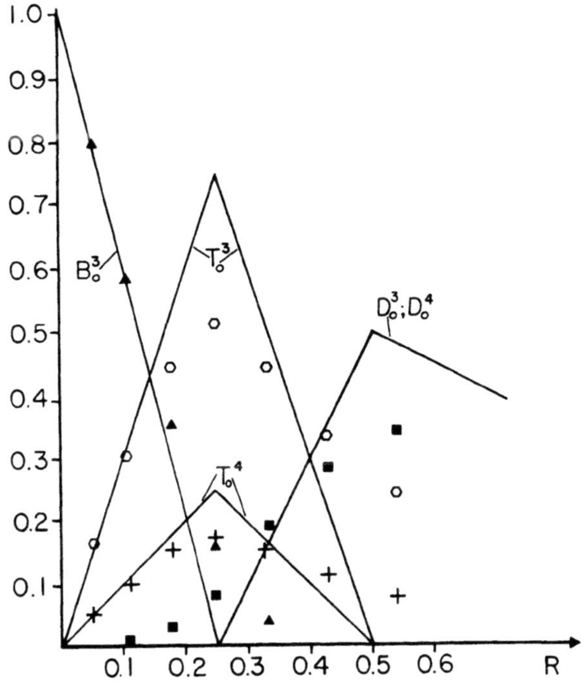

Figure 5: Fractions of B^3 (▲), T^3 (○), D^3 and D^4 (■) and T^4 (+) plotted as a function of R (=x/1-x), x = molar fraction of sodium oxide. The straight line segments are plotted for the case where the lever rule is obeyed (see text).

units. This combination is not seen in crystalline sodium tetraborate (18) or α sodium triborate (19), so it is reasonable to say that site 2 arises only from diborate BO_4 units.

According to Krogh-Moe's model, boroxol groups will predominate at low sodium content. For pure B_2O_3 glass, site 3 is the only site observable (see reference 9). Also, Raman studies (14,15) indicate that the boroxol group is predominant at low sodium content, while X-ray studies (16) indicate that pure B_2O_3 glass consists mostly of boroxol rings. Therefore the assignment of site 3 to BO_3 units in boroxol groups is straightforward. The non-zero asymmetry parameter for site 3 can be attributed to the non-equivalence of the 3 oxygens surrounding the boron atom (two oxygens are in the ring itself, while the other is outside the ring; see reference 9).

At 20 molar % sodium oxide, Krogh-Moe's model states that tetraborate groups will predominate. This is reinforced by Raman studies (15). Furthermore, B-11 NMR measurements of crystalline sodium tetraborate (17) indicates that η=0.27 for the BO_3 units in that compound (compared to 0.26 for site 4). One might be tempted to immediately assign site 4 to BO_3 units in tetraborate groups; however, there is a potential problem with this assignment. It is possible that boroxol or diborate borons could also contribute to site 4, since one does not know how the mixture of groups is interconnected. For the present, it will be assumed, however, that site 4 borons come from BO_3 units in tetraborate groups.

Above 30 molar % sodium oxide, diborate groups predominate (14). X-ray diffraction experiments of crystalline lithium diborate (20, 21) indicate that all BO_3 units are connected to 3 BO_4 units (sodium diborate is not appropriate for comparison, since diborate units do not appear in this compound (22)). Site 5 is then assigned to BO_3 units in diborate groups. NMR measurements of crystalline lithium diborate, however, yield a value (23) η= 0.17, which is clearly greater than η for site 5. This is possibly due to one B-O bond length being 0.04 A shorter than the other two B-O bond lengths (21) in lithium diborate, which would increase the asymmetry parameter. If this is the case, the crystalline diborate groups are <u>more distorted</u>, due to crystal packing requirements, than the glassy diborate units.

ASSIGNMENT OF WEIGHTING FRACTIONS

In the above assignments, and in the determination of the proper weighting factors presented in Table III, the following assumptions have been made:

1) "Loose" BO_3 and BO_4 units are neglected.
2) Groups other than those pictured in Figure 4 are ignored.
3) Pentaborate and triborate units are assumed to occur in pairs, and are called tetraborate groups.
4) Site 2 is postulated to consist solely of BO_4 units in diborate groups.
5) Site 4 is postulated to consist entirely of BO_3 units in tetraborate groups.

Additionally, two facts should be noted: 1) D^3=D^4 and 2) T^3=$T^4/3$ (see Table IV for notation). These facts can be seen from Figure 4 b, c, and d.

For x $\leq$ 0.25 (x = molar fraction sodium oxide), the weighting fractions of Table III are determined by varying only a quantity δ, where δ represents the fraction of diborate borons in excess of the fraction expected by the lever rule. In this region, N_4=x/1-x (2). Therefore, any departure from the lever rule must leave N_4 unchanged. This leads to the following equations for B^3, T^3, D^3, T^4, and D^4:

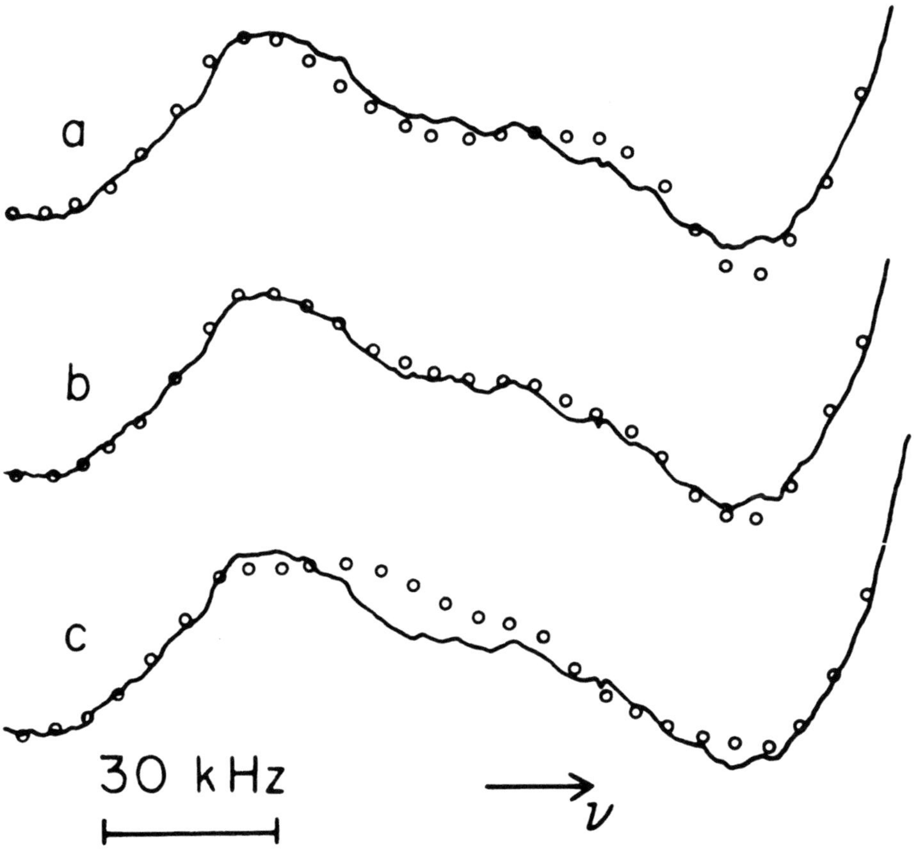

Figure 6. Experimental B-10 derivative of the main feature for 3-coordinated boron atoms (feature A of Figure 2) in the 20 molar % sodium oxide glass (solid line). The simulated spectra (circles) were calculated using the five sites of Table I, and three sets of weighting fractions. The quantity δ is discussed in the text.

Simulation	Site 1	Site 2	Site 3	Site 4	Site 5	δ
a	.19	.06	.12	.57	.06	.06
b	.17	.08	.16	.51	.08	.08
c	.15	.10	.20	.45	.10	.10

$$\left.\begin{aligned} B^3 &= B_o^3 + 2\delta \\ T^3 &= T_o^3 - 3\delta \\ D^3 &= D_o^3 + \delta \\ T^4 &= T_o^4 - \delta \\ D^4 &= D_o^4 + \delta \end{aligned}\right\} \quad x \leq 0.25 \qquad (3)$$

where the subscript o refers to values predicted by the lever rule. Figure 6 shows the experimental spectrum for the x=0.20 glass superimposed upon three computer-simulated traces for different values of δ. As can be seen, Figure 6b represents the best fit, and the simulated spectra are very sensitive to δ.

For $x \geq 0.30$, it was necessary to relax the condition $N_4=x/1-x$, in order to fit the central line and the broad line simultaneously. From Raman studies (15), $B^3=0$ in this region. Again, let $D^3=D_o^3+\delta$. Then

$$\left.\begin{aligned} B^3 &= 0 \\ T^3 &= T_o^3 - 3\delta/2 \\ D^3 &= D_o^3 + \delta \\ T^4 &= T_o^4 - \delta/2 \\ D^4 &= D_o^4 + \delta \end{aligned}\right\} \quad x \geq 0.30 \qquad (4)$$

(Note δ will be negative for these glasses, indicating the "disassociation" of diborate groups and the "creation" of tetraborate groups from the numbers predicted by the lever rule. Also, the number of oxygens will not be conserved by this transformation, since the condition on N_4 is relaxed; this is because non-bridging oxygens have been ignored.)

DISCUSSION

An examination of Figure 5 shows that the fractions B^3, D^3 and D^4 are above the values predicted by the lever rule, while the factions T^3 and T^4 are below the values predicted by the lever rule, in the compositional region 0-25 molar % sodium oxide. Any violation of assumption 5 would tend to decrease T^3 (as well as T^4), so this conclusion is independent of this assumption.

Therefore, in the region 0–25 molar % sodium oxide some of the tetraborate groups predicted by the lever rule have been "disassociated" into diborate groups and additional boroxol groups. Such "disassociation" was predicted by Krogh-Moe (24) in his interpretation of melting point depression data for melts of the sodium borate glass system. He found that the best fit of the experimental data in the region near x = 0.2 occurred when tetraborate, boroxol, and diborate groups were present, and when some of the tetraborate groups, which would be present according to the lever rule, were disassociated into boroxol and diborate groups. Furthermore, Raman studies by Konijnendijk (15) and Bril (25) indicate the existence of boroxol groups at 20 molar % sodium oxide, and infrared measurements by Krogh-Moe (14) indicate the existence of diborate groups at the same composition. Therefore, for the compositional region x=0 to x=0.2, the conclusion can be drawn that the glass is composed of the three principal structural groups (boroxol, tetraborate, and diborate), but that the fractions of these groups in the glasses for which x=0.10, 0.15, and 0.20 do not correspond to the values predicted by the lever rule.

In the region x=0.3 and 0.35, some of the diborate groups predicted by the lever rule have been "changed" into tetraborate groups, though the numbers presented in Figure 5 could be in error. The interpretation of melting point depression data by Krogh-Moe (24) of melts of the sodium borate system in the region of x=1/3 indicates that these two groups (tetraborate and dibora) are present. In addition, infrared (14) and Raman (15) studies indicate that these two groups are the only major components of the glass in this region. In particular, the components of crystalline sodium diborate (the di-pentaborate and the triborate with one non-bridging oxygen) have been ruled out, since the NMR spectra indicated that no observable BO_3 units with one or two non-bridging oxygens were present. Therefore, the tetraborate and diborate groups are the only major components of the glasses for which x=0.3 and 0.35; however, as was mentioned above, the fractions T^3, T^4, D^3, and D^4 do not correspond to values predicted by the lever rule. In particular, the lever rule would predict $T^3 = T^4 = 0$ for the x=1/3 glass. By interpolation, T^3=0.27 and T^4=0.09 at x=1/3, which is clearly not zero. The main consequence of the "changing" of diborate groups into tetraborate groups for the glasses with x=0.3 and 0.35 is that N_4 will fall below the x/1-x line. Figure 7 shows two independent measurements of N_4: 1) using B-11 NMR (shown by the circles (O) of Figure 6), and 2) using B-10 NMR ($N_4 = T^4 + D^4$, shown by the crosses (+) of Figure 6). The solid line of Figure 7 is obtained by summing $T^4_o + D^4_o$ (obtained from the lever rule) of Figure 5. The agreement of both methods is excellent; both methods indicate that the value of N_4 lies below the x/1-x curve for the glasses with x=0.3 and 0.35. Also included in Figure 7 is the theoretical curve for N_4 (shown by the dotted line) from Beekenkamp (26). As can be seen, the experimental data points lie substantially above this theoretical curve.

The conclusions above concerning the fractions T^3, T^4, D^3, and D^4 for the glasses with x=0.3 and 0.35 are highly dependent on assumption 5. Any deviation from this assumption would tend to decrease T^3 and T^4, and increase D^3 and D^4, resulting in a larger value of N_4. However, this deviation could not be too large without creating a disagreement with the B-11 NMR N_4 measurements.

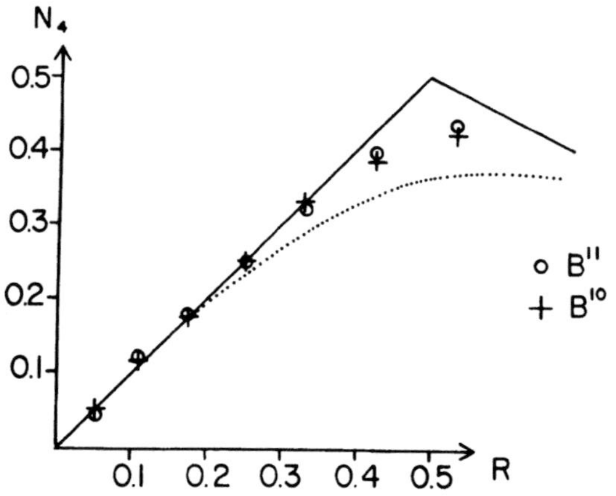

Figure 7. N_4 versus R (=x/1-x) from B-11 NMR (0) and B-10 NMR (+). The straight line is determined by summing $T^4_o + D^4_o$ from Figure 5. The dotted line is from Beekenkamp (25).

REFERENCES

1. A. H. Silver and P. J. Bray, J. Chem. Phys., 29, 984 (1958).
2. P. J. Bray and J. G. O'Keefe, Phys. Chem. Glasses 4, 37 (1963).
3. S. G. Bishop and P. J. Bray, Phys. Chem. Glasses 7, 73 (1966).
4. S. Greenblatt and P. J. Bray, Phys. Chem. Glasses 8, 213 (1967).
5. H. M. Kriz, M. J. Park, and P. J. Bray, Phys. Chem. Glasses 12, 45 (1967).
6. H. M. Kriz and P. J. Bray, J. Non-crystalline Solids 6, 27 (1971).
7. W. Müller-Warmuth, W. Poch, and G. Sielaff, Glastech. Ber. 43, 5 (1970).
8. G. E. Jellison, Jr. and P. J. Bray, Solid State Comm. 19, 517 (1976).

9. G. E. Jellison, Jr., L. W. Panek, P. J. Bray, and G. Rouse, J. Chem. Phys. 66, 802 (1977).
10. G. E. Jellison, Jr., S. A. Feller, and P. J. Bray, accepted for publication in J. Mag. Res.
11. J. F. Baugher, P. C. Taylor, T. Oja, and P. J. Bray, J. Chem. Phys. 50, 4914 (1969).
12. P. C. Taylor, and P. J. Bray, Lineshape Manual 1969 (unpublished).
13. P. C. Taylor, and P. J. Bray, J. Mag. Res. 2, 305 (1970).
14. J. Krogh-Moe, Phys. Chem. Glasses 6, 46 (1965).
15. W. L. Konijnendijk, The Structure of Borosilicate Glasses, Thesis, Philips Res. Repts. Suppl. 1975, No. 1; W. L. Konijnendijk and J. M. Stevels, J. Non-crystalline Solids 18, 307 (1975).
16. R. L. Mozzi and B. E. Warren, J. Appl. Cryst. 3, 251 (1970).
17. C. H. Rhee, J. Korean Phys. Soc. 4, 51 (1971).
18. A. Hyman, A. Perloff, F. Mauer, and S. Block, Acta Cryst. 22, 815 (1967).
19. J. Krogh-Moe, Acta Cryst. B30, 747 (1974).
20. J. Krogh-Moe, Acta Cryst. 15, 190 (1962).
21. J. Krogh-Moe, Acta Cryst. B24, 179 (1968).
22. J. Krogh-Moe, Acta Cryst. B30, 578 (1974).
23. J. F. Baugher, H. M. Kriz, P. C. Taylor, and P. J. Bray, J. Mag. Res. 3, 415 (1970).
24. J. Krogh-Moe, Phys. Chem. Glasses 3, 101 (1962).
25. T. W. Bril, Thesis, Technological University Eindoven, The Netherlands, 1975.
26. P. Beekenkamp, Philips Res. Repts. Suppl. 1966 No. 4.

RANDOM VECTOR STATISTICAL STUDIES OF AMORPHOUS MATERIALS

C. R. Kurkjian, G. E. Peterson and A. Carnevale

Bell Laboratories

Murray Hill, New Jersey 07974

I. INTRODUCTION

The lack of translational symmetry in amorphous solids restricts the amount of information obtainable from direct diffraction studies of glass structure. It also complicates the manner in which the structure can itself be conveniently and concisely described. Because of these difficulties it is necessary to apply all of the available experimental and theoretical techniques to the solution of this problem. The presence of a quadrupole moment at its nucleus allows the variation in symmetry at boron sites to be conveniently studied by nuclear magnetic resonance. This is not the case for silicon and thus an added dimension is given to structural studies of borate glasses. Although glasses in which boron is the major or only glass former are rare, it is a very important ingredient in a large number of commercial glasses. Because of this, and because of the interesting and sometimes anomalous properties shown by borate glasses, their structure continues to be of special scientific and technological interest.

II. HISTORICAL REVIEW OF GLASS STRUCTURE

Table I traces the cyclic character of order-disorder theories of glass structure.[1-14] Although many specific theories and models have been proposed, these order-disorder controversies continue to appear. The random network theory on the one hand, was a rather qualitative model which visualized a statistical assemblage of very small units, e.g., triangles or tetrahedra. On the other hand more ordered models proposed the existence of larger ("crystalline") structural groupings, sometimes as large as 20 to 50 Å.

Table I

Year	"Order"	"Disorder"	Cation Cluster	Two Phase
1852	Leydolt			
1912	Tillotson			
1921	Lebedev			
1930	Randall			
1932		Zachariasen		
1933		Warren		
1936	Valenkov; Porai-Koshits			
1938			Hartleif	
1942			Dietzel	
1953				Prebus
1955				Hoffman
1958				Porai-Koshits
1969		Mozzi; Warren		
1972	Konnert; Karle			

The existence of these two types of theories is indicative of the difficulty in experimentally defining glass structures unambiguously.

This description basically refers to silicate glasses. The history of borate glasses is somewhat different. Firstly, crystalline B_2O_3 is difficult to prepare and thus, interestingly, a determination of the structure of glassy B_2O_3 [15] (e.g., BO_3 triangles) was made before that of crystalline B_2O_3. Rather than an order-disorder controversy, the controversy in simple borate glasses tended to center around the so-called "boron anomaly".[16] This refers to the often-observed maximum or minimum in properties in alkali borate glasses near 15 mole % alkali oxide. Most explanations made use of Biscoe and Warren's[17] finding that BO_3 triangles were converted to BO_4 tetrahedra as alkali was added to B_2O_3. A number of theories were proposed because of the indefiniteness of the extent of the coordination change. NMR results in the early 60's showed that the coordination change continued through the range of the anomaly (up to ~30 mole %) and thus the simple model could not explain the "15% anomaly". To date a completely satisfactory explanation of this anomaly has not been given.

During the 1960's and early 1970's NMR, x-ray and Raman studies on many alkali borate crystals and glasses led to the model described by Griscom in the pre-conference review.[18] This model describes the structure in terms of rather specific proportions of five different crystal-like groupings in the glass. Thus an order-disorder controversy may now appear with regard to borate glasses as well.

Although in 1958 Skatula, Vogel and Wessel[19] and again in 1968 Shaw and Uhlmann[20] described metastable, subliquidous phase-separation in alkali-borate systems, this complication has generally been ignored in structural discussions. In fact this issue has recently[21] been further clouded by the observation of micro-heterogeneities, but not phase separation in alkali borate glasses. The problem must be clarified before the details of borate glass structure can be understood.

III. CONTEMPORARY STUDIES

Various types of spectroscopic techniques have been applied to glasses. These generally yield broad lines compared to crystals. This line broadening is normally qualitatively ascribed to the non-uniqueness of the structural groups and their random arrangement in the glass. In particular, for instance, it was suggested by Silver and Bray[22] at the start of their NMR studies of B^{11} in borate glasses, that the broadness which they observed in the resonance from 3-fold coordinated borons could be accounted for by a distribution in quadrupole coupling constants of the different boron nuclei in the glass. Somewhat later, Kriz and Bray[23] estimated a distribution of coupling constants for B_2O_3 glass by a trial and error method. Peterson et al.[24] found that a distribution of coupling with $\sigma_{\mathcal{Q}}$ = 0.2 MHz would satisfactorily fit the data.

The objectives at the start of our work were two-fold. The first was the development of mathematical techniques for decomposing experimental specta into components in an effort to ascertain whether a unique description of the distribution of coupling constants was possible. Secondly, to apply the latest techniques of quantum chemistry to the calculation of quadrupole coupling constants in B^{11} in various environments. By combining the two studies it was hoped that a unique description of the distribution $\Delta\mathcal{Q}$ together with a knowledge of what geometries were required to give this range of $\mathcal{Q}$'s would allow a rather detailed description of the glass structure to be given.

The quantum chemical calculations and results have been given in reference 25. Details of our ideas in terms of random vectors and the development of the computational techniques are given in the references indicated.[26]

SCF-MO calculations were made using $B(OH)_3$ and BH_3 as models for B_2O_3. Changes in bond angles and bond lengths both in and out of the **plane of the** triangle were made and energies, quadrupole coupling constants and asymmetry parameters were calculated. The results of these calculations are shown in Table II. Only the calculation using charges above and below the original BH_3 molecule produced a large enough difference in $\mathcal{Q}$ to account for that (∿0.2 MHz) estimated by Kriz and Bray[23] and Peterson et al.[24] The possibility of detecting the small changes in $\mathcal{Q}$ represented by the other configurations was considered in a separate paper.[27] In that work it was shown that the solution of the basic equation relating the distribution of the field gradient and the NMR line shape is very unstable with respect to noise;

$$\sigma_E^2 \equiv \sigma_N^2 M^2 \qquad (1)$$

σ_N^2, the noise on the experimental data is related to the noise on the solution σ_E^2 by the magnification factor M. This relation is shown in Fig. 1 for an NMR frequency of 16 MHz and $\mathcal{Q}$ = 2.70 MHz for various resolutions. It can be seen that as one attempts to improve the resolution in the solution, M increases very rapidly and for resolutions greater than ∿0.1 MHz, M is greater than 1 and the noise is greater than the solution.

In a more recent paper, in a culmination of the work on random vectors,[28] biased spectral decomposition was carried out by iterative techniques. These results are shown in Fig. 2. In this work, three "biased" density functions were used as starting points -

(a) Highly random structure

(b) Crystalline-like structure

(c) Partially crystalline, partially random structure.

By means of 43 "building block" spectra, and equations and procedures developed earlier,[26] NMR line shapes were calculated from these density functions. The initial density function was then changed iteratively in order to get a better fit between the "known" and calculated line shapes. Iteration was continued until the fit was better than 0.5%. The final probability density functions for the three different cases are shown in the center column of the figure. It is striking that the PDF's are at once similar in broad features (low resolution) and different in detail (high resolution). Thus some of the original bias is retained in the final solution as fine structure. On the other hand there are clearly two types of structure; a high $\mathcal{Q}$ and a low $\mathcal{Q}$ structure. These are the two peaks normally assigned to 3-fold and 4-fold borons, respectively. It is of importance that we find that the

Table II

Distortion	ΔQ MHz	$\Delta\eta$	M	$2\sigma_E$ for $\sigma_N = .1$/MHz
Boron displacement of ∿.059 Å toward neighboring oxygen as modeled by BH_3	.003	.076	$>10^4$	$>10^3$
Gross bending from planar to tetrahedral as modeled by $B(OH)_3$	.010	0	$\sim 10^4$	$\sim 10^3$
B-O-B angle closure by 10° as modeled by BH_3	.015	.024	$\sim 10^4$	$\sim 10^3$
Oxygen displacement .077 Å away from neighboring boron as modeled by BH_3	.088	.057	14	2.8
Intermolecular electric field effects as modeled* by BH_3	.150	0	2.4	.24

* A computed Mulliken atomic charge distribution of BF_3 was placed 3.12 Å above and below a BH_3 molecule. The BF bonds being parallel to the BH bonds.

relative areas of these two peaks does not change with the initial bias. Thus, apparently a quantitative measure of the relative numbers of 3-fold and 4-fold borons can be made, regardless of their origin. In our example, for instance, N_4 (the ratio of 4-fold to total borons) measured for the final density functions, d, e and f was found to be 0.327, 0.326 and 0.330, respectively.

IV. SUMMARY

By applying up to date computational techniques, rather quantitative tests of a given glass structural model, as well as an assessment of the usefulness and resolution of an experimental tool can be made. The uniqueness of detailed, high resolution models for borate glasses can not be proven using only NMR spectra, but quantitative measurements of N_4 can be made.

*Note added in Proof. After the completion of the above manuscript the following comments appeared in a review paper entitled "The

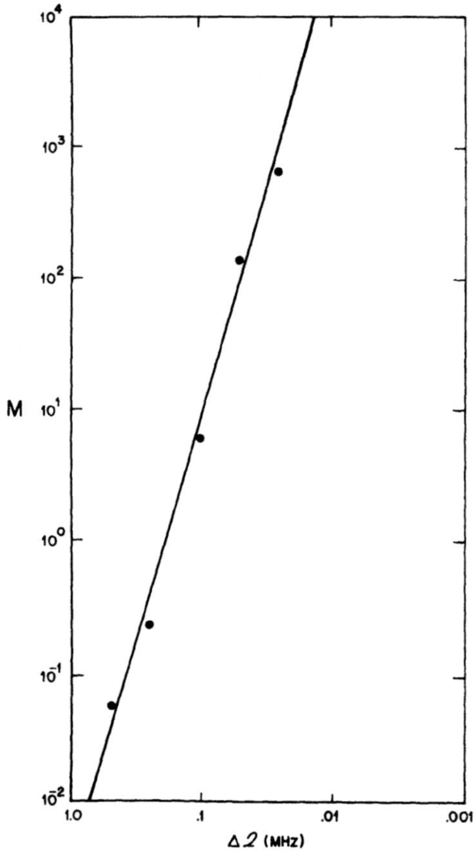

Fig. 1 Plot of the M factor (Eqn. 1) versus resolution.

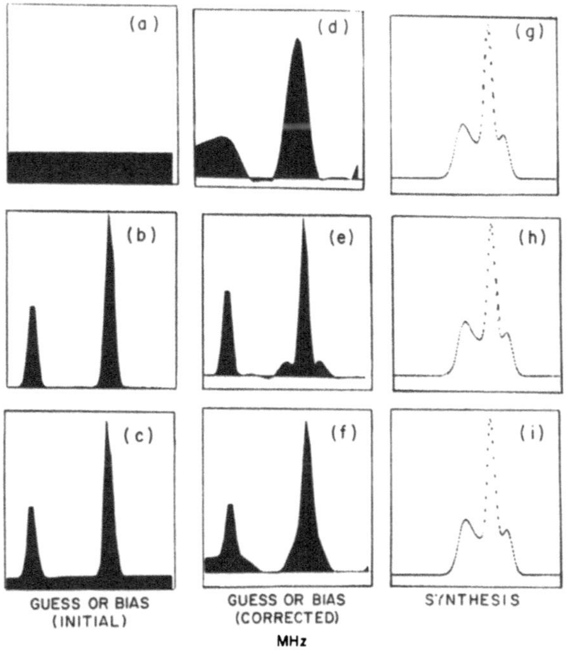

Fig. 2 A family of initial guesses or biases (a,b,c), corrections to these baises (d,e,f) and NMR line synthesis based upon these corrected biases (solid lines - g,h,i). Points represent the synthesized "known" spectra.

Structure of Glass" by E. A. Porai-Koshits, Proc. XI International Congress on Glass, Prague, 1977. We feel they are in essential agreement with three of the main points made in the present paper.

(1) Concerning the cyclic nature of glass research "And as far as according to the laws of Gegel dialectic the development of humanity moves by a spiral, then in 1977 some ideas of more than fifty years standing have suddenly reappeared on the new level of modern mathematical and experimental achievements". This is followed by a comparison of Lebedev's work (1921) with that of Konnert and Karle (1972) and also that of Zachariasen (1932) with that of Bell and Dean (Phil Mag., 25, 1381, 1972).

(2) Concerning the effects of input bias "Investigators who, like Bell and Dean construct and utilize the random network models, come to the conclusion about the validity of Zachariasen's hypothesis. But those investigators who use the quasi-crystalline models find crystallinity or crystal-like regions in glasses".

(3) Concerning experimental resolution "As far as 18 years ago one could see the failure of attempts to extract from the RDC information which, in principle cannot be contained in the experimental intensity curves" (E. A. Porai-Koshits, Vitreous State-Proc. 3rs All-Union Conference, Leningrad, 1960).

REFERENCES

1. . Leydolt, Wein. Akad. Ber., 8, 261, 1852.
2. E. W. Tillotson, Jr., J. Ind. Eng. Chem., 4, 246, 1912.
3. A. A. Lebedev, Trans. State Optical Inst., 2, 57, 1921 (USSR).
4. J. T. Randall, H. P. Rooksby and B. S. Cooper, J. Soc. Glass Tech., 14, 219, 1930.
5. W. H. Zachariasen, J. Am. Chem. Soc., 54, 348, 1932.
6. B. E. Warren, Z. Krist, 86, 349, 1933.
7. N. N. Valenkov and E. A. Porai-Koshits, Nature, 137, 273, 1936. Z. Krist., 95, 195, 1936.
8. G. Hartleif, Z. Anorg. Chem., 238, 353, 1938.
9. A. Dietzel, Z. Electrochem., 48, 9, 1942.
10. A. F. Prebus and J. W. Michener, Ind. Eng. Chem., 46, 147, 1953.
11. L. C. Hoffman and W. O. Statton, Nature, London, 176, 561, 1955.
12. E. A. Porai-Koshits and N. S. Andreev, Dok. Akad. Nauk. USSR, 118, 735, 1958.
13. R. L. Mozzi and B. E. Warren, J. Appl. Cryst., 2, 164, 1969.
14. J. H. Konnert and J. Karle, Nature Phys. Sci., 236, 92, 1972.
15. B. E. Warren, H. Krutter and O. Morningstar, J. Am. Ceram. Soc., 19, 202, 1936.
16. E. J. Gooding and W. E. S. Turner, J. Soc. Glass Tech., 18, 32, 1934.
17. J. Biscoe and B. E. Warren, J. Am. Ceram. Soc., 21, 287, 1938.
18. See NMR refs. these Proc. - D. L. Griscom.
19. W. Skatulla, W. Vogel and H. Wessel, Silikat. Tech., 9, 51, 1968.
20. R. R. Shaw and D. R. Uhlmann, J. Am. Ceram. Soc., 51, 377, 1968.
21. V. V. Golubkov, A. P. Titov and E. A. Porai-Koshits, these Proceedings and Proc. XI Int'l Congress on Glass, Prague, 1977.
22. A. H. Silver and P. J. Bray, J. Chem. Phys., 29, 984, 1958.
23. H. M. Kriz and P. J. Bray, J. Non-Cryst. Solids, 6, 27, 1971.
24. G. E. Peterson, C. R. Kurkjian and A. Carnevale, Phys. Chem. Glasses, 15, 59, 1974.
25. L. C. Snyder, G. E. Peterson and C. R. Kurkjian, J. Chem. Phys., 64, 1569, 1976.
26. This work has been published in a number of articles starting with Phys. Chem. Glasses, 15, 52, 1974 and J. Non-Cryst. Solids, 21, 283, 1976.
27. G. E. Peterson, A. Carnevale and C. R. Kurkjian, "Proc. Symp. on the Structure of Noncryst. Solids", Soc. Glass Tech., Cambridge, 1976.
28. G. E. Peterson, A. Carnevale and C. R. Kurkjian, to be published, Phys. Chem. Glasses.

A GAS PROBE ANALYSIS OF STRUCTURAL TRENDS IN BORON GLASSES

James F. Shackelford

Materials and Devices Research Group and
Department of Mechanical Engineering
University of California
Davis, California

Joseph S. Masaryk

Kaiser Aluminum and Chemical Corporation
Center for Technology
Pleasanton, California

INTRODUCTION

Studies on the structure of boron-containing glasses have been well documented by Griscom (1). The techniques described in those studies are, in general, electromagnetic "probes", e.g. X-ray diffraction, electron spin resonance, etc... An alternate method was suggested in 1962 by Doremus (2), viz. a careful analysis of the solution of relatively small diameter gas atoms or molecules in the glass network structure. More recently, the authors (3) have described a particular version of this method referred to as the "gas probe technique". It is based on a statistical thermodynamic analysis of the gas solution process. A description follows in the next section. In general, the analysis provides values for the binding energy between a dissolved helium atom and the glass network and the vibrational frequency of the atom. This information, in turn, describes the potential energy well for the dissolved atom. The potential well is a direct result of the glass network structure in the immediate vicinity of the dissolved atom, i.e. the solubility site.

The first application of this technique to obtain structural information (3) resulted in good correlation with phase separation observed in alkali silicate glasses. Two systems were studied, viz. sodium silicate and potassium silicate glasses. Phase separation is known to occur in the sodium silicate system. The phase boundaries in this system correspond to clear changes in binding energy and vibrational frequency data trends. These changes were correlated to a significant structural change occurring with increasing Na_2O content. A silica-rich phase was modeled as a random network structure analogous to cristobalite, and a soda-rich phase was modeled as a planar silicate structure analogous to crystalline sodium disilicate. In addition, there was evidence that Na^+ ions produced a net compression of the silica-rich phase network structure, viz., the average diameter of individual solubility sites is reduced by Na_2O additions to SiO_2. Phase separation is not known to occur in the potassium silicate system. There were no sharp breaks in the data trends. The results indicated that the system has a true "stuffed" silicate structure in which K^+ ions fill solubility sites in a cristobalite-like silica network structure. There is a slight compression of the remaining unfilled sites (still accessible to helium atom solution).

THE GAS PROBE METHOD OF STRUCTURAL ANALYSIS

The technique of gas probe structural analysis has been described previously (3). Only a brief review need be given here. The method is based on a statistical thermodynamic analysis of a simple solution system, i.e. helium gas in glass. Using a relatively simple model of an ideal gas for the helium atmosphere above the glass and a simple harmonic oscillator for the dissolved helium atom, one obtains the solubility equation (4)

$$n_s/p = (h^2/2\pi mkT)^{3/2} (kT)^{-1} N_s$$
$$x[e^{-h\nu/2kT}/(1-e^{-h\nu/kT})]^3 e^{-E(0)/RT} \tag{1}$$

where n_s is the number of helium atoms dissolved per cm^3 of glass, p is the pressure of the gas atmosphere, h is Planck's constant, m is the mass of one helium atom, k is Boltzmann's constant, T is the absolute temperature, N_s is the number of solubility sites available per cm^3 of solid, ν is the vibrational frequency of the dissolved helium atom, R is the gas constant and E(0), the binding energy, is the energy of the atom in the lowest quantum state in solution relative to the similar state in the free gas, i.e., removed from the potential field of the solid.

Fitting Eq. (1) to experimental solubility data (n_s/p as a function of T) yields values for E(0) and ν. E(0) is roughly equal to the slope of an Arrhenius plot of ln (n_s/p) versus 1/T. The value of ν is then determined primarily by the magnitude of solubility (n_s/p) at any given temperature. It must be noted that E(0) is not exactly equal to the experimental solution enthalpy (defined as E_s where the experimental data is represented as $(n_s/p) = S = S_o e^{-E_s/RT}$) because there is a non-trivial temperature dependence in the pre-exponential term of Eq. (1), i.e. S_o is not truly constant.

Obtaining values of ν and E(0) as described requires some knowledge of the solubility site density, N_s. A value for N_s has been experimentally obtained for vitreous silica (5). For alkali silicates, the number of sites is assumed to be reduced by the number of alkali-ions added (3), i.e.

$$N_{s,\ \text{silicate}} = (1-2m) \times N_{s,\ \text{vitreous silica}} \tag{2}$$

where m is the mole fraction R_2O in the silicate. A similar assumption will be made in this paper for alkali additions to borate and borosilicate glasses. However, no experimental value for the alkali-free compositions is currently available. Since this paper deals with structural trends in the data, any reasonable value can be used for the alkali-free N_s. In all calculations to follow, the alkali-free N_s used was that for vitreous SiO_2 ($=2.3 \times 10^{21}cm^{-3}$). This value should be a reasonable approximation to the true value for these systems.

E(0) represents the strength of a van der Waals bond between the dissolved helium atom and the glass network. The key structural parameter is ν in that it shows the steepness of the potential well in which the helium atom rests in its solubility site. This is illustrated by the equation for this parabolic well (6)

$$E(r) - E(0) = (2m\pi^2\nu^2)r^2 \tag{3}$$

where r is the distance from the center of the solubility site. Figure 1 illustrates schematically the relationship between decreasing site size, increasing potential well steepness, and increased vibrational frequency.

EFFECT OF THERMAL HISTORY

In vitreous silica, thermal history was found to have no significant effect on helium gas solubility (3). In various boron-containing glasses, some significant effects have been measured (7-9). Individual experimental runs at a given temperature indicate a clear trend of decreasing helium solubility as vitreous B_2O_3 is annealed from the quenched state (e.g., Figure 3 of reference 7). Shelby (7) associated this with the macroscopic observation of increasing glass density upon annealing. Specifically, he concluded that the average interstitial site diameter in the glass network must be decreasing. The use of the gas probe analysis on the experimental data confirms this assumption. Unfortunately, firm conclusions are made difficult because of experimental scatter in the data. The primary reason for the scatter is the short temperature range covered. For a given uncertainty in n_s data, the short temperature range gives a relatively large uncertainty in the slope of the Arrhenius plot of $\ln(n_s/p)$ versus $1/T$. The slope represents approximately $E(0)$. Since ν is directly coupled to $E(0)$ in the curve fitting technique of the gas probe analysis, scatter in both parameters results. Fortunately, data is available for quenched and annealed sodium-borate glasses of various compositions (7,8) giving an indication of the average effect of annealing.

Figure 2 shows the resulting trends in $E(0)$ and ν for various sodium borate glasses. Comparison is made between glasses in the quenched state (cast in room-temperature graphite molds from a 1200°C melt) and the annealed state (slow cooled after 30 minutes at the glass transition temperature). There is too much scatter in the $E(0)$ data to distinguish any changing trend for composition or thermal history. A single average value adequately describes all data. The trend in the ν data indicates a distinction between the quenched and annealed states. Although no significant composition effect is indicated for the short range studied (consistent with the results of the next section), the average ν for all annealed glasses ($3.21 \times 10^{12} s^{-1}$) is clearly higher than that for all quenched glasses ($3.01 \times 10^{12} s^{-1}$). Inspection of Eq. (3) and Fig. 1 indicates that the solubility site radius (or diameter) can be considered to be inversely proportional to the helium atom vibrational frequency. Since vibrational frequency increases upon annealing, the interstitial site diameter is decreasing. The values obtained from Fig. 2 indicate that the average interstitial site diameter shrinks about 6% upon annealing. Due to the experimental scatter, this number must be considered rather approximate. It can be compared to the observed decrease (1.5%) in bulk density for vitreous B_2O_3 upon annealing (10). This indicates that the interstitial site may undergo a much greater structural relaxation than the glass network as a whole. It should be re-emphasized that a wider temperature range for the individual glasses studied would yield more precise structural parameters and provide a more precise picture of interstitial site shrinkage.

EFFECT OF COMPOSITION

As noted in the introduction, composition was found to have a significant effect on helium solubility (and, therefore, glass structure) in sodium and potassium silicate systems. For comparison, data is available on a wide variety of boron containing glasses from Shelby (11) and Altemose (12).

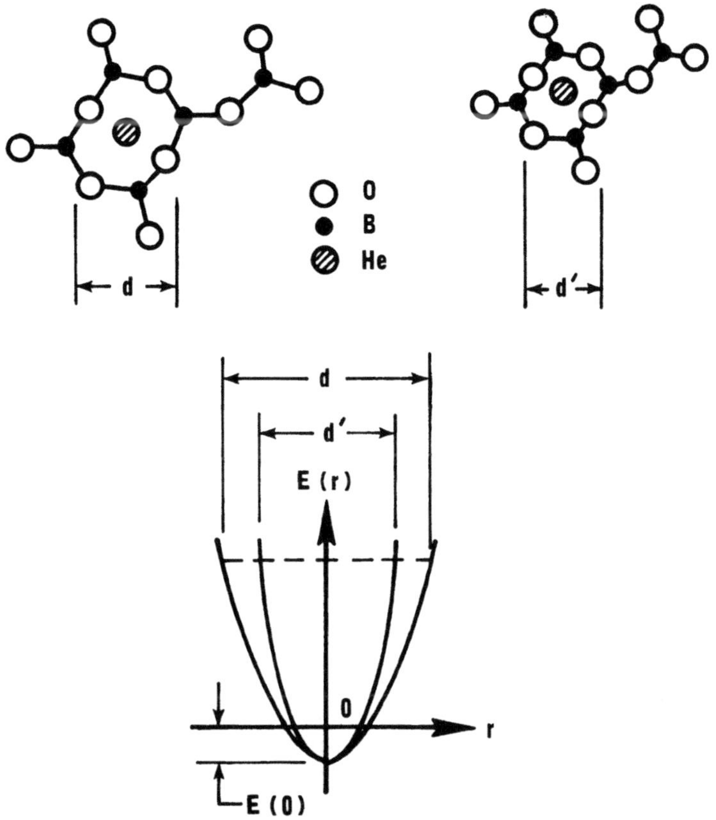

Figure 1. Relationship between decreasing solubility site size and increasing potential well steepness (and, therefore, increasing vibrational frequency, ν, as shown by Eq. (3)).

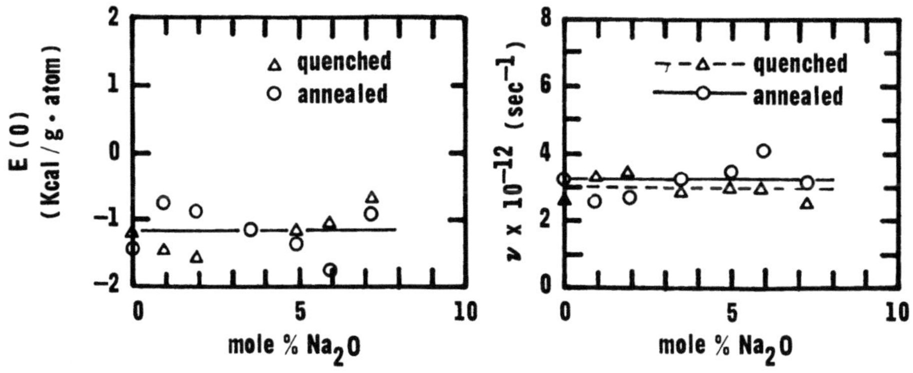

Figure 2. Trends in E(0) and ν for quenched and annealed sodium borate glasses.

Shelby (11) reported helium solubility data for R_2O-B_2O_3 glasses with R = Li, Na, K, Rb, Cs and additions as high as 32.6 mole %. Again because of a limited temperature range studied, E(0) and ν data obtained by curve-fitting are highly scattered. In spite of this, no system shows any sharply changing trend such as that observed for sodium-silicate (3). All data follows essentially one curve similar to the slight upward trend seen in potassium-silicate glasses (3). The results of the gas probe analysis are shown in Figure 3. The scatter in the E(0) data is again quite large and only a horizontal line is shown to indicate a mean value. For the ν data, the scatter is also large. However, there is a substantial number of data points and a linear least squares fit shows a slight upward trend with increasing R_2O concentration indicating a gradual shrinking of the average interstitial site diameter. Specifically, the data indicates a 13% decrease in site diameter for a 30 mole % R_2O addition. This compares with the 6% decrease noted for annealing of similar glasses and a 7% decrease for adding 30 mole % R_2O in the K_2O-SiO_2 system (3).

Altemose (12) reported helium solubility data in a similar study of a series of borosilicate glasses. The basic composition was a 78.1:20.7:1.2 molar ratio of SiO_2:B_2O_3:Al_2O_3. Akali additions were made in 2 and 7 mole % amounts with the exception of Na_2O which was added in various quantities up to 13.4 mole %. Two general features of this data are to be noted. First, the data exhibits less scatter and trends are more apparent. Second, experimental enthalpies of solution are consistently positive. This is somewhat surprising in light of the results of the authors and Shelby, i.e. the enthalpy of solution is usually slightly negative. However, since the primary interest here is in the relative change of vibrational frequency and not absolute values, no judgement is required about the validity of the sign of enthalpy. The results of the gas probe analysis are shown in Figure 4. The E(0) data shows a slight downward trend with alkali addition. More important is the trend in ν data which shows a slight upward trend corresponding to shrinking of the interstitial site diameter with alkali addition. Although this repeats the trend found above for alkali borate glasses, the magnitude of increase in ν corresponds to a site diameter decrease of 29% for an Na_2O addition of only 10 mole %. This is a substantially higher figure than those found for the alkali borates or potassium silicates. The only system indicating a similar magnitude dimensional change was sodium silicate in which a 20 mole % Na_2O addition produced a 25% decrease in average interstitial site diameter. The data at 2 mole % R_2O additions indicate a slight increase in ν with increasing molecular weight (larger radius) alkali ions. While it is reasonable that larger ions might produce a greater compression of the network, the trend is less clear for the 7 mole % additions. Obviously more data is required to clarify the true effects.

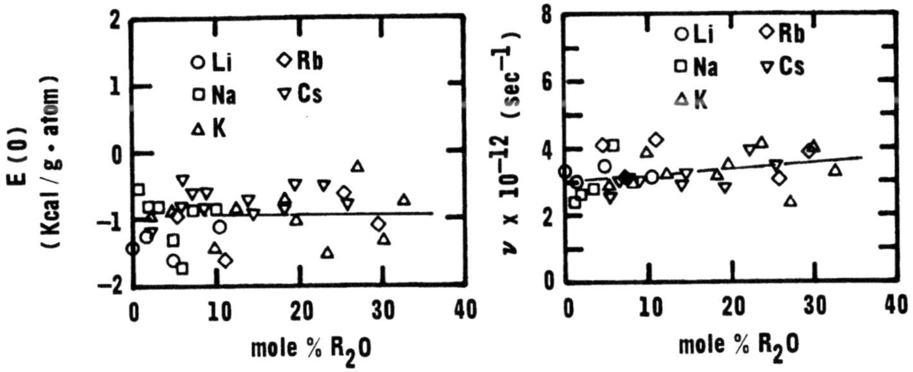

Figure 3. Trend in E(0) and (ν) for alkali borate glasses.

SUMMARY AND CONCLUSIONS

A gas probe analysis of glass structure involves a statistical thermodynamic description of helium solubility giving values for binding energy, E(0), and vibrational frequency, ν, for the dissolved helium atom. Changes in ν occurring as a result of changing thermal history or composition indicate changes in the glass network geometry in the vicinity of the helium atom, i.e. an interstitial solubility site. Analyses of key data in the literature dealing with boron containing glasses yield three main results:

1. Thermal history has a slight effect on network geometry in vitreous B_2O_3 and sodium-borate glasses. The average interstitial site diameter is decreased approximately 6% upon going from a quenched to an annealed state.

2. Composition has a greater effect on network geometry in various alkali-borate glasses. The average solubility site diameter is compressed approximately 13% upon addition of 30 mole % R_2O to vitreous B_2O_3. Experimental scatter in the data does not permit a distinction between the effect of the various R^+ species.

3. Composition appears to have an even greater effect on structure in alkali borosilicate glasses. The average solubility site diameter is compressed approximately 29% upon addition of only 10 mole % R_2O to an SiO_2:B_2O_3:Al_2O_3 glass with molar ratios of 78.1:20.7:1.2. There is a slight indication that the compression of the solubility site may be increased by increasing R^+ radii.

In the first two results, experimental scatter prevents more precise statements. In the third result, an anomaly in the experi-

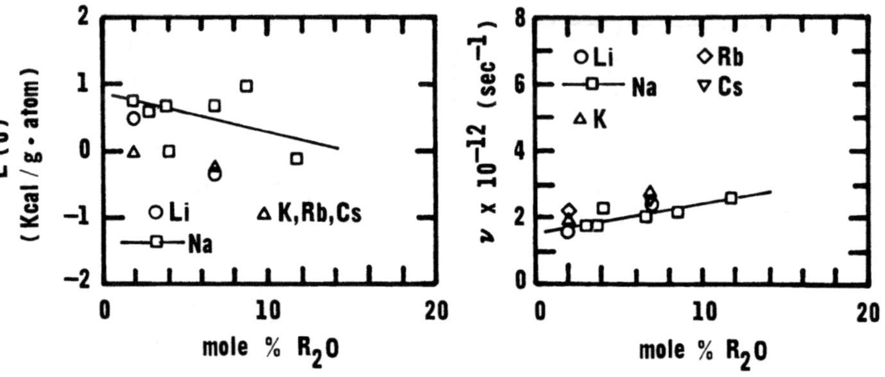

Figure 4. Trend in E(0) and ν for alkali borosilicate glasses.

mental data (positive enthalpies of solution) requires some caution in making comparisons to the first two.

A few concluding remarks are in order. First, more experimental data on more gas-in-glass systems is desirable to allow more attempts at obtaining structural information in this way. Even more useful to the current work would be more precise data (over a wider range of temperature) for the systems already studied. Equally useful would be high pressure gas solubility data of the type available for vitreous silica (5). This gives an experimental value for N_S, the total density of solubility sites, which would allow more quantitative structural information to be obtained. Second, little has been said about the precise geometry of the interstitial solubility site. Purposely, only general, averaged dimensional trends have been implied. A logical future direction of this work would be to relate various structural models for the site to calculated values of E(0) and ν for a dissolved helium atom. These calculations could be compared to the values obtained in this paper by curve-fitting. Specific structural models can be obtained from the structural "building block" information provided by electromagnetic probe techniques as summarized by Griscom (1). Basically, alkali additions yield a continuous addition of BO_4 tetrahedra to the network of BO_3 triangles with these two groups combining to form various polyborate groups as the fundamental building blocks of the glass structure. At this point, the gas probe analysis indicates that the local structure of the solubility site is changing only slightly as a result of thermal history or composition changes. While the functional groups in the local structure may be changing substantially, their net potential field as sensed by the dissolved helium atom is changing only slightly in a smooth fashion. It is interesting to note that the only system which has not exhibited a gradual increase in ν with increasing R_2O content is sodium silicate (3). The sharply changing trend in that data was associated with phase separation and two distinctly different structures. The rather smooth trend in the alkali borate data occurs in systems also known to phase separate (1). This would indicate that structural changes associated with any phase separation that occurred in the glasses used in the solubility experiments were such that the potential field for the dissolved helium atom was changed only slightly. Once again, more precise experimental values might allow distinctions to be made. Finally, it should be noted that the results for the borosilicate glasses indicated a similarly gradual change in the net potential field sensed by the helium atom as R_2O was added. This is noteworthy due to the inevitable complexity of combining silicate and borate structural building blocks even though the compositions chosen were not expected to phase separate.

REFERENCES

1. D. L. Griscom, "Borate Glass Structure", pp. 1-128, this volume.

2. R. H. Doremus, Phys. Chem. Glasses 3 (1962) 127.

3. J. F. Shackelford and J. S. Masaryk, to be published in Ceramic Microstructures 76, 6th International Materials Symposium, Berkeley, CA., 1976.

4. J. F. Shackelford, P. L. Studt, and R. M. Fulrath, J. Appl. Phys. 43 (1972) 1619.

5. J. E. Shelby, J. Appl. Phys. 47 (1976) 135.

6. T. L. Hill, Introduction to Statistical Thermodynamics (Addison-Wesley, Reading, MA., 1963).

7. J. E. Shelby, J. Appl. Phys. 44 (1973) 4588.

8. J. E. Shelby, J. Appl. Phys. 45 (1974) 2536.

9. J. E. Shelby, J. Non-Cryst. Solids 14 (1974) 288.

10. P. B. Macedo, Mechanical and Thermal Properties of Ceramics, National Bureau of Standards special publication No. 303 (U.S. GPO, Washington, D.C., 1969) p. 169.

11. J. E. Shelby, J. Appl. Phys. 44 (1973) 3880.

12. V. O. Altemose, J. Amer. Ceram. Soc. (1973) 1.

HIGH-TEMPERATURE BORATE LIQUIDS: PHYSICAL PROPERTIES OF GLASS-FORMING COMPOSITIONS

E. F. Riebling

Chemical Technology Laboratory, Atlantic Richfield Hanford Company,* Richland, Washington 99352

INTRODUCTION

A pure network liquid such as molten SiO_2 is essentially a random three-dimensional arrangement of corner shared SiO_4 tetrahedra. In contrast, the orthosilicate composition (33 mole % SiO_2) contains a completely depolymerized mixture of SiO_4^{-4} anions. Any composition between these two extremes must therefore contain a relatively complex mixture of partially polymerized anions. Practice has shown that physical-chemical properties of such mixtures are highly dependent upon structure and composition.

Many of the structurally sensitive techniques that were originally developed to examine crystalline solids have also been successfully extended to amorphous oxides. Thus, X-ray, infrared, and NMR have all been used to develop a more sophisticated picture of the polymeric nature of alkali borate, silicate, germanate, and phosphate glasses [1]. However, only with the case of binary phosphate glasses has it been possible to identify both the type (chain, ring, sheet) and relative concentration of each polyanion in a given distribution [2]. The fragment distributions agreed quite well with predictions based on condensation polymer theory. While binary silicate glasses close to the orthosilicate composition have been partially characterized [3, 4], extensive solvent/glass interactions [5, 6] have inhibited efforts to characterize polyanion distributions in higher silica content glasses and crystals [7-10].

*Presently Rockwell Hanford Operations

In contrast, the depolymerizations that accompany modifier additions (between 15 and 50 mole %) to B_2O_3 and GeO_2 create a change of coordination for a fraction of the network forming cations [11-13]. The resultant polyanions can therefore contain mixtures of two different types of polyhedra (BO_3 and BO_4 or GeO_4 and GeO_6). Hence, solvent/glass interactions and polyanion separation/identification procedures can be rather different from those thus far encountered with silicate and phosphate glasses. An initial approach to this problem involves the preparation of binary and ternary borate or germanate glasses that are likely to contain relatively simple polyanion distributions. Such distributions can be obtained by adding large amounts of either a network former or a modifier that does not create a coordination change. Special physical-chemical property techniques can be developed to aid the search for such glasses. Examples of progress in this area include, but are not limited to, glasses in the $Sb_2O_3 \cdot B_2O_3 \cdot GeO_2$ [14], $Bi_2O_3 \cdot GeO_2$ [15], and $Tl_2O \cdot Bi_2O_3 \cdot GeO_2$ [16] systems.

Another approach to the problem of glass characterization is based upon the fact that the available evidence confirms the absence of significant polyanion or polyhedral coordination distribution changes within the glass transformation range [13, 17]. In some cases, the physical-chemical properties of a high-temperature liquid can be either more structure sensitive or more obtainable than the corresponding glass properties. For example, either microphase separation or partial crystallization can occur on cooling and make it difficult to achieve a one phase glass for the exact compositions of interest.

Relatively fluid oxide liquids such as the alkali borates can be hydroscopic, corrosive, and volatile. Also, the properties of viscous network liquids such as pure B_2O_3 can be quite sensitive to impurity concentration levels. Hence, compositional control is essential. However, precise physical-chemical studies of the above types of liquids have been completed and the results used to gain a greater insight into the nature of the glass state. Most of what follows will of necessity be brief and restricted to borate liquids. Information concerning silicates and germanates will be used only for comparative purposes.

NETWORK BORATES

Most ionic liquids expand and become less viscous at higher temperatures. However, temperature can alter the structure of covalently bonded liquids in a number of ways. Thus, although the viscosity of water-free B_2O_3 decreases with increase of temperature, it does so in such a fashion as to suggest the existence of a more complex flow mechanism above 1300° C [18]. Also, the density and expansivity changes suggest a gradual tightening of the arrangement of BO_3 polyhedra at temperatures within the 1100° to 1400° C range. Several types of B-O regroupings are possible. However, it is still not at all certain if the two phenomena are related to the same type of B-O regrouping.

Small SiO_2 or GeO_2 additions (0 to 20 mole %) to molten B_2O_3 increase both the density and viscosity, but decrease the expansivity [19]. The actual deviations from ideal mixing behavior suggest the presence of a nearly normal O-B-O solvent network that contains widely separated silicon or germanium atoms. Similar additions of Al_2O_3 increase the density and expansivity, but decrease the viscosity [18]. The aluminum atoms thus appear to occur as widely separated network modifiers.

Larger additions of either SiO_2 or GeO_2 (up to 60 mole %) to molten B_2O_3 create significant property deviations from ideal mixing [19]. Gross disruptions of the O-B-O solvent are apparently accompanied by microclustering of SiO_4 or GeO_4 tetrahedra at the 40 mole % SiO_2 and 60 mole % GeO_2 compositions. Thus, the higher field strength Si^{+4} appears to be more effective in altering the O-B-O structure. However, electron microscopic examination of the resultant glasses did not reveal actual phase separation.

ALKALI BORATES

For a simple crystalline oxide, one usually encounters an either/or situation with respect to the stability of a given coordination polyhedron. Thus, either tetrahedra or octahedra are the stable species for a given cation at a fixed temperature and pressure [20]. Examples include the pressure dependent cristobalite to stishovite transformation for SiO_2 and the temperature dependent rutile to quartz-like transformation for GeO_2.

The presence of a second type of cation represents a change of electronic environment (composition, oxidation state, etc.) for the original cation. In such a situation, different coordination polyhedra of a given cation can readily coexist as part of a thermodynamic equilibrium mixture. Thus, BO_3 trigonal planar polyhedra and BO_4 tetrahedra, as well as GeO_4 tetrahedra and GeO_6 octahedra, coexist in crystalline alkali borates and germanates respectively. The ratio of BO_4/BO_3 and GeO_6/GeO_4 is quite specific because of (a) the different structural role played by each type of polyhedron, (b) the fixed composition of a given compound, (c) the fixed pressure and temperature, and (d) the fixed geometry of the unit cell.

A gross structure change upon melting, such as from molecular to ionic bonding, is usually characterized by abnormal heats, entropies, and temperatures of fusion. In the absence of any such evidence for a given compound, there is no reason why polyhedral mixtures characteristic of the crystal may not exist in the liquid. The absence of the constraining electrostatic force field of the crystalline lattice may introduce a slight temperature dependence of the BO_4/BO_3 and GeO_6/GeO_4 ratios. However, an equilibrium mixture of polyhedra should still prevail in the liquid. Further, there is no thermodynamic reason, other than phase separation, incongruent melting behavior, and the like, why such mixtures of coordination polyhedra should not be retained during rapid cooling to obtain a glass. This should be particularly true of relatively viscous liquids that are rapidly quenched. There is no convincing evidence to suggest that the glass transformation (Tg) region is one of gross atomic level structure change. If anything, the density and expansivity evidence [17], together with the relatively small heats of glass transformation, suggest the existence of only minor structure changes within the Tg region.

Thus, it was not entirely unexpected that the addition of alkali oxide to molten GeO_2 was found to induce a partial change of coordination for germanium from tetrahedral to octahedral [11] and that similar conclusions could be drawn for the corresponding alkali germanate glasses [17]. A most convincing piece of evidence in these studies was the smaller than normal partial molar volume exhibited by GeO_2 in these mixtures. Higher coordination polyhedra can represent a more efficient mode of cation/oxygen packing. Also, octahedra can be packed more efficiently, via edge sharing, than tetrahedra. What was surprising was that these studies had not been attempted earlier.

Subsequent work with molten alkali borates revealed a pronounced melt densification that accompanied the known viscosity decrease [12]. The partial molar volume decrease exhibited by B_2O_3 in such liquids also suggested a more efficient packing of O-B-O polyhedra. The deviation from ideal mixing at the 40 mole % Na_2O composition (Figure 1) is of the magnitude expected for an equimolar mixture of BO_4 tetrahedra and BO_3 polyhedra. This conclusion agreed with that reported for an NMR study of alkali borate glasses [21]. Both studies also reported a complete return to some form of trigonal coordination for boron at higher alkali contents.

In spite of our present knowledge of the glass transformation range and also of coordination changes that can occur for network forming cations, occasional efforts are put forth which ignore some of what is known for oxide liquids, glasses, and crystals. Such attempts therefore imply, if only indirectly, that significant polyhedral distribution changes occur within the glass transformation range. Several of these attempts will be briefly analyzed below.

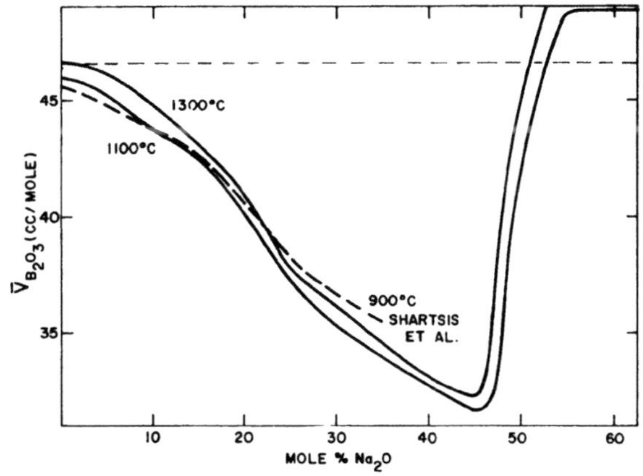

Fig. 1. Partial molar volume of B_2O_3 isotherms for binary sodium borate melts.

Non-bridging oxygen atoms are going to exist if an ionic modifier is added to a covalent network former, regardless of the exact mode of depolymerization (disturbed network, island, discrete anion model with silicates versus coordination change, discrete anion model with borates and germanates). However, some of the physical property changes that accompany the addition of alkali oxide can be highly dependent upon the specific network former. This suggests that given physical property changes can be rather sensitive indicators of a particular mode of network depolymerization. Thus, molten silica can apparently accomodate 10 mole % M_2O without undergoing significant disruption. The supporting evidence includes the nearly ideal partial molar volume of SiO_2 and the absence of an expansivity increase. Larger amounts of modifier successively create a disturbed network, silica-rich "islands," and discrete anions as non-bridging oxygen atoms are gradually introduced [22]. In contrast, the partial molar volumes of molten and amorphous GeO_2 and B_2O_3 decrease significantly, with simultaneous expansivity increases, as modifier is added. Therefore, it should be readily apparent that non-bridging oxygen atoms are not completely responsible for the property changes observed with borate and germanate melts.

However, the "specter" of non-bridging oxygen atoms continues to be raised for borates and germanates, usually in such a way as to subtly cast doubt on the coordination change model [23]. Equally obfuscating are statements to the effect that structural models cannot be proved by physical property information [23]. Unfortunately, such semantics tend to cast doubt on all that has gone before, theoretical as well as experimental. Modern physical science is based on a vast body of information and ideas, not all of which are readily accepted by every scientist. While "blinders" are to some extent necessary, they must be used with caution to avoid myopia. Thus, in trying to develop a proper appreciation of borate structures, one should consider all of the information that is available for liquid, glass, and crystalline borates, germanates, and silicates.

More recently, a viscosity study of sodium borate melts concluded that the results could be explained without resorting to a change of coordination for boron [24]. The significant viscosity decrease that accompanies initial modifier additions was instead rationalized in terms of the appearance of progressively smaller, BO_3-containing, sheet-like discrete polyanions. Unfortunately, such a model implies a gross structure change in the glass transformation range (BO_4 tetrahedra do exist in alkali borate glasses) and therefore casts doubt on the validity of several of the established "milestones" of borate glass science [1].

Thus, if the above model is correct, one or more of the following situations would arise. Borate tetrahedra exist in the glass, but not in the liquid (this would require that a significant polyanion population shift occurs in the glass transformation range). Borate tetrahedra do not exist in either the glass or liquid (this would imply that most of the NMR, density, infrared, etc., measurements and related conclusions are in error). There is no significant evidence to support the first possibility for a binary borate or any other oxide system. There is very little likelihood that the second possibility represents atomic-level reality because many of the same physical property changes (viscosity, density, etc.) are known for alkali germanate liquids and glasses [11, 17, 25, 26]. Mixtures of discrete, sheet-like germanate anions have not been reported or postulated for such liquids, glasses, and crystals. Hence, the sheet-like model for molten borates represents another example of a failure to consider all of the available liquid, glass, and crystalline information, not just for borates, but also for germanates, silicates, etc.

TERNARY BORATES THAT CONTAIN A SECOND NETWORK FORMER

Silica-rich compositions that contain less than 40 mole % M_2O in the $Na_2O{\cdot}B_2O_3{\cdot}SiO_2$ system have been extensively studied because of the various practical applications of borosilicate glasses. Also, silica-rich borosilicate-type glasses are of current interest for possible use as the long-term storage media for selected radioactive wastes that remain from nuclear weapons production [27]. The atomic level structure of any glass directly influences the mechanical, chemical, and processing properties such as strength, durability, and annealing characteristics. Hence, an appreciation of the atomic level structure of such glasses will greatly facilitate the final selections of compositions and processing conditions. A brief examination of some salient features of these high-temperature borosilicate liquids is therefore warranted. Equally important phenomena such as micro- and/or macro-phase separation are not within the scope of this paper.

A density study of the relatively fluid B_2O_3-rich ternary liquids in the $Na_2O \cdot B_2O_3 \cdot SiO_2$ system concluded that BO_4 tetrahedra exist in most compositions that contain between 15 and 45 mole % Na_2O [12]. For example, the concentration of BO_4 tetrahedra was estimated to be highest (about 50 percent of the boron atoms) for ternary melts close to the 40 mole % Na_2O composition in the sodium borate system. In contrast, only about 25 percent of the boron atoms appeared to be tetrahedrally coordinated in the ternary melts with B/Si ratios close to 2.00.

A recent viscosity and NMR examination of several silica-rich glasses in the above system confirmed the presence of significant concentrations of BO_4 tetrahedra [28]. An appreciation of the equilibrium control of the BO_4/BO_3 ratio was also presented. For example, in a series of silica-rich glasses that contained 15 mole % Na_2O, the fraction of tetrahedrally coordinated boron atoms was proportional to the silica content. This confirmed the earlier prediction, based on a viscosity and density study of $B_2O_3 \cdot SiO_2$ liquids and glasses [19], that the boron atoms could be forced to adopt a tetrahedral configuration in silica-rich compositions.

The nature of similar liquids and glasses that contain less than 40 mole % Na_2O in the $Na_2O \cdot B_2O_3 \cdot GeO_2$ system was also examined [13]. The objectives were to examine the compositional stability of BO_4 tetrahedra and the influence of the $BO_4 \rightleftarrows BO_3$ equilibrium on the stability of GeO_6 octahedra. Only the former phenomenon will be briefly discussed at this time.

Initial substitutions of GeO_2 for B_2O_3 in sodium borate compositions produced viscosity increases that were attributed to the potential network forming role of GeO_4 tetrahedra under such conditions. Both the molar volume and expansivity results also suggested a significant concentration of BO_4 tetrahedra at most ternary compositions. Therefore, there is a high degree of similarity for the structure of B_2O_3-rich liquids and glasses in the ternary silicate and germanate systems.

The expansivities of the above ternary borosilicate and borogermanate liquids were also found to be highly sensitive indicators of structural change. In this technique, compositions that contain the same expansion coefficient are joined to form iso-expansivity contours on a ternary composition diagram (Figure 2). Only the liquid phase iso-expansivity contour patterns were readily

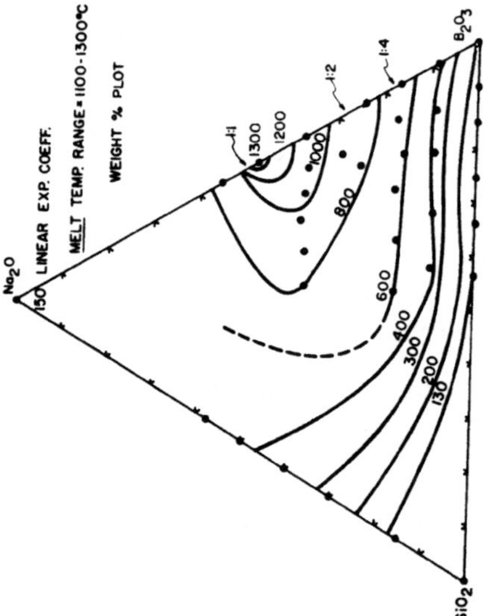

Isoexpansion coefficient contours for $Na_2O \cdot SiO_2 \cdot B_2O_3$ melts.

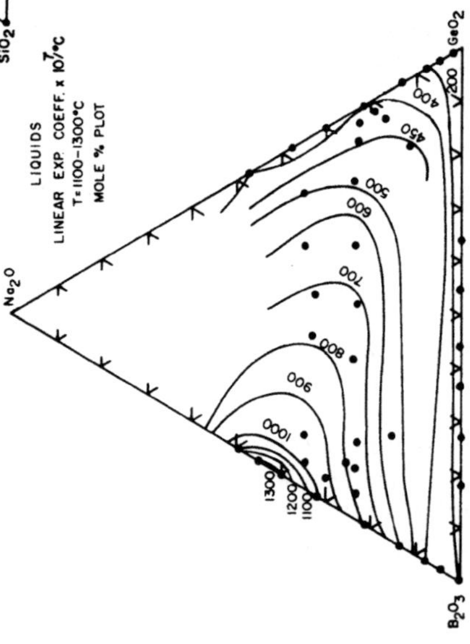

Fig. 2. Iso-expansion coefficient contours for melts in the $Na_2O \cdot GeO_2 \cdot B_2O_3$ system.

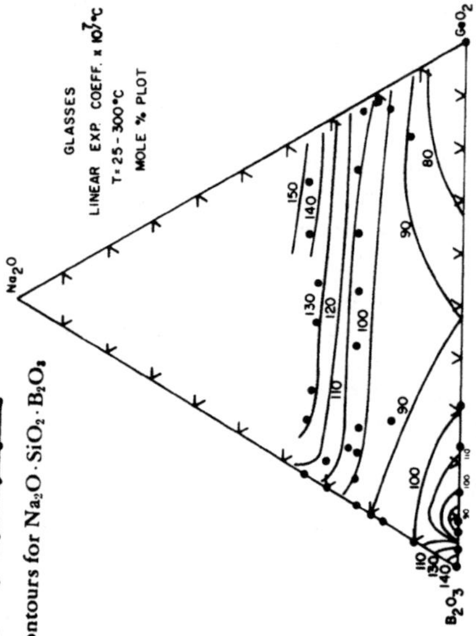

Iso-expansion coefficient contours for glasses in the $Na_2O \cdot GeO_2 \cdot B_2O_3$ system.

associated with the presence of BO_4 tetrahedra. The absence of a similar correlation for the corresponding glasses was not unexpected. There are known structure-sensitive factors that influence the liquid expansivity to a much greater extent [17, 29].

Crystalline Sb_2O_3 has a double-chain polymeric structure that appears to be retained in the liquid and glass states. Visually estimated viscosities of GeO_2-rich liquids in the $Sb_2O_3 \cdot B_2O_3 \cdot GeO_2$ system suggested that moderate concentrations of Sb_2O_3 depolymerize the GeO_2 network more effectively than does B_2O_3, but less effectively than does an alkali oxide [14]. However, extensive depolymerization of either a Ge-O-Ge or a B-O-Ge network appeared to require at least 10 to 15 mole % Sb_2O_3. This was confirmed by the density and infrared changes that were observed for the corresponding glasses [14]. For example, Figure 3 shows the compositional dependence of the transmission minima for the main Ge-O and B-O vibrations. Thus, at least 30 mole % B_2O_3 can be added to GeO_2 without causing a significant frequency shift for ν_{Ge-O}. Such a situation can reflect the absence of gross depolymerization of GeO_2 (condensed or polymerized polyhedra can possess higher stretching frequencies than their isolated counterparts). The linear increase of ν_{B-O} confirms the retention of a network for all $B_2O_3 \cdot GeO_2$ compositions.

In contrast, the addition of Sb_2O_3 to GeO_2 causes a substantial shift of ν_{Ge-O} to lower frequencies. While 30 mole % Sb_2O_3 causes a frequency shift that is comparable to that caused by 30 mole % alkali oxide additions, the absence of relatively large volume deviations ruled out a coordination change for the Ge atoms. A distribution of smaller Ge-O-Sb species seemed more likely. For B_2O_3, the substitution of just 50 percent of the B atoms by Sb appeared to create a ν_{B-O} shift that is twice that observed for the substitution of 99 percent of the B atoms by Ge atoms. That suggested that Sb_2O_3 causes extensive depolymerization of the B-O-B network.

A technique had been previously developed to more clearly illustrate the relation between infrared frequency and the extent of cross-linking in ternary oxide glasses [30]. Isofrequency contours for a given vibration (B-O, Ge-O, etc.) can be developed by connecting ternary compositions that exhibit the same frequency for that vibration. Such contours can, but not necessarily, connect glasses of similar structure because more than one depolymerization mechanism can cause a frequency shift. Figure 4 depicts these ν_{Ge-O} contours for GeO_2-rich glasses in the $Sb_2O_3 \cdot B_2O_3 \cdot GeO_2$ system. The bent ν_{Ge-O} patterns of Figure 4 may partially reflect the more limited network-forming ability of Sb in a B-O-Ge glass compared to that in a Ge-O-Ge glass. On the other hand, GeO_2-rich B-O-Ge networks may simply be more susceptible to scission by

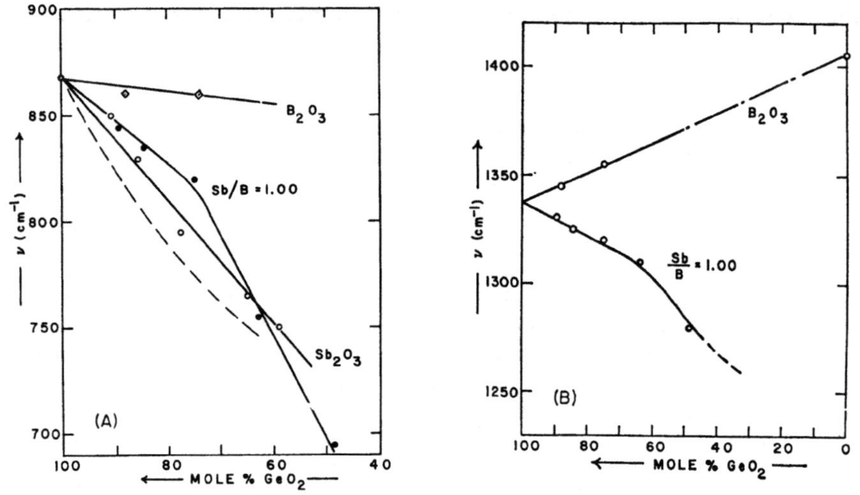

Fig. 3. *(A)* **Frequency of Ge–O vibration vs mol% GeO_2 for Sb_2O_3-GeO_2, B_2O_3-GeO_2, and Sb/B = 1.00 germanate glasses. Alkali germanate data** *(dashed line)*. *(B)* **Frequency of B–O vibration vs mol% GeO_2 for B_2O_3-GeO_2 and Sb/B = 1.00 germanate glasses.**

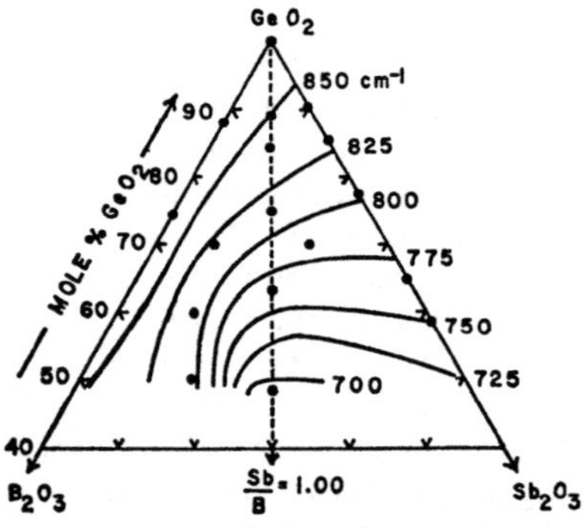

Fig. 4. Constant-wave-number contours for Ge–O vibration in GeO_2-rich Sb_2O_3-B_2O_3-GeO_2 glasses.

Sb atoms. In any event, it was possible to readily associate these observations with the differing structural roles of Sb_2O_3 (O-Sb-O chains) and B_2O_3 (O-B-O network).

CONCLUSIONS

It was concluded that several experimental approaches will have to be used if we are to develop a better and more useful understanding of the polymeric constitution of oxide glasses. The state of knowledge for borates was selectively reviewed and used as an example to emphasize the need to consider all of the information that is available for other similar materials such as germanates and silicates. The narrow approach of examining just the properties of the amorphous solids themselves was shown to be insufficient. Instead, a broad approach that also involves a consideration of the corresponding crystals and high-temperature liquids was shown to be superior. It was also concluded that if BO_4 tetrahedra exist in borate crystals, then they may also exist in similar borate high-temperature liquids and their corresponding glasses.

REFERENCES

1. D. L. Griscom, "Borate Glass Structure," pp 1-128, this volume.

2. A.E.R. Westman, in "Modern Aspects of the Vitreous State," Vol. 1, J. D. Mackenzie, ed., Butterworths, London (1960) chapter 4.

3. C. R. Masson, I. B. Smith, and S. G. Whiteway, Canadian J. Chem. 48 (1970) 1456.

4. P. C. Hess, Geochimica et Cosmochimica Acta 35 (1971) 289.

5. D. A. Crerar and G. M. Anderson, Chemical Geology 8 (1971) 107.

6. R.W. Luce, R. W. Bartlett, and G. A. Parks, Geochimica et Cosmochimica Acta 36 (1972) 35.

7. J. Gotz and C. R. Masson, J. Chem. Soc. A1970 (1970) 2683.

8. C. R. Masson, Polymer Preprints 13 (1972) 792.

9. B. R. Currell, H. G. Midgley, and M. A. Seaborne, J. Chem. Soc., Dalton 1972 (1972) 490.

10. B. R. Currell, H. G. Midgley, and M. A. Seaborne, Nature, Physical Science 236 (1972) 108.

11. E. F. Riebling, J. Chem. Phys. 39 (1963) 3022.

12. E. F. Riebling, J. Amer. Ceram. Soc. 50 (1967) 46.

13. E. F. Riebling, P. E. Blaszyk, and D. W. Smith, J. Amer. Ceram. Soc. 50 (1967) 641.

14. E. F. Riebling, J. Amer. Ceram. Soc. 56 (1973) 303.

15. E. F. Riebling, J. Mater. Sci. 9 (1974) 753.

16. E. F. Riebling, J. Mater. Sci. 10 (1975) 1565.

17. E. F. Riebling, J. Amer. Ceram. Soc. 51 (1968) 143.

18. E. F. Riebling, J. Amer. Ceram. Soc. 49 (1966) 19.

19. E. F. Riebling, J. Amer. Ceram. Soc. 47 (1964) 478.

20. E. F. Riebling, in "Phase Diagrams, Materials Science and Technology," Vol. 3, A. M. Alper, ed., Academic Press, New York (1970), chapter 7.

21. P. J. Bray and J. G. O'Keefe, Phys. Chem. Glasses 4 (1963) 37.

22. J. O'M. Brockris, J. W. Tomlinson, and J. L. White, Trans. Faraday Soc. 52 (1956) 299.

23. J. E. Shelby, J. Amer. Ceram. Soc. 57 (1974) 436.

24. G. H. Kaiura and J. M. Toguri, Phys. Chem. Glasses 17 (1976) 62.

25. E. F. Riebling, J. Chem. Phys. 30 (1963) 1889.

26. E. F. Riebling and S. D. Gabelnick, J. Electrochem. Soc. 112 (1965) 822.

27. M. J. Kupfer and W. W. Schulz, Bull. Amer. Ceram. Sco. 56 (1977) 338.

28. M. P. Brungs and E. R. McCartney, Phys. Chem. Glasses 16 (1975) 48.

29. E. F. Riebling, in "Proceedings of the 12th Symposium on the Art of Glassblowing," American Scientific Glassblowers Society (1968), p. 42.

30. E. F. Riebling, J. Mater. Sci. 7 (1972) 40.

KINETICS OF VOLATILIZATION OF SODIUM BORATE MELTS

M. Cable

Department of Ceramics, Glasses and Polymers
Sheffield University, Northumberland Road, S10 2TZ
England

INTRODUCTION

The high temperature chemistry of simple glasses is notoriously complex: apart from the experimental difficulties commonly found at temperatures above 1000°C, chemical equilibria often are strongly concentration-dependent even at low concentrations. In such circumstances it is very attractive to consider evaluating the concentration and temperature dependence of chemical activities by studying the vapor in equilibrium with a melt (provided that the experimental difficulties can be overcome). The simplest experiment that may give useful information related to this is to measure weight loss and analyse the residue, thus determining the average composition of the volatile material. This kind of experiment is also of direct interest to the glassmaker for the insight it can give into problems of inhomogeneities produced by volatilization, corrosion of superstructure and regenerators by vapors and atmospheric pollution problems. The next refinement is to examine and analyze the vapor, then to design an experiment which ensures that the melt and vapor are in equilibrium.

It is obvious that kinetic effects may be important when the vapor does not have the same composition as the melt and diffusion of the volatile species in the melt may be expected to play an important role. A considerable number of studies of silicate glass melts[1,2,3,4,5] have shown that assuming instantaneous achievement of the final equilibrium between melt and vapor at the interface and control entirely by diffusion in the melt does not correctly describe the kinetics of volatilization of Na_2O, Li_2O, K_2O or PbO. Apart from failing to describe the kinetics correctly, this model cannot allow for two important effects which are well established: the sensitivity of the rate of loss to pressure[3] and velocity of flow of the atmosphere.[6] However it has been possible to produce a slightly modified model which can describe the data accurately and which can include the effects of pressure and flow of the atmosphere.

As boric oxide is one of the most volatile materials normally used in glassmaking it is worth seeing whether the same model can describe its loss from borosilicate and borate melts. These systems have the additional interesting feature that the volatile material is not a single virtually pure oxide as in all the cases referred to above. Fortunately the theory only requires that the composition of the volatile material is constant, not that it must be a particular compound.

BORATE MELTS:PUBLISHED WORK

Although not extensively investigated, several authors[7,8,9,10] have estimated the composition of the vapor lost from various alkali borate melts by analyzing the initial and residual melts. Some differences might be expected from the difficulties of determining boron accurately by traditional chemical methods and also from different arbitrary choices of time or other important parameters affecting kinetics. Figure 1 shows a representative selection of these results and it can be seen that there are significant differences between different authors for the same composition in some cases. All authors agree that vapor and melt are identical in composition for

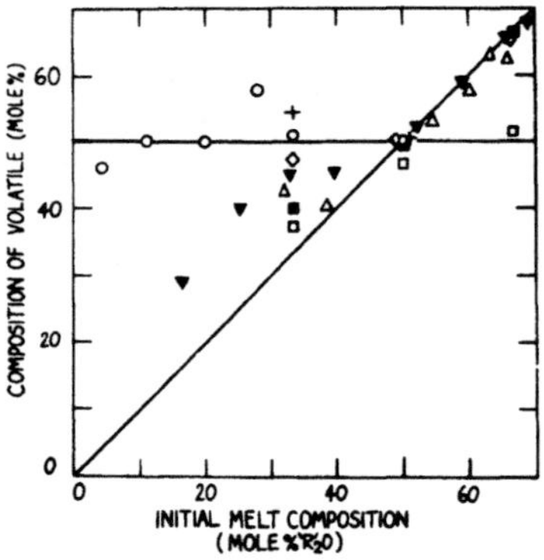

Fig. 1. Average composition of material volatilized from alkali borate melts according to various authors. □ Li_2O (1400°C)[8]; ◇ K_2O (1300°C)[8]; ■ Na_2O (1300°C)[8]; + Na_2O (1300°C)[7]; Δ K_2O (1200°C)[9]; ▼ Na_2O (1200°C)[9]; ○ Rb_2O (900-1300°C)[10].

all metaborates ('R'BO_2) so that diffusion in the melt is not necessarily involved and one would expect loss in weight to be proportional to time. The reasonably recent data of Adams and Quan[10] suggest that all Rb_2O-B_2O_3 melts with $4 \leqslant Rb_2O \leqslant 50$ (mole%) lose $RbBO_2$ but the other fairly recent data on sodium and potassium borates indicate more complex behaviour:between about 15 and 45 mole% alkali oxide the vapor appears to be distinctly richer in alkali than the melt, implying diffusion of 'Na_2O' or 'K_2O' through the melt whilst above 50 mole% alkali melt and vapor seem to be of almost identical composition with the melt making diffusion unnecessary.

These factors might be expected to affect the kinetics and an appropriate kinetic model might help to quantify and interpret the results. Unfortunately no detailed studies of kinetics of volatilization have been published. Kolykov[11] made such studies of B_2O_3, twenty-five sodium borates and forty-three sodium borosilicates at 1200°C but his observations continued only for eight hours nor does his paper give the data in sufficient detail for them to be re-examined. By a fortunate coincidence an unpublished thesis by Lawton[12] contains a detailed study of the Na_2O-B_2O_3 system for 0 to 34.3 mole percent Na_2O for times up to 100 hours at 1000°C, 50 hours at 1100°C and 20 hours at 1200°C. The main purpose of this paper is to examine these data.

THEORETICAL MODEL

If a melt contained a concentration C_o (g cm^{-3}) of some volatile species and the interface reached its final equilibrium concentration of $C_\infty = 0$ very rapidly, one would expect the loss to be controlled by diffusion in the melt and the results to be described by[3]

$$M = \frac{2}{\sqrt{\pi}} (C_o - C_\infty) \sqrt{Dt}, \tag{1}$$

where $\underline{M}$ is the weight loss (g cm^{-2}), $\underline{D}$ is the effective binary diffusivity (cm^2 s^{-1}) and $\underline{t}$ is time (s). This relation would hold even if diffusivity were concentration-dependent in which case the diffusivity would be

$$D = \frac{1}{C_o - C_i} \int_{C_i}^{C_o} D(c)\,dc, \tag{2}$$

where $\underline{C_o}$-$\underline{C_i}$ is the range of concentrations involved.

Data for volatilization from glass melts do not usually fit this model but can be described by introducing a new boundary condition at the gas-melt interface. The flux (j) into the gas is now defined by

$$j = \alpha(C_i - C_\infty); \tag{3}$$

where α is a mass transfer coefficient (cm s^{-1}) and C_i the concentration in the melt at the interface x=0. It is still assumed that transport by diffusion occurs inside the melt. This model leads to the loss being given by[13]

$$M = \frac{C_o - C_\infty}{h}\left[\exp(h^2Dt)\,\mathrm{erfc}(h\sqrt{Dt}) - 1 + \frac{2}{\sqrt{\pi}}h\sqrt{Dt}\right] \tag{4}$$

where $h=\alpha/D$ and the interface concentration is obtained from

$$\frac{C_i - C_\infty}{C_o - C_\infty} = \exp(h^2Dt)\,\mathrm{erfc}(h\sqrt{Dt}). \tag{5}$$

These equations assume that the melt is deep enough to behave as a semi-infinite body. A melt of finite depth can also be treated.[14] It has been shown that this finite model successfully describes some data on the loss of PbO from lead crystal[4]. If the vapor happens to have the same composition as the melt so that diffusive mass transfer is unnecessary, equation (3) at once leads to

$$M = \alpha(C_o - C_\infty)t \tag{6}$$

but this situation can still be described by equation (4) if $D \gg \alpha$.

Unfortunately the Boltzmann transformation does not apply here, as it does to equation (1), and equations (4) and (5) are not valid if α or D is concentration dependent (even if h is still constant). A recent paper by Cable and Cardew explored some of the complications that may arise on this account in a system showing behaviour similar to sodium silicate melts. It was found that equation (4) could be fitted very well to data for a system with strongly concentration-dependent properties but that the meanings of the parameters α^* and D^* obtained in this way could have at least two different interpretations. These were:-

Case I

Data restricted to relatively short times ($h^*\sqrt{D^*t} < 1$), $\alpha^* \simeq \alpha$ for C_o, D^* merely a curve fitting parameter and outside the range $D(C_o)$ to $D(C_i)$.

Case II

Data extend to longer times ($h^*\sqrt{D^*t} > 1$),

$$\alpha^* = \frac{h}{C_o - C_i}\int_{C_i}^{C_o} D(c)\,dc \tag{7}$$

and

$$D^* = \frac{1}{t} \int_0^t D_i dt. \tag{8}$$

Values of α^* and D^* obtained by fitting equation (4) to experimental $\underline{M}$, $\underline{t}$ values, thus need to be approached with caution.

DATA FOR SODIUM BORATES

Lawton[12] made experiments with approximately 3.8 g samples of glass in small platinum capsules of about 4.9 cm^2 surface area, which were heated in a horizontal tube furnace. The main sillimanite tube was about 38 cm long and 8.5 cm diameter but the samples were placed inside another sillimanite tube about 27 cm long and 5.0 cm internal diameter; the volume of the chamber containing the samples was about 0.54 litre. The stopper at the rear of the furnace was sealed into place to reduce, as far as possible, accidental varying currents in the furnace atmosphere. No deliberate gas flow was introduced.

Twelve different glasses, including B_2O_3 were studied being weighed every two hours at 1200°C, every 5 or 10 hours at 1100°C and every 2 or 20 hours at 1000°C according to magnitude of the loss. No set of data had fewer than five pairs of $\underline{M}$, $\underline{t}$ data and many had ten observations.

To use the theory just described it is necessary to express the initial concentration of volatile material in the appropriate units, here g cm^{-3}, and this implies knowledge of the composition of the volatile material and volume-composition-temperature relations for the system. As Lawton analyzed all his initial melts and the residues after the longest times it is easy to deduce the average composition of the volatile material. In the following analysis it is assumed, as a first order approximation, that the composition of the vapor is independent of time and given by this average value. The density data needed to convert these values to the appropriate units were obtained from the work of Riebling.[16] Table 1 summarizes the most important data for the experiments done at 1200°C. Values of $\underline{M}$ have been corrected for loss of platinum from the capsules.

Figure 2 shows Lawton's results for the average composition of the material lost at all temperatures. Even though the largest loss was only about 13% of sample weight, this represents a significant change in composition of the melt during the experiment, in some cases, and plotting average melt composition is better than using initial melt composition. When the data are plotted in this way all points lie in a narrow band giving only slight evidence that the vapor composition varies with temperature. So plotting the data also shows reasonable agreement with the results of Tamura et alia[9] for sodium borates. From figures 1 and 2 it is now possible that

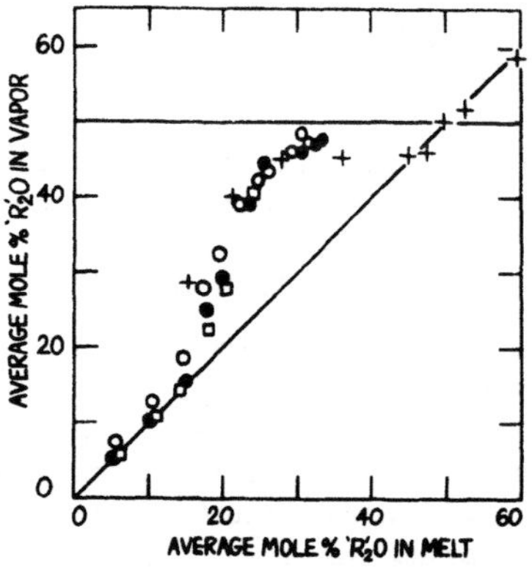

Fig. 2. Lawton's data for average conposition of vapor lost from sodium borates at 1000°C (□), 1100°C (●) and 1200°C (○). Some of the data of Tamura et al. for 1200°C are shown by +.

Table 1. Summary of Experiments Done at 1200°C with Values of α^* and C^* Obtained.

Compositions (mole% Na_2O)								
Melt (initial)	(average)	Vapor (average)	C_o ($g\ cm^{-3}$)	M($mg\ cm^{-2}$) 2h	20h	$\alpha^* \times 10^6$ ($cm\ s^{-1}$)	$D^* \times 10^6$ ($cm^2\ s^{-1}$)	$h\sqrt{D^*t}$ (max.)
0	0	0	1.478	12.1	-	1.45	- †	4.3×10^{-4}
5.61	5.47	7.42	1.204	12.2	126	1.69	- *	0.304
10.6	10.46	12.6	1.428	13.6	130	1.51	41.4	0.063
14.8	14.50	18.6	1.414	14.8	142	1.53	5.52	0.173
18.2	17.3	28.0	1.182	16.5	160	1.92	24.9	0.103
20.8	19.4	32.5	1.172	21.0	184	3.03	4.46	0.099
24.7	22.5	39.0	1.169	32.3	236	4.64	2.94	0.674
27.9	24.8	42.4	1.217	36.2	294	5.07	5.30	0.566
29.3	26.1	43.4	1.250	41.3	307	5.80	4.25	0.470
33.1	29.3	46.2	1.331	59.3	363	8.12	2.46	1.39
35.0	30.5	48.6	1.333	66	396	11.9	2.44	1.89
36.9	31.6	47.1	1.453	98	506	16.9	2.38	2.95

† $D = 8.07 \times 10^{-1}$ * $D = 1.47 \times 10^{0}$

the behaviour falls into three regions:

1) 0 to ca. 15% Na_2O: vapor and melt have almost the same composition but the vapor may be slightly richer in Na_2O at the highest temperatures.

2) 15 to 50% Na_2O: the volatile material is distinctly richer in soda than the melt. The largest difference in composition occurs close to 30% Na_2O but the vapor does not go beyond $NaBO_2$.

3) Above 50% Na_2O: the volatile material and the melt have the same composition as far as the data extend.

These interesting results must be related in some way to the structure and chemical properties of the melts.

Figure 3 shows the weight loss data for a representative selection of the experiments at 1200°C. It can be seen that the relation appears to be exactly linear for melts with up to about 15% Na_2O, in accordance with equation (6), but becomes distinctly curved at higher alkali contents where diffusion in the melt becomes necessary.

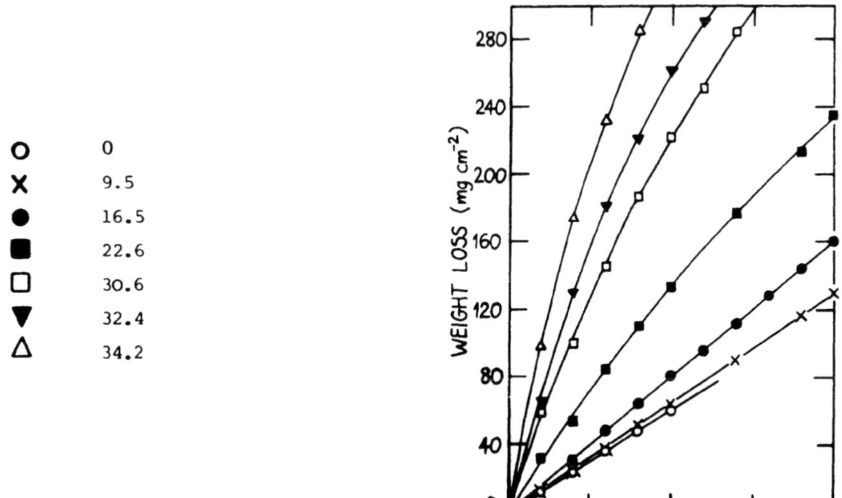

Fig. 3. Typical examples of Lawton's loss data obtained at 1200 C. The initial compositions were (mole % Na_2O).

ANALYSIS OF THE DATA

All the data have been analyzed by a computer program which rearranges equation (4) and uses the M, t data to calculate h for every set of observations when an arbitrary choice of $h\sqrt{D}$ is made. Next $dh/d\sqrt{t}$ is calculated and stored. A new value of $h\sqrt{D}$ is then selected and the cycle repeated until $dh/d\sqrt{t}$ is very close to zero. The method has been described in more detail elsewhere[4]. The known values of h and $h\sqrt{D}$ are then used to recover α^* and D^*. Examination of calculated values of M and the original data shows an excellent fit in all cases and no clear evidence of consistent trends in the discrepancies; differences greater than 2% are rare and they generally lie within limits attributable to experimental errors. The equation thus is an excellent description of the data. The values of α^* and D^* obtained for 1200°C are included in Table 1 which also gives the maximum values of $h^*\sqrt{D^*t}$.

It is clear that values of α^* follow an internally consistent trend and that the program recognizes the cases in which diffusion in the melt is formally unnecessary by giving values of D^* about 10^5 greater than in the other cases (implying $D^* \to \infty$). The results at 1000 and 1100°C show very similar trends in values of α^* which again, appear to be reliable and all of these results are plotted in figure 4. Although experiments continued to longer times at lower temperatures the maximum values of $h^*\sqrt{D^*t}$ are nearly all less than 0.10 at 1000°C and only exceed 0.20 for compositions with more

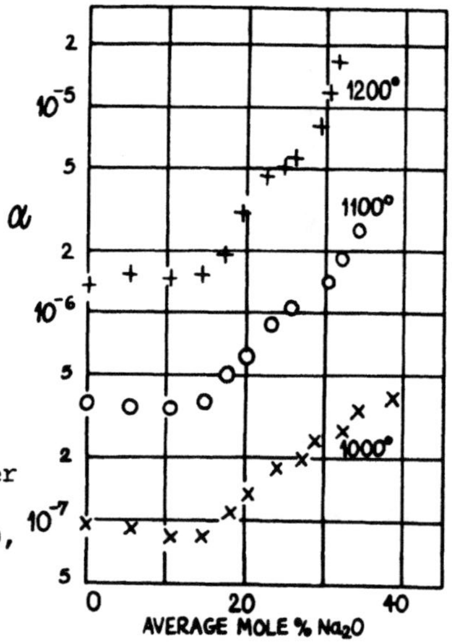

Fig 4. Logarithmic plots of mass transfer coefficient α vs. the average compositon of the melt during the experiment at 1000, 1100 and 1200°C.

than 25% Na_2O at 1200°C. The values of diffusivity (where diffusion in the melt is necessary) lie in the range 9×10^{-9} to 5×10^{-8} $cm^2 s^{-1}$ at 1000°C and 9×10^{-8} to 3×10^{-7} at 1100°C but are not considered reliable for two reasons: the form of $\alpha^*(C)$ indicates that properties are concentration dependent and the times do not, anyway, cover a sufficient range to give reliable values of diffusivity. These strictures fortunately do not apply to the values of α^*.

The theoretical model for concentration-dependent behaviour in such a system examined by Cable and Cardew[15] suggests that the values of D* may easily be distinctly too low to represent any real diffusivity and D* is only a curve fitting parameter. As other reliable diffusivities for mass transfer processes in alkali borate melts are remarkably scanty it is not proposed to discuss the diffusivity values further. In these circumstances it is probable that the form of $\alpha^*(C_{av})$ is very close to the true form of $\alpha(C_o)$.

THE SIGNIFICANCE OF α

The meaning of the mass transfer coefficient (α) is clearly defined in the mathematical sense but a plausible physicochemical model is needed before any attempt is made to relate it to other data on melt structure, chemical activities, and so on. As the size and shape of the containing vessel and of the furnace chamber itself, also gas flow rate and pressure, can affect rate of loss it is natural to consider whether transport through the gas phase might not determine the flux from the sample in some conditions - most obviously those where melt and volatilized matter have the same composition and diffusion in the melt is unnecessary.

The basic postulate of the theoretical model is that

$$j_{iM} = j_{iG} = \alpha(C_{iM} - C_{\infty M}) \tag{9}$$

where subscripts M and G refer to melt and gas phase respectively. If one uses the concept of a boundary layer to redefine j_i in both melt and gas, the first two terms in equation (9) may be written

$$\frac{D_M(C_o - C_i)_M}{\delta_M} = \frac{D_G(C_i - C_\infty)_G}{\delta_G} \tag{10}$$

and it is expected that δ_M is determined by diffusion alone but convection may have a very important effect on δ_G. When first considering such problems with glass melts it is usual to assume that transport in the gas phase cannot limit the process because $D_G \gg D_M$ but this is an insufficient condition unless C_M and C_G are of similar magnitudes. From Table 1 it can be seen that reasonable estimates are $(C_o - C_i)_M \simeq 0.5$ g cm^{-3}, $D \simeq 5 \times 10^{-6}$ $cm^2 s^{-1}$. On the gas side $D_G \simeq 1.8\ cm^2\ s^{-1}$ (calculated by standard methods for B_2O_3 in air[17]) but 1 atm. of air represents only $2.4 \times 10^{-4}\ g\ cm^{-3}$; if the vapor pressure of the volatile species is much less than 1 atm. $\underline{C}_{iG}$ may be very low. An effective vapor pressure of 10^{-5} atm. and a molecular weight of 65 would give $\underline{C}_{iG} \simeq 4.5 \times 10^{-9}\ g\ cm^{-3}$. In such a situation one may easily find cases for which $D_M(C_o - C_i)_M >> D_G(C_i - C_\infty)_G$ despite the very large differences in $\underline{D}$. Transport in the gas would then be of negligible influence only if $\underline{\delta}_G << \underline{\delta}_M$. A satisfactory test of this model requires information that is not available but a preliminary test may be made for B_2O_3.

The vapor pressure of pure B_2O_3 is known fairly accurately and given by[18]

$$\log P(\text{atm}) = 8.921 - \frac{20381}{T(^\circ K)} \tag{11}$$

From this the equilibrium interface concentration of B_2O_3 vapor at 1200°C should be about $5.5 \times 10^{-9}\ g\ cm^{-3}$. Equations (9) and (10) now lead to

$$\alpha C_{oM} = \frac{D_G C_{iG}}{\delta_G} \tag{12}$$

assuming that $C_\infty \to 0$. All these quantities but $\underline{\delta}_G$ are available and give

$$\delta_G = \frac{D_G C_{iG}}{\alpha C_{iM}} = 4.6 \times 10^{-3}\ \text{cm}. \tag{13}$$

Even if the atmosphere over the melt is not deliberately set in motion, some convection is likely to occur. If the boundary layer thickness given by equation (13) could be produced by a velocity of flow likely to occur in this way, the model could be accepted. For forced convection over a flat slab in laminar conditions the standard analysis gives, for a slab 1 cm long,

$$\begin{aligned} \delta_G &\simeq 3\ Sc^{1/3}\ Re^{1/2} \\ &\simeq 3\ D^{1/3}\ \nu^{1/6}\ U^{-1/2} \end{aligned} \tag{14}$$

where $\underline{\nu}$ is the kinematic viscosity of the gas which here is about 2.3. The necessary gas flow velocity then proves to be about $\underline{U} = 100\ m\ s^{-1}$, which is absurd.

However, an important flaw can be seen in the argument. The melts were made from H_3BO_3 nor was the air dried so that both melt and atmosphere contained enough water to make it likely that the true evaporation process should be described not by

$$B_2O_3(\ell) \rightarrow B_2O_3(g) \tag{15}$$

but instead as

$$B_2O_3(\ell) + H_2O \rightarrow 2HBO_2(g). \tag{16}$$

For the water vapor content of typical United Kingdom air the vapor pressure of HBO_2 at 1200°C is likely to be about 5 x 10^{-3} atm which would give $C_{oG} \simeq 1.4 \times 10^{-6}$ g cm^{-3}. Repeating the calculations now gives $\delta_G \simeq 1.2$ cm and $U \simeq 17$ cm s^{-1}. This value of the velocity is about five times what seems immediately acceptable but, given the assumptions made, within the bounds of possibility. Experiments using well defined flow conditions and identifying the species in the vapor are needed for a rigorous test.

It is suggested that the mass transfer coefficient may reasonably be considered to represent

$$\alpha = \frac{D_G}{\delta_G}\frac{C_{iG}}{C_{iM}} = \frac{D_G}{\delta_G}\frac{1}{K_H}, \tag{17}$$

where K_H would represent the Henry's law constant for a liquid behaving in a regular manner. It is thus tentatively proposed that the form of $\alpha(C)$ indicates the composition dependence of K_H at constant temperature; different species in the vapor will, in principle, alter D_G and δ_G but the changes will be very small. Because the composition of the material evaporated varies considerably with composition $\alpha(C)$ must not be thought to indicate directly the change of activity of a specific component with melt composition.

The temperature coefficient of α likewise does not directly give the energy of evaporation of the volatile material but the properties of the vapor are very weakly dependent on temperature and this assumption is likely to be acceptable. All sets of data shown in figure 4 give good straight lines when plotted as log α against 1/T. Up to 15% Na_2O the values lie between 49.3 and 56.3 k cal $mole^{-1}$ but steadily increase with soda content and reach 73 kcal $mole^{-1}$ at 35% Na_2O. The best values for the heats of sublimation are[19] 57.9 ±1 kcal $mole^{-1}$ for HBO_2 about 87 kcal $mole^{-1}$ for B_2O_3 and 77.0 for $NaBO_2$. These values tend to support the argument that HBO_2 was the major species lost from 'pure' B_2O_3 and are consistent with the vapor lost from the melts richest in Na_2O being equivalent to $NaBO_2$, but agreement of activation energies alone is insufficient evidence to prove such assumptions.

CONCLUSION

The theoretical model previously used to interpret volatilization data for other systems which lose only one component has been found to apply equally well to similar experiments with $Na_2O-B_2O_3$ melts even though both components are volatile. This model can be interpreted as involving mass transport by diffusion in the melt, equilibrium between melt and vapor at the interface and transport by both diffusion and convection in the gas. It is suggested that the form of $\underline{\alpha}(\underline{C})$ indicates the composition dependence of the Henry's law constant for the system but that values of diffusivity obtained in the present work may only be curve fitting parameters.

The model may be used to estimate volatilization losses from glass melts and to understand how various factors such as flow of the atmosphere can affect the process. Few, if any, aspects of glass melting can at present be understood at a deeper level than that of macroscopic parameters such as viscosity, density, diffusivity or surface tension. Nevertheless data such as those showing the change in composition of the volatile matter with melt composition (figure 2) or the composition and temperature dependence of $\underline{\alpha}$ (figure 4) invite attempts to produce atomic models to interpret the information. The first inflexion on figures 2 and 4 does not correspond to any obvious feature of the phase diagram[20] but the composition of the vapor appears constant and almost equal to $NaBO_2$ over the whole of the range in which $NaBO_2$ is the primary crystalline phase.

Although there is an abundance of tools for investigating glass structures near room temperature and extensive data exist on borate glasses, few of these lend themselves to direct use at high temperatures. Some of those that do are best used on vapors. One tends to assume that low temperature observations may be extrapolated to much higher temperatures. This volume contains a number of papers on borate glass structures; it is a particularly rigorous test to ask whether the information they contain can successfully interpret the observations reported here.

REFERENCES

1. Terai, R. and Ueno, T., J.Ceram.Assoc.Japan, 1966, 74,283-94.
2. Matoušek, J. and Hlaváč, J., Glass Technol., 1971, 12, 103-6.
3. Cable, M. and Chaudhry, M.A., Glass Technol., 1975, 16,125-34.
4. Cable, M., Apak, C. and Chaudhry, M.A., Glastechn. Ber., 1975, 48, 1-11.
5. Cable, M., Apak, C. and Chaudhry, M.A., Glastechn. Ber., 1975, 48, 127-34.
6. Barlow, D.F., Proceedings VII International Congress on Glass, Brussels 1965. Vol. I, Paper No. 19.
7. Cole, S.S. and Taylor, N.W., J.Amer.Ceram.Soc., 1935, 18,82-5.

8. Solomin, N.V., Proceedings, Conference on the Structure of Glass, Leningrad, 1953. Consultants Bureau Translation 1958, pp 181-3.

9. Tamura, Y., Oishi, Y. and Hamano, Y., Proc. Symposium on Glass Melting, Brussels 1958. Union Scientifique Continentale du Verre, Charleroi. pp 543-56.

10. Adams, C.E. and Quan, J.T., J.Phys.Chem., 1966, 70, 331-340.

11. Kolykov, G.A., Proceedings, Conference on the Structure of Glass, Leningrad, 1953. Consultants Bureau Translation 1958, pp 184-92.

12. Lawton, A. Volatilization of Glasses at High Temperatures. Sheffield University, 1942.

13. Carslaw, H.S. and Jaeger, J.C. Conduction of Heat in Solids. Clarendon Press, Oxford. 2nd edition, 1959, pp 70-73.

14. Carslaw, H.S. and Jaeger, J.C. Op. cit. pp. 121-5.

15. Cable, M. and Cardew, G.E., Chem. Engng Sci., 1977, 32,535-41.

16. Riebling, E.F., J.Amer.Ceram.Soc., 1967, 50, 46-52.

17. Bird, R.B., Stewart, W.E. and Lightfoot, E.N., Transport Phenomena, John Wiley, New York, 1960.

18. Hildenbrand, D.L., Hall, W.F. and Potter, N.D., J.Chem.Phys., 1963, 39, 296-301.

19. JANAF Thermochemical Tables. Dow Chemical Company Midland, Michigan, 1965.

20. Morey, G.W. and Merwin, H.E., J.Am.Chem.Soc., 1936, 58,2252-71.

VISCOUS FLOW IN BINARY BORATE MELTS

Charles J. Leedecke	Clifton G. Bergeron
Ceramics Dev. Div. I	Dept. of Ceramic Eng. and
Organization 5845	Materials Research Laboratory
Sandia Laboratories	University of Illinois
Albuquerque, NM 87115	Urbana, IL 61801

INTRODUCTION AND OVERVIEW

The rheological properties and structure of vitreous B_2O_3 and borate glasses have been the subject of several investigations because of their unusual behavior. For example, the isothermal viscosity of alkali borate glasses has been found to decrease to a minimum and subsequently increase to a maximum as a modifier oxide is added to B_2O_3 (1,2). The effect has been termed the "boron anomaly" and attempts have been made to explain this from a group theory approach or by phase separation.

Nuclear magnetic resonance (NMR) studies have shown that boron coordination changes from three to four fold as the alkali oxide content in binary borate compositions is increased up to 33 mole percent (3,4). Phase separation is not considered responsible for this effect (3,5). From the results of infrared studies (6) on crystals and glasses of the same composition, coupled with the NMR results, Krogh-Moe devised a series of structures which were prominent in certain composition regions. A recent Raman study (7) further corroborated the findings of Krogh-Moe. The maximum in the viscosity isotherms could then be attributed to an interplay between structures containing BO_4 units and BO_3 units.

In these same systems liquid-liquid phase separation at temperatures near 600°C has been observed. Furthermore, these immiscibility gaps extend beyond several stoichiometric compounds having stable crystalline states (8). Alkaline earth borates have a considerable immiscibility gap predominantly at low alkaline earth concentrations preventing the formation of a homogeneous vitreous system (9,10).

Numerous models have been proposed to describe the viscosity-temperature relationship in glass forming systems. The absolute reaction rate theory (11), the free volume theory (12-14), as well as others have been derived. However, the studies (15-17) have shown that no theory can adequately describe the viscosity over large temperature regions. Nevertheless, within certain temperature-viscosity regions in specific instances the models can be applied with some degree of success. For example, Kumar (18-19) has suggested that the free volume theory presents an adequate description of the high temperature flow behavior in binary borate melts.

This review correlates viscosity results obtained in selected binary borate compositions with structural changes which occur with changes in temperature and composition. In addition, parameters associated with the free volume theory and reaction rate theory will be presented for the compositions.

EXPERIMENTAL PROCEDURES

Sample Preparation

The glasses used for the molar volume and viscosity measurements were prepared from reagent grade raw materials. Boric acid (H_3BO_3) and sodium carbonate (Na_2CO_3), barium carbonate ($BaCO_3$), or potassium carbonate (K_2CO_3) were used as the starting materials. The appropriate quantities of these materials were fused in a platinum crucible at temperatures ranging from 900°C to 1000°C. To ensure homogeneity and to remove excess water, the melts were stirred with a platinum rod and dry air was bubbled through the melts for one hour. The melts were quenched on water-cooled stainless steel. Resultant compositions of the glasses are presented in Table I.

Viscosity Measurements

The viscosity of the selected sodium, barium and potassium borate compositions was determined with the use of a variable torque viscometer (Brookfield Engineering Laboratories, Inc.) in conjunction with a platinum-rhodium spindle designed by Tiede (20). The melts were contained in platinum-rhodium crucibles approximately 8 cm high and 4 cm in diameter. The temperature of the melt was measured with a Pt-Pt10%Rh thermocouple, which had previously been calibrated with an NBS standard thermocouple, contained within the spindle. The viscosity apparatus was calibrated with an NBS 711 standard glass (lead-silica glass) at high viscosities and

TABLE I

Melt Compositions[+]

Glass	Metal Oxide	Mole % B_2O_3
NB15R	Na_2O	85.30
NB20R	Na_2O	79.94
NB24R	Na_2O	75.99
BB18R	BaO	81.86
BB20R	BaO	80.08
BB23R	BaO	76.61
BB27R	BaO	72.70
KB20R	K_2O	79.70
KB33R	K_2O	66.71

[+] Wet chemical analysis - Coors Spectro Chemical Laboratory, Golden, CO

viscosity oils at lower viscosities prior to the measurements on the borate melts.

Viscosity values were obtained at temperatures above and below the liquids. The highest viscosity values obtained in the sodium and barium borate systems were determined by the limits of the viscometers since no crystallization problems were encountered. In the potassium borate system, however, the onset of crystallization, indicated by a rapid increase in viscosity with time, dictated the limits.

Molar Volume Measurements

The molar volume of each melt was determined by the counter-balance sphere technique developed by Riebling (21.). The apparatus consisted of a platinum sphere suspended from the core of an LVDT by a 0.5 mm platinum wire. The LVDT used in the measurement had a linear range which exceeded the distance that the platinum sphere would travel in the melt. The balance used to apply various offsets was capable of 0.1 mg incremental changes in weight. The melts were contained in the same crucibles used in the viscosity determinations. The temperature was monitored with a Pt-Pt10%Rh thermocouple in contact with the crucible. The thermocouple was calibrated against another located in the central region of the melt.

For low viscosity values, $\eta < 10$ poise, the equilibrium weight of the sphere in the liquid could be measured directly. At larger viscosities, the indirect technique was required due to the viscous drag on the sphere and the inertia of the balance.

At each temperature at least five positive and five negative offsets were used. Any irregularity in the displacement versus time plots was indicative of such problems as crystallization or bubbles on the platinum sphere.

RESULTS AND DISCUSSION

The temperature dependence of the viscosity in the binary borate compositions investigated is presented in Figure 1. The viscosity data cannot be described by either an Arrhenius equation (11) or a Vogel-Fulcher Tamman (VFT) equation (22), defined by Equation 1, over the entire temperature region.

$$\log \eta = A + \frac{B}{T-T_O} \quad , \qquad (1)$$

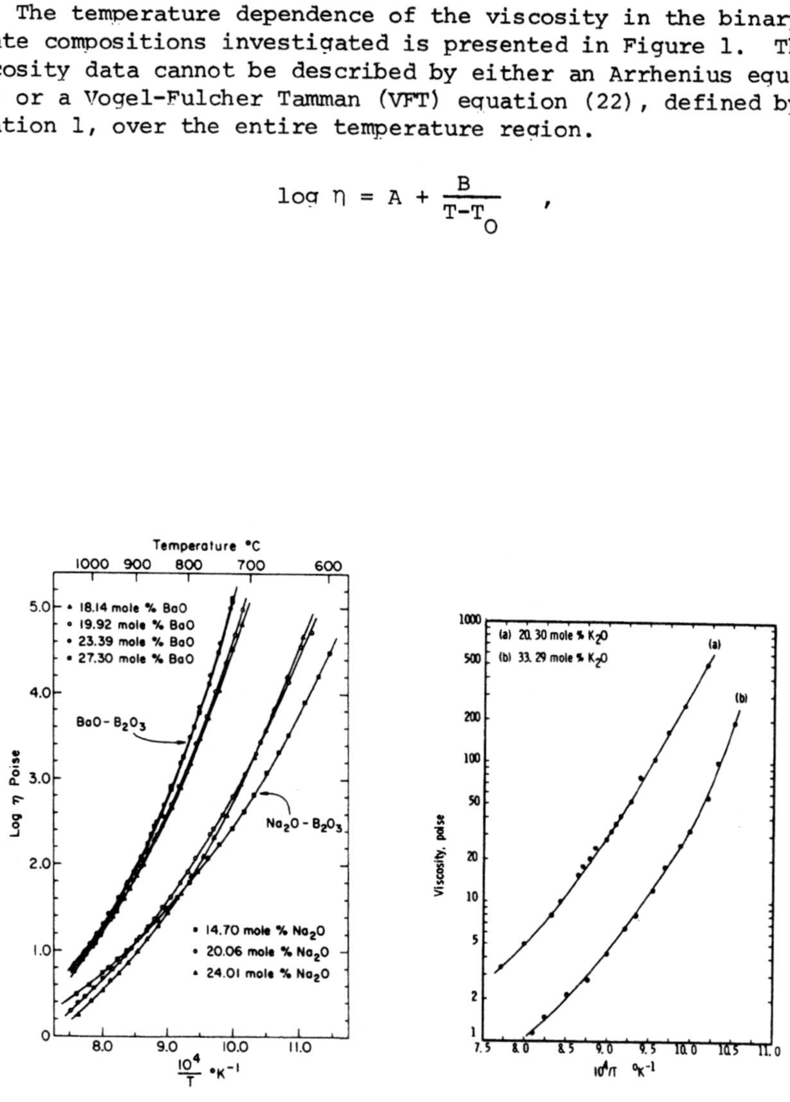

A B

Figure 1. The temperature dependence of the viscosity in binary borate systems. (A) $Na_2O-B_2O_3$ and $BaO-B_2O_3$ (B) $K_2O-B_2O_3$.

where A, B, T_0 are constants and T is the temperature in ^oC. The Arrhenius equation is not applicable over a very wide temperature range, suggesting a temperature dependent activation energy for viscous flow. Figure 2, showing compositions containing approximately 20 mole percent modifier oxide, indicates that the VFT equation is applicable over a much wider temperature region but, as previously found in other glass forming melts (15,16), it overestimates the viscosity of the binary borate melts in the region of high viscosities.

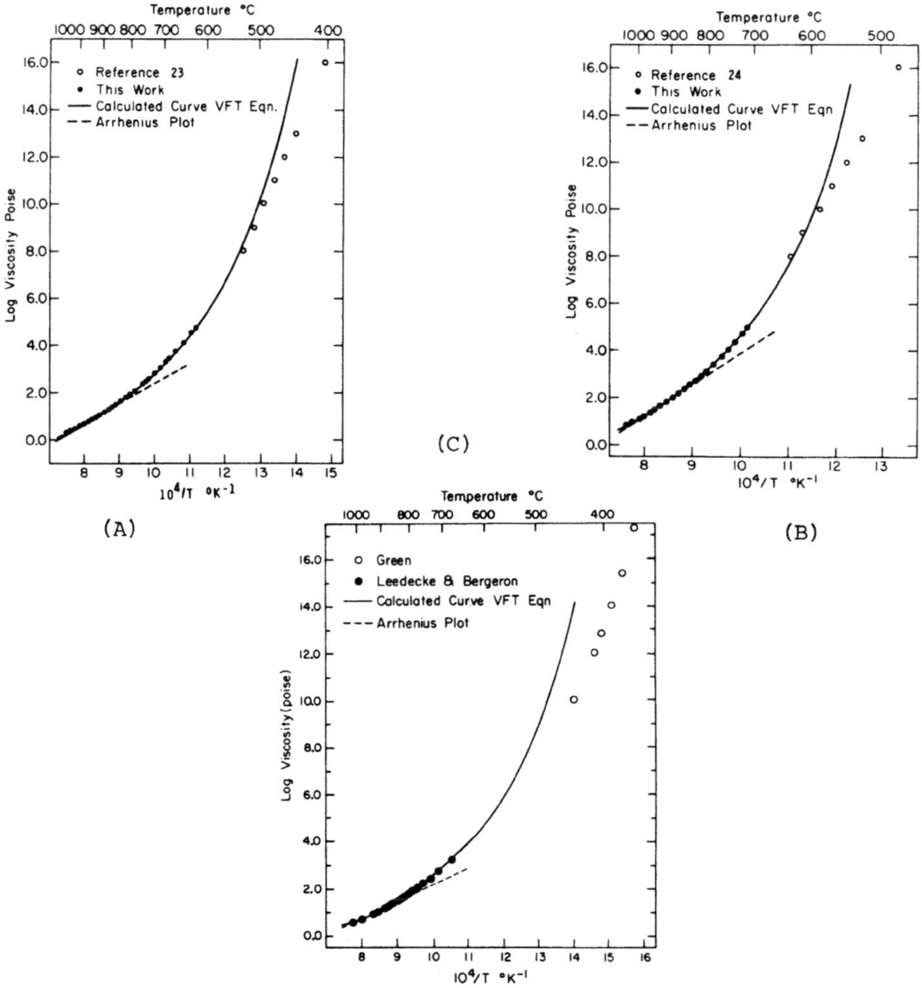

Figure 2. Viscosity-temperature relation for binary borates (A) NB20R, (B) BB20R, (C) KB20R. Solid curve is the VFT equation with constants listed in Table IV. Low temperature data are from Nemilov (23), Romanov and Nemilov (24), and Green (25), respectively.

The viscosity-composition curves (Figure 3) have an interesting feature in the region of low modifier oxide content in binary borate melts. At temperature well above the liquidus, increasing modifier content decreases the viscosity, a situation encountered in many glass forming systems. However, at intermediate and low temperatures the viscosity isotherms reach a maximum somewhere between 20 and 30 mole percent modifier, depending upon the temperature. This occurrence has been explained in terms of two opposing phenomena (1): a) the breakdown of the coordinated structure with the formation of singly bonded oxygens which weakens the structure and causes a decrease in the viscosity and b) the formation of four fold coordinated borons which strengthens the structure and causes an increase in the viscosity. The first phenomenon dominates in the 0 to 5-10 mole percent modifier region. On further additions the second phenomenon, the formation of groups with four fold coordinated borons, dominates up to some concentration where it becomes inactive and the network breakdown effect is again dominant. This compositional effect is shown for some alkali borates over a wide composition region in Figure 3. The type of structures present in the various composition regions which are responsible for the behavior are listed in Table II and the schematic representation of these groups has been presented earlier in the volume (26).

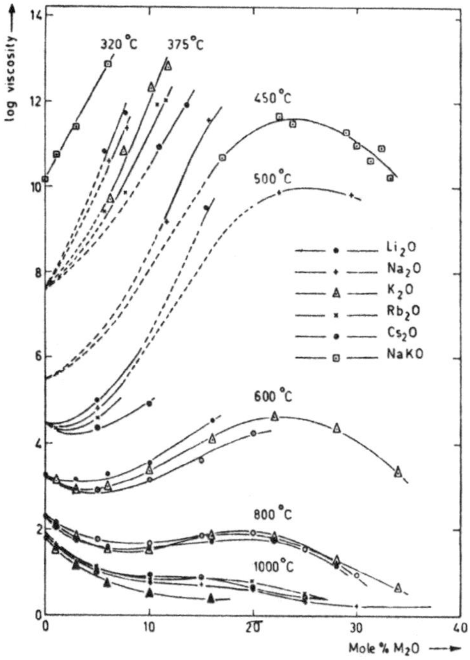

Figure 3. Isothermal viscosity curves for alkali borate glasses. (After Ref. 2)

TABLE II

Borate Groups Present in Various Composition Regions (from Ref. 7)

0-20 mole % R_2O	20-35 mole % R_2O	35-50 mole % R_2O
Boroxol groups	Tetraborate groups	Diborate groups
Tetraborate groups	Diborate groups	Metaborate groups
Loose BO_3 triangles	Loose BO_3 triangles	Pyroborate groups
Loose BO_4 tetrahedra	Loose BO_4 tetrahedra	Orthoborate groups
		Loose BO_3--three bridging oxygens
		Loose BO_3--one non-bridging oxygen

The variation of viscosity, η, with temperature has been expressed in terms of the reaction rate theory according to Eq. 2 (11).

$$\eta = Nh/V_m \exp(\Delta G_\eta^{\neq}/RT) \qquad \text{or} \qquad (2)$$

$$\eta = Nh/V_m \exp(\Delta H_\eta^{\neq}/RT) \exp(-\Delta S_\eta^{\neq}/R)$$

where $\Delta G_\eta^{\neq}$ is the free energy of activation for viscous flow, $\Delta H_\eta^{\neq}$ is the activation enthalpy, $\Delta S_\eta^{\neq}$ is the activation entropy, h is Planck's constant, N is Avogadro's number, and V_m is the molar volume. From Eq. 2 $\Delta G_\eta^{\neq}$ was estimated with a knowledge of the viscosity-temperature and the molar volume-temperature relationships, Figures 1 and 4 respectively. By assuming $\Delta S_\eta^{\neq}$ to be constant and V_m a linear function of temperature Eq. 2 can be reduced to

$$\eta = C/T \exp(\Delta H_\eta^{\neq}/RT) \qquad (3)$$

where C is a constant. $\Delta H_\eta^{\neq}$ was thus estimated from a plot of $\ln(\eta T)$ versus $1/T$ such as the one shown in Figure 5. The result was straight line regions representing different $\Delta H_\eta^{\neq}$ values. $\Delta H_\eta^{\neq}$ represents an average height of a potential barrier and is not meant to imply a single relaxation time.

A possible explanation for the change in activation enthalpies at a particular temperature may be in the changes which occur in the structure of the melt as it is cooled. The degree of association of the melt is expected to increase during cooling; a change if this sort is suggested by the break in the molar volume-temperature plot at about 700°C in the Na_2O-B_2O_3 system of Figure 4.

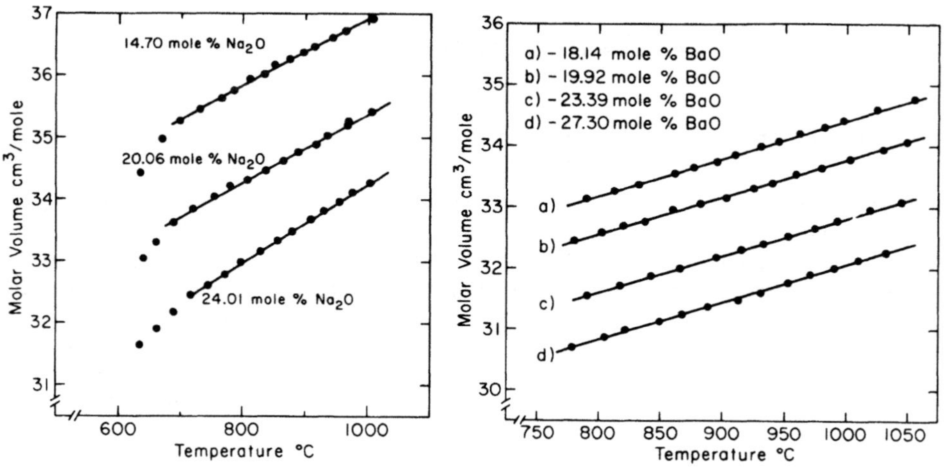

(A) (B)

Figure 4. Temperature dependence of the molar volume in the (A) Na_2O-B_2O_3 and (B) BaO-B_2O_3 systems.

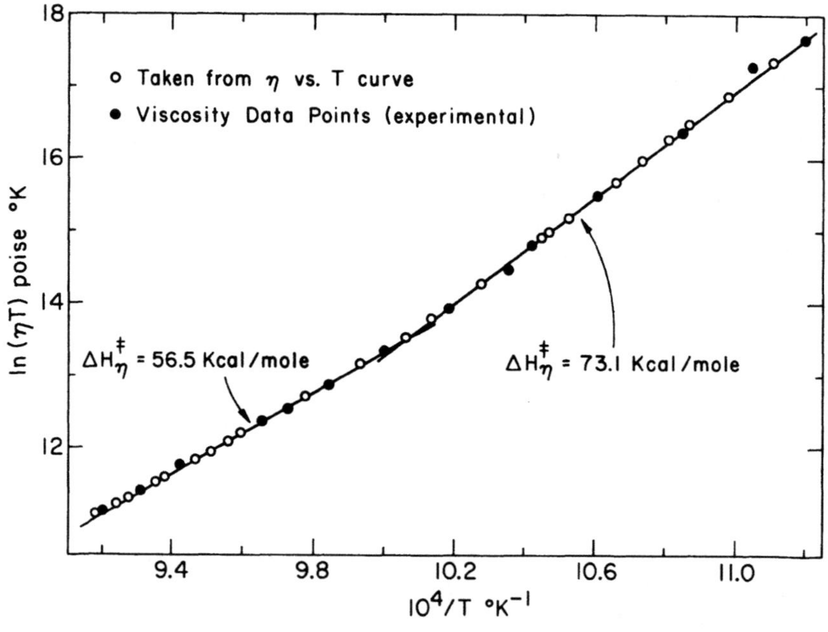

Figure 5. An estimation of the activation enthalpy for viscous flow for a composition 20.06 mole % Na_2O-79.94 mole % B_2O_3.

Similar changes were noted in the viscosity plot of Figure 5 and in a related study on the growth of $Na_2B_8O_{13}$ in Na_2O-B_2O_3 melts (27). As the structural units in the melt become larger the ease with which viscous flow can occur becomes less and the energy barrier to movement accordingly increases. The enthalpy of activation for viscous flow may be considered to be the sum of the enthalpy required to activate a molecule for flow and the enthalpy to create a nearby hole in the liquid into which the molecule may jump.

$\neq$ $\Delta S^{\neq}_{\eta}$, obtained from the thermodynamic expression $\Delta G^{\neq}_{\eta} = \Delta H^{\neq}_{\eta} - T\Delta S^{\neq}_{\eta}$, was in all cases positive and is indicative of the presence of a complex entangled structure. $\Delta S^{\neq}_{\eta}$ changes in the same temperature region as $\Delta H^{\neq}_{\eta}$ and the increase in $\Delta S^{\neq}_{\eta}$ at the lower temperatures is consistent with the concept of an association of individual groups. A summary of the values is presented in Table III.

TABLE III

Activation Parameters Estimated from Reaction Rate Theory

Composition	$\Delta G^{\neq}_{\eta}$ cal/mole	=	$\Delta H^{\neq}_{\eta}$ cal/mole	-	$T\Delta S^{\neq}_{\eta}$ esu	Temperature °C	$\Delta G^{\neq}_{\eta}$ at 800°C
NB15R	$\Delta G^{\neq}_{\eta}$	=	44,200	-	T(14.0)	688-1042	29,200
	$\Delta G^{\neq}_{\eta}$	=	68,200	-	T(39.9)	599-688	
NB20R	$\Delta G^{\neq}_{\eta}$	=	56,500	-	T(25.3)	720-1058	29,400
	$\Delta G^{\neq}_{\eta}$	=	73,100	-	T(42.6)	620-720	
NB24R	$\Delta G^{\neq}_{\eta}$	=	59,300	-	T(29.2)	721-1038	28,000
	$\Delta G^{\neq}_{\eta}$	=	83,300	-	T(53.4)	631-721	
BB18R	$\Delta G^{\neq}_{\eta}$	=	67,500	-	T(30.8)	798-1055	34,400
	$\Delta G^{\neq}_{\eta}$	=	94,500	-	T(56.0)	714-798	
BB20R	$\Delta G^{\neq}_{\eta}$	=	71,400	-	T(34.1)	816-1038	34,800
	$\Delta G^{\neq}_{\eta}$	=	96,400	-	T(57.7)	711-816	
BB23R	$\Delta G^{\neq}_{\eta}$	=	79,500	-	T(39.9)	795-1033	36,700
	$\Delta G^{\neq}_{\eta}$	=	106,000	-	T(65.5)	726-795	
BB27R	$\Delta G^{\neq}_{\eta}$	=	83,700	-	T(44.9)	793-1047	35,500
	$\Delta G^{\neq}_{\eta}$	=	108,400	-	T(67.8)	728-793	
KB20R	$\Delta G^{\neq}_{\eta}$	=	36,100	-	T(7.6)	814-1022	29,900
	$\Delta G^{\neq}_{\eta}$	=	47,900	-	T(18.3)	677-814	
KB33R	$\Delta G^{\neq}_{\eta}$	=	29,700	-	T(5.5)	741-1049	23,800
	$\Delta G^{\neq}_{\eta}$	=	66,200	-	T(41.3)	634-741	

Based on one mole of glass

In a study where activation energies for viscous flow were obtained from reaction rate theory by analyzing the most probable relaxation times, it was suggested that the width of the distribution in relaxation times is a measure of the microheterogeneities in the melt (28,29). In binary borate compositions the microheterogeneities or microstructure may result from liquid-liquid phase separation (8) or BO_4 and BO_3 group formation (30). In a study in the K_2O-B_2O_3 system (28) the maximum in the activation energy around 20 mole % K_2O was suggested as the cause of the maximum in the most probable relaxation time. A tightly bound structure due to compound formation or the BO_4 structures of Bray (3) and Beekenkamp (30) was said to be responsible for this occurrence.

In a structural model relating microstructure to viscous flow and structural relaxation properties of molten oxide glasses, the broadening of relaxation times was attributed to difference in flow properties between the phases or groups making up the microstructure and on the size of the groups or phases. The effect is present only with large viscosity or relaxation time differences when the groups are on the order of 25-50 Å (29).

An alternative theory, the free volume theory, has been applied by Kumar (18,19) to the high temperature flow behavior in borate systems. The free volume theory models the flow process in terms of a hard sphere model. According to the work of Cohen and Turnbull (14) the viscosity-temperature data in this treatment should fit an equation of the form

$$\log(\eta/T^{1/2}) = \log A + \gamma V^* \log e/V_f \qquad (4)$$

where γ is an overlap factor, V^* is the volume of a flow unit, A is a constant and V_f is the free volume calculated from

$$V_f = M\ (1/\rho_T - 1/\rho) \qquad (5)$$

where M is the molecular weight based on one mole of glass and ρ is the density at room temperature and at a temperature T. The slope of a log $(\eta/T^{1/2})$ versus $1/V_f$ plot is related to the size of a flow unit, V^*. The results are shown in Figure 6 and provide support for the applicability of the free volume treatment in the high temperature region for the alkali borates. The failure of this approach in the binary barium borate compositions, suggested by the unreasonably small flow units, may be associated with phase separation.

The parameters used to describe the temperature dependence of the viscosity using the VFT equation and the calculated free volume quantities are presented in Table IV. In Table IV the size of the flow unit is represented as the number of oxygen ions

(A)

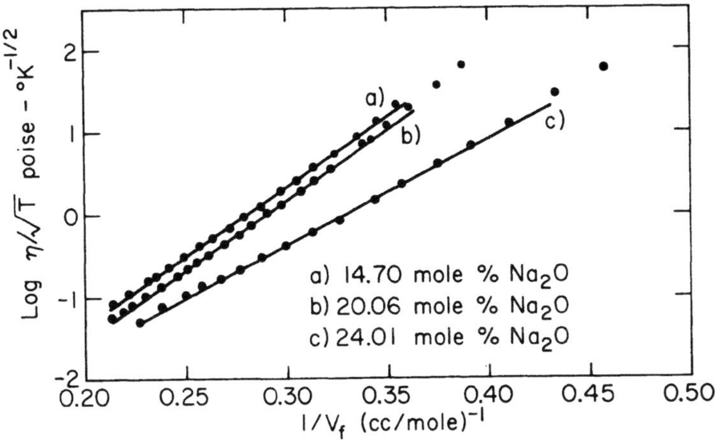

(B)

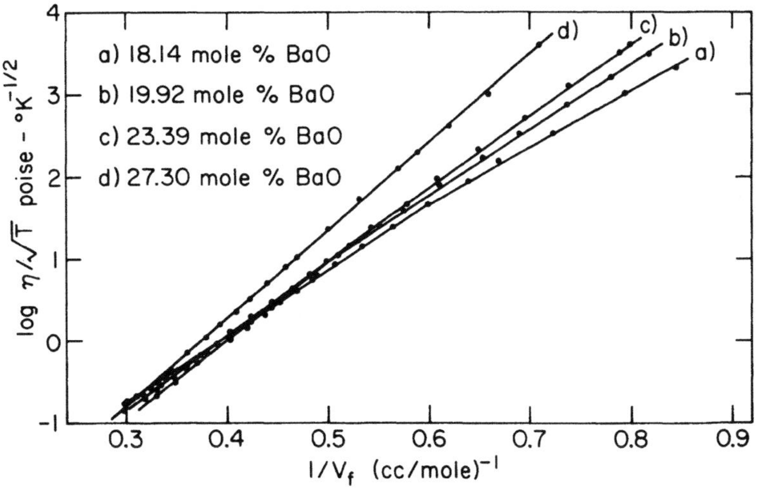

Figure 6. Relationship between log $(\eta/T^{1/2})$ and $1/V_f$ for (A) sodium borate and (B) barium borate compositions.

present in the unit, γn_o, since the ionic radius of the O^{2-} ion is much larger than that of the cation. The reference temperature, T_a, which is the temperature at which the viscous flow data departs significantly (defined as 0.3 in $\log_{10}\eta$ (15)) from the VFT curve, provides the lower limit to the usefulness of the free volume model in these binary borates. The free volume concept is thus unable to describe transport in the region of high viscosities. From the data in Figure 2 it also appears that another flow process, an easier one, must become dominant in this region. However, the results of the free volume approach in the high temperature region suggest that the flow units for these binary borate melts are relatively large ionic clusters.

TABLE IV

Parameters Used to Describe the High Temperature Viscous Flow Behavior in Binary Borate Melts

Mole % Modifier	A	B	T_o °C	T_g °C	γn_o	T_a °C	η at T_a
14.70 Na_2O	-2.103	1883	313.8	385	5.59	---	---
20.06 Na_2O	-2.804	2269	321.5	444	5.54	527	10^8
24.01 Na_2O	-2.503	1830	376.0	460	4.26	---	---
18.14 BaO	-2.926	2365	408.2	511	2.71	---	---
19.92 BaO	-2.018	2412	408.6	521	2.88	597	10^9
23.39 BaO	-3.000	2280	444.0	551	3.11	---	---
27.30 BaO	-3.132	2344	441.8	562	3.70	---	---
20.30 K_2O	-2.300	1951	321.0	400	6.11	533	$10^{6.5}$
33.29 K_2O	-1.433	774	465.4	---	5.14	---	---

T_g: $\eta = 10^{13}$ poise

CONCLUSIONS

The high temperature flow behavior of selected compositions in the Na_2O-B_2O_3, BaO-B_2O_3, and K_2O-B_2O_3 systems was obtained with a variable torque viscometer. The results were with previous low temperature data to represent the viscosity-temperature relationship over a wide range of values. A free volume analysis of the data indicated that this approach gives an appropriate description of the flow behavior at elevated temperatures in the alkali borates; however, at lower temperatures it overestimates the difficulty of the flow process. The failure of the approach in the BaO-B_2O_3 system was attributed to phase separation. In addition to the free volume parameters, estimates of $\Delta S^{\neq}_{\eta}$ and $\Delta H^{\neq}_{\eta}$

were obtained using the Eyring model. The large positive values obtained for $\Delta S_{\eta}^{\neq}$ were suggestive of entangled structures in the liquid. Observations on the variation in viscosity with composition were in agreement with present ideas on the structure of borate glasses.

ACKNOWLEDGMENT

This work was supported in part by the National Science Foundation under Contract NSF-DMR-76-01058.

REFERENCES

1. L. Shartsis, W. Capps and S. Spinner, J. Am. Ceram. Soc. 36 (1953) 319.

2. T. A. M. Visser and J. M. Stevels, J. Non-Crystalline Solids 7 (1952) 376.

3. P. J. Bray and J. G. O'Keefe, Phys. Chem. Glasses 4 (1963) 37.

4. S. Greenblatt and P. J. Bray, Phys. Chem. Glasses 8 (1967) 213.

5. S. E. Svanson, E. Forslind and J. Krogh-Moe, J. Phys. Chem. 66 (1962) 174.

6. J. Krogh-Moe, Phys. Chem. Glasses 6 (1965) 46.

7. W. L. Konijnendijk and J. M. Stevels, J. Non-Crystalline Solids 18 (1975) 307.

8. R. R. Shaw and D. R. Uhlmann, J. Am. Ceram. Soc. 51 (1968) 377.

9 E. M. Levin and H. F. McMurdie, J. Am. Ceram. Soc. 32 (1949) 99.

10. L. W. Coughanour, L. Shartsis, H. F. Shermer, J. Am. Ceram. Soc. 41 (1958) 324.

11. S. Glasstone, K. Laidler and H. Eyring, Theory of Rate Processes (McGraw-Hill, New York, 1941).

12. A. K. Doolottle, J. Appl. Phys. 22 (1951) 1471.

13. M. L. Williams, R. F. Landel and J. D. Ferry, J. Am. Chem. Soc. 77 (1955) 3701.

14. M. H. Cohen and D. Turnbull, J. Chem. Phys. 31 (1959) 1164.

15. M. Cukierman, J. W. Lane and D. R. Uhlmann, J. Chem. Phys. 59 (1973) 3639.

16. W. T. Laughlin and D. R. Uhlmann, J. Phys. Chem. 76 (1972) 2317.

17. A. J. Barlow, J. Lamb and A. J. Matheson, Proc. Roy. Coc. 292A (1966) 322.

18. S. Kumar, Phys. Chem. Glasses 4 (1963) 106.

19. S. Kumar, Phys. Chem. Glasses 6 (1965) 147.

20. R. T. Tiede, J. Am. Ceram. Soc. 42 (1959) 537.

21. E. F. Riebling, Rev. Sci. Instr. 34 (1963) 568.

22. G. S. Fulcher, J. Am. Ceram. Soc. 8 (1925) 339.

23. S. V. Nemilov, Inorganic Matls. 2 (1966) 300.

24. N. V. Romanov and S. V. Nemilov, Inorganic Matls 6 (1970) 1160.

25. R. Green, J. Am. Ceram. Soc. 25 (1942) 83.

26. D. L. Griscom, "Borate Glass Structure," pp 1-128, this volume.

27. C. J. Leedecke and C. G. Bergeron, to be published in Phys. Chem. Glasses.

28. R. Scully, J. H. Simmons and P. B. Macedo, J. Non-Crystalline Solids 12 (1973) 18.

29. J. H. Simmons and P. B. Macedo, J. Res. Nat. Bureau Stand. 75A (1971) 175.

30. P. Beekenkamp, Physics of Non-Crystalline Solids ed. J. A. Prinis (North-Holland Publishing Co., Amsterdam, 1965).

IMMISCIBILITY AND MICROSTRUCTURE IN AMORPHOUS BORATES

Thomas P. Seward, III

Research and Development Laboratories

Corning Glass Works, Corning, N.Y. 14830, USA

I. INTRODUCTION

Immiscibility in glassforming oxide systems has been known for a long time. Perhaps the first recorded systematic studies were those of Guertler [1], working with binary borate glasses, and Grieg [2], with binary silicates. Both authors observed stable immiscibility in alkaline earth and transition metal oxide-containing melts. Guertler concluded that the binary alkali borates were completely miscible.

In recent times, much attention has been given to the subject of metastable immiscibility in borate, silicate, and borosilicate systems. Many melts which had been previously believed to be completely miscible were found to exhibit two-liquid regions at subliquidus temperatures.

Recent review articles by Levin [3], James [4], and Uhlmann and Kolbeck [5] have collectively covered most aspects of the subject of immiscibility in glassforming systems. Other useful discussions can be found in References [6] and [7]. The present chapter will not attempt to be as comprehensive as those reviews, but will focus on areas of special interest to the author and will hopefully supplement the previous writings. References have been confined to boron-containing systems whenever practical; and many valuable references have regrettably been omitted because of space limitations.

By immiscibility we mean that, over some range of temperature and composition, a glass melt tends to separate into two distinct liquid phases. In terms of a phase diagram, the system can display stable or metastable immiscibility, depending on whether the two-liquid region extends to temperatures above the liquidus or remains entirely below it. Although all systems showing stable immiscibility also have a metastable extension below the liquidus [6], many systems show only metastable immiscibility.

In the region of stable immiscibility, the system can exist as two liquids for an indefinite period of time. However, in the metastable region the system ultimately tends toward the thermodynamically stable crystalline state and can exist as two liquids for only the period of time permitted by the crystallization kinetics. The tendency for crystallization can be reduced or halted by cooling below the glass-transition temperatures of both separated phases, whereby the glassy phases become kinetically stable.

Some glassforming systems may thermodynamically favor separation into two amorphous phases only at temperatures below the glass transition of the homogeneous melt. In such cases, the separation is kinetically unattainable and is never observed.

In this chapter, the term "phase separation" refers to the process by which a single-phase melt separates into two liquid phases. A phase-separated glass is thus a two-phase glass formed from a melt which has undergone phase separation.

In the literature, the term "miscibility gap" sometimes refers to the entire two-liquid region on a phase diagram and sometimes to the composition range over which a stable two-liquid region intersects the liquidus curve. Here, to avoid confusion, it is used only in the latter sense.

Phase-separated glasses often show an unusual microstructure which is strongly influenced by the viscosities of the separating phases, their relative volume fractions, and the time available for separation to occur. At one extreme, characterized by very fluid melts in stable immiscibility regions, the melt will form two liquid layers, the more dense layer on the bottom, if given sufficient time. Often a few minutes is all that is required.

On the other hand, in a moderately viscous melt (10^7 to 10^{13} Poise) phase separation generally occurs on a very fine scale, often of the order of 10 to 10^4 Å. The microstructure usually consists of spherical particles of the minor (lower volume fraction) phase imbedded in a matrix of the major phase. However, if the phase volumes are approximately equal, the phases often form two interpenetrating, internally connected phases. Coarsening of these microstructures into two liquid layers is hindered by the slow kinetics and, if the separation is metastable, by the eventual intervention of crystallization. Yet, it is the slow kinetics often found in metastable immiscibility regions which make them most suited for controlled studies of the phase-separation process.

II. BORATE GLASS IMMISCIBILITY

A great many binary borates show stable and/or metastable immiscibility regions. These include alkali metal, alkaline earth, transition metal, rare earth, and heavy metal borates. While the composition limits of most stable immiscibility gaps have been measured, the detailed shapes of the two-liquid regions have been determined in relatively few cases.

The metastable immiscibility boundaries in the binary alkali borates (Li, Na, K, Rb and Cs) were studied by Shaw and Uhlmann [8] with the aid of electron microscopy. They determined the consolute temperatures (maximum temperature of immiscibility) to lie between 570° and 660°C, with the corresponding compositions at about 16 mole % for the Na_2O system and about 10% for the others. While the actual composition range of detectable immiscibility depends on the system chosen, in no case did they find the alkali-rich end point to exceed about 24 mole % alkali oxide.

The stable barium borate immiscibility boundary was determined by Levin and Cleek [9] as part of their investigation of the ternary BaO-B_2O_3-SiO_2 system. The consolute point was found to lie at 1225°C and about 15 wt % BaO.

A number of workers have studied the shape of the immiscibility region in the binary lead borate system by observing the onset of visible opalescence [10-13]. The boundary has a very broad maximum extending from about 8 to 41 wt % at 750°C, with the consolute temperature at about 780°C.

A pronounced sigmoidal liquidus shape and certain physical property data suggest the presence of a metastable two-liquid region in the binary borosilicate system [14]. In an early study, Rockett and Foster [15] found no evidence of immiscibility even after holding a $3SiO_2 \cdot 2B_2O_3$ sample at 650°C for one month. Later, using chemical activity data derived from the liquidus curve (see Section III), Charles and Wagstaff [16] predicted the location of a very broad two-liquid region at temperatures below about 520°C. After a one-month heat treatment at these temperatures, the presence of immiscibility at three separate compositions was verified.

The immiscibility characteristics of ternary borates depend in great measure on whether the third component is another network former or a network modifier. In general, two modifier oxides with strong tendencies toward immiscibility will lead to a two-liquid region extending smoothly, like a tunnel, across the B_2O_3-rich corner of the ternary phase diagram. The addition of a more miscible modifier to a binary system of high immiscibility leads to a decrease in the extent of immiscibility. This was verified by Shartsis et al [17] for alkali oxides added to $RO-B_2O_3$ type binary systems.

Geffcken and Faulstich [18] have compiled data for the glass-forming regions in a number of rare earth oxide borate systems. Geffcken showed homogeneous glass formation to be limited by phase separation at the B_2O_3-rich corner of a number of ternary systems including those containing lanthanum oxide in combination with Ca, Sr, Ba, Zn and Cd oxides and those containing calcium oxide in combination with Sr, Ba, Zn and Cd oxides. The boundaries plotted by these authors give some indication of the extent of immiscibility in these systems.

The ternary borate whose immiscibility region has been most completely studied is $PbO-Al_2O_3-B_2O_3$. Zarzycki and Naudin [19] studied the boundary by observing the onset of opalescence using a platinum hot stage and a low power optical microscope. Addition of Al_2O_3 to the binary $PbO-B_2O_3$ first decreases then increases the tendency for immiscibility. A saddle point was found at 80 wt % B_2O_3, 15% PbO and 5% Al_2O_3 at about 660°C.

Several of the ternary alkali and alkaline earth borosilicate systems have been studied in some detail. Sastry and Hummel [20] defined compositional limits of immiscibility in the lithium borosilicate system, although no boundary isotherms for the two-phase region were determined.

The sodium borosilicate system, because of its commercial value (see Section VIII), has been rather extensively investigated by a number of workers including Hood and Nordberg [21], who delineated the two-phase region of optimum leaching; Molchanova [22], who produced a full curve enclosing the opalescence region; and Haller et al [23], who characterized the entire immiscibility region. The general shape of the two-phase region is that of a central, elongated dome running parallel to the $B_2O_3-SiO_2$ boundary. The dome peak occurs at about 5 wt % Na_2O, 70% SiO_2, 25% B_2O_3, and about 760°C. At lower temperatures, this dome extends to intersect the $B_2O_3-SiO_2$ binary surface and to meet immiscibility regions entering from the Na_2O-SiO_2 and $Na_2O-B_2O_3$ binary surfaces, forming pronounced troughs where they meet.

The potassium borosilicate system has received much less attention than its soda analog, but a similar region of optimum leachability has been shown [21].

The barium borosilicate two-liquid region has been studied by Levin and Cleek [9]. The boundary surface appears as a tunnel extending from a stable immiscibility region at the $BaO-B_2O_3$ binary, passing through the liquidus surface and intersecting the binary $BaO-SiO_2$ surface. This metastable intersection ha been studied by Seward et al [24].

In order to characterize phase-separation behavior within an immiscibility region, not only must one define the immiscibility surface but also the tie lines for the system. In a binary system, phase separation is confined to the plane of the binary surface; but in a ternary (or more complicated) system, it is not apparent from the phase diagram into what end compositions a given starting composition will separate. In particular, if an arbitrary binary cut is taken through a multi-component phase diagram, the tie lines are not likely to lie in the plane of this cut, unless the cut happens to lie very close to one of the binary end members.

The tie lines in the sodium borosilicate system have been studied by a number of workers [23,25-30].

Because of the fine scale of the phase-separated microstructures and the difficulties involved in coarsening them to large sizes, wet chemical analysis of the separated phases is generally difficult. Indirect methods for composition determination have been used which involve measurement of volume fractions from electron micrographs with certain assumptions about the densities of the phases [e.g., 23,30]. Information from leaching studies [25,26] has also been employed. More recently, Scholes and Wilkinson [27] used electron microprobe chemical analysis of the separated phases and Mazurin and Streltsina [29] derived the tie lines from measurements of glass-transition temperatures.

For a phase-separated glass there are two glass transitions, one corresponding to each of the separated phases. (There can be three or more glass transitions if the separation takes place in a multi-phase region, or if some phases secondarily phase separate into regions of different composition.) Mazurin et al [28] suggested that tie lines in two- (or three-) phase regions could be determined from the glass-transition temperature of the softer of the two separated phases as found by thermal expansion measurements or differential thermal analysis (DTA). Since the compositions of the two phases in equilibrium along any given tie line do not change (only the relative volumes change), the glass-transition temperatures of the two phases should also remain constant along the tie line. (Conversely, all experimentally determined lines along which the glass transitions remain constant need not be tie lines.)

The statement--glass-transition temperatures are constant along a tie line--does not imply that a single glass transition can be assigned for a two-phase glass. Nor does it imply that the glass transition of a homogeneous quenched (not yet phase-separated) glass would be constant across the immiscibility region.

Using thermal expansion measurements to determine glass-transition temperatures, Mazurin and coworkers [28,29] derived several tie lines for the sodium borosilicate system. They claim better agreement with the results of Scholes and Wilkinson [27] than with those of other workers [23,25,26,30]. However, the variation between the tie line directions of all the investigations is significant.

The accuracy to which the tie line direction can be determined in any experiment should be considered. For instance, indirect tie line determinations based on electron microscopy depend on how accurately the volume fractions are measured and how accurately the isotherms describing the immiscibility region are known. To demonstrate this effect, Mazurin and Streltsina [29] drew tie lines differing by more than ±45° rotation from a selected line and calculated that volume differences of only ±5 volume percent should be attained. They cite data of Burnett and Douglas [31] to show that ranges of experimental error for volume determinations can be ±15 volume percent.

The technique suggested by Mazurin has its own difficulties. For example, it loses its effectiveness in regions where the glass-transition temperature varies slowly with composition. Composition changes which occur during quenching can, of course, affect the accuracy of measurements made by any of the techniques.

Zarzycki and Naudin [19] determined two tie lines in the PbO-Al_2O_3-B_2O_3 system. The system was sufficiently fluid to allow the microstructure to be coarsened for wet chemical analysis. They also determined the tie lines from glass-transition temperatures measured by DTA and found good agreement between the results of the two techniques.

It should be noted at this point that in ternary phase diagrams three-liquid immiscibility regions are theoretically possible. Haller et al [23] pointed out the possible location of such a three-phase region in the sodium borosilicate system. There is no strong experimental evidence of such an occurrence, although secondary phase separation (see Section V) has sometimes been mistaken for such three-liquid immiscibility.

III. ORIGINS OF IMMISCIBILITY

The physical-chemical origins of liquid immiscibility lie in the tendency of certain liquids to separate into two (or more) distinct phases, together more energetically favored (on the basis of bond energies and geometric configurations) than the homogeneous mixture. Counteracting this tendency is a thermal disordering effect; the increased thermal energy present at high temperatures increases the likelihood of attaining the energetically less favored configurations of the homogeneous mixture. These effects are most conveniently discussed in thermodynamic terms. One considers a free-energy function which consists of an enthalpic (energy) term and an entropic (thermal disordering) term. A one- or two-phase system is obtained, depending on which minimizes the free energy of the system.

Most of the early explanations of phase separation centered around geometric or bond strength arguments. The geometric arguments were helpful to explain, and in some cases predict, those composition endpoints which result during phase separation. The bond strength arguments indicated the relative degree to which immiscibility would extend to high temperatures despite the homogenizing entropic effects. Such crystal chemical arguments were often used to explain the compositional extents of stable miscibility gaps; however, any such application can be misleading since the gap depends not only on the liquid unmixing tendency but also on the relative stability of the crystalline phases involved.

The geometric and bond strength arguments cannot explain the detailed variation of the two-phase boundary on a phase diagram. A quantitative thermodynamic treatment requires detailed knowledge of the liquid-phase, free-energy curves as a function of temperature and composition. Although such information is not generally available, Charles [32] was able to account for the location of the immiscibility boundary in several binary silicate systems using thermodynamic activity data derived from the liquidus curve shape. (The analysis utilizes melting point depression calculations and an assumption of the constancy of the partial molar heats of solution of silica as a function of temperature.) His derived activity data was supplemented by volatility, cell potential, and other thermodynamic data where available. A similar treatment was applied to the B_2O_3-SiO_2 system by Charles and Wagstaff [16]. They were able to predict the location of the immiscibility region, and they experimentally confirmed three points within that two-phase region. No other workers appear to have applied this approach to borate immiscibility, perhaps due to a lack of sufficient thermodynamic data.

In a subsequent study [33], Charles sought to determine what physical-chemical effects lead to the endothermic enthalpic term in the free-energy functions for phase-separable binary silicate melts, and also what is the nature of the entropic term. He concluded that the principal contribution to the entropy of mixing is the large number of configurations available to oxygen (bridging, non-bridging and free) in the homogeneous melt. He also demonstrated (from free energy of solution data for FeO, MnO and CaO and from melting point depression calculations for BaO and several alkali oxides) that although the heat of mixing (enthalpy) term is endothermic for the transition metal oxide systems, it is exothermic for the alkali metals and alkaline earth oxides. Since an endothermic term is required for immiscibility, Charles concluded that the partial molar heat of solution of silica, at high silica concentrations, must be endothermic. He argued the endothermic effect to be the elastic strain energy taken up when the silica network is forced to bend or flex during solution. The extent of the bending and the magnitude of the effect would depend on the ionic field strength of the modifier cations.

Haller et al [34] interpreted the endothermic solution effects of silica to imply a tendency for complex structural units of silica, larger than a single silica tetrahedra, to exist in the high silica melts. Macedo and Simmons [35] extended these arguments to borate glasses.

To quantitatively explain immiscibility in systems for which there is a lack of thermodynamic data, an alternative approach can be taken, namely: Hypothesize a model, devise a free-energy equation to fit it, and from this derive the two-phase boundary and even the locus of the spinodal points. An approach of this sort has been applied by Haller et al [34] to binary silicates and by Macedo and Simmons [35] to binary borates. Starting with a regular solution model, they account for the asymmetry of the two-phase boundary curves by choosing complex molecules as the mixing species corresponding to the respective endpoint compositions. To adjust the relative sharpness or flatness of the boundary at its maximum, those workers added a second entropy term to account for an increase or decrease in the number of states (configurational or vibrational) available in the homogeneous melt.

Macedo and Simmons fitted the published experimental two-phase boundary curves for the binary alkali borates (except for Na_2O-B_2O_3) and lead borate using $(B_2O_3)_5$ and $R_xO \cdot 3B_2O_3$ or $R_xO \cdot 4B_2O_3$ molecular units. To fit the rather flat-topped two-phase boundary for PbO-B_2O_3, a relatively large positive additional entropy of mixing term had to be added. This implies an increased availability of states which was proposed as due to the possibility of tetrahedral coordination of lead with oxygen in the mixed melt.

This approach is more applicable to fitting known immiscibility data than predicting the location of two-phase regions in the absence of such data. It should be emphasized that a reasonable fit to the data does not ensure uniqueness of the model. For instance, as Haller pointed out [34], any multiple applied to both the molecular unit sizes would yield the same analytic results. Other explanations for the required additional entropy of mixing term could be offered. While this work is consistent with the thesis that complex glass-forming molecular units are present in the melt, it should not be taken as direct confirming evidence.

Extensions of these approaches to the calculation of immiscibility regions in ternary or more complicated glasses have yet to be reported.

IV. EXPERIMENTAL TECHNIQUES

There are a number of experimental techniques which can be employed to study phase separation in glasses. Direct optical techniques, such as visual inspection or petrographic microscope examination, are useful to study the limits of stable two-phase regions. If the system tends to develop opalescence (because of relatively large particle sizes and/or refractive index differences) the metastable phase boundaries can be studied by such direct observation. X-ray diffraction can be used to verify that the observed opalescence is not due to crystallization.

Measurements of the angular dependence of light scattering are usually only useful in microstructure studies when the scale of the separated phases is comparable to, or greater than, a wavelength of light.

Electron microscopy (scanning, replica or direct transmission) can be useful in determining phase boundaries when direct visual evidence, such as opalescence, is absent. These techniques also give evidence of the microstructure details and can give insight into the phase-separation process. However, since quenched samples are generally examined, care must be taken to ascertain which structural effects are characteristic of the conditions being studied and which developed during the quench.

Of the three electron microscope techniques, direct transmission is most capable of resolving fine structure (on the order of 10 Å), but generally only separated phases with large density differences provide sufficient contrast for detailed study. Etching techniques can improve the microstructure contrast by producing corresponding thickness variations, but with increased chances of introducing misleading structural alterations or other artifacts.

When operated in a dark field mode, the electron microscope can also detect and give information about the shapes and orientations of small crystalline regions. Electron diffraction from selected sample areas can yield information about atomic pair distributions and, if crystals are present, about their structure and d-spacings. Zarzycki and Naudin [36] have used an optical transform method and spatial (correlation) filtering techniques to obtain information about three-dimensional microstructure spacings within bulk PbO-Al_2O_3-B_2O_3 glasses from two-dimensional transmission electron micrographs.

Small angle X-ray scattering (s.a.x.s.) can give indirect evidence concerning electron density fluctuations within the glass or melt. In two-phase systems, if the differences in electron density between the phases are sufficiently large, information about differences in composition, particle sizes and shapes and interfacial areas can be inferred. S.a.x.s. is particularly applicable to spinodal decomposition studies since the linearized theory of spinodal decomposition and s.a.x.s. theory are both formulated in reciprocal space. Because the technique is indirect, unique interpretations of the data are not always possible; and because of the electron density difference requirement, the techniques are not applicable to all phase-separated glasses.

Indirect evidence about phase separation can also be determined from property measurements such as viscosity, electrical conductivity, gas permeability and leaching rates (see Section VI).

V. PHASE-SEPARATION PROCESSES

A melt cooled into a two-phase region may enter the region of metastability, the region of instability, or both in sequence, depending on the initial melt composition and the cooling rate. Compositions near one side of a two-phase region generally enter the region of metastability on cooling. There phase separation proceeds by nucleation and growth of the minor phase (the phase characteristic of the opposite end of the tie line).

For melts of the consolute composition (corresponding to the maximum immiscibility temperature) the liquid enters the unstable region directly from the stable region. As a consequence of the instability, there exists no classical nucleation barrier to the formation of a new phase. The separation may begin by long range, low amplitude composition fluctuations which grow with time. This continuous process of phase separation is often referred to as "spinodal decomposition," because it takes place within the boundaries defined by the spinodal points.

Cahn and Charles [14] were the first to treat such a phase separation process in glassforming systems on a theoretical, mathematical basis. In its simplest (linearized) form, the theory predicts that, at a fixed temperature, there exists a critical wavelength of zero growth such that fluctuations of shorter wavelength will not grow but those of longer wavelength will. It further predicts a wavelength of maximum growth rate. A consequence of such growth is that two interpenetrating phases develop with a scale approximately that of the wavelength of maximum growth rate. When composition differences become large, the linearized theory is no longer valid and more involved solutions are required.

For compositions near consolute, the system may pass sequentially through the metastable into the unstable temperature range. Under the right conditions nuclei could form at the upper temperatures and act as sites from which wavelike composition fluctuations could spread at the lower temperatures.

The time evolution of the microstructure, in later stages, is strongly controlled by the relative viscosities and volume fractions of the separated phases, the glass-glass interfacial energies, and the diffusion coefficients. Structures consisting of two interpenetrating phases can coarsen for considerable periods of time without losing their connectivity; the scale increases and the net interfacial area decreases [e.g., 37]. The coarsening process is sometimes interrupted by crystallization to a more stable phase before spheroidization of the glassy microstructure occurs.

Because the continuous decomposition theory predicts the establishment of two interpenetrating phases, the existence of such microstructures in glass has often been taken as evidence of such a spinodal decomposition mechanism having been involved. However, as pointed out by Haller [38], interconnected microstructures can develop by a sequence of nucleation and subsequent growth of droplets to the extent that they contact each other. The interconnectivity of each phase results from the relatively fixed positions of the nuclei and the nearly equal volume fractions of the resulting phases.

Experimental confirmation that intersecting growth can yield continuous interpenetrating structures has been found in silicate and borosilicate systems [30,37,39,40]. Perhaps the greatest difficulty in experimentally demonstrating the continuous spinodal decomposition mechanism has been obtaining quenched glasses of sufficient homogeneity that the initial stages of the decomposition could be followed.

Experimental studies of the course of phase separation in borate-based glasses have been concentrated in the $PbO-B_2O_3$ and $PbO-Al_2O_3-B_2O_3$ systems. This is in part due to the convenient range of phase-separation temperatures and to the large density, electron density, and refractive index differences between the resulting phases.

Zarzycki and Naudin [12] used direct transmission electron microscopy and s.a.x.s. techniques to study phase separation in the classical nucleation and growth region of the $PbO-B_2O_3$ system. For melt compositions of 1 and 2 wt % PbO they demonstrated that particle diameters obey a $(\text{time})^{1/3}$ law during the later stages (Ostwald ripening).

In subsequent studies, Zarzycki and Naudin [36] and others [41,42] added Al_2O_3 to the $PbO-B_2O_3$ system. This suppressed the immiscibility temperatures to a viscosity range where the separation could be more carefully controlled and which would permit the quenching of a more acceptably homogeneous starting structure. Scattering curves for the earliest stages of decomposition show distinct maxima whose intensities increase with time, and whose positions remain relatively unchanged. This agrees with predictions of the continuous decomposition theory that certain wavelengths are favored to grow with time. However, essentially no crossovers (corresponding to the critical wavelength of zero growth) were observed in the scattering data at larger angles, indicating that, contrary to the linearized theory, short wavelength fluctuations do not die out. An explanation of this, and other discrepancies, has been offered in terms of thermally activated composition fluctuations [43].

For longer times, corresponding to structure coarsening, Zarzycki and Naudin [36] found the characteristic wavelengths of the microstructure to increase in proportion to $(\text{time})^{1/3}$. This indicates similar kinetics for the coarsening of interconnected structures as for ripening of droplet phases.

No equivalent s.a.x.s. studies have been made of a nucleation, growth and interconnection process in a heavily nucleated system. Without such data, it is not clear that s.a.x.s. studies can uniquely distinguish between the two mechanisms, even in the earliest stages.

If the later-stage microstructure is indeed independent of the initial separation mechanism, then from a practical point of view the question of the initial mechanism may have little importance, however interesting it may be from a purely scientific point of view.

Another subject worthy of mention in this section is that of supercritical fluctuations. For compositions near the center of an immiscibility region and at temperatures just above the coexistence surface, the free energy, as a function of composition, is quite flat. Hence, sizeable thermally activated composition fluctuations are possible. If the refractive index varies significantly with composition, these fluctuations may give rise to visible opalescence. This effect, known as critical opalescence, is similar to that found at the critical point in a liquid-gas system.

Zarzycki and Naudin [19] studied supercritical fluctuations in the $PbO-Al_2O_3-B_2O_3$ system using high temperature s.a.x.s. and found the Debye theory of correlation lengths to be applicable. The correlation range, L_D, is determined directly from the s.a.x.s.; and the range of molecular forces, ℓ, is obtained from a plot of $(1/L_D)$ vs $(T-T_c)$. For a glass of composition 77 wt % B_2O_3, 18% PbO, 5% Al_2O_3, $\ell \approx 10$ Å. Although it is difficult to relate ℓ to structural entities, it would presumably be related to the nearest distances of the heavier metal ions, Pb and Al, which produce the s.a.x.s. Zarzycki and Naudin compared their experimental ℓ values to the Pb-Pb and Al-Al distances (calculated for the range of glass compositions studied, assuming the ions were uniformly distributed in the melt) and found ℓ to agree well with the shorter of the two distances, switching from the Pb-Pb distance at high lead concentrations to the Al-Al distance at high alumina levels.

The final aspect of the phase-separation process to be discussed, secondary phase separation, has been observed by a number of workers including [24] and [44]. If a melt is held for some time at a given temperature within an immiscibility region, two phases, of compositions given by the appropriate tie line endpoints, develop. If the temperature is then lowered, each of these phases can separate into two new phases whose compositions are determined by the tie lines appropriate for each parent phase at that new temperature. This secondary phase separation is usually more prone to occur in the more fluid of the two parent phases.

VI. PHYSICAL AND CHEMICAL PROPERTIES

Physical and chemical properties of a phase-separated glass tend to vary continuously between those of the individual phases in equilibrium, as the average composition is varied across an immiscibility region. Physical properties such as density, heat capacity, molar volume, thermal expansion and elastic modulus tend to depend primarily on the compositions and relative volumes of the two separated phases, but not strongly on the microstructure, although some microstructural dependence can be detected. They tend to vary gradually as a function of composition. Shaw and Uhlmann have discussed some of these properties [45].

On the other hand, viscosity, transport properties such as electrical conductivity, ion diffusion and gas permeability, and chemical properties such as acid etching rates, depend very much on microstructure. They are often found to change rather abruptly over the narrow composition range where the connectivity of one or the other phase ceases. Also, for a given composition, these properties tend to vary with the degree of coarsening of the microstructure; that is, they are time-dependent at temperatures where the phase separation can progress.

For example, consider the viscosity isokomes for the PbO-B_2O_3 system shown by Zarzycki and Naudin [12] or Mazurin et al [46]. The glass is more fluid (isokomes at the lower temperatures) toward the B_2O_3-rich side of the diagram and more viscous toward the PbO-rich side. Rather abrupt changes occur where a decreased interconnectivity of the PbO-rich phase might be expected on the basis of volume fraction considerations.

As an example of the time variability of properties, Simmons et al [47] reported, for a sodium borosilicate glass at a temperature 200°C below the immiscibility temperature, a viscosity increase of five orders of magnitude over a time interval of 10^5 minutes.

The various transport properties are usually controlled by the more highly conducting phase as long as it remains three-dimensionally continuous. A-C dielectric loss shows interesting structure-dependent effects [48]. Some excellent chemically durable glasses, such as certain borosilicate PYREX® brand products, show a loss of chemical durability upon development of the two-phase microstructure with extended heating.

VII. DEFORMED MICROPHASES

The microstructure of phase-separated glasses can be altered by mechanical deformation. For example, spherical second-phase droplets will deform during bulk glass deformation, provided the forces generated by viscous flow are sufficient to overcome the restoring forces due to interfacial tension [49]. Droplet deformation will be more rapid if the droplet viscosity is comparable to, or less than, that of the matrix.

Seward [49] produced an aligned array of high aspect ratio, needle-like particles by stretching a phase-separated 40 wt % PbO-60% B_2O_3 glass at temperatures near the softening point. Subsequent annealing caused the needles to disintegrate into rows of spherical particles via a necking process which began near the needle tips.

Such arrays of aligned elongated particles can give rise to unusual optical properties, among which are birefringence and light guiding. The birefringence can have two causes. First, inherent differences in thermal expansion produce a net anisotropic strain field. The resulting stress birefringence cannot be annealed out, except by changing the microstructure. Second, a "shape birefringence" arises from the differences in refractive indices of the phases since, for non-spherical particles, the relative field concentration in particles and matrix depends on the polarization direction of the light [50].

Certain highly elongated phase-separated glasses, when viewed along the tensile axis, behave like coherent fiber optic image transfer plates [50,51]. In these glasses the particles are of lower refractive index than the matrix. The light guiding is seen as taking place in the matrix with internal reflections occurring at the particle/matrix interfaces.

VIII. PRACTICAL APPLICATIONS

There are many practical applications of liquid immiscibility in borate glasses. Light scattering by certain phase-separated borosilicate glasses is sufficiently strong that white opal tableware or art glass objects can be produced.

Certain phase-separated alkali borosilicate glasses can be leached to remove the alkali borate-rich phase. Consolidating the resulting porous glass structure at moderately high temperatures produces a dense, transparent, high silica (often near 96%) glass [21]. This process is used to make many of the Corning Glass Works VYCOR® brand products.

The porous materials, such as Corning Code 7930, produced by leaching phase-separated borosilicate glasses have been used in many specialized applications including dessicants, molecular sieves, salt bridges, beads for chemical chromatography, and high surface area supports for immobilization of enzymes. The pore sizes required for these various applications can be controlled, in part, by the thermal history prior to leaching.

Porous glasses have been backfilled with various materials for a number of applications. Stuffing with metals has produced granular superconductors; resistance heating elements and thermometers have resulted from impregnating the pores with carbon.

Reflectance losses of certain phase-separated borosilicate glasses can be drastically reduced by selective etching of both the low-silica and silica-rich phases [52]. A graded refractive index profile is developed on the surface which has the advantage of reducing reflectance losses over a broad visible and infrared wavelength region.

Leaching and backfilling techniques have also been employed in the preparation of optical waveguide materials [53].

IX. CONCLUSIONS

It is apparent that our understanding of silicate, borate and borosilicate immiscibility has expanded considerably during the past two decades as a result of some very significant experimental and theoretical work. For the near future, I suspect that the areas which need, and will get, the most attention are those involving the thermodynamic treatment of immiscibility and the practical applications of phase-separated glasses. I also believe that continued work with deformed two-phase glasses will lead to new, and perhaps exciting, applications.

REFERENCES

[1] W. Guertler, Z. Anorg. Chem. 40 (1904) 225.
[2] J.W. Grieg, Am. J. Sci., Ser. 5 13 (1927) 1; ibid., 133.
[3] E.M. Levin, in: Phase Diagrams, Vol. III, edited by A.M. Alper (Academic Press, New York, 1970) 143.
[4] P.F. James, J. Mater. Sci. 10 (1975) 1802.
[5] D.R. Uhlmann and A.G. Kolbeck, Phys. Chem. Glasses 17 (1976) 146.
[6] T.P. Seward, III, in: Phase Diagrams, Vol. I, edited by A.M. Alper (Academic Press, New York, 1970) 295.
[7] H. Rawson, Inorganic Glass Forming Systems (Academic Press, London, 1967) esp. chs. 7 and 8.
[8] R.R. Shaw and D.R. Uhlmann, J. Am. Ceram. Soc. 51 (1968) 377.
[9] E.M. Levin and G.W. Cleek, J. Am. Ceram. Soc. 41 (1958) 175.
[10] R.F. Geller and E.N. Bunting, J. Res. Natn. Bur. Stand. 18 (1937) 585.
[11] D.J. Liedberg, C.C. Ruderer and G.C. Burgeron, J. Am. Ceram. Soc. 48 (1965) 440.
[12] J. Zarzycki and F. Naudin, Phys. Chem. Glasses 8 (1967) 11.
[13] J.H. Simmons, J. Am. Ceram. Soc. 56 (1973) 284.
[14] J.W. Cahn and R.J. Charles, Phys. Chem. Glasses 6 (1965) 181.
[15] T.J. Rockett and W.R. Foster, J. Am. Ceram. Soc. 48 (1965) 75.
[16] R.J. Charles and F.E. Wagstaff, J. Am. Ceram. Soc. 51 (1968) 16.
[17] L. Shartsis, H.F. Shermer, and A.G. Bestul, J. Am. Ceram. Soc. 41 (1958) 507.
[18] W. Geffcken, Glastechn. Ber. 34 (1961) 91; and M. Faulstich, ibid., 102.
[19] J. Zarzycki and F. Naudin, J. Non-Cryst. Solids 5 (1971) 415.
[20] B.S.R. Sastry and F.A. Hummel, J. Am. Ceram. Soc. 42 (1959) 81.
[21] H.P. Hood and M.E. Nordberg, U.S. Patent 2,106,744 (1938); and U.S. Patent 2,215,039 (1940).
[22] O.S. Molchanova, Steklo Keram. 14 (1957) 5.
[23] W. Haller, D.H. Blackburn, F.E. Wagstaff and R.J. Charles, J. Am. Ceram. Soc. 53 (1970) 34.
[24] T.P. Seward, III, D.R. Uhlmann and D. Turnbull, J. Am. Ceram. Soc. 51 (1968) 278.
[25] R.J. Charles, J. Am. Ceram. Soc. 47 (1964) 559.
[26] T.L. Tran, Verres Refract. 19 (1965) 416.
[27] S. Scholes and F.C.F. Wilkinson, Discuss. Faraday Soc. 50 (1970) 175.
[28] O.V. Mazurin, M.V. Streltsina and A.S. Totesh, Phys. Chem. Glasses 10 (1969) 63.
[29] O.V. Mazurin and M.V. Streltsina, J. Non-Cryst. Solids 11 (1972) 199.
[30] G.R. Srinivasan, J. Tweer, P.B. Macedo, A. Sarkar and W. Haller, J. Non-Cryst. Solids 6 (1971) 221.

[31] D.G. Burnett and R.W. Douglas, Phys. Chem. Glasses 11 (1970) 125.
[32] R.J. Charles, J. Am. Ceram. Soc. 50 (1967) 631.
[33] R.J. Charles, Phys. Chem. Glasses 10 (1969) 169.
[34] W. Haller, D.H. Blackburn and J.H. Simmons, J. Am. Ceram. Soc. 57 (1974) 120.
[35] P.B. Macedo and J.H. Simmons, J. Res. Natn. Bur. Stand. 78A (1974) 53.
[36] J. Zarzycki and F. Naudin, J. Non-Cryst. Solids 1 (1969) 215.
[37] T.H. Elmer, M.E. Nordberg, G.B. Carrier and E.J. Korda, J. Am. Ceram. Soc. 53 (1970) 171.
[38] W. Haller, J. Chem. Phys. 42 (1965) 686.
[39] T.P. Seward, III, D.R. Uhlmann and D. Turnbull, J. Am. Ceram. Soc. 51 (1968) 634.
[40] P.F. James and P.W. McMillan, Phys. Chem. Glasses 11 (1970) 50; ibid., 64.
[41] A.F. Craievich, Phys. Chem. Glasses 16 (1975) 133.
[42] G.R. Srinivasan, R. Colella, P.B. Macedo and V. Volterra, Phys. Chem. Glasses 14 (1973) 90.
[43] H.E. Cook, Acta. Metall. 18 (1970) 297.
[44] E.A. Porai-Koshits and V.I. Averjanov, J. Non-Cryst. Solids 1 (1968) 29.
[45] R.R. Shaw and D.R. Uhlmann, J. Non-Cryst. Solids 1 (1969) 474; ibid., 5 (1971) 237.
[46] O.V. Mazurin, G.P. Roskova and V.P. Kluyev, Discuss. Faraday Soc. 50 (1970) 155.
[47] J.H. Simmons, P.B. Macedo, A. Napolitano and W. Haller, Discuss. Faraday Soc. 50 (1970) 191.
[48] R.J. Charles, J. Am. Ceram. Soc. 49 (1966) 55.
[49] T.P. Seward, III, J. Non-Cryst. Solids 15 (1974) 487.
[50] T.P. Seward, III, in: Non-Crystalline Solids, edited by G.H. Frischat (Trans Tech Publications, Aedermansdorf, Switzerland, 1977) 342.
[51] L.J. Randall and T.P. Seward, III, U.S. Patent 3,870,399 (1975).
[52] T.H. Elmer and F.W. Martin, Bul. Am. Ceram. Soc. 56 (1977) 366.
[53] P.B. Macedo and T.A. Litovitz, U.S. Patent 3,938,974 (1976).

CRYSTAL GROWTH KINETICS IN BINARY BORATE MELTS

C. G. BERGERON

Department of Ceramic Engineering and Materials Res. Lab.

University of Illinois at Urbana-Champaign, Urbana 61801

INTRODUCTION

The rate of crystal growth from undercooled melts depends upon a number of factors: (1) the magnitude of the free energy difference between the melt and crystal, (2) the concentration of sites suitable for molecular attachment at the crystal surface, (3) the rate at which growth units can be transported across the melt-crystal interface, (4) possible orientation requirements at the melt-crystal interface, and (5) the availability of crystallizable species at the interface. Most crystal growth data can be represented reasonably well by the standard equation for crystal growth [1]:

$$U = fD''/a_o[1 - \exp(V\Delta G_V)/RT] \qquad (1)$$

where U = growth rate in cm/sec,
f = fraction of sites at interface suitable for atomic or molecular attachment,
D'' = diffusion coefficient for transport across the melt-crystal interface,
a_o = molecular diameter,
V = molar volume,
ΔG_V = free energy change per unit volume accompanying the transformation from melt to crystal,
R = gas constant,
T = absolute temperature.

In analyzing crystal growth data, it is frequently assumed that the molecular motions involved in transporting molecules or atoms across the melt-crystal interface are similar to those for viscous flow in the bulk melt. Hence, D" in equation (1) is assumed to be inversely proportional to the melt viscosity, η, i.e., $D'' = b/\eta$. The temperature dependency of f (the fraction of sites which are suitable for attachment on the crystal surface) is related to the growth mechanism and can be examined by writing equation (1) in terms of a reduced growth rate, U_R:[2]

$$U_R = U\eta/[1 - \exp(V\Delta G_V/RT)] = (b/a_o)f. \qquad (2)$$

For the standard models of crystal growth, a plot of U_R vs undercooling would be expected to yield a horizontal line if growth occurs by a continuous or 'normal' mechanism in which the crystal surface contains a large fraction of sites suitable for molecular attachment. For the screw dislocation model, a plot of U_R vs undercooling is expected to yield a line of positive slope passing through the origin; if the growth rate is controlled by the rate at which surface nuclei are formed, a curve of positive slope passing through the origin is anticipated.

The linear growth rates have been determined for a number of binary borate crystals [3-11] in which the melts were of the same composition as that of the crystal. In the majority of the systems reported, high purity materials (less than 150 ppm impurities) were used to make the glasses. In all cases, the crystals were observed to grow at rates which were independent of time, suggesting that steady-state conditions prevailed during the measurements. Additional data such as melt viscosity, enthalpy of fusion, heat capacity, and molar volume have been determined for many of these systems.

In the present paper, the crystallization kinetics of nine binary borate crystals are compared and analyzed in an effort to ascertain the probable growth mechanisms and the principal factors which influence the growth rate.

GROWTH RATES

Figure 1 shows the growth rates plotted as a function of undercooling. The growth rates vary from 2 microns/sec for PbB_4O_7 to about 3000 microns/sec for $Li_2B_4O_7$. It would seem that an explanation for this large disparity in growth rates might be found by comparing some of the known properties of the crystals and their melts. Table I lists some of the pertinent properties. The entropies of fusion, in terms of cal/cm^3/°, vary from 0.11 for $Cs_2B_6O_{10}$ to 0.43 for PbB_4O_7 and SrB_4O_7. The driving force term $[1 - \exp(V\Delta G_V/RT)]$ in equation (1) can be evaluated using the approximation $\Delta G_V = -\Delta S\Delta T$. A comparison of this term as a function of undercooling for the nine

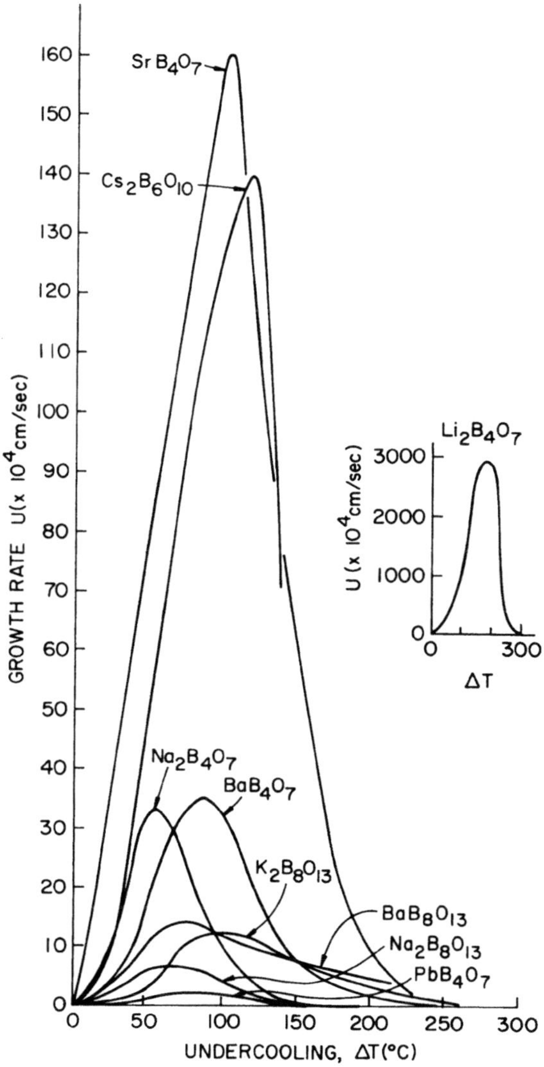

Fig. 1. Growth rate as a function of undercooling.

TABLE I. Properties of Binary Borates

Crystal	M.P. °C	Xtal system	Density $\frac{g}{cm^3}$	ΔH Fusion $\frac{Kcal}{mole}$	ΔH Fusion $\frac{Cal}{cm^3}$	ΔS Fusion $\frac{Cal}{mol \times °C}$	ΔS Fusion $\frac{Cal}{cm^3 \times °C}$	U_{max} $\frac{cm}{sec \times 10^4}$	$\eta_{U_{max}}$ Poise	$\Delta H^{\ddagger}_{\eta}$ (Kcal/mole) High T	$\Delta H^{\ddagger}_{\eta}$ (Kcal/mole) Low T
$PbO \cdot 2B_2O_3$	775	Ortho	5.22	31.5	455	30	0.43	2	60	45	38
$SrO \cdot 2B_2O_3$	997	Ortho	4.01	33.4	551	26.3	0.43	154	50	60	82
$BaO \cdot 2B_2O_3$	910	Mono	3.57	20.7	249	17.5	0.21	35	322	71	99
$Li_2O \cdot 2B_2O_3$	917	Tetr	2.44	28.8	415	24.1	0.35	3000	3	32	46
$Na_2O \cdot 2B_2O_3$	742	Ortho	2.37	18.1	213	18	0.21	33	250	60	93
$Cs_2O \cdot 3B_2O_3$	837	Ortho	3.39	17.2	119	15.5	0.11	140	230	44	81
$Na_2O \cdot 4B_2O_3$	816	Mono	2.32	31.3	213	29	0.20	7	230	56	73
$K_2O \cdot 4B_2O_3$	857	----	2.14	30.1	173	26.6	0.15	12	150	37	51
$BaO \cdot 4B_2O_3$	889	Ortho	2.91	27.6	186	24	0.16	15	550	71	91

crystal systems is shown in Fig. 2. The slowest growing crystal (PbB_4O_7) exhibits the largest value for the driving force. However, the values differ by no more than a factor of three between the different systems.

If the fluidity of the melt is an appropriate measure of the rate of transport across the melt-crystal interface, a comparison of the melt viscosities may provide some insight. Figure 3 shows the viscosities as a function of undercooling and in Table I the viscosities at the temperature of maximum growth rate are listed. It can be seen that at U_{max} the viscosities of the borates of Na, K, Cs, and Ba are similar in magnitude (150-550 poises) and that the viscosities of the borates of Sr and Pb are considerably less (50-60 poises) while that of Li is only a few poises. It is apparent that the product of the driving force and the fluidity does not provide a clear correlation with the measured growth rates in these systems.

GROWTH MECHANISMS

If the growth rate is limited by the availability of sites suitable for molecular attachment, the temperature dependence of the reduced growth rate, U_R, can provide information about the probable growth mechanism in terms of the standard models of crystal growth. Morphological studies [11,12] have suggested that $Na_2B_4O_7$, PbB_4O_7, and $Na_2B_8O_{13}$ grow by the screw dislocation mechanism. The plots of U_R vs ΔT for these three systems are shown in Figs. 4, 5, and 6, in which the undercooling was calculated using the crystal-melt interface temperatures measured by Herron [13]. The plots do not exhibit the linear relationship anticipated for a screw dislocation mechanism but instead suggest a relationship characteristic of a surface nucleation model. If the data conform to a surface nucleation model a plot of $\ell n(U\eta)$ vs $1/T\Delta T$ would be expected to yield a line. Such plots resulted in curves for all three crystals; Fig. 7 (PbB_4O_7) shows a typical plot.

For the screw dislocation model, the fraction of preferred growth sites, f, is given by [14]:

$$f = (a_o \Delta S \Delta T)/4\pi\gamma V_m) \simeq A\Delta T \tag{3}$$

where ΔS is the entropy of fusion, γ is the solid-liquid interfacial energy, and V_m is the molar volume. The growth equation then becomes:

$$U = A\Delta T(D''/a_o)[1 - \exp(-\Delta S\Delta T/RT)] \tag{4}$$

D" can be written in the activated state formulation as:

$$D'' = a_o \nu \exp(-\Delta G_U^{\ddagger}/RT) \tag{5}$$

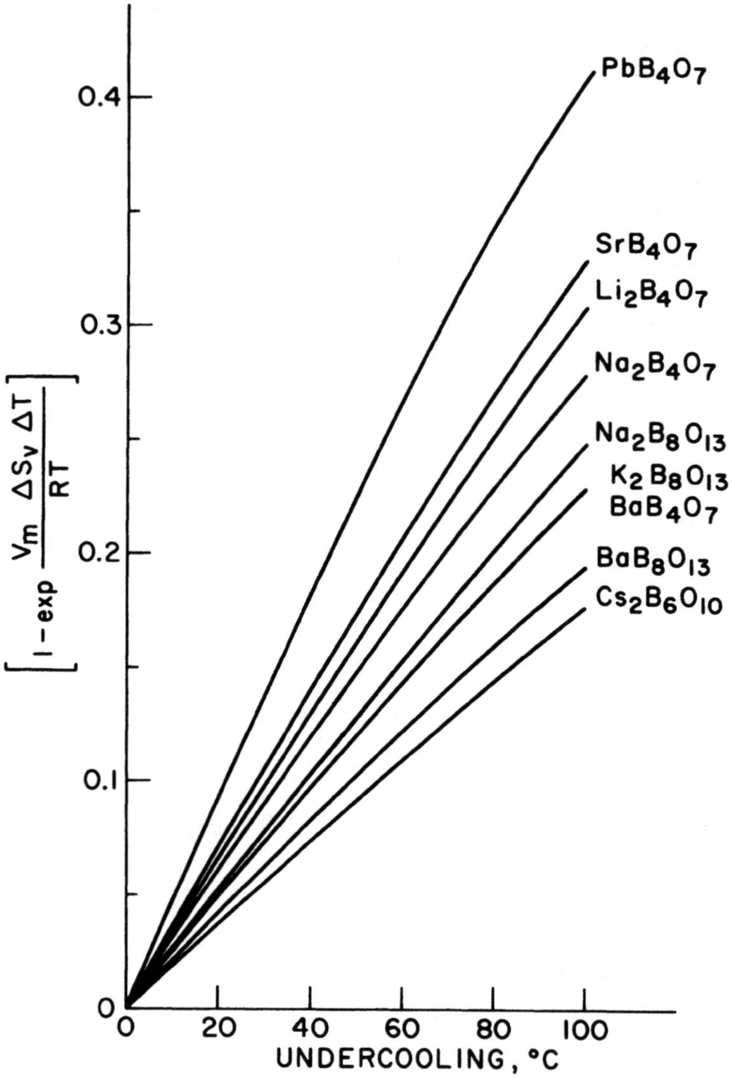

Fig. 2. Driving force for growth vs. undercooling.

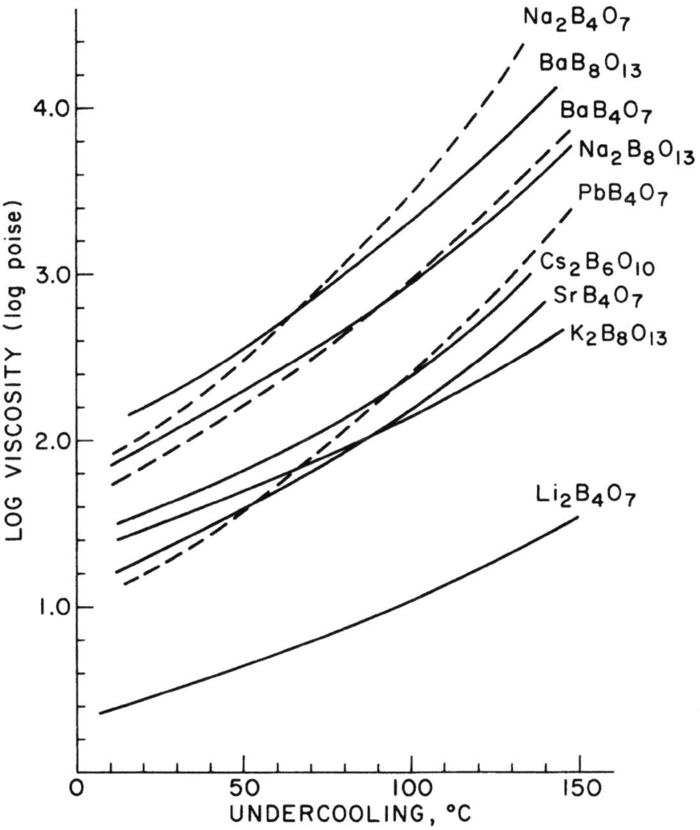

Fig. 3. Log viscosity vs. undercooling.

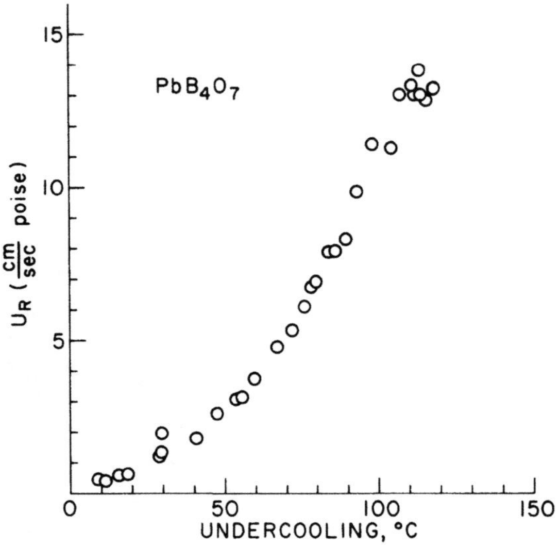

Fig. 4. Reduced growth rate vs. undercooling for PbB_4O_7.

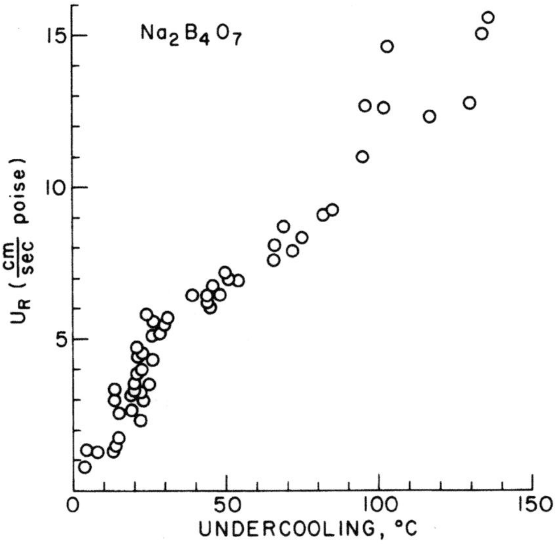

Fig. 5. Reduced growth rate vs. undercooling for $Na_2B_4O_7$.

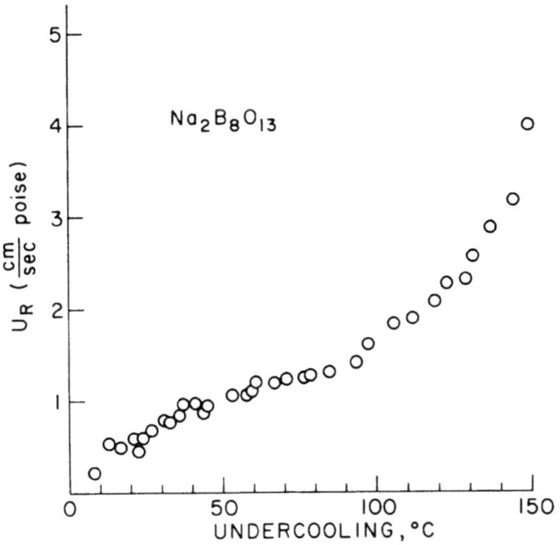

Fig. 6. Reduced growth rate vs undercooling for $Na_2B_8O_{13}$

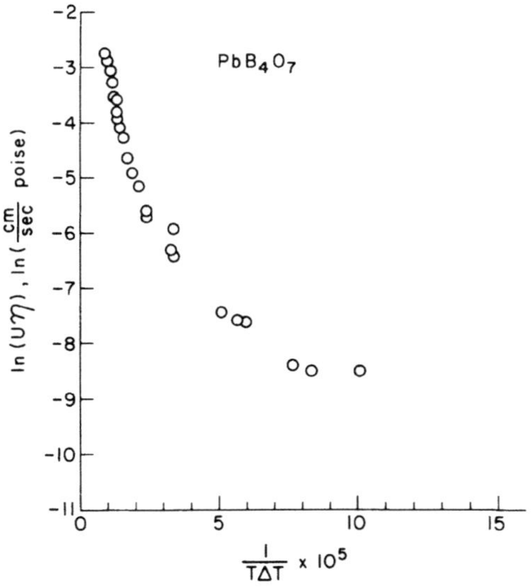

Fig. 7. ℓn (growth rate × viscosity) vs 1/T for PbB_4O_7.

where ν = vibration frequency and $\Delta G_U^{\ddagger}$ is the free energy of activation for transport across the interface. Kumar [15] has shown that if the entropy of activation for transport, $\Delta S_U^{\ddagger}$ is assumed independent of temperature, the growth rate (assuming a screw dislocation model) can be written as:

$$U = A\Delta T a_o \nu \exp(\Delta S_U^{\ddagger}/R)\exp(-\Delta H_U^{\ddagger}/RT)[1 - \exp(-\Delta S\Delta T/RT)] \quad (6)$$

and a plot of $\ln\{U/T\Delta T[1 - \exp(-\Delta S\Delta T/RT)]\}$ vs 1/T should yield a linear relationship. The value of $\Delta H_U^{\ddagger}$ can be calculated from the slope of the line. Plots of this type for PbB_4O_7, $Na_2B_4O_7$, and $Na_2B_8O_{13}$ are shown in Figs. 8, 9, and 10, where, in each case, two linear segments are observed, suggesting a reasonable fit to the model. The change in slope occurs at a temperature corresponding approximately to the inflection point on the low temperature side of the U vs T curve. The enthalpies of activation calculated from the slopes of these lines can be compared with the corresponding values calculated for viscous flow. An estimate of the enthalpy of activation for viscous flow, $\Delta H_\eta^{\ddagger}$, can be made from the Eyring equation [16]:

$$\eta = Nh/V_m \exp(\Delta H_\eta^{\ddagger}/RT) \exp(-\Delta S_\eta^{\ddagger}/R) \quad (7)$$

where η is the coefficient of viscosity, N is Avagadro's number, V_m is the molar volume, and $\Delta S_\eta^{\ddagger}$ is the entropy of activation for viscous flow. Since V_m is essentially a linear function of temperature (Fig. 11), a plot of $\ln(\eta T)$ vs 1/T can be used to calculate $\Delta H_\eta^{\ddagger}$ which is proportional to the slope. Plots of this type are shown in Fig. 12. In Table II the calculated values of $\Delta H_\eta^{\ddagger}$ for the high and low temperature portions of the graphs are compared with the values of $\Delta H_U^{\ddagger}$ for corresponding temperature ranges. In all three cases the values of $\Delta H_\eta^{\ddagger}$ and $\Delta H_U^{\ddagger}$ are significantly different. Thus it is not surprising that the U_R vs ΔT plots for these crystals show a non-linear relationship despite the apparent fit to a screw dislocation model; $\Delta H_U^{\ddagger}$ and $\Delta H_\eta^{\ddagger}$ would have to be approximately equal in order for the U_R vs ΔT plot to be linear. A curve of decreasing slope would occur for $\Delta H_\eta^{\ddagger} < \Delta H_U^{\ddagger}$, and a curve of increasing slope for $\Delta H_\eta^{\ddagger} > H_U^{\ddagger}$. For the borate glasses, perhaps a more appropriate test for a fit to the screw dislocation model would be a plot of $\ln\{U/T\Delta T[1 - \exp(-\Delta S\Delta T/RT)]\}$ vs 1/T.

The plots of U_R vs ΔT for the crystals SrB_4O_7, BaB_4O_7, $Li_2B_4O_7$, $Cs_2B_6O_{10}$, $K_2B_8O_{13}$, and BaB_8O_{13} yield curves similar in shape to that shown in Fig. 13 for $K_2B_8O_{13}$, which suggests a surface nucleation growth mechanism. A plot of $\ln(U\eta)$ vs $1/T\Delta T$ should therefore exhibit a linear relationship if the data are consistent with the surface nucleation model. Figure 14 shows such a plot for $K_2B_8O_{13}$; instead of the linear relationship expected, two distinct curved segments are seen. Curves similar to this were obtained for SrB_4O_7, BaB_4O_7, and BaB_8O_{13}. The standard surface nucleation model for growth is based upon the

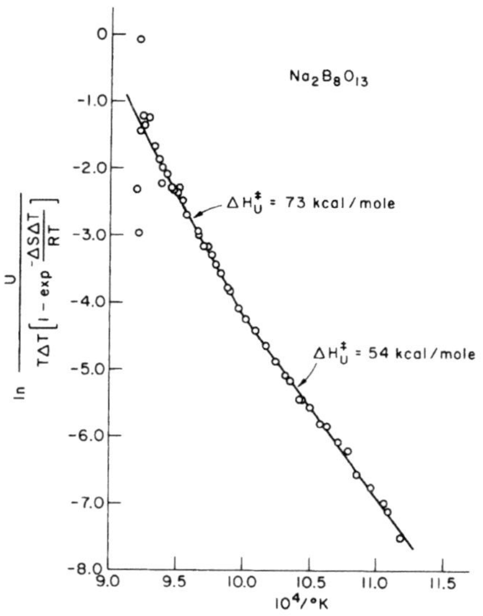

Fig. 8. $\ell n[U/T\Delta T(1 - \exp\frac{-\Delta S\Delta T}{RT})]$ vs 1/T for $Na_2B_8O_{13}$.

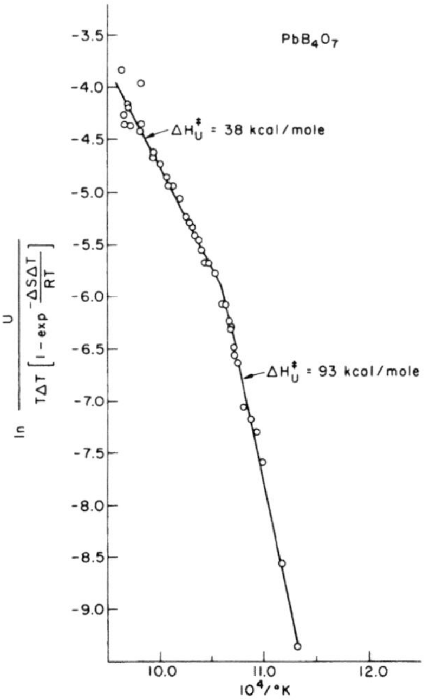

Fig. 9. $\ell n[U/T\Delta T(1 - \exp\frac{-\Delta S\Delta T}{RT})]$ vs 1/T for PbB_4O_7.

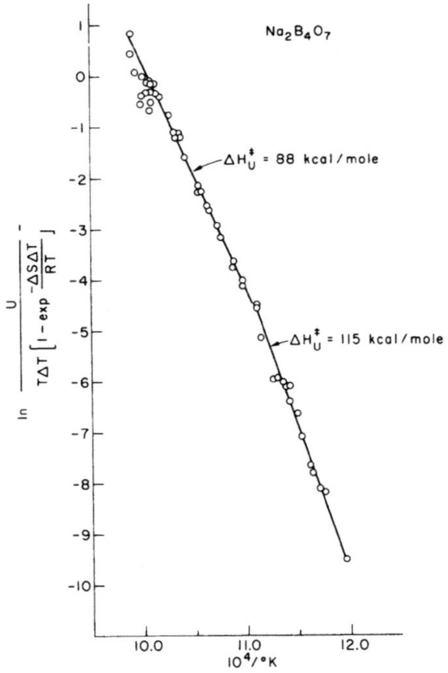

Fig. 10. $\ln[U/T\Delta T(1 - \exp\frac{-\Delta S\Delta T}{RT})]$ vs 1/T for $Na_2B_4O_7$.

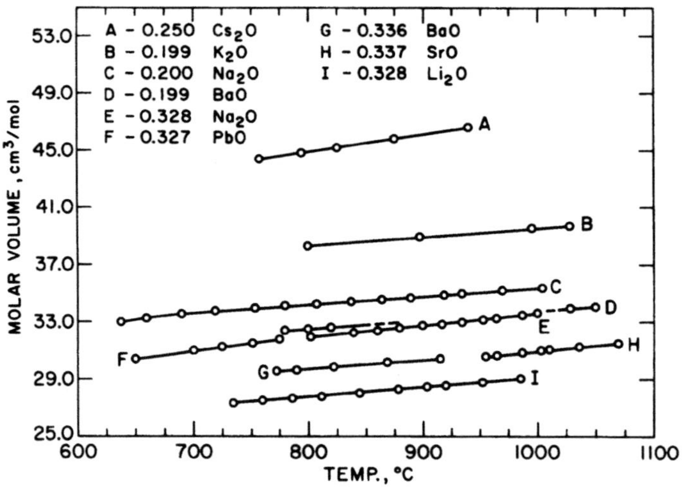

Fig. 11. Molar volume vs temperature for binary borate melts.

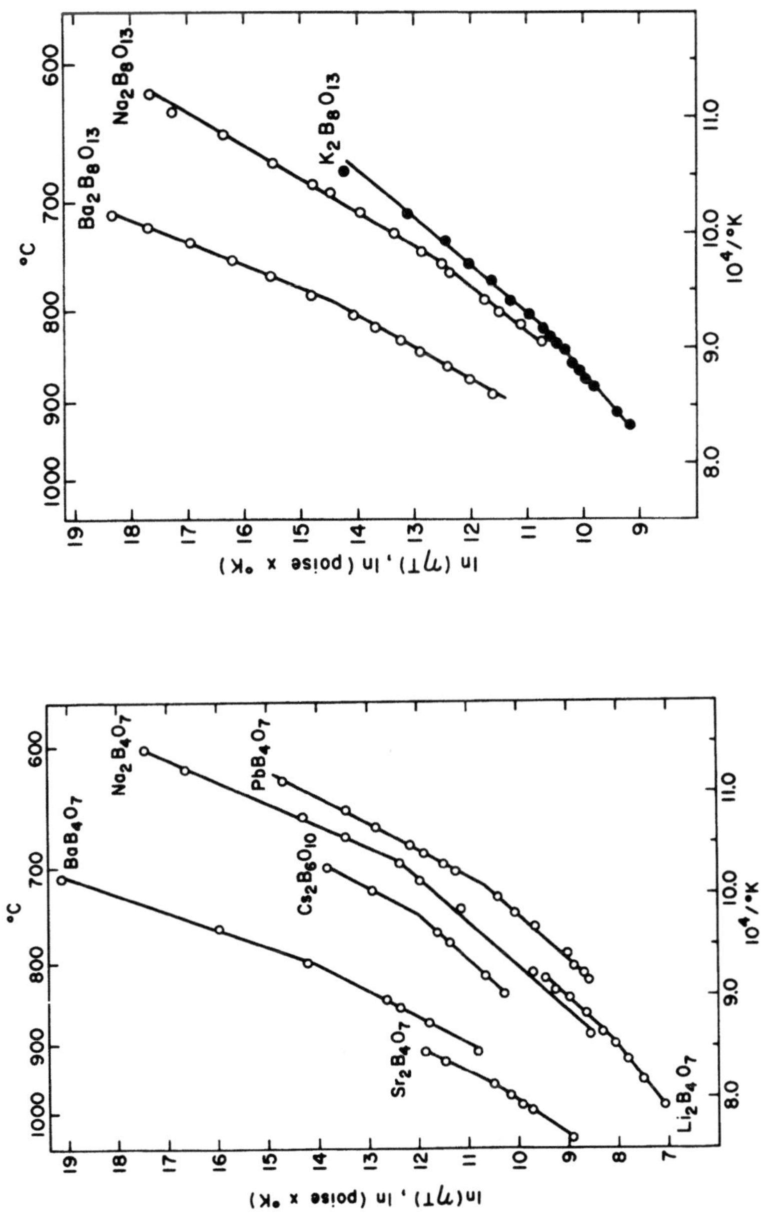

Fig. 12. ℓn(viscosity × temperature) vs 1/T for binary borate melts

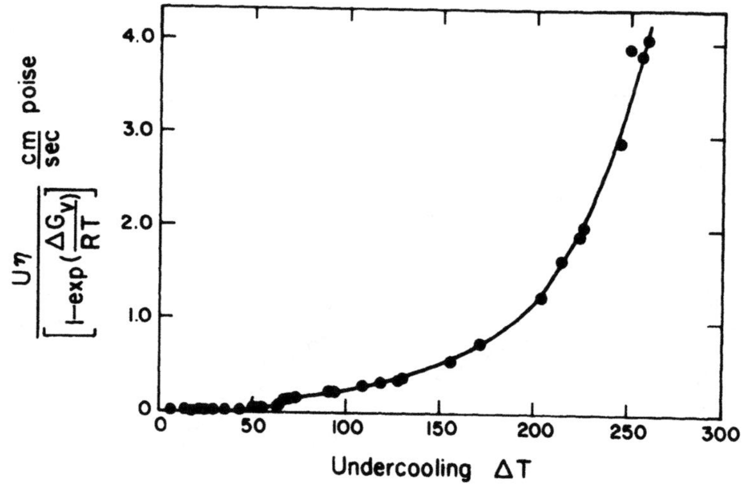

Fig. 13. Reduced growth rate vs. undercooling for $K_2B_8O_{13}$ (after ref. 8).

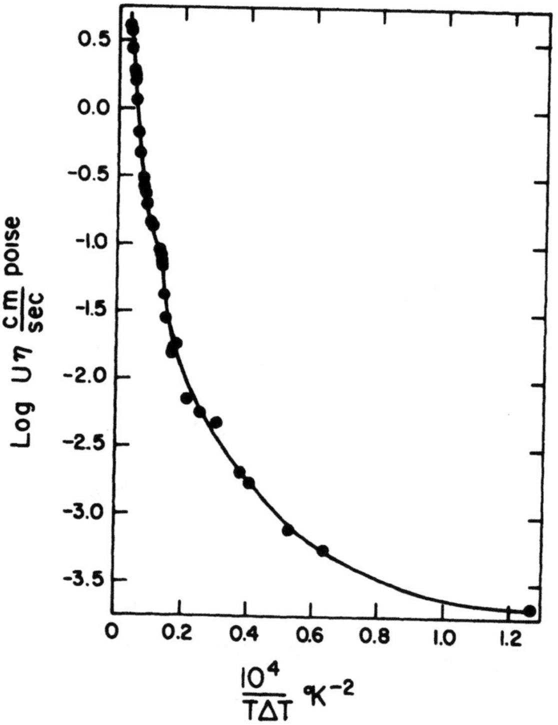

Fig. 14. Log (growth rate × viscosity) vs $1/T\Delta T$ for $K_2B_8O_{13}$ (after ref. 8).

TABLE II. Comparison of $\Delta H_U^{\ddagger}$ and $\Delta H_\eta^{\ddagger}$ for Na- and Pb-Borates

Crystal	Temp. Range	$\Delta H_\eta^{\ddagger}$	$\Delta H_U^{\ddagger}$
$Na_2B_4O_7$	742-685	60	88
	685-625	93	88
	625-550	93	115
$Na_2B_8O_{13}$.	816-731	56	73
	731-720	56	54
	720-620	73	54
PbB_4O_7	775-725	45	38
	725-670	73	38
	670-610	73	93

assumption of an equilibrium size and shape for the nucleus, i.e., a pillbox one atomic diameter high and of radius r. However, the work of Binsbergen [17] on a computer simulation of embryo growth has shown a high probability for the formation of two-dimensional embryos which deviate strongly from the equilibrium shape, although the temperature dependency of the nucleation frequency remains the same as that derived from classical theory. Regardless of the shape of the surface nucleus, the work of formation, $\Delta\phi$, can be expressed as the sum of an interfacial free energy and a volume free energy:

$$\Delta\phi = A_e\gamma + V_e\Delta G_v \tag{8}$$

where A_e = area of embryo edge, γ = interfacial energy per unit area, V_e = volume of the embryo, and ΔG_v = the free energy change per unit volume for the liquid to crystal transition. The temperature dependency of ΔG_v is expected to be dominant; therefore, as shown by Leedecke and Bergeron [8], a plot of $\ln\{U \exp(\Delta G_\eta^{\ddagger}/RT)/U_o[1 - \exp(\Delta G_v/RT)]^{2/3}\}$ vs $(\Delta G_v/T)$ (where U_o is a constant) would be expected to yield a straight line if the data are consistent with a surface nucleation mechanism (large crystal model). Such a plot, shown in Fig. 15 for $K_2B_8O_{13}$, yields two straight-line segments intersecting at about 70° of undercooling. Plots of similar shape were obtained for the crystals SrB_4O_7, BaB_4O_7, and BaB_8O_{13}. The change in the slope may be an indication of the temperature range in which the transition to transport-rate controlled growth is occurring.

The growth data for $Cs_2B_6O_{10}$ have shown [7] a reasonable fit to the standard surface nucleation model; however deviation from the linear relationship was observed at large undercoolings in the $\ln(U\eta)$ vs $1/T\Delta T$ plot. To date the growth rate data of $Li_2B_4O_7$ have

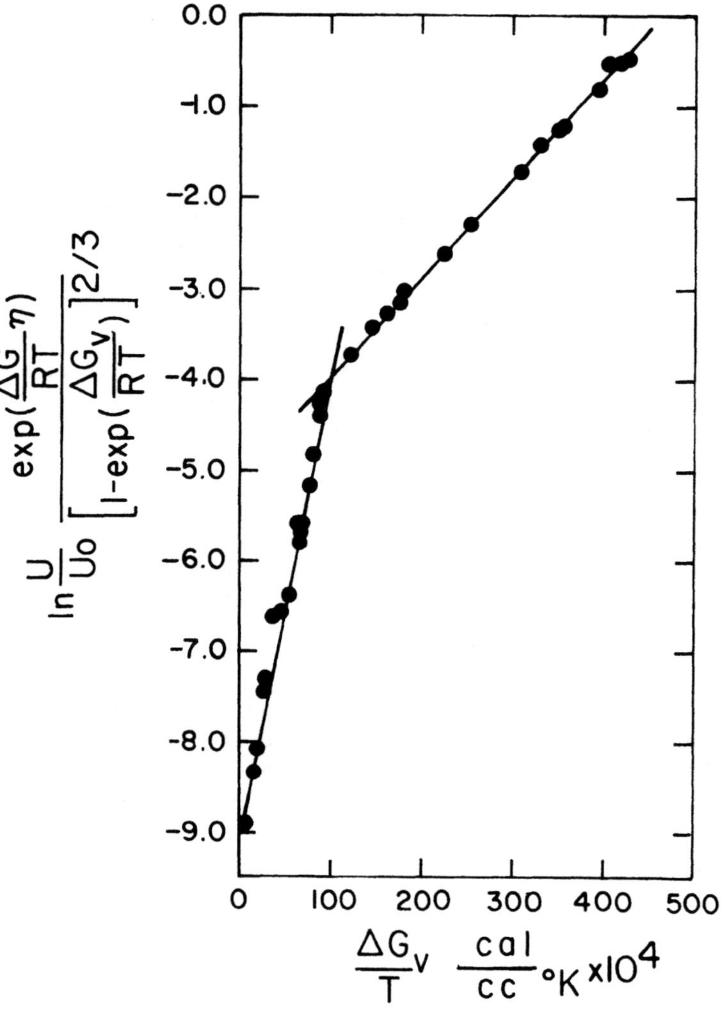

Fig. 15. Modified reduced growth rate vs $\Delta G_v/T$ for $K_2B_8O_{13}$ (after ref. 8).

not been amenable to analysis because of the difficulty of determining the undercooling. The very large growth rates of $Li_2B_4O_7$ cause a rise in temperature at the interface which Herron [13] has shown to be more than 150°C.

SUMMARY

In terms of the standard models of crystal growth, the growth data for $Na_2B_4O_7$, $Na_2B_8O_{13}$, and PbB_4O_7 gave a best fit to the screw dislocation model. Calculation of the enthalpy of activation for transport of molecular species across the melt-crystal interface, assuming a screw dislocation mechanism, resulted in values which were significantly different in magnitude from those calculated for viscous flow in the bulk melt.

The growth rate data for SrB_4O_7, BaB_4O_7, BaB_8O_{13}, and $K_2B_8O_{13}$ indicated a best fit to a surface nucleation model provided the configuration of the nucleus on the crystal surface was not restricted to the theoretical equilibrium size and shape.

The maximum growth rates for these crystals cover a broad range, yet the melt viscosities at the respective temperatures of maximum growth rate are relatively similar except for that of $Li_2B_4O_7$. It suggests that the viscosity of the melt does not necessarily reflect the true transport rate at the interface or, perhaps, that the orientation requirements at the interface may be rate limiting in some instances.

It is recognized that the standard growth models represent a simplification of a complex process and that more than one attachment mechanism may be operating concurrently. Nevertheless, the growth rate data exhibit temperature dependencies which appear to fit the models remarkably well.

ACKNOWLEDGEMENT

This work was supported in part by the National Science Foundation under Grant NSF DMR-72003026.

REFERENCES

[1] D. Turnbull and M. Cohen, in: Modern Aspects of the Vitreous State, Vol. 1, Ed. J. D. MacKenzie (Butterworth, London, 1960) pp. 38-62.

[2] K. A. Jackson, D. R. Uhlmann, and J. D. Hunt, J. Crystal Growth 1(1967) 1.

[3] R. J. Eagan, J. P. DeLuca, and C. G. Bergeron, J. Am. Ceram. Soc. 53(1970) 214.
[4] J. A. Laird,and C. G. Bergeron, J. Am. Ceram. Soc. 53(1970) 482.
[5] S. R. Nagel and C. G. Bergeron, Advances in Nucleation and Crystallization in Glasses, eds. L. Hench and S. Freiman (American Ceramic Society, Columbus, OH 1972) 183.
[6] S. R. Nagel and C. G. Bergeron, J. Am. Ceram. Soc. 57 (1974) 129.
[7] A. J. Marlor, H. S. A. Kumar, and C. G. Bergeron, Phys. Chem. Glasses 16(1975) 108.
[8] C. J. Leedecke and C. G. Bergeron, J. Crystal Growth 32(1976) 327.
[9] S. R. Nagel, L. W. Herron, and C. G. Bergeron, J. Am. Ceram. Soc. 60(1977) 172.
[10] C. J. Leedecke and C. G. Bergeron, Phys. Chem. Glasses (to be published).
[11] C. J. Leedecke, Ph.D. Thesis, University of Illinois, Urbana 1977.
[12] H. S. A. Kumar and C. G. Bergeron, J. Crystal Growth 22(1974) 58.
[13] L. W. Herron, Ph.D. Thesis, University of Illinois, Urbana 1977.
[14] W. B. Hillig and D. Turnbull, J. Chem. Phys. 74(1956)914.
[15] H. S. A. Kumar, Ph.D. Thesis, University of Illinois, Urbana 1974.
[16] H. Eyring, S. Glasstone and K. J. Laidler in: The Theory of Rate Processes (McGraw-Hill, New York, 1941).
[17] F. L. Binsbergen, J. Crystal Growth 13/14(1972) 44.

VISCOELASTIC RELAXATION IN B_2O_3

P.B. Macedo, C.J. Montrose, C.T. Moynihan and C.C.Lai

Vitreous State Laboratory, Catholic University of America

Washington, DC 20064

INTRODUCTION

The relative ease with which its viscosity can be measured over a wide range of temperature (260-1600°C) has made B_2O_3 one of the materials most often used to test viscosity theories. It exhibits two Arrhenius regions: a high temperature region between 800-1400°C and a low temperature one, 260-350°C. In a previous publication[1] we have shown that none of the theoretical models for viscosity thus far proposed account for this behavior; not only are they inadequate in an exact quantitative sense, but they also fail to explain the general trends. This inadequacy is quite a general one, having been noted in many types of liquids including silicates, molten salts, aqueous solutions and organic liquids. Recognizing the difficulties that other investigators had encountered, our approach to trying to understand viscous flow was to ask a set of fundamental questions followed by an intensive experimental investigation. In this way we hoped that the data would naturally guide the theoretical development.

The first question we raised was with respect to the size of the liquid region involved in structural relaxation. If a liquid sample is cut in half, the viscosity of each half is still the same. How many times can the liquid be cut in half before the viscosity changes? When do the hydrodynamic equations fail? Since it is difficult to measure the viscosity of a sample with dimensions smaller than 1000A, we took advantage of phase segregation near a liquid/liquid immiscibility to subdivide the liquid into small pieces. In a series of papers which were summarized in a recent publication[2] it was demonstrated that for borosilicate glasses structural relaxation may demand cooperative motions of molecules within a region of perhaps 50A radius. There are a great many atoms on such a sphere! None of the previous theories took into account so many particles; their shortcomings are therefore not surprising.

This complexity had been suspected before, especially by those researchers who observed that in most liquids shear stress does not relax exponentially but exhibits a spectrum of relaxation times. Complicated interaction between molecules was presumed to be the origin of this behavior. Several models involving the diffusion and relaxation of order parameters were proposed, and while they provided some insights, the fundamental dynamical behavior remained a mystery. The order parameter was a convenient way to cover up our ignorance of the number of bonds, angles, number of vacancies, etc. (In fact it played a very similar role to that of the fictive temperature.)

A parameterized fit to a featureless viscosity versus temperature curve is in general not a sensitive test of a complicated theory of viscosity. However, if one requires the simultaneous fit of the temperature dependence of viscosity and the distribution of relaxation times, one can significantly increase the sensitivity of the test. The measurement of the distribution of relaxation times, that is, characterization of the structural relaxation, is not only very time-consuming, but requires the use of many techniquies. Here we will compile data taken for short times (less than 10^{-6} s) by ultrasonics; for intermediate times (10^{-6} to 1 s) by digital clipped auto-correlator light scattering; for times of order of one minute by rate heating specific heat measurements; and for long times in excess of several minutes by annealing experiments.

Even though in all cases the relaxation of the B_2O_3 structure is observed, the results of the different techniques can be compared to the observations of a group of people viewing a play from different vantage points: they will observe different details.

ULTRASONICS

By measuring the propagational parameters of shear and longitudinal waves, one can calculate the frequency dependence of the elastic constants. In a series of papers [3] measurements were made from 30KHz to 10GHz of longitudinal propagation and, over a more limited range, of shear propagation.

The ultrasonic data can be used to calculate either a compliance (e.g., compressibility) or a modulus. The data was analyzed in the modulus representation. The most remarkable feature was the shear modulus relaxational behavior between 600°C and 900°C. Here a fundamental question could be asked: Is there a relationship between non-Arrhenius behavior of the shear viscosity and the spectrum of relaxation times?

In Fig. 1, the logarithm of the shear viscosity is plotted vs. reciprocal absolute temperature. Over a 600°C interval the data fall on a straight line. This is the Arrhenius region. Below 800°C the behavior becomes non-Arrhenius.

Consider now the temperature dependence of the relaxation spectrum. Above 800°C the spectrum is single and the relaxation times τ follow an Arrhenius curve with the same slope as the viscosity. At lower temperatures the spectrum is not single and is symmetric in $\ln \tau$. A Gaussian distribution was chosen to represent the spectrum or the distribution. The distributions at various temperatures are shown in Fig. 2. Considering the uncertainty in the data, one cannot differentiate between a delta function and that Gaussian labeled 850°C (i.e., $\underline{b} > 2.50$ is effectively a single relaxation). However, the difference between the 850°C and 650°C distribution is easily observed in the moduli data. Since the function is symmetric in $\ln \tau$ but not in τ, the average time $\bar{\tau}$ is not the same as the most probable time τ'. The average times are indicated by the arrows in Fig. 2; they are displaced to longer times at lower temperatures. The average time $\bar{\tau}$ departs from the Arrhenius line in the same temperature region as the viscosity (see Fig. 1). This is expected since the viscosity is proportional to $\bar{\tau}$. The most probable time τ' falls on the Arrhenius line even in the non-Arrhenius region, indicating that even when the viscosity becomes non-Arrhenius, the average activation energy has not changed. This result leads to a rather significant conclusion concerning the apparent distribution of activation energies.

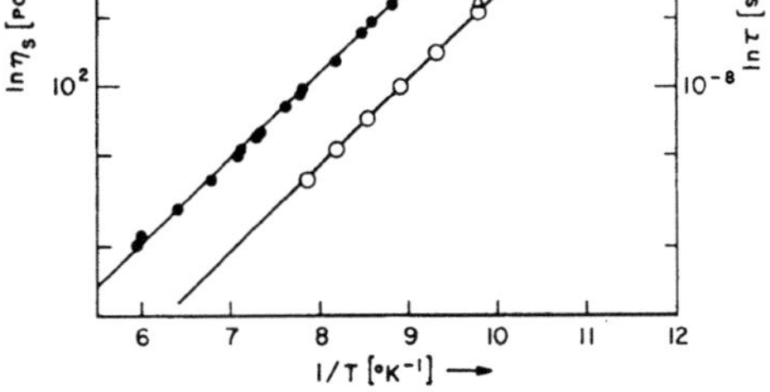

Fig. 1. Arrhenius plot of shear relaxation time and viscosity.

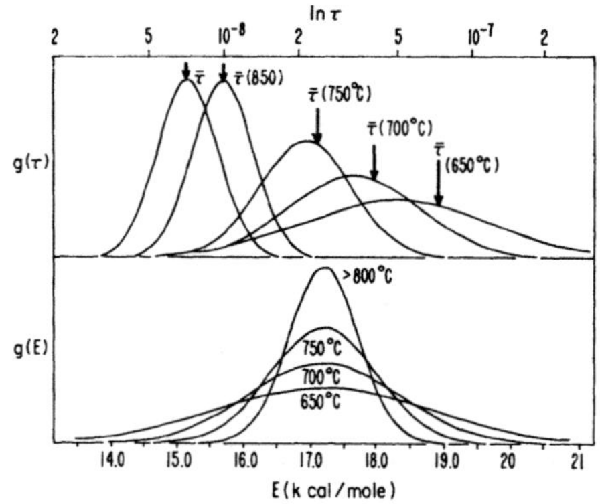

Fig. 2. Plot of the relaxation time spectrum and the activation energy distribution.

By using Eyring's rate equation

$$\tau = A \exp (E/RT) \tag{1}$$

where A is a constant, E the activation energy, T the temperature in K, and R the gas constant. Assuming that A is the same constant for all the relaxation spectrum, one can calculate the distribution of activation energies. The fraction of relaxation mechanisms with time τ_i between $\ln \tau_i - \frac{1}{2} d \ln \tau$ and $\ln \tau_i + \frac{1}{2} d \ln \tau$ is

$$g_S(\tau_i)\, d \ln \tau = g_E\,(E_i) dE \tag{2}$$

where $g_E(E_i)dE$ is the fraction of mechanisms with activation energy between $E_i - \frac{1}{2}\, dE$ and $E_i + \frac{1}{2}\, dE$. Thus, the activation energy E_i is given by

$$E_i = RT(\ln \tau_i - \ln A) \tag{3}$$

$A = -11.336$ (with units of $\ln s^{-1}$). Fig. 2 shows $g(E_i)$ for the various temperatures. Note that $g(E_i)$ is a Gaussian in E, not in ln E. Thus the average and most probable values coincide. Also the average value of the energy $\bar{E}$ is temperature independent.

Clearly the cause of the non-Arrhenius region is not a higher average activation but rather the appearance of a distribution of energies, some of which are larger than $\bar{E}$. Not only have higher activation energies appeared, but also lower ones. Any theory accounting for non-Arrhenius behavior must also account for this surprising fact.

The present theories for non-Arrhenius behavior generally depend on some increase in cooperative behavior or a loss in certain diffusional degrees of freedom as the temperature is lowered. These are typified by the work of Ree et al.,[4] Adam and Gibbs[5], and Davies and Matheson[6]. These theories lead to an increase in apparent activation energies. It is not clear how any of these can lead to smaller activation energies.

LIGHT SCATTERING

The next most pronounced feature in the temperature dependence of the viscosity is the low temperature (260-350 °C) Arrhenius region found in a subsequent study[1]. It raised the question: Is the low temperature Arrhenius region associated with single relaxation? Even though the annealing measurements of Boesch et al.[7] indicated this was not so, full interpretation of these results was only possible after the light scattering results of Lai et al.[8] became known Using a digital clipped auto-correlator, the correlation function for polarized light was measured.

Experimental correlograms ($|G_n(t)|^2$ vs t) were obtained at ten temperatures in the range 300 °C to 475 °C. For these experimental conditions it has been shown[9] that the measured $G_n(t)$ is just the isothermal structural relaxation function. As a result we expect the average relaxation time

$$<\tau> = \int_0^\infty dt\, G_n(t) \tag{4}$$

to be simply proportional to the viscosity and indeed this is observed to be the case. The more interesting aspect of the result comes in examining the actual shape of the function (typical results are shown in Fig. 3) for 10^{-6}s $\leqslant$ t $\leqslant$ 1 s vs. temperature. It was found that at each temperature the results were acceptably described by

$$G_n(t) = \exp[(-t/\tau_o)^\beta] \tag{5}$$

where τ_o and β are characteristic (temperature dependent) parameters. If this is regarded as the superposition of exponential decay functions, i.e.

$$G_n(t) = \int_0^\infty d\tau f(\tau) e^{-t/\tau}$$

then the average relaxation time $<\tau> = \int_0^\infty dt G_n(t) = \int_0^\infty d\tau f(\tau)\tau$ and the width of the distribution $<\Delta\tau^2> \equiv <\tau^2> - <\tau>^2$, where $<\tau^2> = \int d\tau\, \tau^2 f(\tau)$ can be given in terms of the parameters τ_o and β[10]

$$<\tau> = (1/\beta)\Gamma(1/\beta)\tau_o.$$

$$<\Delta\tau^2> = [(1/\beta)\Gamma(2/\beta) - (1/\beta^2)\Gamma^2(1/\beta)]\tau_o^2 \tag{6}$$

The experimental results for $<\Delta\tau^2>$ as a function of temperature are

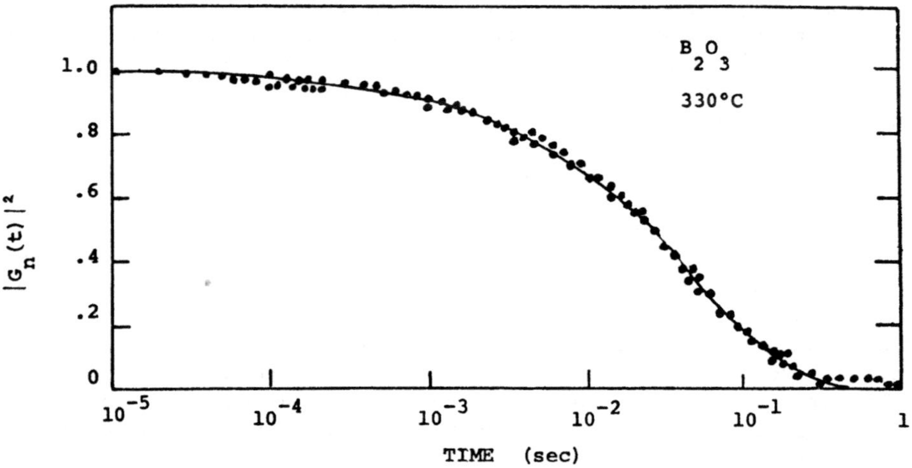

Fig. 3. A typical correlogram showing $|G_n(t)|^2$ vs. t. The data points are a superposition of 7 runs with varying bin sizes. The solid curve is calculated from Eq. 5 with τ_o= .126 sec and β= .66.

presented in Fig. 4. Also shown is the apparent activation energy E governing the temperature variation of $\langle\tau\rangle$ through an Arrhenius formula: $\langle\tau\rangle \propto \exp(E/RT)$. Clearly there is a close correlation between the two temperature dependences: when E is constant, $\langle\Delta\tau^2\rangle$ is also constant.

ANNEALING EXPERIMENTS

At the time the annealing experiment[7] was performed, it was recognized that the correlation function was related to the isothermal compliance, rather than to the adiabatic modulus as in the ultrasonic experiments. However, because of problems associated with nonlinearity, the exact shape of the correlation function was unknown. With the correlation shape as determined by light scattering we were able to derive an expression for the crossover experiment in terms of β and τ_o[11]. The spectrum width β was very sensitive to crossover depth. Thus we were able to confirm that the spectrum is temperature independent in the lower Arrhenius region.

RATE HEATING EXPERIMENTS

Now that we have measured the spectrum of relaxation times over the entire temperature range of interest, what additional questions can be asked which might throw light onto the phenomena of viscous flow? If one displaces a sample sufficiently far from equilibrium, its approach to equilibrium will be non-linear.

The non-linearity can be handled using Narayanaswamy's method[12] by replacing the reduced time $\xi = t/\tau$ with the integral

$$\xi \equiv \int_0^t dt' (\tau_e^{\ddagger}/\tau) \quad (7)$$

$$= \int_0^t dt' \exp\left[\frac{x\Delta h^*}{R}\left(\frac{1}{T^{\ddagger}} - \frac{1}{T}\right) + \frac{(1-x)\Delta h^*}{R}\left(\frac{1}{T^{\ddagger}} - \frac{1}{T_f}\right)\right] \quad (8)$$

where $\tau_e^{\ddagger}$ is the equilibrium relaxation time at reference temperature $T^{\ddagger}$, x is a non-linearity parameter, and Δh^* is the activation enthalpy for structural relaxation. DeBolt et al.[13] have shown that using the linearized ξ of Eq. (7) with correlation function of Eq. (5) and the Boltzmann superposition principle, one can properly analyze rate cooling and heating relaxation data. Fig. 5 gives the experimental specific heat for B_2O_3 at a heating rate of 10 K/min following a rate cool through the transition region at 10 K/min. This data is reduced by taking into account the temperature dependence of the glass-like and liquid-like specific heats, Cpg and Cpe. The reduced data for several cooling rates together with the computer fits using the above theory are presented in Fig. 6. The fit is excellent and very sensitive to the relaxation parameters. This data permits us to ask another question: How do the correlation functions for enthalpy, $G(\tau_H)$, and volume, $G(\tau_V)$, compare?

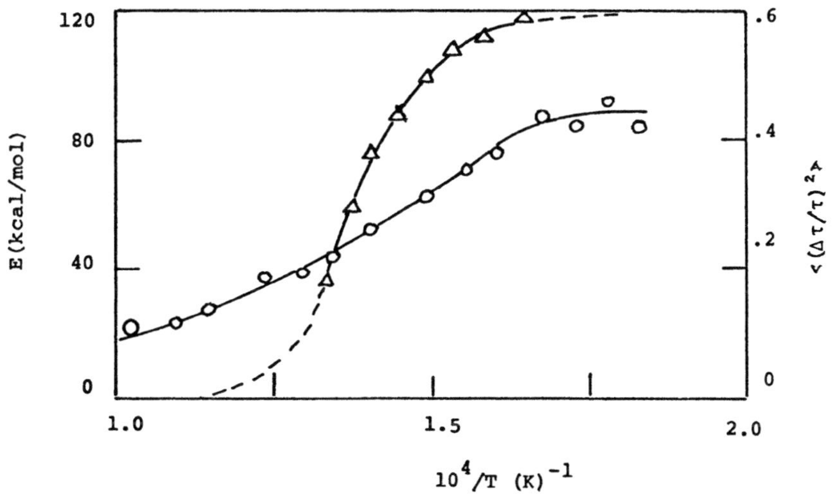

Fig. 4. The activation energy E (-o-) and the distribution width $<(\Delta\tau/\tau)^2>$ (-Δ-) plotted versus reciprocal temperature for B_2O_3.

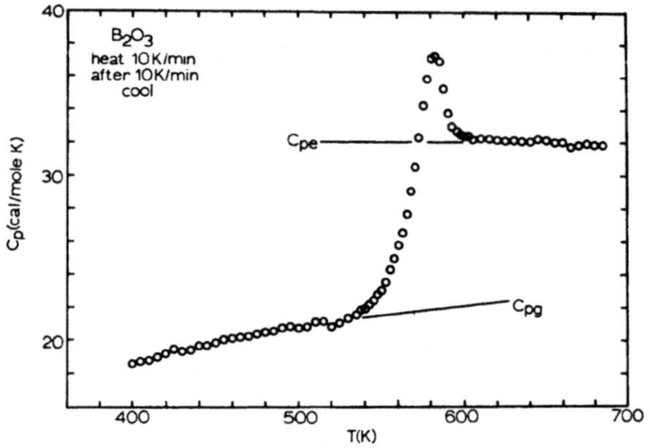

Fig. 5. Heat capacity vs. temperature for B_2O_3 at a heating rate of 10 K/min following rate cool through transition region at 10 K/min.

Note that in case of the ultrasonic longitudinal modulus the correlation function could be represented by a symmetric log Gaussian spectrum of relaxation times, while the light scattering longitudinal compliance correlation function was represented by the distribution function given in Eq. (5), which is skewed toward short times[10]. This difference was ascribed to the difference between a modulus and a compliance representation. In this case both correlation function ($G(\tau_V)$ and $G(\tau_H)$) are in the compliance representation and are both skewed. Fig. 7 shows the comparison. The correlation functions are different.

DEPOLARIZED LIGHT SCATTERING

Using the auto-correlator described above, one can measure the correlation function for light whose polarization is rotated upon scattering. This rotation of polarization is due to molecular asymmetries. As can be seen in Fig. 8, we again observe different correlation functions for the same B_2O_3 structural relaxation.

Phenomenologically this behavior may be explained assuming that the τ_i can be associated on a one-for-one basis with microscopic order order parameters, z_i, that change during structural relaxation. It may then be shown that for the correlation function $\phi_p(t-t_1,t)$, the corresponding weighting coefficients, g_{ip}, are given by[14]

$$g_{ip}=[(\partial p/\partial z_i)_T(\partial z_i/\partial T)_e]/[(\partial p/\partial T)_e-(\partial p/\partial T)_g] \tag{9}$$

where the subscripts g and e refer as usual to glass and equilibrium liquid. Thus, it follows that the correlation functions for two properties, e.g. H and V, will not be the same unless

$$(\partial H/\partial z_i)_T/(\partial H/\partial z_j)_T = (\partial V/\partial z_i)_T/(\partial V/\partial z_j)_T \qquad (10)$$

for all pairs of order parameters z_i and z_j. There is no a priori reason to expect Eq. (10) to be true. In terms of the independent-order-parameter model of structural relaxation this would be analogous to requiring that the enthalpy changes for a number of

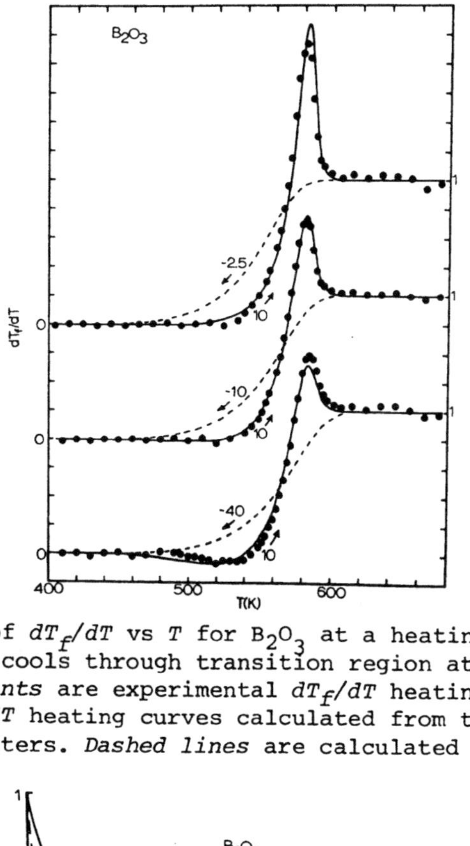

Fig. 6. Plot of dT_f/dT vs T for B_2O_3 at a heating rate of 10 K/min following rate cools through transition region at rates shown (K/min) on figure. *Points* are experimental dT_f/dT heating curves, and *solid lines* are dT_f/dT heating curves calculated from the best-fit A, Δh^*, x, and β parameters. *Dashed lines* are calculated dT_f/dT cooling curves.

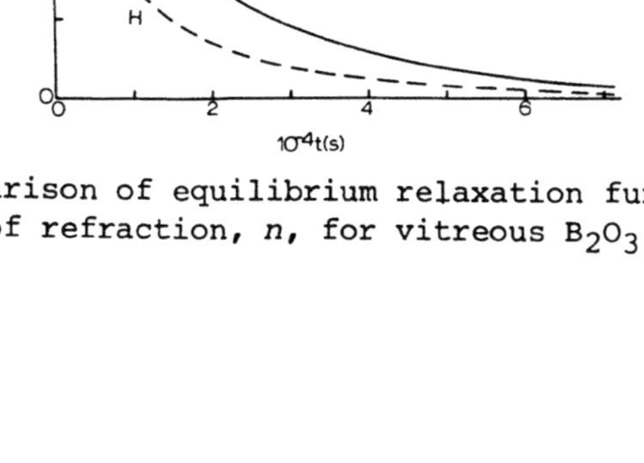

Fig. 7. Comparison of equilibrium relaxation functions for enthalpy, H, and index of refraction, n, for vitreous B_2O_3 at 536.4 K.

independent chemical reactions be proportional via the same factor to their respective volume changes.

Although convenient for purposes of a phenomenological description of structural relaxation, the microscopic order parameters implied may have no counterpart in reality. Rather, structural relaxation may involve a cooperative, inherently nonexponential re-arrangement of the average configuration of the liquid. Those properties which are more affected by the early stages of this process will then appear to relax faster than those which are more affected by the later stages. For instance, the enthalpy changes occurring during the approach to equilibrium of a network glass following an increase in temperature could be considered to arise from the breaking of some of the network bonds, whereas the volume changes result from rearrangement of the structure into less densely packed configurations. In these terms Fig. 7 might be interpreted as implying that, for B_2O_3, the bond-breaking steps occur on the average earlier than the structural rearrangements causing the volume changes. In fact if Eq. (10) held the Prigogine-Defay ratio which is

$$\Pi = \Delta C_p \, \Delta\kappa_T / T \, V \, (\Delta\alpha)^2,$$

would be equal to 1,[15] where $\Delta C_p = C_{pe} - C_{pg}$, and similarly for

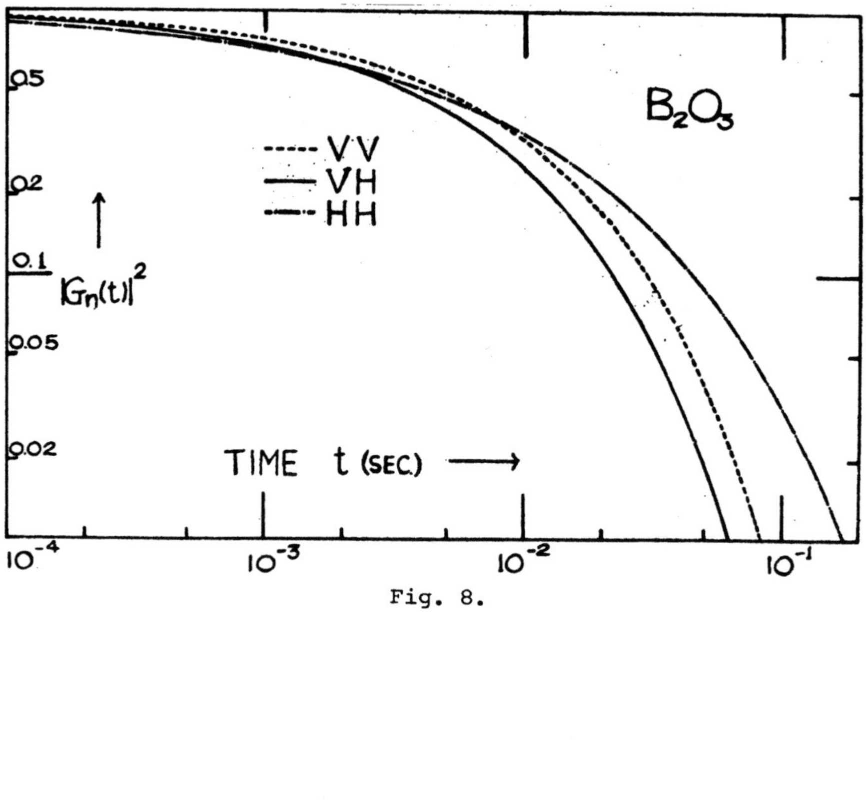

Fig. 8.

isothermal compressibility K_T and thermal expansion coefficient α. Leidecker et al.[16] have investigated Π for B_2O_3. Its temperature dependence is shown in Fig. 9. Π is not equal to one. Here the thermodynamic parameters give us an even greater insight into the differences between the volume and enthalpy correlation functions. Leidecker et al. were able to describe the thermodynamics of B_2O_3 with a three state model. In state 1 the structural elements are bonded and have the lowest volume and energy. In state 2 the bonds are broken, giving a higher energy but not much additional volume. Finally in state 3 there is a major increase in volume without much additional energy. This is consistent with the explanation given above for the fact that volume relaxation times in response to temperature changes are longer than those for enthalpy.

NON-LINEARITY PARAMETER

The non-linearity parameter x in Eq. (8) is generally found to be approximately 0.5 for network glasses[12,13].. A possible reason for this may lie in a consideration of the temperature dependence of the activation free energy for viscous flow. Many investigators have postulated models involving temperature dependent activation enthalpies, but these approaches often suffer from inadvertantly violating the thermodynamic requirement

$$\left(\frac{\partial \Delta h^*}{\partial T}\right)_P = T\left(\frac{\Delta s^*}{\partial T}\right)_P$$

That is, a temperature dependent activation enthalpy Δh^* implies also a temperature dependent activation entropy Δs^*. It is more convenient to construct a model in terms of a temperature dependent Gibbs free energy of activation. Fig. 10 illustrates such a model for a network glass such as B_2O_3. We assume here that at low temperatures, where the network is totally bonded, the activation free energy is given by an expression of the form

$$g = h_1 - Ts_1 \tag{11}$$

At high temperatures, where the network is totally unbonded, the activation free energy is

$$g = h_2 - Ts_2 \tag{12}$$

At intermediate temperatures the two straight lines implied by Eqs. (11) and (12) must join smoothly, as shown in the figure. Note that the apparent activation enthalpy Δh^* at any temperature is the intercept with the g axis of a tangent drawn to the g vs. T curve. Hence in the region of the inflection point of the curve in Fig.10 the equilibrium liquid viscosity will exhibit an unusually high apparent activation enthalpy, which we have designated Δh^*_e, and will

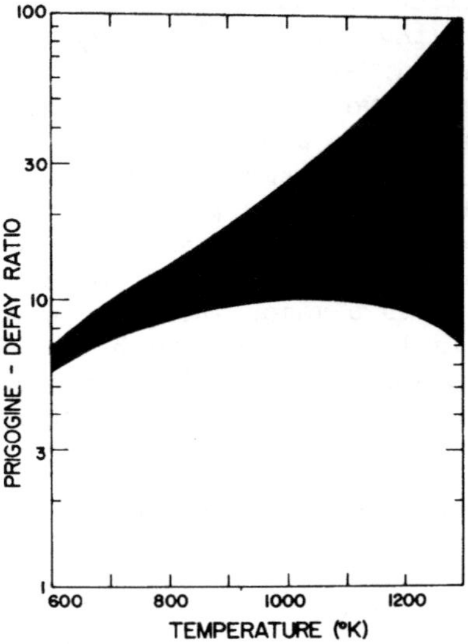

Fig 9. Prigogine-Defay ratio plotted against temperature for B_2O_3. The shaded region shows the expected uncertainty resulting from experimental error and the need to extrapolate ΔC_p. The dashed curve is calculated from the proposed three-state model.

also appear to be Arrhenius. For B_2O_3 we presume that this region lies in the glass transition range around temperature T_g.

The simplest analytical curve that has the approximate shape of the g vs T curve in Fig. 10 is a Lorentzian:

$$g = \frac{A}{1+(\frac{T_f}{T_g})^2} - BT \tag{13}$$

The first term expresses the structure dependence of g and is therefore a function of the fictive temperature, T_f. The second term is the normal temperature dependence of a free energy contained in the entropy term. We may now calculate the activation enthalpy as a function of T and T_f:

$$\Delta h^* = g - T(\frac{\partial g}{\partial T})_P \tag{14}$$

$$= A[1+(\frac{T_f}{Tg^2})^2 + 2(\frac{T_f}{Tg})^2\frac{dT_f}{dT}]/[1+(\frac{T_f}{Tg})^2]^2$$

For the equilibrium liquid the condition $dT_f/dT = 1$ applies, so that at temperature Tg the apparent activation enthalpy is

$$\Delta h^*_e = A$$

For the glass the condition $dT_f/dT = 0$ applies, that is, the struc-

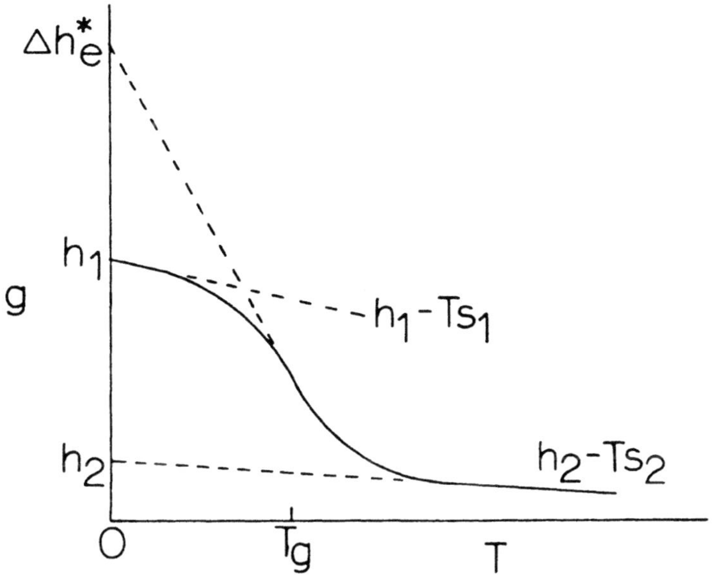

Fig. 10. Temperature dependence of activation free energy g of a network glass.

ture measured in terms of T_f is constant, so that at temperature T_g the apparent activation enthalpy is

$$\Delta h^*_g = A/2$$

In the glass transition region the non-linearity parameter x, via Eq. (8), is

$$x = \Delta h^*_g / \Delta h^*_e = 0.5,$$

in agreement with experiment.

* * * * *

Acknowledgement. The research described herein was supported by contracts from the Office of Naval Research.

REFERENCES

1. P.B. Macedo and A. Napolitano, J. Chem. Phys., 49, 1887 (1968).
2. R. Mahoney, G.R. Srinivasan, P.B. Macedo, A. Napolitano, and J.H. Simmons, Phys. Chem. Glasses, 15, 24 (1974).
3. J. Tauke, T.A. Litovitz, and P.B. Macedo, J. Am. Ceram. Soc., 51, 158 (1968); P.B. Macedo and T. A. Litovitz, Phys. Chem. Glasses, 6, 69 (1965).
4. T.S. Ree, T. Ree, and H. Eyring, Proc. Natl. Acad. Sci. U.S., 48, 501 (1962).
5. G. Adam and J.H. Gibbs, J. Chem. Phys., 43, 139 (1965).
6. D.B. Davies and A.J. Matheson, J. Chem. Phys., 45, 1000 (1966).
7. L. Boesch, A. Napolitano, and P.B. Macedo, J. Am. Ceram. Soc., 53, 148 (1970).
8. P.B. Macedo, C.J. Montrose, and C.C. Lai, Proc. 10th Int. Cong. on Glass, 11-68, 1974.
9. C. Demoulin, C.J. Montrose and N. Ostrowsky, Phys. Rev. A., 9, 1740 (1974).
10. C.T. Moynihan, L.P. Boesch, and N.L. Laberge, Phys. Chem. Glasses, 14, 122 (1973)
11. C.T. Moynihan et.al., Ann. N.Y. Acad. Sci., 279, 15 (1976).
12. O.S. Narayanaswamy, J. Am. Ceram. Soc., 54, 491 (1971).
13. M.A. DeBolt, A.J. Easteal, P.B. Macedo, and C.T. Moynihan, J. Am. Ceram. Soc., 59, 16 (1976).
14. C.T. Moynihan, A.J. Easteal, M.A. DeBolt, and J. Tucker, J. Am. Ceram. Soc., 59, 12 (1976).
15. P.K. Gupta and C.T. Moynihan, J. Chem. Phys., 65, 4136 (1976).
16. H.W. Leidecker, J.H. Simmons, T.A. Litovitz, and P.B. Macedo, J. Chem. Phys., 55, 2028 (1971).

GLASS FORMATION IN BORATE SYSTEMS

L. D. Pye, V. D. Fréchette and D. E. Rase

New York State College of Ceramics, Alfred University

Alfred, New York

INTRODUCTION AND OVERVIEW

Glass formation achieved by quenching a melt is a fundamental material property associated with a number of independent variables: composition, melt volume, type of quench, quench temperature and melt history[1]. Collectively, these variables combine to yield a quench rate which may or may not be sufficient for a given melt to suppress crystallization when cooled, and thus transform to the vitreous state through a reversible increase of viscosity. When assessing a tendency towards glass formation, two related factors are a) one's method (capability) of detecting crystallization and b) how one chooses to define a glass. The first, of course, is an experimental description; regarding the second, an appropriate working definition for the present text is "a transparent or opaque solid composed of a single or multiple non-crystalline phases(s)".

It will be shown below that the adoption of this definition* allows one to treat glass formation as a sub-topic in the overall subject area of phase transformation. This allows assessment of glass formation in terms of such transformations as melting, liquid-liquid immiscibility, and of course, the glass transition itself. In turn, the related concepts of phase equibria, free energy

*The "working" nature of this definition is emphasized. For a more complete discussion of this matter see "The Vitreous State; Definition and Nomenclature", L. D. Pye (to be published).

differences, driving forces and energy barriers may be invoked as required.

The present work attempts to discuss glass formation in borate systems from the above viewpoint. To this end, glass formation of pure B_2O_3 is discussed first, followed by illustrations of the glass transition in multicomponent borate system where both stable and metastable liquid immiscibility have been reported. Finally, new results on glass formation in the GeO_2-B_2O_3-SiO_2 system are presented.

THE GLASS TRANSITION IN PURE B_2O_3

It is generally recognized that the Gibbs free energy, G, of any phase of a system may be described by

$$G = H - TS \qquad (1)$$

where H, T, and S have their usual meaning. Because of the dominance of the (TS) term, it is expected that G will fall with increasing temperature and indeed this is the case[2]. Noting that the free energy difference between two phases is zero at their coexistence temperature, it is relatively easy to apply this equation in a schematic manner to a glass forming compound such as B_2O_3. For this compound, only one stable crystalline phase, α - B_2O_3, is known to exist below its melting point ($T_m \sim 460^oC$) at one atmosphere. Assuming a glass transition temperature, T_g, of about 280^oC, we are led to Fig. 1. Note that the free energy curves are continuous through T_m; thus the direct transformation (1) at 490^oC from α - B_2O_3 to liquid B_2O_3 is indeed possible. This is sometimes referred to as superheating. In a similar manner, transformation (2) represents crystallization of liquid B_2O_3; its absence prior to the transition to the vitreous state at T_g, is indicative of supercooling. In this approach, the vitreous state is regarded as having a somewhat higher free energy than the liquid state below T_g.

The temperature interval, $\Delta T \sim 180^oC$ is a fundamental property of liquid B_2O_3 as is the ratio $T_g/T_m \sim 2/3$ (see reference 5). ΔT represents that range of temperatures where 1) there is a negative free energy difference between liquid B_2O_3 and α - B_2O_3, e.g. a driving force exists for a phase change, and 2) there is sufficient atom mobility to allow the transformation to occur. Below T_g, this transformation is not anticipated because of low atom mobility even though a driving force is present. In the above sense,

B_2O_3 is similar to other prominent glass forming oxides such as SiO_2 and GeO_2 in that all have a measurable T_g, T_m and therefore a ΔT. However, unlike SiO_2 and GeO_2, transformation (2) of Fig. 1 has never, to the authors' knowledge, been observed for pure B_2O_3 at one atmosphere. Thus, this compound exhibits a strong tendency towards glass formation.

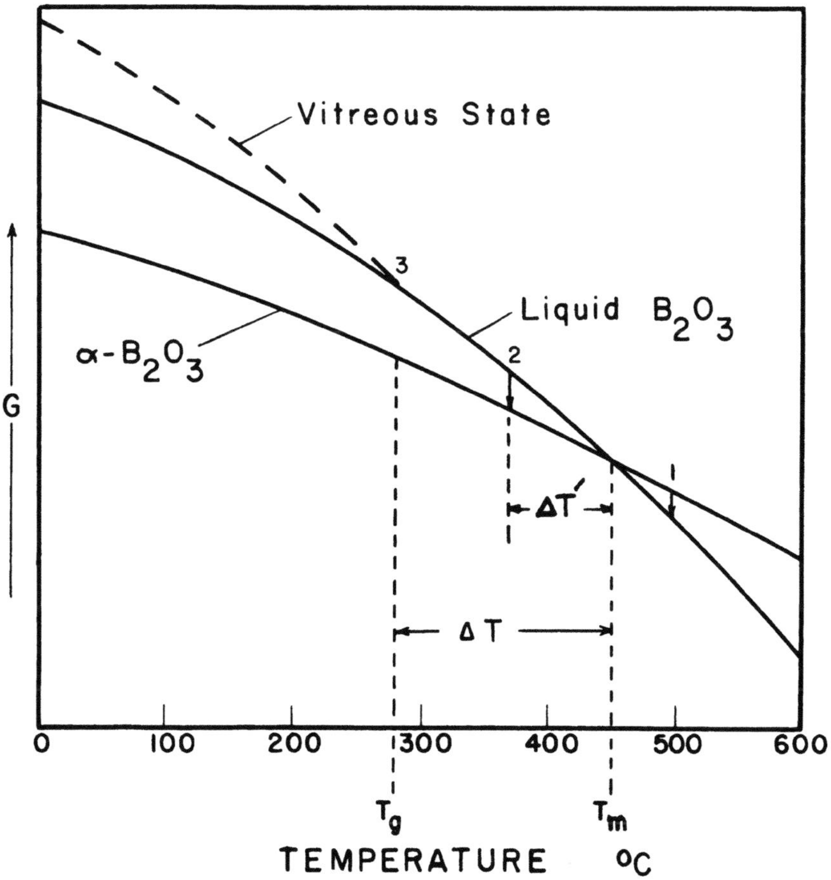

Figure 1. Hypothetical free energy-temperature diagram for B_2O_3. T_g is strongly dependent on water content (ref. 3). Transformation (1) was observed by Mackenzie and Clausen for α-B_2O_3 that had been prepared by high pressure methods (ref. 4).

In view of the emphasis of these proceedings on borate glasses, it is appropriate to examine in some detail the various properties of several oxides which influence transformation (2) in hopes of establishing why it has never been observed in B_2O_3. Two lines of examination are possible: 1) a kinetic inquiry and 2) a crystal-chemical examination of this phase change. The latter is regarded as somewhat more general and is therefore emphasized below. To this end, a variety of information has been assembled for B_2O_3, GeO_2, and SiO_2 (Table I). These compounds all form glasses; however, both GeO_2 and SiO_2 exhibit transformation (2). Thus a comparison of the properties of these oxides which influence the crystallization of their melts is especially valuable. In the way of normalizing this comparison, a potential melt crystallization temperature, T_c, was calculated for each oxide and is defined as

$$T_c = \frac{T_m + T_g}{2} \quad (2)$$

Thus, T_c is located one-half way between the respective glass transition and melting temperature. In Table I, the properties of the liquid phases for each oxide correspond to T_c. The driving force for this transformation was calculated for each compound using the Hoffman[6] relationship:

$$\Delta G = \frac{\Delta H \cdot \Delta T' \cdot T_c}{T_m^2} \quad (3)$$

This driving force represents the volume free energy difference between the liquid and crystalline phases of these oxides at T_c, and it is this quantity which appears in the kinetic equations describing this transformation[7]. $\Delta T'$ is defined as $T_m - T_c$ (see Fig. 1).

In perusing the data of Table I, it is evident that the driving force for liquid B_2O_3 to crystallize at T_c is slightly greater than GeO_2 and about four times larger than SiO_2. On this basis, crystallization of liquid B_2O_3 should occur much more readily than liquid SiO_2. Yet, as noted above, the converse is observed experimentally. From the viscosity data of Table I, and assuming the usual inverse relationship between this property and crystal growth, one would again expect that liquid B_2O_3 should crystallize

TABLE I†

ITEM	OXIDE: B_2O_3	GeO_2	SiO_2 ††
1. T_m	$\sim 460^oC$	1115^oC	1710^oC
2. T_g	~ 280	550	1170
3. T_c	~ 370	835	1440
4. ΔT	· 90	280	270
5. ΔG(Kcal/mole)	0.57	0.42	0.14
6. LIQUID PHASE			
a. Viscosity @ T_c ($Nsec/m^2$)	$\sim 6 \times 10^6$	$\sim 1.0 \times 10^7$	$\sim 1.0 \times 10^9$
b. Activation Energy for viscous flow (Kcal/mole)	78	90	125
c. Thermal Expansion coefficient ($X\ 10^7/^oC$	~ 380	~ 300	~ 10
7. BOND STRENGTH (Kcal/mole) e.g. B-O, Ge-O, Si-O	100	108	106
8. % DENSITY INCREASE	+ 33	+ 28	+ 5.0

† Data compiled from ref. 5-14.

†† Vitreosil

readily. Once more, this is not the case. Moreover, the very high thermal expansion coefficient of liquid B_2O_3 implies a relatively open-structured liquid capable of the high degree of atomic rearrangement requisite to crystallization. Again, the direction of increasing expansion coefficient does not correlate well with the ease in which these oxides crystallize. From all of these viewpoints, probably the most fundamental anomaly encountered in borate glasses is the reluctance of B_2O_3 to crystallize below its melting temperature.

To explain these observations, we are left to consider some of the remaining factors of Table I: Bond strength, % density increase during crystallization, and the relative values of T_c. It is seen that the density increase is greatest for B_2O_3 (albeit only slightly larger than GeO_2) and is suggestive of a larger atomic re-arrangement during crystallization than that encountered for SiO_2. At the same time, the respective cation-oxygen bond strengths are nearly the same for all compounds. The crucial factor, then, underlying crystallization in these compounds may be the amount of energy available to break the cation-oxygen bond and form the more dense crystalline phase. A measure of this energy is the value of T_c for each compound. For B_2O_3, T_c is about 40% of that for GeO_2 and about 25% that of SiO_2. Therefore, the amount of energy at T_c for liquid B_2O_3 is significantly less than that for GeO_2 and SiO_2. Hence the reluctance of liquid B_2O_3 to re-arrange itself into a crystalline form of higher density is somewhat more understandable. This relationship between low melting (crystallization) temperatures and ease of glass formation among various compounds (including B_2O_3) was first suggested by Rawson[15]. It serves to re-enforce a long-held experimental observation: to increase glass stability, decrease the thermodynamic melting temperature.

The significance of a low crystallization temperature was also recognized by Turnbull[16]. The latter author suggested that the glass forming tendency is greater whenever a reduced thermodynamic crystallization temperature, τ_m, is small. This quantity is defined as

$$\tau_m = \frac{K \cdot T_c}{H_v} \tag{4}$$

where K is the Boltzman constant, and H_v is the molecular heat of vaporization. This equation is based on the theory of corresponding states.[17, 18]

For GeO_2, B_2O_3 and SiO_2 it is expected that H_v should be about the same for all of these compounds in view of their similar cation-oxygen bond strengths. With this assumption, τ_m should be snallest for B_2O_3, intermediate for GeO_2, and largest for SiO_2, that is, τ_m should vary in direct relation to T_c. Thus B_2O_3 should exhibit the strongest tendency towards glass formation or conversely, the weakest inclination towards crystallization. Alternatively, T_c in equation (4) can be replaced by $\Delta H / \Delta S$ (the molar heats of fusion and entropy) providing these quantities are nearly independent of temperature. That is:

$$\gamma_m = \frac{K \Delta H}{H_v \Delta S} \tag{5}$$

Here γ_m has real meaning since it may be interpreted in terms of bond energies and atomic rearrangement associated with crystallization. Turnbull and Cohen[19] have emphasized that these quantities encompass the Zachariasen[20] conditions for glass formation.

GLASS TRANSITION IN MULTI-COMPONENT BORATES

In this section, we wish to examine glass formation in multicomponent borate systems. In so doing, an additional difficulty is encountered in that many borate melts tend to decompose into two stable liquids when cooled from a single liquid region. This tendency represents a potential for another phase transformation to take place in addition to that of crystallization. It is important that this phenomenon be understood in relationship to glass formation and glass transition. Accordingly, Figures 2 and 4 have been constructed and are discussed in some detail below.

In Figure 2, the information presented is for a specific composition, C_o, in a binary system. As in Fig. 1, the free energy curves are continuous through the melting and liquidus temperatures. Here, transformation (1) represents a partial crystallization of a single phase liquid. Transformation (2) again represents partial crystallization only in this case, a metastable two-liquid melt structure is by-passed. That this is not always required is illustrated by transformation (3). To achieve a pure vitreous state, e. g. the transformation of a single liquid phase into the vitreous state at (4), it is clear that both transformation (2) and (3) must be by-passed. For other compositions in this binary system, the intersection of the two-liquid curve with the single liquid curve takes place below the glass transition temperature. For this case, the development of two liquids is not expected. Therefore glass formation for these compositions is concerned only with by-passing transformation (1).

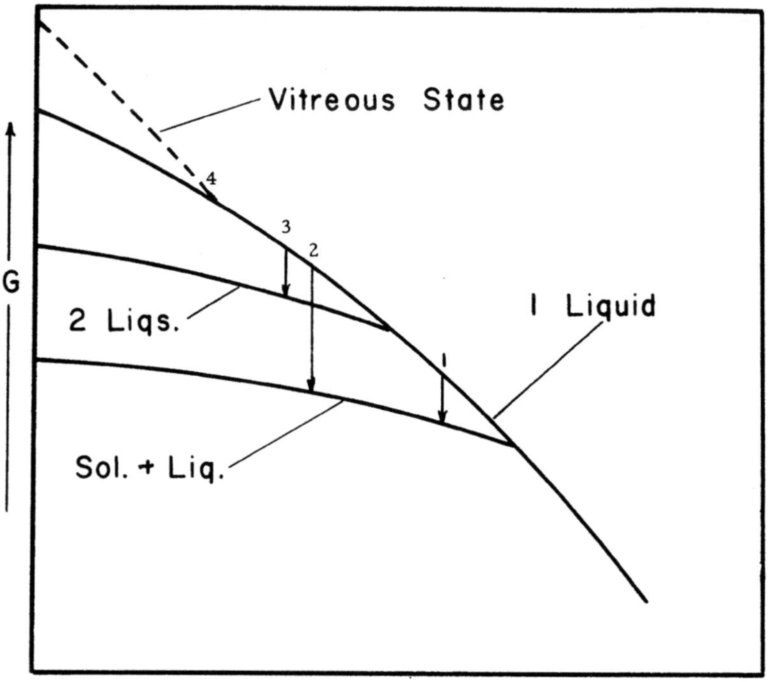

Figure 2. Hypothetical free energy-temperature diagram for a potential glass forming composition, C_o, in a binary or ternary system with metastable liquid immiscibility.

This state of affairs can be illustrated by a number of binary and ternary borate systems. As representative of the former, we turn our attention to the Na_2O - B_2O_3 system. Here glass formation has been reported over the range of 0 - 40% Na_2O in one study[21]. From another it is reported that a metastable miscibility gap intersects the glass transition-composition curve at approximately 8 and 24 % Na_2O[22]. From the above discussion we conclude that over the composition range 8-24% Na_2O, the formation of a single phase glass can be obtained only through the by-passing of transformation (2) and (3) of Fig. 2. Similarly, glass formation in the ranges of 0-8% and 24-40% Na_2O is achieved by by-passing the development of a crystalline phase such as is described in Fig. 1. In this system, the development of a discernible amorphous microstructure is only possible in the range of 8-24% Na_2O.

An important ternary system showing similar behavior is the Na_2O - B_2O_3 - SiO_2 system. Figure 3 shows the limits of glass formation in this system as reported by Abe[23] and the estimated region of metastable liquid immiscibility as reported by Haller, et al[24]. Although the glass transition temperatures are not known for all glass forming compositions outlined in this diagram, to a first approximation homogeneous glasses are most likely produced in the clear area between limits of glass formation and the boundaries of the immiscibility regions. That is, glasses in such areas are unstable only with respect to a crystalline phase during their formation. As before, single phase glasses prepared in the region showing metastable liquid immiscibility, are formed only through by-passing transformations (2) and (3) of Fig. 2.

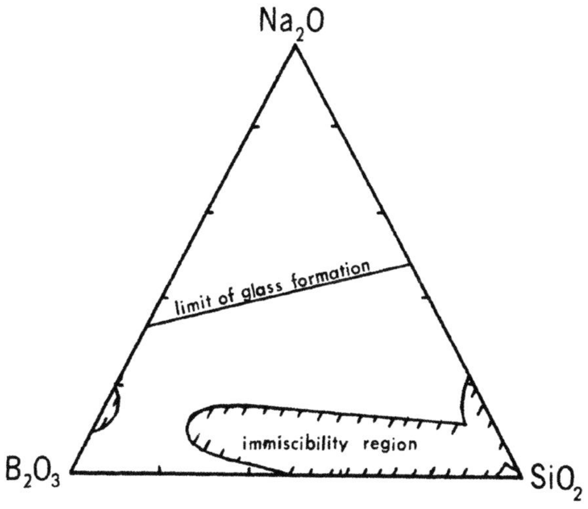

Figure 3. Limits of glass formation and metastable liquid immiscibility in the Na_2O-B_2O_3-SiO_2 system (from ref. 23 and 24).

A slightly different situation is encountered when the two-liquid region in a binary system appears as part of the stable equilibrium phase diagram. This condition is described in Figure 4. Note that transformation (1) again represents the decomposition of a single phase liquid into two, stable liquids. If transformation (1) is by-passed, this decomposition could occur at (2) followed by a conversion to a partly crystalline melt, e. g. transformation (4).

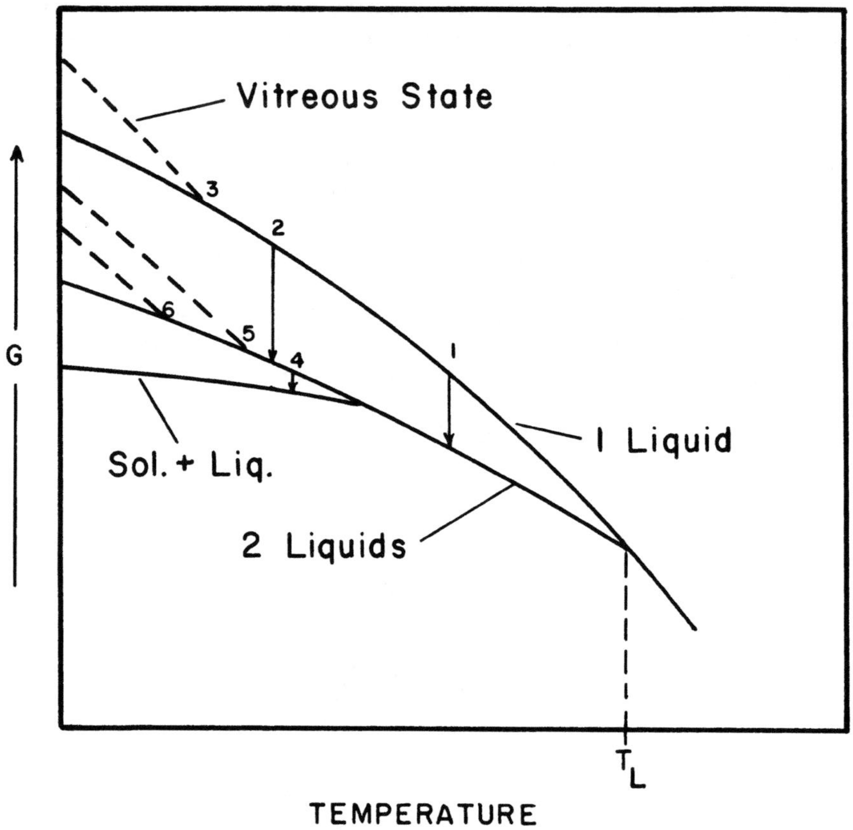

Figure 4. Hypothetical free energy-temperature diagram for a potential glass forming composition, C_O, in a binary or ternary system with stable liquid immiscibility.

Alternatively, the single liquid could transform into a solid and liquid directly via transformation (2) in a manner similar to that described in Fig. 2. However, should two liquids form first, as in transformation (2), then each liquid can in theory be cooled and converted to their respective vitreous states at (5) and (6). Again, a pure vitreous state, e.g., a completely single phase glass, could be achieved for this composition only if both transformations (1) and (2) are by-passed. Thus there are strong similarities between Figures 2 and 4. Indeed, metastable two-liquid formation is always encountered in systems where the equilibrium phase diagram contains a region of stable liquid immiscibility. The converse is not true.

There are, of course, many borate systems that possess a region of stable liquid immiscibility (see references 24-27). Most are characterized by a region of glass formation for compositions rich in B_2O_3. One of the best illustrations of glass formation in such instances is to be found in the PbO - SiO_2 - B_2O_3 system. Here, quenched melts whose compositions lie within the known immiscibility region for this system are entirely non-crystalline. However, it is observed that transformation (1) of Figure 4 nearly always takes place during quenching. This is revealed by the opaqueness of these glasses (see Figure 5) which arises from the presence of a clearly discernible amorphous microstructure. In fact, it is extremely difficult to prevent transformation (1) of Fig. 4 from occurring. On this basis, glass formation in this system is clearly much more involved then simply saying that crystallization has failed to occur.

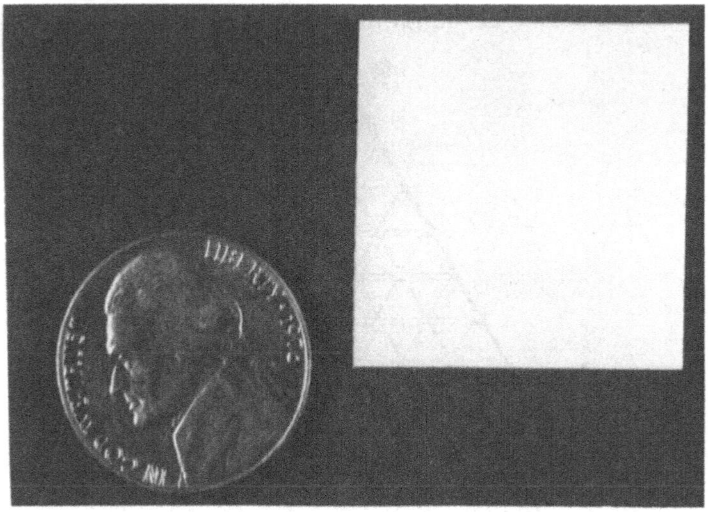

Figure 5. Illustration of a non-crystalline, opaque solid developed through a transformation of a single phase liquid into two liquids during quenching (see transformation (2) in Fig. 4. Composition of this solid - 50% PbO (weight), 30% B_2O_3 and 20% SiO_2.

GLASS FORMATION IN THE B_2O_3-SiO_2-GeO_2 SYSTEM

With the exception of P_2O_5, extensive glass formation has been reported whenever B_2O_3 is joined with other glass forming oxides, e. g. SiO_2, GeO_2, Sb_2O_3 and TeO_2. Because of the fairly intense commercial interest in glasses containing B_2O_3, GeO_2 and SiO_2, an investigation to determine the area of glass formation in this ternary system was undertaken. About 5 gram batches within this system were prepared mixing the dried oxides of each component. About 3 mg of each batch was sealed in platinum capsules to prevent volatilization during melting. All compositions studied were melted in a resistance furnace for two hours at 1450°C ∓ 15° C. The capsules were then quenched to room temperature and their contents examined by optical microscopy techniques.

Glasses formed easily from liquids quenched according to the above. At 1450°C, the system is characterized by extensive liquid immiscibility including the binary system GeO_2-SiO_2. Immiscibility was not observed in the bounding GeO_2-B_2O_3 binary. The refractive indices varied from 1.46 to 1.60. The stable coexistence of crystalline phases was not observed in any of the compositions studied. Our results are summarized in Fig. 6. Current research efforts are designed to complete the definition of phase equilibrium in the entire condensed ternary system.

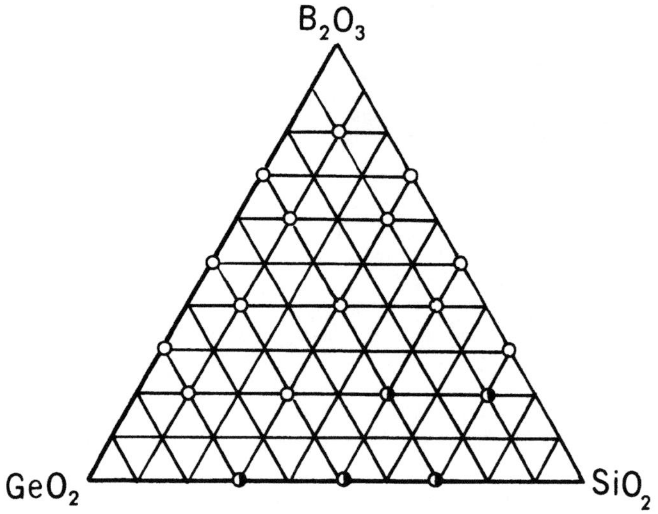

Figure 6. Glass forming compositions in the B_2O_3-GeO_2-SiO_2 system. Single phase glasses - O , bi-phase glasses - ◐ .

SUMMARY

From the foregoing, it is apparent that glass formation forms an integral part of the large subject area of phase transformation. The reluctance of pure B_2O_3 to crystallize compared to GeO_2 and SiO_2, seems to correlate best with its low potential crystallization temperature, T_c. In multi-component systems, compositional regions can be identified where formation of a single phase glass is possible. Similarly, for other compositions, glass formation can be achieved only with a concomittant development of an amorphous microstructure.

ACKNOWLEDGEMENTS

The authors acknowledge the help of Ms. Marina Pascucci in the preparation of the batches used in the glass formation studies.

REFERENCES

1. Pye, L. D., Introduction to Glass Science, Plenum Press, New York, 1972, 1-39.
2. Pye, L. D., Treatise on Materials Science and Technology, Vol. II, Academic Press, New York, 1977, 151-94.
3. Poch, V. S., Glastechn. Ber. 37 (1964) 533.
4. MacKenzie, J. D., Claussen, W. F., J. Am. Ceram. Soc. 44 (1961) 79.
5. Sakka, S., MacKenzie, J. D., J. Non-Cryst. Solids, 6 (1971) 145.
6. Hoffman, J. D., J. Chem. Phys., 29 (1958) 1192.
7. Uhlmann, D. R., Hays, J. F., Turnbull, D., Phys. Chem. Glasses 8 (1967) 1.
8. MacKenzie, J. D., J. Chem. Phys. 25 (1956) 88.
9. Macedo, P. B., Napolitano, A., J. Chem. Phys. 49 (1968) 1887.
10. Napolitano, A., Macedo, P. B., Hawkins, E. G., J. Am. Ceram. Soc. 48 (1965) 613.
11. Napolitano, A., Macedo, P. B., J. Res. NBS 72A (1968) 425.
12. Sun, K. H., J. Am. Ceram. Soc. 30 (1947) 277.
13. Fontana, E. H., Plummer, W. A., Phys. Chem. Glasses, 7 (1966) 139.

14. Brückner, R., J. Non-Cryst. Solids, 5 (1970) 123.
15. H. Rawson, Proc. IV Int. Cong. on Glass (1957) 62.
16. Turnbull, D., Symposium on Nucleation and Crystallization, The Am. Ceram. Soc. (1962) 75.
17. Guggenheim, E. A., J. Chem. Phys., 13 (1945) 253.
18. De Boer, J., Physica, 14 (1948) 139.
19. Turnbull, D. and Cohen, M. H., Modern Aspects of the Vitreous State, I, Butterworth Inc., Washington D. C., 1960, 75.
20. Zachariasen, W. H., J. Am. Chem. Soc., 58 (1932) 3841.
21. Becherescu, V. D., Cristea, V. and Menessy, I., Glastechr. Ber. 42 (1969) 91.
22. Shaw, R. R. and Uhlmann, D. R., J. Am. Ceram. Soc., 51 (1968) 377.
23. Abe, T., J. Am. Ceram. Soc., 35 (1952) 284.
24. Haller, W., Blackburn, D. H. Wagstaff, F. E. and Charles, R. J., J. Am. Ceram. Soc., 53 (1970) 34.
25. Levin, E. M., Physics Chem. Glasses, 7 (1966) 92.
26. Levin, E. M. and Block, S., J. Am. Ceram. Soc., 40 (1957) 95.
27. Shartsis, L., Shermer, H. F. and Bestul, A. B., J. Am. Ceram. Soc., 41 (1958) 507.

ELECTRICAL AND DIELECTRIC PROPERTIES OF BORATE GLASSES

R. K. MacCrone

Rensselaer Polytechnic Institute

Troy, New York 12181

ABSTRACT

Some common features of the "well known" dielectric relaxation in oxide glasses are discussed.

A structural model of glasses which may be generally appropriate to account for electric properties is suggested. Indirect experimental evidence for the model is briefly described.

In this paper we will be only concerned with the first relaxation that occurs below the softening temperature in all glasses containing a single modifying ion; that is, the well known dielectric relaxation due to e.g. sodium ion migration. This relaxation is also mechanically active, and gives rise to an analogous mechanical relaxation. We will not discuss relaxations due to surface polarization and electrode effects, or the mixed alkali effect.

INTRODUCTION

A convenient starting point for discussion are the well known Debye (1) equations

$$\varepsilon' = \varepsilon_\infty + (\varepsilon_s - \varepsilon_\infty) \frac{1}{1+\omega^2\tau^2} \qquad (1)$$

and

$$\varepsilon'' = (\varepsilon_s - \varepsilon_\infty) \frac{1}{1+\omega^2\tau^2} \qquad (2)$$

where $\varepsilon^* = \varepsilon' + i\varepsilon''$ is the relative complex dielectric constant.

These equations describe the electrical response of a dilute assembly of dipoles embedded in an insulating medium, whose dynamical response is governed by a single relaxation time τ. Specifically, the relaxation time τ describes the rate of dipolar orientation following the application of a dc electric field at time $t = 0$:

$$P(t) = P(\infty)\,(1-e^{-t/\tau}) = P(\infty)\,\phi_p(t) \qquad (3)$$

The quantity $(\varepsilon_s-\varepsilon_u)$ is called the static limiting relaxation strength and is given in a first approximation by an equation of the form

$$\varepsilon_s-\varepsilon_\infty = \frac{C}{kT} \qquad (4)$$

The constant C contains parameters describing the concentration and magnitude of the elementary dipoles and also takes into account the local electric field at the dipole site in terms of the average macroscopic electric field E.

Equations (1) and (2) result from the definition of the dielectric constant $\varepsilon^* = D/E$. Alternatively, one may define the electric modulus $M^* = E/D$. The choice of ε or M to represent a system is entirely arbitrary and may be chosen for convenience. The components of ε and M are related

$$\varepsilon' = \frac{M'^2}{M'^2 + M''^2} \qquad (5)$$

$$\varepsilon'' = \frac{M''}{M'^2 + M''^2} \qquad (6)$$

and conversely.

The observed relaxation in the glass is much wider than predicted from equation (2). Thus, it is customary to write

$$\varepsilon'' = (\varepsilon_s-\varepsilon_\infty)\int_0^\infty \frac{N(\tau)\ \omega\tau\ d\tau}{1+\omega^2\tau^2} \qquad (7)$$

where $N(\tau)$ is the relaxation time distribution function subject to the condition $\int_0^\infty N(\tau)\,d\tau = 1$. The details of the distribution function $N(\tau)$ reflect the processes taking place within the glass. It is thus expected to be a function of glass composition, thermodynamic parameters, temperature, and other variables which are involved in the relaxations.

THE RELAXATION TIME DISTRIBUTION FUNCTION

Although a wide distribution of relaxation times is involved in $N(\tau)$, it turns out that there are many features of $N(\tau)$ which are common to many glass systems, provided the mole fraction of the migrating species exceeds $\sim$ 15 mole %. We describe some of the features of $N(\tau)$ appropriate to glasses containing such a "high" concentration of mobile species.

(i) Temperature Dependence

The first variable to consider is that of temperature, and to inquire what inferences may be drawn from the temperature dependence of the distribution function, or alternatively the temperature dependence of the dispersion.

A relaxation whose width decreases with increasing temperature is characteristic of a process whose distribution or relaxation times is due to a distribution of activation energies. Conversely, a relaxation whose width is temperature independent is characteristic of a process whose distribution of relaxation times is due to a distribution in the pre-exponential factor τ_o only, and not in the activation energy Q.

This inference is based on the assumption that the relaxation times are thermally activated. Given each τ to be of the form $\tau_i = \tau_o \, exp \, Q_i/kT$, then as $T \to \infty$ $\tau \to \tau_o$: that is τ_i will tend to a value independent of the quantity Q_i. If τ_o is the same for all the dipolar entities, one expects $N(\tau)$ to narrow with increasing temperature (to become a δ-function at $T = \infty$) and the width of the relaxation to narrow accordingly.

For the relaxations of interest in glasses, there is abundant evidence to indicate that the distribution of relaxation times is not due to a distribution in activation energy, as in the example above, but mainly due to a distribution in the pre-exponential factor τ_o.

This is illustrated in Fig. 1, taken from the paper by Charles, (2) in which is superimposed some measurements of

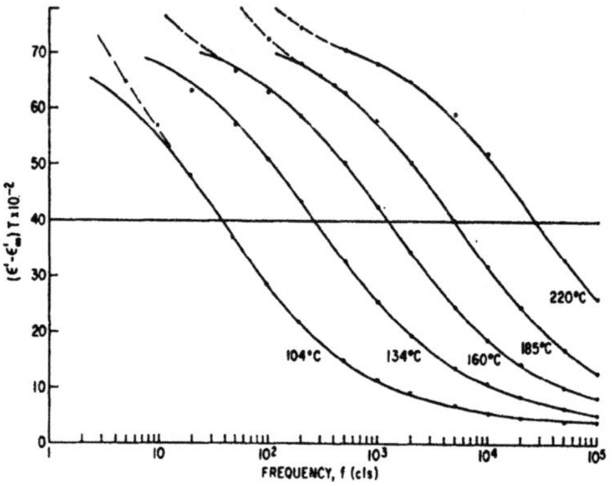

Fig. 1. Temperature-corrected dispersion of the dielectric constant for a lime glass.

Taylor.(3) The temperature independence of the maximum value of $d\varepsilon'/d\omega$ indicates very little distribution in the value of the activation energy. Heroux (4) also found no evidence for a distribution in relaxation times in lithium borate glasses. The independence of the relaxation time distribution function on energy has been investigated numerically by Boesch and Moynihan,(5) with the conclusion that only a small distribution of activation energies is involved in a wide distribution of relaxation times.

The homologous anelastic dispersions induced by stress have been shown by Moore and Day (6) to behave similarly as far as the temperature dependence is concerned, i.e. to display only a weak temperature dependence of $N(\tau)$.

Small polaron hopping also gives rise to dielectric relaxation, and in glasses the dispersions are remarkably similar to the ionic dispersions discussed above. The temperature dependence of this relaxation time distribution at short relaxation times has been investigated in detail,(7,8) and been found to be essentially temperature independent. This is shown by the data shown in Fig. 2.

Fig. 2 is a Cole-Cole plot of ε' vs. ε'' of a sequence of specimens. In this representation, the slope of the curve as $\varepsilon'' \to 0$ is a measure of the width of the distribution. The data in Fig. 2 derived from different specimens of varying concentrations of transition metal ions, at various ratios of the two different valence states, i.e. various valence concentration ratios. The data was taken at various frequencies and temperatures. As can be seen, there is no observable dependence of this slope on temperature.

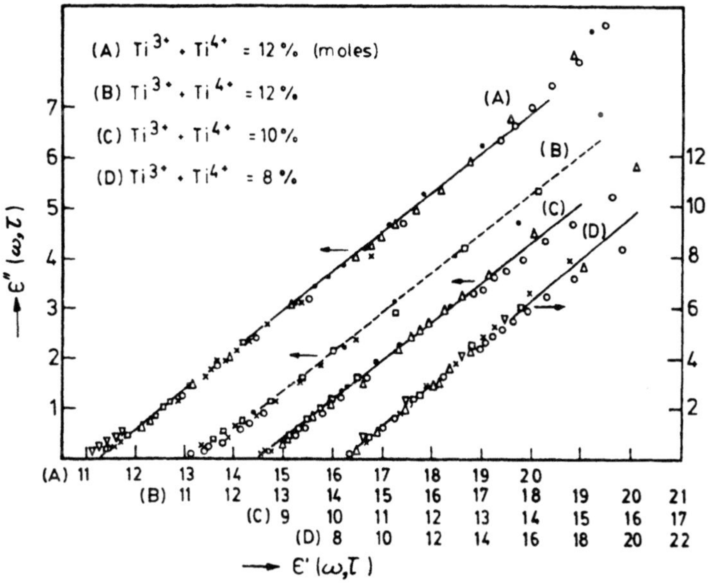

Fig. 2. Cole-Cole plots for various glasses, $w\tau \to \infty$. (A), (C), (D): selective τ, various w; (B): selective w, various τ

The temperature independence of the results (no distribution in Q) at least at the shorter relaxation time regime is remarkable.

(ii) Shape of the Dispersion

In glasses, the dispersion is not symmetric about the maximum in loss and is skewed towards the higher frequencies. This is well shown by the data of Hakim and Uhlmann (9) for a comprehensive series of alkali glasses. These data show the characteristics of relaxation time distribution function, a non-symmetrical function with a tail extending into the short relaxation time regime. [The distribution function was found to be essentially independent of temperature.]

This asymmetry, with tail extending to higher frequencies, is shown also by the data of refs. (2,5,13), for example, and by the electronic relaxation previously discussed.

(iii) The ac and dc Activation Energies

The previous paragraph established that the dielectric relaxation may be described by a reasonably unique value for the activation energy Q. The value of Q is found from the temperature dependence of the frequency for at which ε" or M" is a maximum and assuming a relation of the form

$$2\pi f_m \tau = \omega_m \tau_o \, exp\,(Q/kT) = \text{constant.}$$

Dc conductivity measurements on the same glasses give the same values for the conductivity activation energy in many cases.

The interrelation between the ac dispersion and dc conductivity is even deeper, as first pointed out by Isard (10) for an inhomogeneous glass, but later found to be true for homogeneous glasses as well; namely

$$\sigma = \varepsilon_o \, \varepsilon_\infty \, \omega_m \tag{8a}$$

or

$$\sigma = \varepsilon_o \, \Delta\varepsilon \, \omega_m \tag{8b}$$

The remarkable validity of the above relation is shown in Fig. 3 taken from the work of Nakajama.(11) The data shown include glasses whose relaxations are due to both ion motion and electron motion. This data is by no means comprehensive.

The close relationship between dielectric relaxation and the dc conductivity has received attention, and explanations have been advanced from a structural basis.

For example, Nakajama has analyzed the electrical response of a zig-zag path, and obtained equation (8b) for a pathway of simple geometry. The basic idea is that carriers will tend to those regions of the curved pathway with lower potential energy due to the applied electric field.

DIPOLAR RELAXATION AND CONDUCTIVITY RELAXATION

The previous discussion has been based on the tacit but very basic physical assumption; namely, that the dc conductivity is separate, physically distinct, from the conductivity resulting from the retarded response of the dipoles. This is manifested clearly in the way the data was analyzed:

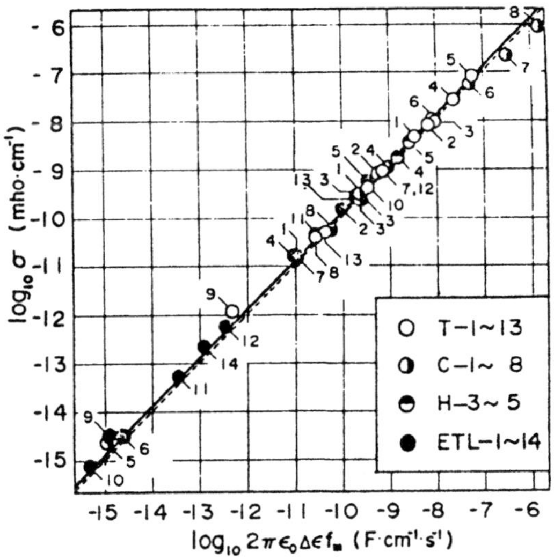

Fig. 3a. Correlation between dc conductivity and dielectric parameters for alkali-metal-containing glasses.

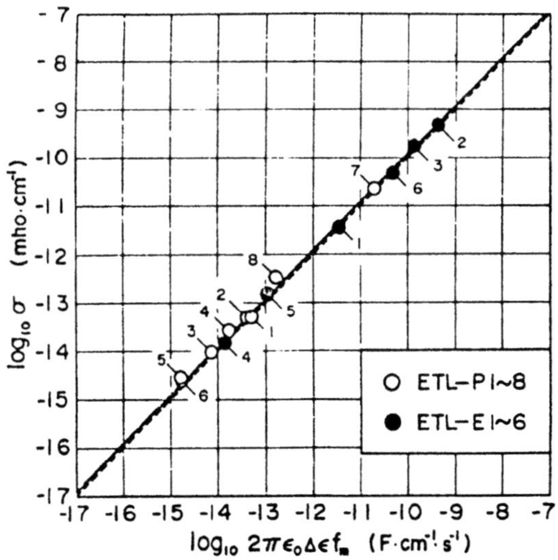

Fig. 3b. Correlation between dc conductivity and dielectric parameters for nonalkali-metal glasses.

$$\varepsilon'' = \frac{1}{\varepsilon_o \omega} (\sigma_{ac} - \sigma_{dc}) \tag{9}$$

where σ_{ac} is the conductivity measured at $\omega = \omega$ and σ_{dc} is the conductivity measured at $\omega = 0$. This is certainly a legitimate procedure provided the dipolar process exists independently of a dc conductivity process with a single relaxation time (see later). However, the very close similarity between the ac and dc activation energies, and the Isard relation, suggest that this is not the case in many glasses. The ac and dc conductivities and the maximum in ε'' may in fact be manifestations of the same process.

In the dipolar relaxation, the dc conductivity is ignored, and the electric field is assumed to follow the equation

$$E(t) = E(o) + [E(\infty) - E(o)]\ \phi_p(t) \tag{10}$$

after a charge q is placed on the plates of a condenser with the material as dielectric.

In the conductivity relaxation case, the dc conductivity is not ignored, and the electric field is assumed to follow the equation

$$E(t) = E(o)\ \phi_E(t) \tag{11}$$

after a charge q is placed on the condenser plates.

In the dipolar relaxation case, the electric field decays to the steady value $E(\infty)$; in the conductivity relation case, the electric field decays to the steady value 0. In both cases, of course, $\phi(t) \to 0$ as $t \to \infty$.

Equation (10) leads to the Debye equations for $\phi(t) = 1 - e^{-t}/\tau$. Equation (11) leads to a dispersion in the dielectric constant only if there is a distribution of "conductivity relaxation times τ_σ". Litovitz, Macedo, Moynihan, et al (12,13) have shown that

$$\Delta\varepsilon = \frac{1}{M_s}\left[\frac{\langle\tau_\sigma^{\ 2}\rangle}{\langle\tau_\sigma\rangle^2} - 1\right] \tag{12}$$

and

$$\sigma_{dc} = \frac{\varepsilon_o}{M_s} / \langle\tau_\sigma\rangle \tag{13}$$

The similarity of equation (13) to equations (8) should be noted.

THE SHAPE OF THE DISTRIBUTION FUNCTION

Considerable computation effort (14) has revealed that the decay function

$$\phi_E(t) = exp\ [-(t/\tau_o)^\beta]$$

describes the observed dispersions remarkably well.

To the author's knowledge no explicit microscopic model leading to the above decay function has been advanced. Williams et al (15) have shown that the above decay function closely approximates the dispersion expected from Glarum (16) defect diffusion relaxation, a model in which a relaxation step takes place only after some defect has diffused to the site.

At this point in the discussion, it might be inferred that the basic phenomenology had been established. All that remains is to determine some suitable microscopic model. However, there is one important variable that has been omitted, namely concentration of the mobile species. We shall describe some results as a function of composition.

DISPERSIONS AS A FUNCTION OF COMPOSITION

The activation energies for dc conductivity and ac conductivity have been measured in the $BaO-B_2O_3-SiO_2$ glasses containing $TiO_2-Ti_2O_3$ as a function of titanium composition.(17) The results are shown in Fig. 4. As can be seen, the dc activation energy is considerably higher than the ac activation energy, but significantly approaches the lower ac value as the concentration increases to about 12-15 mole % TiO_2. The ac activation energy is even more interesting, in that it also decreases with increasing Ti concentration, but reaches a fixed value at a lower concentration, namely ∿ 6 mole %. The dc activation energy clearly is tending towards this lower activation energy.

At lower concentrations it is possible to observe ac effects which are different to dc effects. As the concentration of relevant species increases, the two effects merge together, and it apparently makes no sense, finally, to distinguish between ac conductivity as being different in origin to dc conductivity.

If 15 mole % TiO_2 glass had been studied, it is suggested that the observations would be compatible with the characteristic behavior described previously: the activation energies for ac and dc conductivities would be the same, the distribution in activation energy would be narrow, the relaxation would be skewed towards

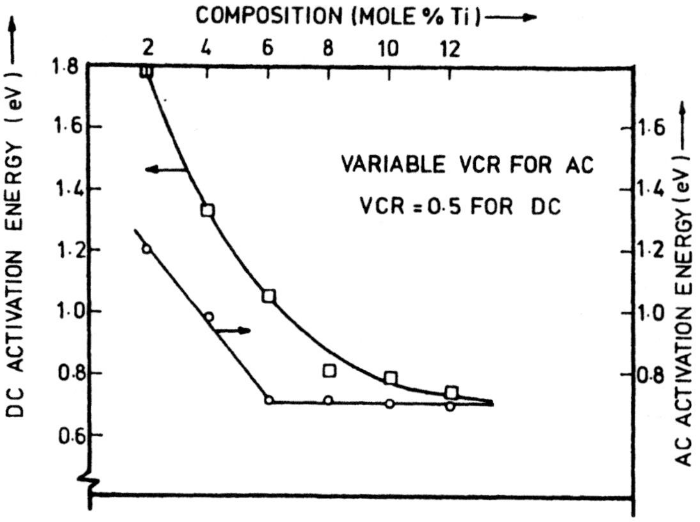

Fig. 4. ac and dc activation energies as a function of TiO_2 concentration.

higher frequencies, and be described by the decay function $\phi(t) = exp\ [-(t/\tau_o)^\beta]$ with some appropriate value of β.

GLASS STRUCTURAL CONSIDERATIONS

dc and ac measurements reflect the spatial distance over which charge may be transported before becoming blocked in some way. ac measurements will involve much shorter distances than dc measurements. As the concentration of the mobile species increases, the activation energies decrease, reaching a lower limit. The ac conductivity reaches this lower limit at a lower concentration than the dc conductivity.

A very natural explanation of these observations is to suppose the existence of microscopic inhomogeneity regions or pathways or chains in which the species move more easily. We suppose that the structure of these pathways approaches some characteristic structure rather rapidly with increasing concentration of mobile species. We also suppose that the number and length of these characteristic pathways or chains likewise increase with increasing concentration.

At "very low" concentrations of mobile species, many of the

pathways are short in length and ill-formed. Most pathways are not connected to others. Both the ac and dc conductivities are relatively low, and the activation energies decrease with increasing concentration.

At intermediate concentrations many of the elementary pathways have assumed a characteristic structure, but interconnectivity between the chains or pathways is still relatively weak. In this composition range, the ac activation energy has now attained the (low) characteristic value. The dc activation energy is larger arising from the relatively poor connections between some of the elemental chains. (A "poor connection" is a region of lower than average concentration of mobile species, requiring a slightly larger activation energy.)

At the higher concentrations, the interconnectivity between the characteristic segments is extensive. Many pathways are essentially unbroken. The ac and dc activation energies are equal.

The inhomogeneity envisaged above is not on a scale that may be described as phase separation. The scale envisaged is much finer, something more like a tracery of tree like structures.

The structure proposed in the three regimes of concentration are schematically shown in Fig. 5.

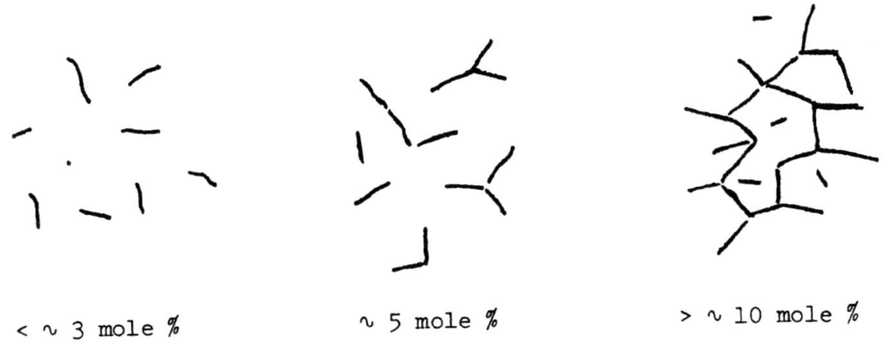

Fig. 5. Schematic diagram of glass structure as a function of composition.

Truncated Chain Model

The chain model described in the preceding paragraph has been considered by Tomandl,(18) Anderson and MacCrone,(7) Aiken and MacCrone,(19) and by Pollak and Pohl (20) in connection with dielectric loss in polymers. When all chains are interconnected, a model similar to that of Nakajama is obtained.

The dynamics are described by

$$\frac{df_i}{dt} = -2\,\alpha_i f_i + \alpha_{i-1} f_{i-1} + \alpha_{i+1} f_{i+1}$$

for any site i within the chain, where f_i is the probability that the site i is occupied and the α_i are the rates at which a particle in the site i will jump to the neighboring site. The equations governing the sites at the end of the chain, that is i = 1 and i = n, are

$$\frac{df_1}{dt} = -\alpha_1 f_1 + \alpha_2 f_2$$

$$\frac{df_n}{dt} = \alpha_{n-1} f_{n-1} - \alpha_n f_n$$

respectively.

In matrix notation, the equations are

$$\begin{pmatrix} \dot{f}_1 \\ \dot{f}_2 \\ \cdot \\ \cdot \\ \cdot \\ \cdot \\ \dot{f}_n \end{pmatrix} = \begin{pmatrix} -\alpha_1 & \alpha_2 & 0 & 0 & 0 & \cdot & \cdot & \cdot \\ \alpha_1 & -2\alpha_2 & \alpha_3 & 0 & 0 & \cdot & \cdot & \cdot \\ 0 & \alpha_2 & -2\alpha_3 & \alpha_4 & 0 & \cdot & \cdot & \cdot \\ & & & & & & & \\ & & & & \alpha_{n-1} & & & -\alpha_n \end{pmatrix} \begin{pmatrix} f_1 \\ f_2 \\ \cdot \\ \cdot \\ \cdot \\ \cdot \\ f_n \end{pmatrix}$$

The truncated chain model implies that hopping between any two adjacent sites is the same for all sites, i.e. $\alpha_1 = \alpha_2 = \alpha_3$ etc.

The dynamical response of this system may be found from a standard normal mode solution. The eigenvalues are given by

$$\lambda_m = 2\alpha\,[\cos \pi \frac{m-1}{n} - 1]$$

while the eigenvectors are given by

$$S_{mj} = \frac{1}{|S_m|} \cos\,[\pi \frac{(m-1)(2j-1)}{2n}].$$

The relaxation mode with m = 2 has a relaxation time $\tau_2 = n^2/f_o = n^2\tau_o$ where τ_o is the jump between sites. The relaxation strength of this mode $R_{m=2} \alpha n^2d^2$, where nd is the length of the chain.

The relaxation times for $m \geq 2$ are smaller than τ_2, and the corresponding relaxation strengths are very much smaller. Thus the dielectric dispersion due to a chain or pathway with n members does in fact give rise to a tail on the high frequency side. However, the dispersion is not sufficiently wide to account for the observed relaxations without introducing a distribution in chain lengths.

Various distributions in chain length have been suggested

$$G(n) = A/n^2$$

$$G(n) = Ae^{-n/n*}$$

and have been found to give reasonable agreement with observed dispersions.(7,8)

Alternatively, one may deduce the distribution of chain lengths from the decay of the polarization. Some elegant work along these lines has been done by Tomandl.(18) A distribution obtained by him is shown in Fig. 6.

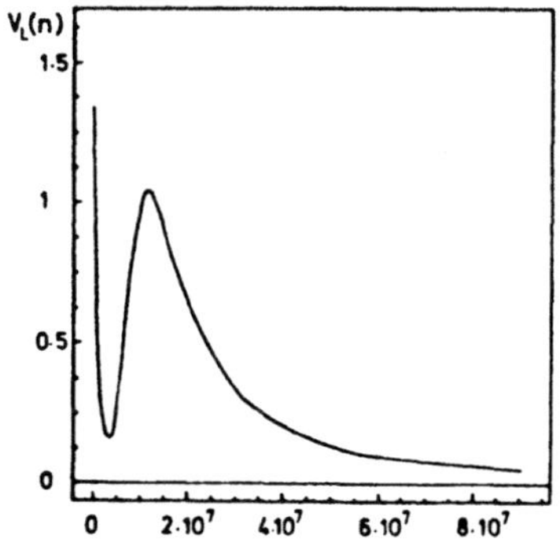

Fig. 6. $V_1(n)$ for glass 1.

CONCLUSION

The thesis presented here is that the modifying and mobile ions in glass are not distributed randomly in the glass. On the contrary, it is proposed that these ions are grouped together even at low concentration of modifying ion, say 1-5 mole %. The spatial extent of these groupings is below the limit of distinct phases.

This model ascribes the electrical conduction to short pathways whose composition and structure become characteristic at a fairly low concentration of mobile species, and whose interconnectivity increases with mobile ion concentration. The model is able to account in a very natural way for

(i) ac and dc activation energies becoming equal above a critical composition,

(ii) the empirically observed relation between the dc conductivity and the frequency at which ε'' is a maximum, and

(iii) the wide distribution of relaxation times not due to a distribution in activation energies.

In conclusion, it is therefore appropriate to briefly consider the indirect evidence in support of the hypothesis. There are several:

(i) Structural Studies

If a large fraction of the conducting species are in well defined regions, then it would be expected that a characteristic inter-ion distance might be detected by X-ray or neutron diffraction measurements. This characteristic distance should be insensitive to the composition of the modifier and/or conducting species.

Many structural studies have been made, in which a characteristic distance between the modifying ions was found much shorter than expected were these ions distributed at random. In most cases the authors did not emphasize their findings strongly enough.

Among these studies, we cite here only that of Krogh-Moe (21) and Krogh-Moe and Jurin.(22)

(ii) Dipole-Dipole Interactions

The line width of a nuclear resonance will be broadened by dipole-dipole interactions at sufficiently low temperatures. The extent of the broadening depends upon distance between the ions, and in the first approximation depends upon the inverse third power.

i.e. $\Delta\nu \propto \dfrac{1}{R^3}$

Thus, line widths would be expected to increase with increasing concentration, as $\bar{R}$ decreases. A line width proportional to the concentration would be a first guess. NMR line width studies indicate that the line width is in fact only a very weak function of composition in silicate and borate glasses. In alkali-silicates, for example, Mosel et al (23) in fact suggest the existence of pairs of alkali ions, or "regions of higher than average alkali contents"; in our parlance these are the conductive segments.

(iii) Exchange Interactions

The exchange interaction between spins $J_{ij}\, S_i\, S_j$ effects the shape of a spin resonance line. The effect of the exchange interaction is to cause the line to narrow in the center and broaden in the wings. The exchange integral J_{ij} is a sensitive function of distance.

Fig. 7. is the normalized esr resonance of Ti^{+3} in our glasses. The shape in this line is insensitive to composition from 2-10 mole % which we take as indicative of a characteristic fixed structure, independent of composition. [It might be argued that the line shape is characteristic of the distribution of environments of isolated Ti^{+3} ions. However, the line widths here are narrower than the line widths observed in very dilute glasses by Peterson et al.(24)]

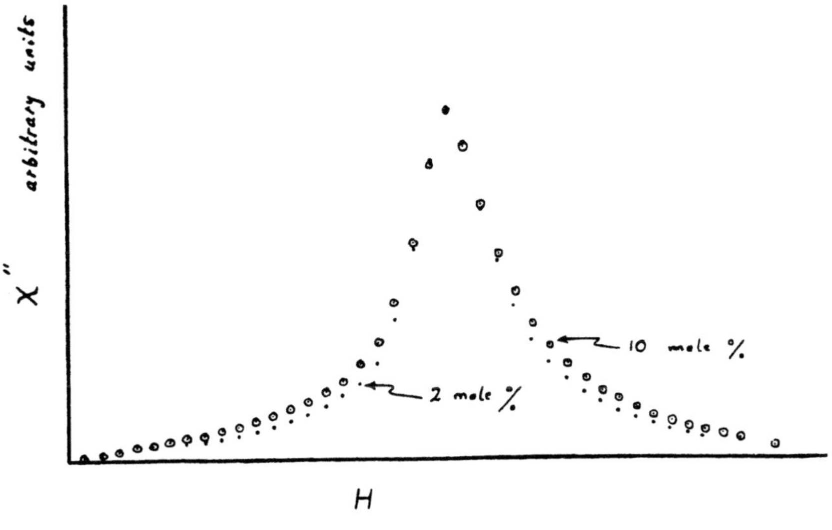

Fig. 7. Normalized esr of glasses containing 2 and 10 mole % TiO_2.

(iv) Charge Transfer Absorption

An electron may be optically excited from a state localized on one atom to a state localized on a neighboring atom. In our particular case, we are interested in the transition

$$Ti^{+3} + Ti^{+4} \rightarrow Ti^{+4} + Ti^{+3}$$

Such charge transfer has been studied in glasses and aqueous solutions and is known to peak in the uv.

This absorption has been studied and indicates a very non-linear dependence on Ti^{+3} concentration. The behavior will be described in detail elsewhere.(25)

POSTSCRIPT

Other approaches to this problem have been taken. We refer those interested to two: that of Doremus,(26) based on space charge effects, and that of Tomozawa,(27) based on a modified Debye-Hückel model.

ACKNOWLEDGEMENTS

This work was supported by National Science Foundation Grant No. GH 34548, for which we are most grateful.

REFERENCES

1. H. Frohlich, Theory of Dielectrics, Oxford University Press (1958).

2. R. J. Charles, J. Appl. Phys. 32, 1115 (1961).

3. H. E. Taylor, J. Soc. Glass Tech. 41, 350 (1957); 43, 124 (1959).

4. L. Heroux, J. Appl. Phys. 29, 1639 (1958).

5. L. P. Boesch and C. T. Moynihan, J. Non-Cryst. Solids 17, 44 (1975).

6. D. W. Moore and D. Day, Phys. Chem. Glass 12, 75 (1971).

7. R. A. Anderson and R. K. MacCrone, J. Non-Cryst. Solids 14, 112 (1974).

8. B. Rawal and R. K. MacCrone, in The Physics of Non-Crystalline Solids, 4th International Conference, Clausthal-Zellerfeld, 1976, Ed. G. H. Frischat, Trans. Tech. Publications, 1977, p. 248.

9. R. M. Hakim and D. R. Uhlman, Phys. Chem. Glasses 14, 81 (1973).

10. J. O. Isard, Proc. I.E.E., Supl. No. 22, 109B, 440 (1962).

11. T. Nakajima, 1971 Annual Rep., Conference on Electrical Insulation and Dielectric Phenomena, National Academy of Science, p. 168.

12. T. A. Litovitz and P. B. Macedo, in Amorphous Materials, Eds. R. W. Douglas and B. Ellis, Proc. of Int. Conference on the Physics of Non-Crystalline Solids at Sheffield, 1970, p. 251.

13. V. Provenzano, L. P. Boesch, V. Volterra, C. T. Moynihan and P. B. Macedo, J. Amer. Ceram. Soc. 55, 492 (1972).

14. C. T. Moynihan, L. B. Boesch and N. L. Laberge, Phys. Chem. Glasses 14, 122 (1973).

15. G. Williams, D. C. Watts, S. B. Dev and A. M. North, Trans. Farad Soc. 67, 1323 (1971).

16. S. H. Glarum, J. Chem. Phys. 33, 1371 (1960).

17. B. Rawal, Ph.D. Thesis, Rensselaer Polytechnic Institute (1977).

18. G. Tomandl, J. Non-Cryst. Solids 14, 101 (1974).

19. J. Aiken and R. K. MacCrone, in "Mass Transport in Solids", Eds. A. R. Cooper and A. H. Heuer, Plenum Press, New York.

20. M. Pollak and H. A. Pohl, J. Chem. Phys. 63, 2980 (1975).

21. J. Krogh-Moe, Phys. Chem. Glasses 3, 208 (1962).

22. J. Krogh-Moe and H. Jurin, Phys. Chem. Glasses 6, 30 (1965).

23. B. D. Mosel, W. Müller-Warmuth and H. Dutz, Phys. Chem. Glasses 15, 154 (1974).

24. G. E. Peterson and A. Carnevale, J. Chem. Phys. 56, 4868 (1972).

25. B. Rawal and R. K. MacCrone, to be published.

26. R. Doremus, in Glass Science, J. Wiley and Sons, 1973, p. 204.

27. M. Tomozawa, in Treatise on Materials Science, Ed. H. Herman, Academic Press, to be published.

CHEMICAL STABILITY OF BORON CONTAINING GLASS ENAMELS WITH SPECIAL REFERENCE TO LEAD RELEASE

A. Paul and D. Cooke

Department of Ceramics, Glasses and Polymers

University of Sheffield, England

The use of low softening point, coloured glasses as applied decoration on different types of glass articles is popular and widespread. Enamel colours which are permanently fused to the glass surface have the advantage of being uniformly glossy and durable.

Conventional glass enamels are lead borosilicate glasses to which heat-stable pigments have been added. Lead borosilicates are used because they have softening points below those of the articles to which they are applied; they also have low thermal expansion coefficients which match quite well with the expansion coefficients of most commercial soda lime silica glasses, thus preventing crazing and peeling on cooling. Besides, lead based enamels show brilliance, lustre and a smooth finish which are due to their low softening points and viscosities. When lead is excluded from an enamel, the softening point rises sharply and the enamel does not acquire the same gloss and smoothness.

Lead, like many other metals, is a trace element in the human system, but it is not essential, and in sufficient quantities is toxic. Symptoms of lead poisoning include severe abdominal cramps, loss of appetite, constipation, anaemia, fatigue, headaches, motor-nerve paralysis and encephalopathy. This raises a problem with lead based enamels. On the one hand the public must be protected against possible toxic effects, and on the other hand the enamels must look attractive and be economical to manufacture and apply.

The chemical resistance of enamels is important, not only because of the harmful effects of lead on health as described above, but also because a nondurable enamel soon loses its gloss and hence its attractiveness. In order to understand and thereby to control the toxic metal release from enamels, a systematic study of the effect of composition on chemical resistance was undertaken, and here we report a part of the results with special reference to lead release.

A large part of the $PbO-B_2O_3-SiO_2$ system involves liquid-liquid immiscibility (shown in Figure 1) and has been studied by various workers (1-4). In fact most of the commercial glass enamels are phase separated.

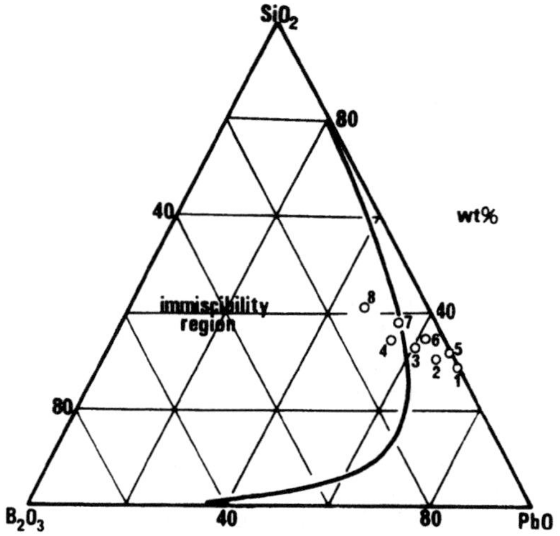

Glass number	Molar composition
1	$60SiO_2,40PbO$
2	$60SiO_2,35PbO,5B_2O_3$
3	$60SiO_2,30PbO,10B_2O_3$
4	$60SiO_2,25PbO,15B_2O_3$
5	$65SiO_2,35PbO$
6	$65SiO_2,30PbO,5B_2O_3$
7	$65SiO_2,25PbO,10B_2O_3$
8	$65SiO_2,20PbO,15B_2O_3$

Figure 1: Phase diagram of the $PbO-B_2O_3-SiO_2$ system showing glass compositions studied.

In general the possible microstructures of two-phase enamels may be broadly divided into two categories:

a) The "Two-framework" structure where both phases are continuous, and

b) The "Droplet" structure where one phase forms discrete regions inside the other, matrix phase.

The overall durability of the two-framework type is determined by the less durable phase, whereas the resistance of the two-phase glass of the droplet type is determined primarily by the durability of the matrix phase.

Usually, one phase is more durable and the other less durable than the initial glass before phase separation. Where the phase separation process leads to the formation of droplets of the unstable phase inside the stable matrix phase, the durability of the glass increases as a whole; this usually happens in the Pyrex type of borosilicate glass. In all other cases the separation process decreases the durability of the glass. The more pronounced the separation process, the greater is the change in durability.

The compositions of glasses reported in this paper are shown in Figure 1. 500 g of glass were melted at a time, and the melts were quenched by pouring on a cold steel plate. Parts of the quenched samples were subsequently heat-treated at different temperatures for further phase separation. All the samples were made into standard grains (-30, +52 mesh) (5) and extracted using different pHs at 70°C for 4 hours in the dark. During extraction 1 g of glass grains was taken with 50 ml of buffered solution. All the leached glass constituents in solution were estimated by the conventional methods.

Figures 2 and 3 show the amount of lead extracted from glasses 1 to 8 at various pHs; the extraction results are expressed as a percentage of the total lead present in the virgin glass. These results show that less lead is extracted from the glasses in the series containing 65 mol% silica than from their opposite numbers (same B_2O_3 content) in the series containing 60 mol% silica below pH 7 but more lead is extracted above pH 7. The quantity of lead extracted increases as the acidity or alkalinity of the solution is increased and remains relatively low around the neutrality point and in feebly alkaline solutions.

Figures 4 and 5 show the pH dependence of silica extraction from these glasses. It is clear that the quantity of silica extracted increases as the alkalinity of the solution is increased, particularly around pH $\sim$ 12. This is associated with the ionization of silicic acid and is a common feature of all silicate glasses in aqueous solutions. In parallel with the lead extraction at a given pH, less silica is extracted from the glasses in the series containing 65 mol% silica than that from glasses containing 60 mol% silica.

To show the effect of B_2O_3 on lead and silica extraction, some of the results of Figures 2-5 are replotted in Figures 6 and 7 as a function of B_2O_3 content of the glass. It is clear that the lead and silica extraction increases dramatically as the B_2O_3 content of the glass increases.

Figures 8 and 9 show the effect of heat-treatment on lead and silica extraction from glasses containing 65 mol% silica. These results show that lead as well as silica extraction increases with the time of heat-treatment and that the increase became greater as the B_2O_3 content of the glass increases.

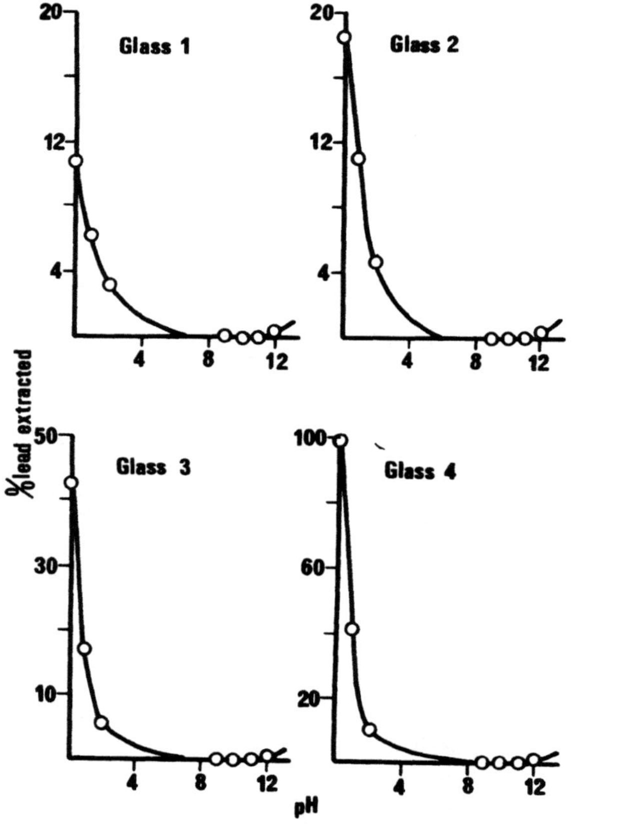

Figure 2: Effect of pH on lead extraction from glasses 1 to 4

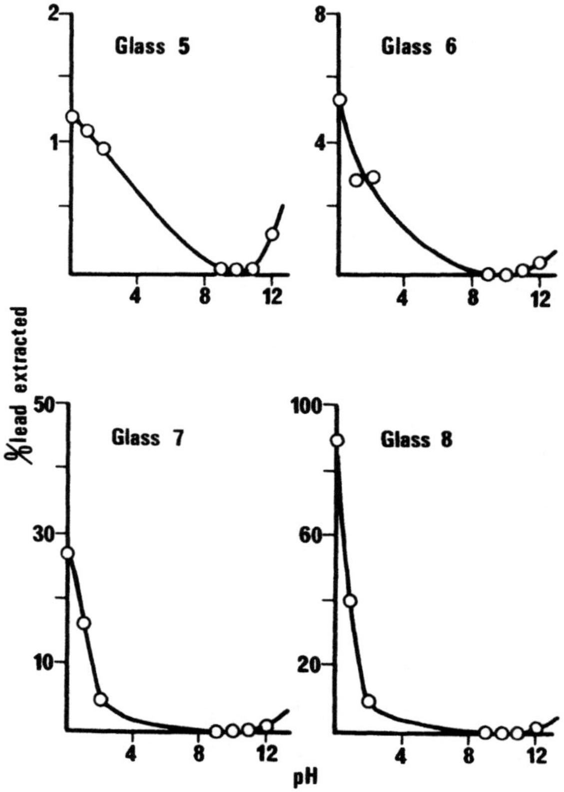

Figure 3: Effect of pH on lead extraction from glasses 5 to 8

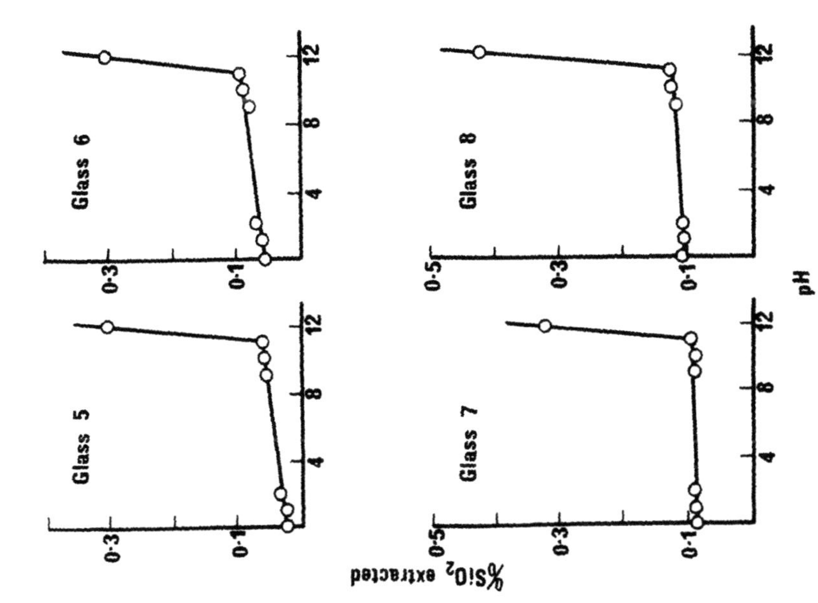

Figure 5: Effect of pH on silica extraction from glasses 5 to 8

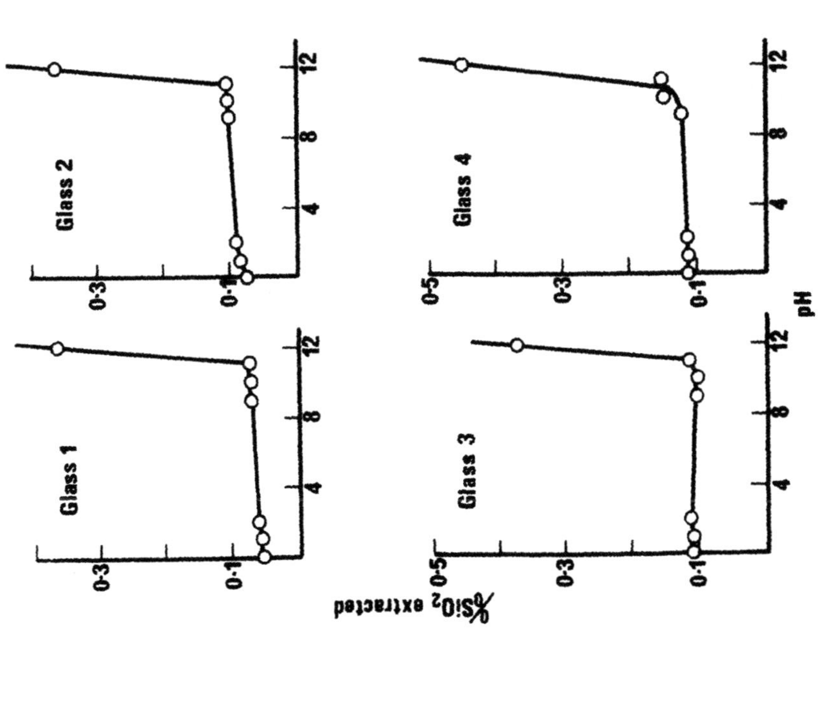

Figure 4: Effect of pH on silica extraction from glasses 1 to 4

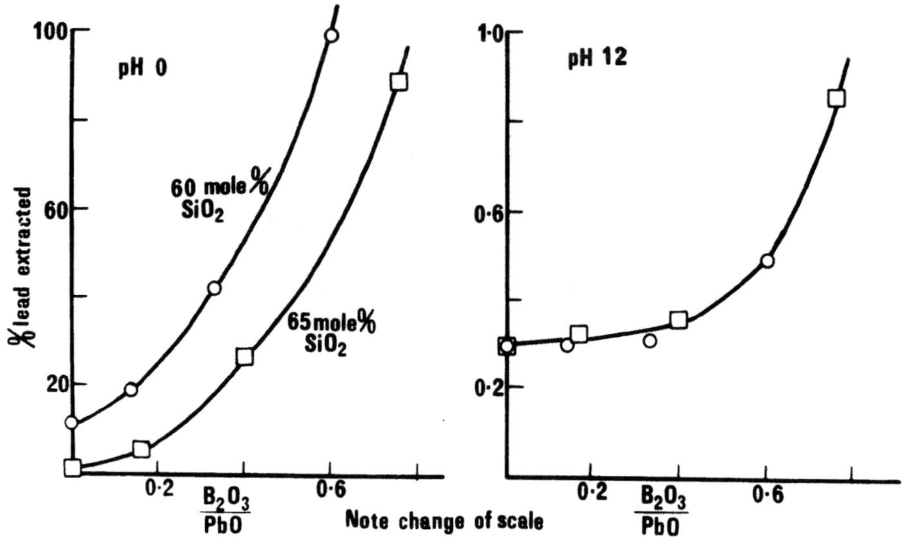

Figure 6: Effect of the addition of B_2O_3 on lead extraction

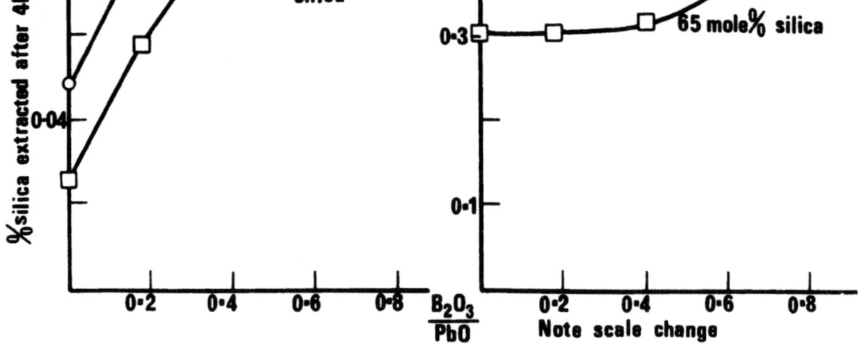

Figure 7: Effect of the addition of B_2O_3 on silica extraction

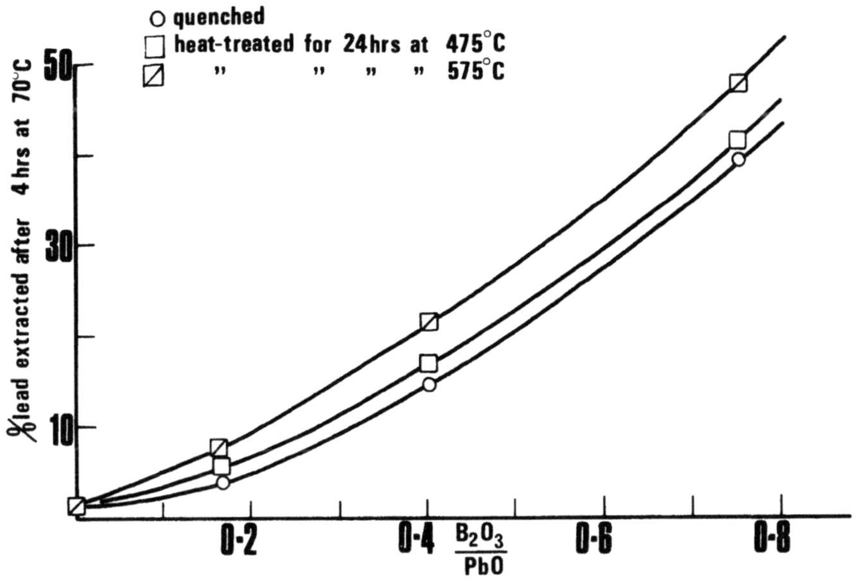

Figure 8: Influence of heat-treatment on lead extraction at pH = 1

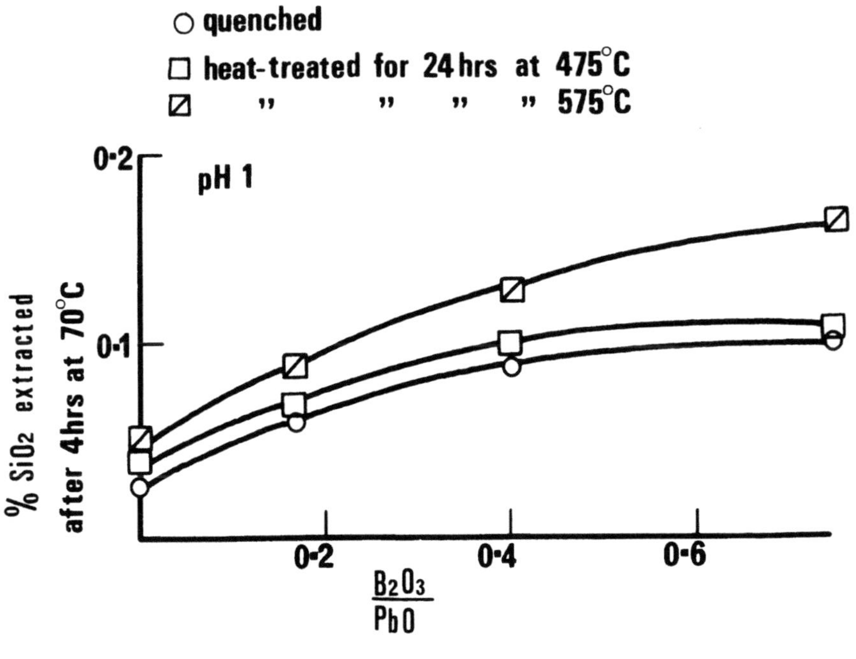

Figure 9: Influence of heat-treatment on silica extraction at pH = 1

Immiscibility is known to exist in all the four systems: PbO-B_2O_3 (6), PbO-SiO_2 (4), B_2O_3-SiO_2 (3) and PbO-B_2O_3-SiO_2 (2). An examination of the PbO-B_2O_3-SiO_2 phase diagram (Figure 1) shows that as B_2O_3 is added to a PbO-SiO_2 glass the tendency towards immiscibility increases. Figure 10(a) shows the electron micrograph of a quenched sample of glass 8, the composition of which lies inside the immiscibility region; clearly two finely interconnected phases may be observed. Figures 11(a) and 11(b) show the microstructures of glasses 8 and 7 respectively after heat-treatment. In both cases the microphases increase considerably in size as the heat-treatment temperature increases. Comparing figures 10(a) and 11(a) with the results of figures 8 and 9 it appears that it is the growth of the microphases that causes the soluble phases to become more accessible and hence increases extraction rates.

In spite of its cost common glass enamels contain 2-5 wt% Li_2O. Figure 12 shows the effect of the addition of Li_2O on lead extraction. The deterioration in durability of binary alkali silicate glasses as the alkali content of the glass is increased is well-known (7), and it can be seen that this is also the case in the Li_2O-PbO-SiO_2 glasses studied in this investigation, i.e. the durability decreases as the lithia content increases. In the case of lead borosilicates however the addition of Li_2O presents a completely different picture. Addition of 5 mol% Li_2O to the glass with a PbO:B_2O_3 ratio of 1:0.4 results in the durability of the glass improving by a factor of about 2. The addition of 10 mol% Li_2O in this glass also increases the durability but not by as much as 5% Li_2O does. Examination of ion beam thinned thin sections of these lithium lead borosilicates under the electron microscope shows that phase separation has been completely eliminated. The anomalous effect of the increase in durability by addition of Li_2O can thus be explained. From Figure 12 it would appear that the increase in durability due to the elimination of phase separation is greater than the decrease due to the addition of Li_2O and also that there is an optimum amount of Li_2O necessary to eliminate phase separation. Amounts greater than this optimum again lead to a decrease in durability.

Figure 13 shows the activity of PbO in various Li_2O-B_2O_3 melts at 1000°C. It may be readily seen that as the Li_2O/B_2O_3 ratio increases, the activity curves are consistently displaced towards more positive values. Thus addition of more than the critical amount of Li_2O is to be avoided.

In principle the chemical stability of a glass towards aqueous solutions may be considered to be a function of the activity of the component oxides in the glass, and the stability of the individual oxides towards hydration, ionization and complexation in aqueous solution. With the available thermodynamic data it is possible to

Glass 8 Glass 7

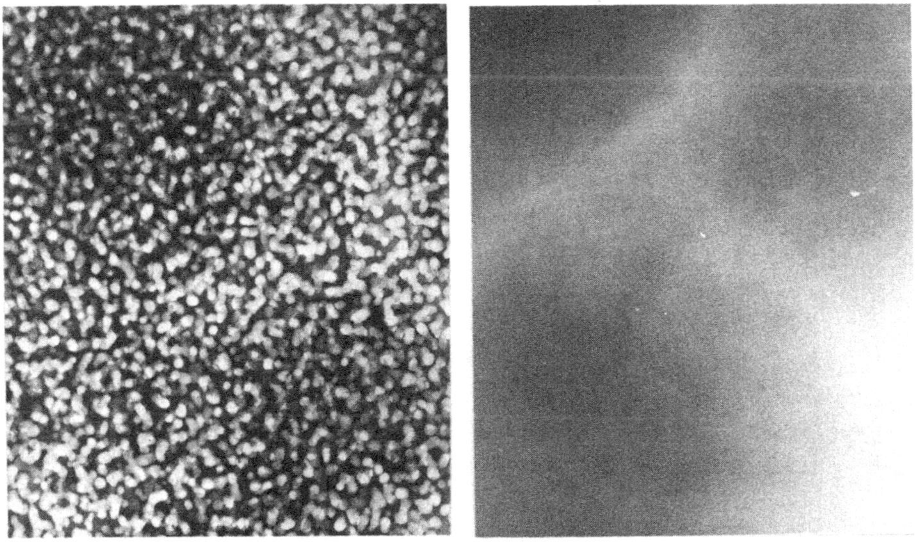

(a) (b)

Figure 10: Electron micrograph of quenched glasses

Glass 8 Glass 7

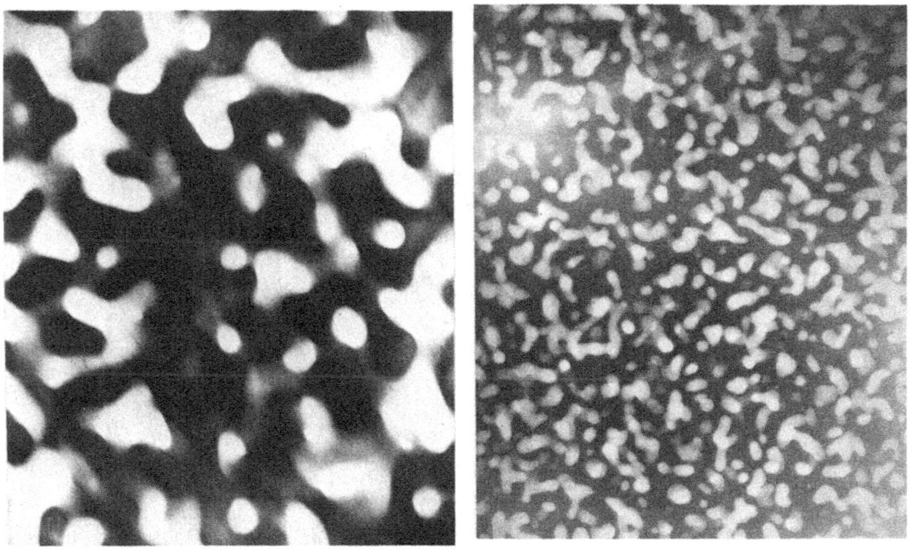

Figure 11: Electron micrograph of glasses head-treated at 575°C for 24 hours

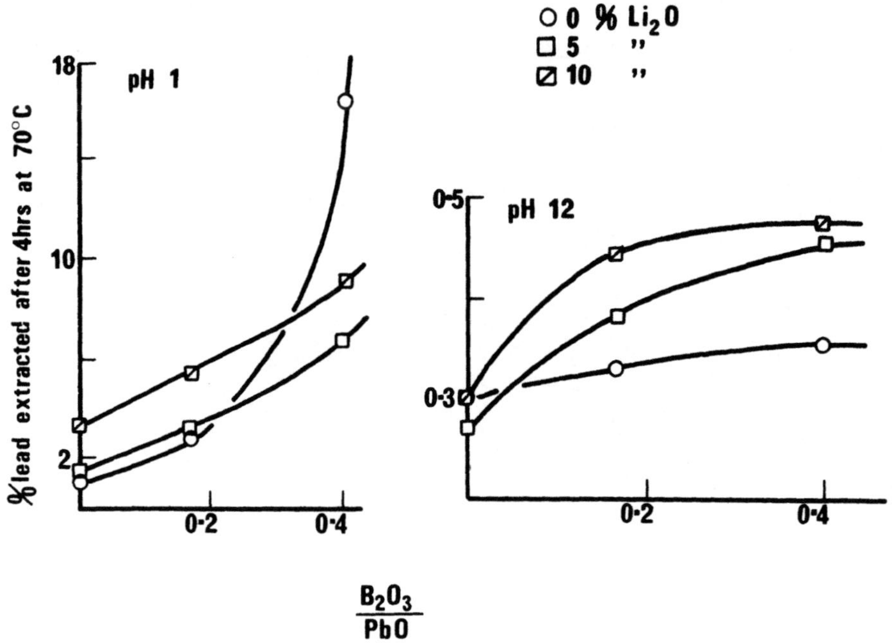

Figure 12: Influence of the addition of Li_2O on lead extraction

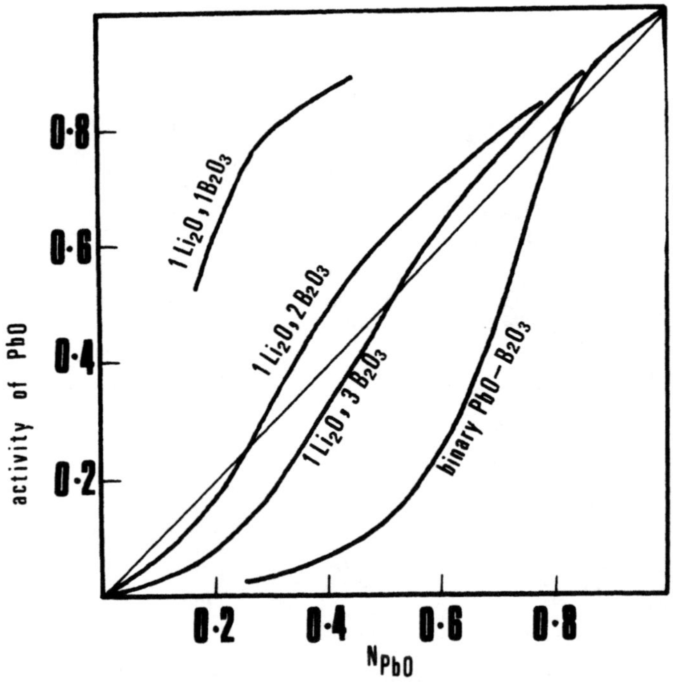

Figure 13: Activity of PbO in some lithium borate melts at 1000°C

TABLE I

THERMOCHEMICAL DATA FOR REACTIONS OF SiO_2, PbO AND B_2O_3 WITH WATER AT 25°C

Reaction	ΔG° (kcal/mol)	log K	Relationship
$SiO_2(glass) + H_2O(l) \rightleftharpoons H_2SiO_3$	+ 5.590	- 4.098	$\log(H_2SiO_3) = -4.098$
$H_2SiO_3 \rightleftharpoons HSiO_3^- + H^+$	+13.640	-10.000	$\log(HSiO_3^-) = -14.098 + pH$
$HSiO_3^- \rightleftharpoons SiO_3^= + H^+$	+16.360	-11.994	$\log(SiO_3^=) = -26.092 + 2pH$
$PbO(c) + H_2O(l) \rightleftharpoons Pb(OH)_2$	+ 1.140	- 0.836	$\log(Pb(OH)_2) = -0.836$
$Pb(OH)_2 + H^+ \rightleftharpoons PbOH^+ + H_2O$	- 7.820	+ 5.736	$\log(PbOH^+) = 4.9 - pH$
$Pb(OH)_2 + 2H^+ \rightleftharpoons Pb^{2+} + 2H_2O$	-18.590	+13.629	$\log(Pb^{2+}) = 12.793 - 2pH$
$Pb(OH)_2 \rightleftharpoons HPbO_2^- + H^+$	+19.600	-14.369	$\log(HPbO_2^-) = -15.205 + pH$
$B_2O_3(glass) + 3H_2O(l) \rightleftharpoons 2H_3BO_3$	- 9.930	+ 7.28	$\log(H_3BO_3) = 3.64$
$H_3BO_3 \rightleftharpoons H_2BO_3^- + H^+$	+12.570	- 9.216	$\log(H_2BO_3^-) = -5.576 + pH$
$H_2BO_3^- \rightleftharpoons HBO_3^= + H^+$	+17.340	-12.713	$\log(HBO_3^=) = -18.289 + 2pH$
$HBO_3^= \rightleftharpoons BO_3^{3-} + H^+$	+18.810	-13.79	$\log(BO_3^{3-}) = -32.079 + 3pH$
$4H_3BO_3 \rightleftharpoons HB_4O_7^- + H^+ + 5H_2O$	+ 9.080	- 6.66	$\log(HB_4O_7^-) = -6.66 + pH$
$4H_3BO_3 \rightleftharpoons B_4O_7^= + 2H^+ + 5H_2O$	+21.350	-15.65	$\log(B_4O_7^=) = -15.65 + 2pH$

calculate the various energy changes associated with these processes, and from them the stability of the glass under various conditions of acidity and alkalinity can be judged. Since these data are usually available only for 25°C and one atmosphere pressure, the following discussion refers to that temperature and pressure only.

The various aqueous reactions of the major component oxides (SiO_2, PbO and B_2O_3) in the present glasses are listed in Table I, and the stability diagrams for the different oxides, calculated from the standard free energy data (given in Table I) are shown in Figures 14, 15 and 16.

The very limited solubility of silica in neutral or acidic aqueous solutions (except HF) is one of the main factors in determining the corrosion resistance of silicate glasses. From Figure 14 it can be seen that the solubility of silica, in terms of H_2SiO_3 in the solution, is independent of the pH, but that in the presence of alkali, when the pH of the solution increases above 10, additional silica in the form of silicate ions ($HSiO_3^-$ and $SiO_3^{=}$) passes into the solution. In fact one can divide the pH in this system into three zones as is shown in Figure 14. It can be seen that in the first zone (pH < 10), the minimum solubility is represented by the undissociated silica (hydrated form), the soluble portion (H_2SiO_3) being $\sim 8 \times 10^{-5}$ molar; this species predominates between pH = 0 and 9. In the second zone (pH = 10 to 12) most of the silica which passes into the solution is due to the formation of $HSiO_3^-$. In the third zone (pH > 12), the $SiO_3^{=}$ ion predominates in the solution. At pH > 9 the total silica solubility increases exponentially with the alkalinity of the solution. A satisfactory qualitative correspondence between Figure 14 and the experimental results of Figures 4 and 5 can be seen.

The stability of PbO in aqueous solutions of different pH is shown in Figure 15. Hydration of PbO is small. In the acidic range (pH < 6.8) lead dissolves as Pb^{2+} and $PbOH^+$, the activity of the former is always much greater than that of $PbOH^+$. In the alkaline range lead forms $HPbO_2^-$ and the activity of $HpbO_2^-$ becomes greater than that of $Pb(OH)_2$ only above pH > 14.5. Thus PbO in a silicate glass is expected to increase the alkaline durability, whereas the acidic durability should decrease. This is indeed the fact as can be seen from Figure 1 and 2 where lead extraction from glasses 1 to 8 is shown.

When boric oxide comes in contact with an aqueous solution, orthoboric acid is formed which then ionizes in several ways depending upon its concentration and the pH of the solution. In relatively dilute solutions, containing less than 0.01 gm atom boron/litre, the boron exists essentially in the form of orthoboric acid and

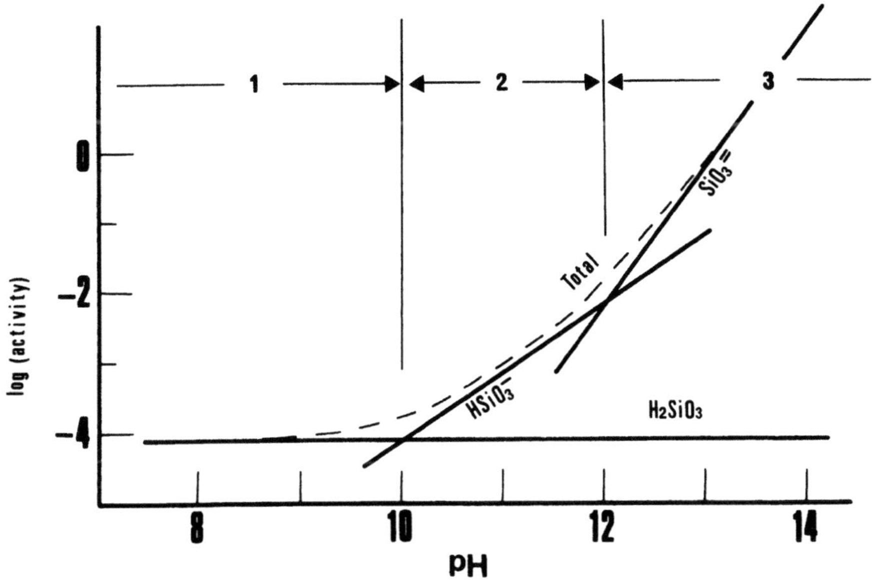

Figure 14: Stability diagram of silica in aqueous solution at 25°C

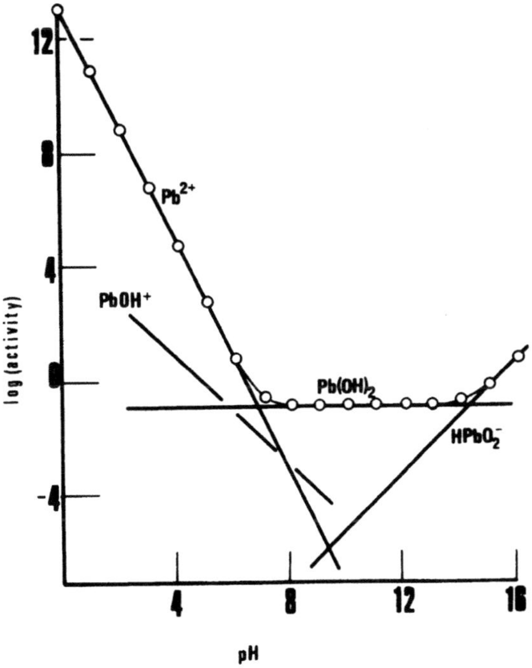

Figure 15: Stability diagram of PbO in aqueous solution at 25°C

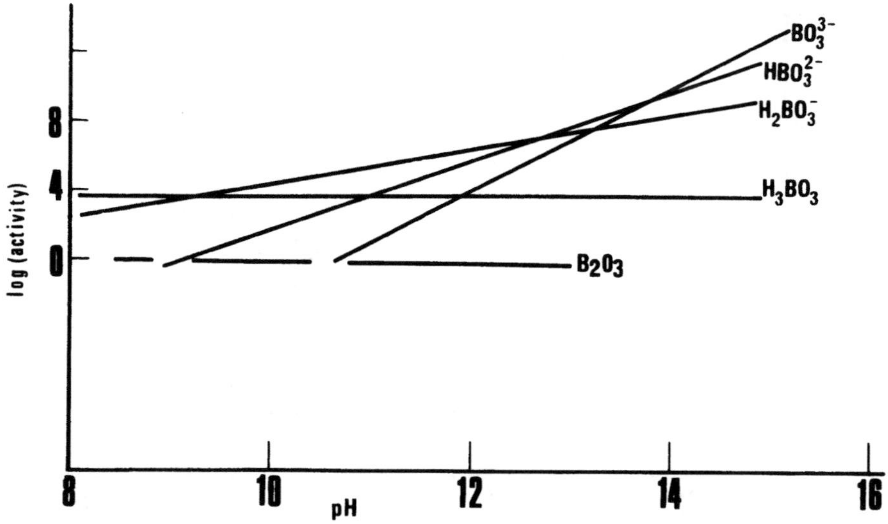

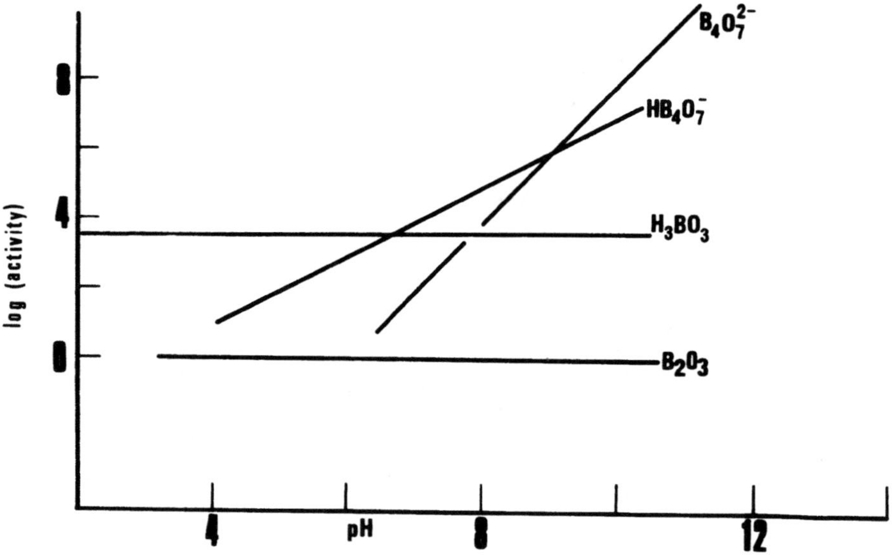

Figure 16: Stability diagram of B_2O_3 in aqueous solution at 25°C
(a) dilute boric acid solution
(b) concentrated boric acid solution

orthoborate ions ($H_2BO_3^-$, HBO_3^{2-} and BO_3^{3-}). For greater concentrations, there is polymerisation with the formation of tetraboric acid and the tetraborate ions ($HB_4O_7^-$ and $B_4O_7^{2-}$).

From the stability diagrams (Figures 16(a) and 16(b)) it can be seen that, for low concentrations of H_3BO_3, below about pH 9 orthoboric acid dissolves mainly as undissociated H_3BO_3; between pH 9 and 12.6 $H_2BO_3^-$ ions are the main species; between pH 12.6 and 13.8 HBO_3^{2-} ions are predominant and above pH 13.8 BO_3^{3-} ions are the predominant species. However, in more concentrated solutions (which are unlikely to be encountered with glasses of reasonable durability) the orthoborate ions do not control the dissolution of H_3BO_3 (Figure 16(b)).

Comparing Figure 16(a) with Figure 14 it can be seen that in aqueous solution the stability of B_2O_3 is very much inferior to that of SiO_2. The first ionization of orthoboric acid predominates at pH $\sim$ 10. There is also a vast difference in hydration characteristics of B_2O_3 and SiO_2. In fact from energetic considerations it can be shown that a glass having a boric oxide activity of as low as $\sim 1.8 \times 10^{-8}$ will have a comparable aqueous stability to fused silica. This is an exceedingly small value (at 25°C) and no such borates are known. Thus it appears that B_2O_3 substitution for SiO_2 in any single phase oxide glass will cause its chemical resistance towards water to deteriorate. However, boron containing glasses with better chemical stability can be obtained by controlling the morphology of the microphases in phase separated systems.

REFERENCES

1. R.F. Geller and E.N. Bunting, J.Res.Natl.Bur.Sts., 23, (1939), 275.

2. D.W. Johnson and F.A. Hummel, J.Amer.Ceram.Soc., 51, (1968), 196.

3. R.J. Charles and F.E. Wagstaff, J.Amer.Ceram.Soc., 51, (1968), 16.

4. P.D. Calvert and R.F. Shaw, J.Amer.Ceram.Soc., 53, (1970), 350.

5. R.F.R. Sykes, Glass Technol., 6, (1965), 178.

6. R.F. Geller, A.S. Creamer and E.N. Bunting, J.Res.Nat.Bur.Std., 18, (1937), 585.

7. V. Dimbleby and W.E.S. Turner, J.Soc.Glass Tech., 10, (1926), 304.

8. A. Paul, J.Mater.Sci., (in press).

9. G.G. Charette and S.N. Flengas, Can.Met.Quart., 7, (1968), 191.

CHEMICAL DURABILITY OF BORATE GLASSES

P. B. Adams and D. L. Evans

Research and Development Laboratories

Corning Glass Works, Corning, N.Y. 14830

INTRODUCTION

The Na_2O-B_2O_3 binary system has been studied extensively. The work of Krogh-Moe and others, summarized by Griscom,[1] has attempted to explain the structural reasons for various property changes that occur as the ratio of Na_2O to B_2O_3 is varied. Many inflections in the curves relating properties and composition seem to occur at about 15-20% Na_2O, the so-called "B_2O_3 anomaly". This has been interpreted as indicative of a structural connectivity maximum. Other information, such as that obtained by nuclear magnetic resonance, suggests that there is a smooth continuous structural change through the system, with BO_3 units converting to BO_4 units as Na_2O is increased up to one to one ratio of Na_2O to B_2O_3.

The Na_2O-B_2O_3 system per se is of little commercial interest to the glass industry. However, if knowledge about it can be related to silicate glasses, e.g., the Na_2O-B_2O_3-SiO_2 ternary system, this becomes useful to the glass manufacturer. Figure 1 shows some of the general features of this ternary system. Note that "anomalous" behavior has been observed in the ternary that seems to correlate with that observed in the binary.

In this paper, chemical durability experiments are described for glasses in the Na_2O-B_2O_3-SiO_2 system that contain between 30 and 90% SiO_2. Isodurs, or lines of constant chemical resistance, are constructed. A theoretical model to account for these is presented. The nature and form of the lines have suggested certain structural configurations which seem to relate to those observed in the Na_2O-B_2O_3 binary. The relevance of these isodurs to commercial glasses is discussed.

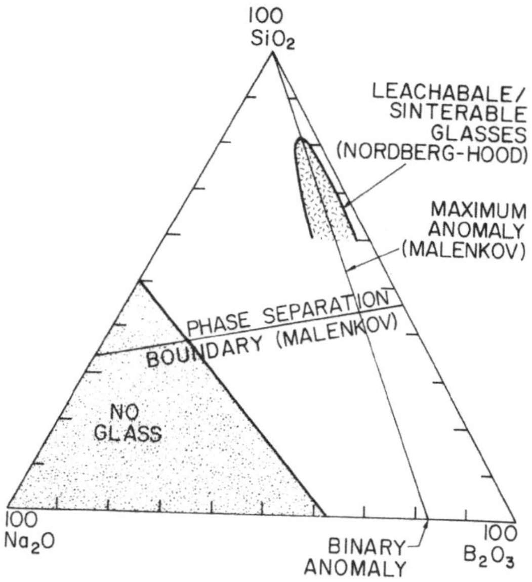

Figure 1. General Features of the $Na_2O-B_2O_3-SiO_2$ Ternary (Composite from Figures Taken from Volf[2]).

The tests were done in water under one set of test conditions. This puts constraints on the use of the data: (1) They apply only to near neutral environments. They cannot be used to predict chemical durability in acid or base solutions. (2) They indicate trends and are not quantitative when relating different levels of attack. The kinetics of dissolution change as the intensity of attack changes and the mechanism of dissolution probably changes from H^+ for Na^+ ion exchange at high SiO_2 contents to hydrolysis of the entire glass network at low SiO_2 contents.

In spite of these constraints and the obvious lack of sufficient data to accurately define the isodurs, the results demonstrate that chemical durability measurements are a useful tool for understanding glass structure.

EXPERIMENTAL

Twenty-seven glasses were prepared in the $Na_2O-B_2O_3-SiO_2$ ternary system in the range from 30 to 90 mole percent SiO_2. The batches were mixed in 1000 gm quantities using 200 mesh Supersil sand, soda ash, and anhydrous boric oxide as batch materials. They were melted in platinum crucibles for 16 hours and then chilled by pouring the molten glass directly into water. The glasses were then analyzed by flame spectroscopy. Table I shows the nominal batched compositions, the melting temperatures employed and the analyzed compositions. Note that those glasses melted at 1550°C lost considerable Na_2O and B_2O_3.

Table I. Nominal Compositions, Melting Temperatures and Analyzed Compositions.

	Nominal Mole %			Melt	Analyzed Mole %[a]		
Glass	SiO_2	Na_2O	B_2O_3	Temp (°C)	SiO_2	Na_2O	B_2O_3
PO	90.0	10.0	0.0	1650	89.3	10.7	0.0
PP	90.0	0.0	10.0	1650	90.0	0.0	10.0
PQ	80.0	20.0	0.0	1650	78.0	22.0	0.0
PR	80.0	10.0	10.0	1650	82.5	8.7	8.8
PS	80.0	0.0	20.0	1650	80.2	0.0	19.8
PT	70.0	30.0	0.0	1650	66.7	33.3	0.0
PU	70.0	20.0	10.0	1650	78.0	16.7	5.3
PV	70.0	10.0	20.0	1650	76.4	6.9	16.7
PW	70.0	0.0	30.0	1650	70.9	0.0	29.1
PX	60.0	40.0	0.0	1550	60.0[b]	40.0[b]	0.0
PY	60.0	30.0	10.0	1550	65.4	30.3	4.4
PZ	60.0	20.0	20.0	1550	71.2	16.9	11.9
QA	60.0	10.0	30.0	1550	63.6	8.1	28.2
QB	60.0	0.0	40.0	1550	61.0	0.0	39.0
QC	50.0	50.0	0.0	1550	-[c]	-	-
QD	50.0	40.0	10.0	1550	59.7	35.5	4.8
QE	50.0	30.0	20.0	1550	65.0	27.3	7.7
QF	50.0	20.0	30.0	1550	67.7	10.4	22.0
QG	50.0	10.0	40.0	1350	50.9	10.3	38.8
QH	50.0	0.0	50.0	1350	51.0	0.0	49.0
QI	40.0	40.0	20.0	1350	42.1	41.0	16.9
QJ	40.0	30.0	30.0	1350	40.8	31.5	27.8
QK	40.0	20.0	40.0	1350	40.3	22.7	37.0
QL	40.0	10.0	50.0	1350	40.0	10.0	50.0
QM	40.0	0.0	60.0	1350	41.0	0.0	59.0
QN	30.0	40.0	30.0	1350	34.7	38.6	26.7
QO	30.0	30.0	40.0	1350	32.6	29.4	37.9

[a]Results have been normalized to 100.0%. The raw analyses totaled 100 ± 1% including LOI, "loss on ignition" which was probably mostly water. Glasses exceeding 0.1% LOI by weight, were PO-0.2, PW-2.7, PZ-0.5, QB-0.9, QI-0.3, QG-1.0, QH-4.5. Some glasses showed a negative LOI, that is, they gained weight when ignited. These were PT-5.0, QD-6.6, QI-5.5, QL-6.5 and QN-28.

[b]There was insufficient sample for analysis. Result shown is the nominal composition.

[c]This sample was at the boundary of the glass forming region so that it totally dissolved when chilled, i.e., plunged into water.

Two sets of glasses having different thermal histories were tested. The first set was as described above, i.e., chilled as rapidly as possible, to minimize phase separation. The second set was annealed for 2½ hours at 600°C. Several glasses from set one, Nos. PU, PY, PZ, QA, QF, were examined by electron microscopy. There was no microstructure present above the 100 A scale.

The samples were prepared for the chemical durability tests by crushing and sieving to the 40-50 mesh size sieve range and then washing to remove fines using standard techniques.[3] One gram samples were then extracted with 25 ml of deionized water at room temperature for 24 hours. The concentration of sodium in the extract solution was determined by flame spectroscopy. For those glasses that contained no sodium, boron was determined.

The durability results were expressed as the ratio of Na_2O (or B_2O_3) extracted to that present originally in the glass. This was computed as follows:

$$R = \frac{P \times F \times V \times 10^4}{C}$$

where

R = ratio of Na_2O (or B_2O_3) extracted to that present originally
P = concentration in extract solution, ppm (μg/ml) of Na or B
F = factor to convert the metal to the oxide
V = volume of extract solution, ml
C = percent of oxide present in the original glass.

Thus, if R = 1, all of the Na_2O was extracted; if R = 0.01, 1% of the total Na_2O originally present was extracted.

For ease in displaying the numbers in the subsequent discussion of results, R is multiplied by 10^4 to provide whole integers. Thus, if $R \times 10^4 = 10{,}000$, all of the Na_2O is extracted; if $R \times 10^4 = 100$, 1% of the total was extracted. Table II shows the results for set one, chilled glass, and set two, annealed glass.

The two sets of samples gave the same chemical durability test results within experimental error except for glasses QA and QG. Both of these fall within the highly phase separable zone defined by Hood and Nordberg and shown in Figure 1. Since the other annealed samples, set 2, did not decrease in durability and since the electron micrographs showed no structure above 100 A, there does not seem to be any gross phase separation in the set 1 samples.

CONSTRUCTION OF ISODURS

Each durability result, expressed as $R \times 10^4$, was plotted on a ternary diagram at the position defined by the composition of the glass for which that result was obtained. Figure 2 shows the

Table II. Chemical Durability Results.

Glass	R x 10^4 Chilled	R x 10^4 Annealed[b]	Glass	R x 10^4 Chilled	R x 10^4 Annealed[b]
PO	23	25	QB	3000[a]	-
PP	2	-	QD	880	1220
PQ	84	62	QE	22	27
PR	2	2	QF	19	15
PS	39	-	QG	770[c]	4400[a]
PT	620	-	QH	2100	-
PU	9	4	QI	4000[a]	-
PV	6	9	QJ	360	400
PW	310	-	QK	340	270
PX	3700[a]	3200[a]	QL	2900[a]	4900[a]
PY	46	36	QM	2000[a]	-
PZ	5	5	QN	5100[a]	-
QA	4[c]	180[c]	QO	770	-

[a]When more than 1/5 of the total has been extracted (R x 10^4 > 2000); the test may not be relevant.

[b]Where no results are shown, no test was made.

[c]Difference between chilled and annealed samples is analytically significant.

data plotted from Table II for set 1 samples. The data all seem consistent except for the value of R x 10^4 = 4 at 63% SiO_2 (glass QA).

Estimated positions for values of R x 10^4 = 10, 100 and 1000 were interpolated between the data points and best fit lines were drawn. Thus, three lines representing a factor of 10 difference in durability between each were constructed. All glass compositions falling on a line have the same chemical durability as measured by the test employed. Obviously there exist other lines, or isodurs, between those shown.

STRUCTURAL IMPLICATIONS OF ISODURS

Figure 3 shows the ternary system with the isodurs at R x 10^4 = 10, 100 and 1000. Note the line ABC. Moving from A to B, the durability increases (the solubility decreases) and is maximized at B. Moving from B to C, the durability decreases. Note the line DBE. This is drawn through B and perpendicular to the tangents to the isodurs at the points of inflection. It represents all compositions for which the durability is maximized at a given SiO_2 content.

These experimental facts suggest that the chemical durability results can serve as a filter for accepting or rejecting structural models. They will be interpreted in as simple a manner as is consistent with what is known about a) the coordination chemistry of boron oxide, and b) the configurations that are either known for crystalline compounds or topologically possible for vitreous compounds, i.e., planar BO_3 and tetrahedral BO_4.

The line DE represents a ratio of Na_2O:B_2O_3 of 46:54 = 0.85:1 (experimental error would probably permit this ratio to vary from 0.7:1 to 1.0:1). The best durability exists at this ratio, whatever the exact number may be. The optimum connectivity of structure must also exist at or near this ratio.

The simplest structural model for these results is one where, at the best durability for minimum silica, virtually all the boron is in BO_4 units analogous to the SiO_4 tetrahedra. The equivalence between chemical and structural equations is

$$SiO_2 = SiO_{4/2}$$
$$B_2O_3 = 2BO_{3/2}$$
$$B_2O_3 + Na_2O = 2BO_{4/2} + 2Na^{+1}.$$

The subscripts indicate the number of oxygens about each cation/number of bonds from each oxygen to each cation.

When $Na_2O > B_2O_3$, non-bridging oxygens are introduced which decrease the overall connectivity of the network. When $Na_2O <$ B_2O_3, both $BO_{4/2}$ and $BO_{3/2}$ units are present and the number of triangular BO_3 units increases as Na_2O decreases. This also systematically decreases the overall connectivity of the 3-dimensional network.

The durability results imply that, so long as invariance of the tetrahedral connectivity of a 3-dimensional network is maintained in the homogeneous glass forming region of the Na_2O-B_2O_3-SiO_2 system, composition changes can be systematically calculated to change other properties without changing durability.

A THEORETICAL MODEL

The approach taken now is to ask whether a 1:1 correspondence exists between any statistical invariant of network topology and the experimental isodurs. Specifically consider only the connectivity of an idealized network (X x Y) generated in terms of:

X = the fraction of all cations that may be either 4- or 3-connected network formers

Y = the fraction of network formers that may be 4-connected.

Oxygen then plays either of 2 roles depending upon the Na_2O: B_2O_3 ratio. It can "create" tetrahedral $BO_{4/2}$ units from triangular $BO_{3/2}$ units or decrease the connectivity of tetrahedral $SiO_{4/2}$ units by the introduction of NBO's via the reaction, $SiO_{4/2}+\frac{1}{2}O=SiO_{(3/2+1/1)}$.

The connectivity-probability logic will be more fully developed in a separate paper. The primary intent here is to find whether a simple physical picture of the connectivity structure invariant correlates reasonably well with durability in Na_2O-B_2O_3-SiO_2 glasses. This means that the structure model must produce the approximate shape of the isodurs. The question to be answered is whether contour lines of constant statistical geometry are similar to isodurs over a wide range of composition. If the answer to this question is yes, then we have a useful theoretical guide for improving durability.

Figure 4 shows two lines for structure invariant values, $X \times Y$ = 0.85 and 0.65 respectively. The data points bracketed by these two lines include experimental $R \times 10^4$ values between 2 and 40.

Three things are clear:

(1) Agreement between the first-order connectivity theory described above and the experiment is good. Durability values, where $R \times 10^4 = 2$, have an $X \times Y=0.825\pm0.005$ while those values which are an order of magnitude higher (20 to 40) are on or very near $X \times Y=0.650$. Note that the structure invariant lines of Figure 4 "fit" the experimental data as well as the estimated isodur contours of Figures 2 and 3. Also note that some anomolous data points in each of the figures have significantly different batched and analyzed compositions (Table I).

(2) There is not enough experimental data to make the model more realistic. In particular, the breadth of the minima must be experimentally determined. Once these are known, the idealized complete chemical order assumed in this model can be modified mathematically to conform to the chemical disorder of real glasses.

(3) There is a sound structural basis for using isodur contours of the ternary system as a useful framework for fitting chemical durability of more complex glasses. This is done in the next section.

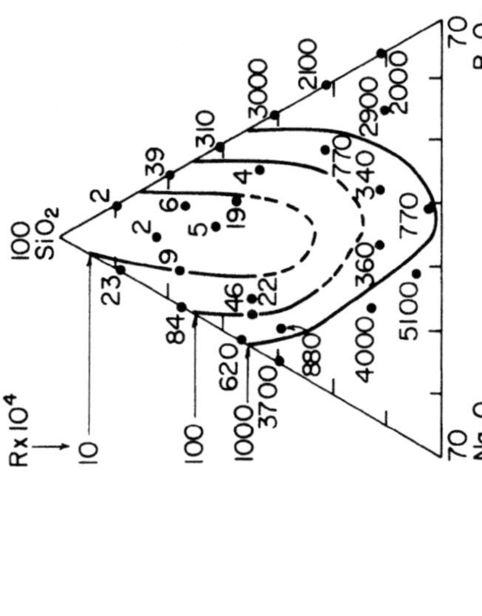

Figure 2. Isodurs in Na_2O-B_2O_3-SiO_2 Ternary System.

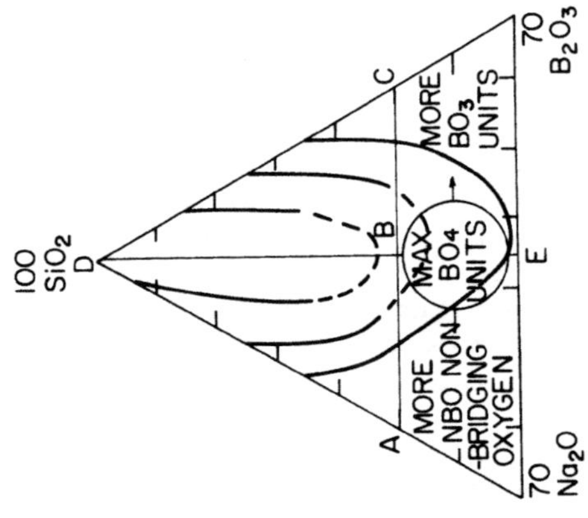

Figure 3. Structural Interpretation of Isodurs.

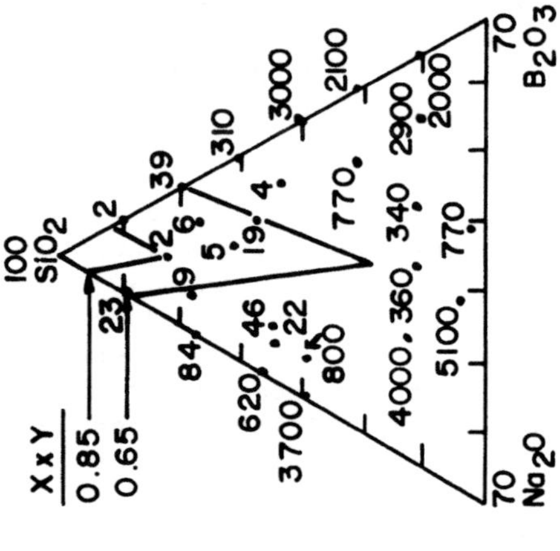

Figure 4. Calculated Connectivity Invariants (XxY) with Data Points shown.

RELEVANCE OF ISODURS TO COMMERCIAL GLASS DURABILITY

Few commercial glasses are simple ternaries. Yet, if the isodurs are to serve a practical purpose, it must be possible to examine commercial glasses in such a framework. Therefore, an attempt has been made to "force-fit" complex commercial glasses into a ternary description.

An empirical method was used which is different from the theoretical approach described in the previous section. "XO_2" is defined as the sum of all oxides that are "network formers" and known to be relatively insoluble in water. XO_2 replaces SiO_2 on the ternary. "R_2O" is defined as the sum of all alkali oxides, those "network modifiers" that can contribute non-bridging oxygen. There is some doubt as to whether the alkaline earths, "RO" should be considered as part of the XO_2 since they generally enhance water durability, or as part of the RO_2 since they can potentially contribute oxygen and thus enhance the formation of BO_4 units. This uncertainty is expressed on the subsequent diagrams as a bar. The point defined by the upper extremity of the bar defines the ternary position if RO is included with XO_2. The point defined by the lower extremity of the bar defines the position if RO is added to R_2O. R_2O (+RO) replaces Na_2O in the subsequent diagrams.

We have examined three categories of commercial glasses: (1) sodium borosilicates that contain minor amounts of other elements, (2) typical optical glasses, and (3) prototype glasses under study for the incorporation of nuclear waste for disposal. The compositions of these glasses are shown in Table III, including the sums for XO_2 without RO included, for R_2O and for RO. Figures 5 through 7 show these compositions plotted on ternary diagrams with the isodurs superimposed and with chemical durability test data indicated.

Figure 5 pictures three commercial sodium borosilicates. The highly phase separable region for sodium borosilicates defined by Hood and Nordberg[2] is shown. The "powder test data" result from a standard test in which crushed glass is extracted by water at 90°C for 4 hours. Results are expressed as % Na_2O extracted from the glass. Note that the results fall in the order predicted by the isodurs.

Figure 6 illustrates complex optical glasses. The weathering ranking represents results of a test in which these glasses were exposed to 98% relative humidity at 50°C. Note that glass having fair resistance ranges from a durability position less than isodur 10 to a position greater than isodur 100 depending on whether RO is included in XO_2 or added to R_2O. Probably the midpoint of the bar is the best position to choose in the framework of these isodurs.

Table III. Compositions (Mole %) of Some Commercial Glasses.

	Borosilicate			Optical			Nuclear Waste			
	1	2	3	4	5	6	7	8	9	10
SiO_2	83.3	76.3	80.5	64.1	54.9	44.1	62.8	44.5	52.5	37.7
Al_2O	1.2	3.8	0.9	6.1	3.7	1.8		0.7		
ZrO_2				1.3		0.4	1.1	1.4	1.1	1.3
TiO_2								2.8		
Fe_2O_3							0.9	1.1	0.9	1.0
Cr_2O_3								0.1		0.1
P_2O_5							1.2	1.6	1.2	1.4
Y_2O_3							0.1	0.1	0.1	0.1
MoO_3							1.2	1.5	1.3	1.4
TeO_2							0.1	0.1	0.1	0.2
La_2O_3					0.3	0.1	0.8	1.0	0.8	0.9
CeO_2							1.7	2.1	1.7	1.9
Pr_6O_{11}							0.1	0.1	0.1	0.1
Nb_2O_3							0.9	1.1	0.8	1.0
Gd_2O_3							0.1	0.1	0.1	0.1
Sm_2O_3							0.1	0.1	0.1	0.1
Na_2O	4.0	7.4	3.5	2.0			13.0	26.9	14.8	9.1
K_2O		0.4				0.1	0.3	0.4	0.3	4.1
Li_2O				6.0						
CaO		0.9			1.4	0.1		2.6		2.2
BaO		0.9		3.0	26.6	28.1	0.3	0.3	0.3	0.8
SrO					0.1		0.3	0.4	0.3	1.2
ZnO					2.5	0.1				21.5
NiO							0.4	0.5	0.4	0.4
PbO			0.5	1.5						
B_2O_3	11.5	10.0	14.8	16.0	10.3	25.2	14.7	10.4	23.3	13.3
XO_2	84.5	80.1	81.4	71.5	58.9	46.4	71.0	58.4	60.7	47.3
R_2O	4.0	7.8	3.5	8.0	0.0	0.1	13.0	27.3	15.1	13.2
RO		1.8	0.5	4.5	30.6	28.4	1.3	3.8	1.0	26.1

Figure 7 shows four "nuclear waste" glasses. (It is interesting to note that the waste loading, i.e., the nuclear waste material incorporated into the glass, represents 25 to 30% by weight but only about 10% by molecular volume.) The corrosion results shown represent the rates predicted for water at 25°C, the prediction having been made from short term tests. Glasses 7, 9 and 10 have corrosion rates in the order predicted by the isodurs, assuming we consider glass 10 to be positioned in the ternary at about the midpoint of the bar. Glass 8 has better durability than predicted, indicating that the empirical isodur model needs some refinement. The previous discussion of structural implications suggests that glasses 8 and 10 might be improved without changing XO_2 by increasing B_2O_3 at the expense of Na_2O to bring the ratio closer to unity.

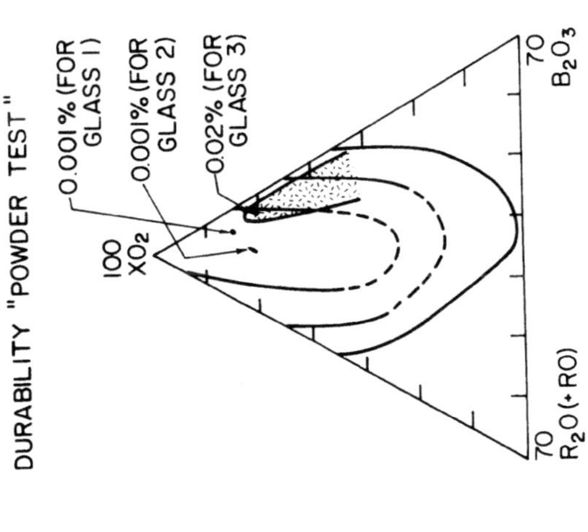

Figure 5. Sodium Borosilicate Glasses.

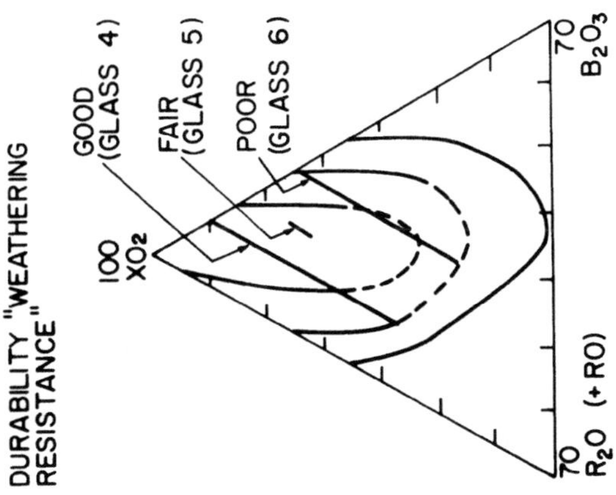

Figure 6. Optical Glasses.

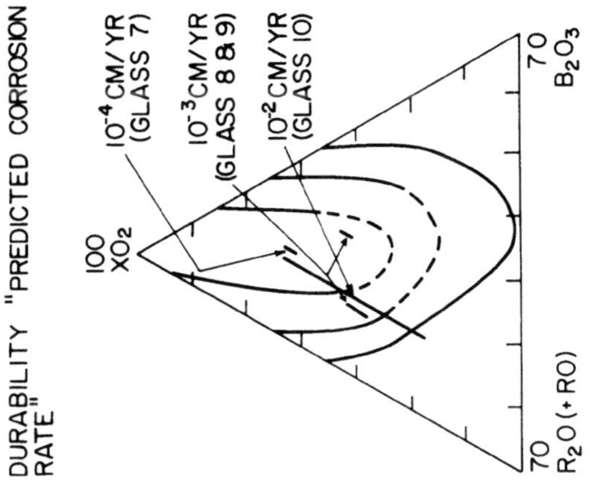

Figure 7. "Nuclear Waste" Glasses.

DISCUSSION AND CONCLUSIONS

Chemical durability measurements are interpreted to be indicative of structural integrity, or connectivity, in the Na_2O-B_2O_3-SiO_2 ternary. The fact that this property is constant for a variety of continuously changing compositions suggests that such compositions have similar structural integrity.

The inflection in the isodurs is interpreted to have particular structural significance. It can be the composition for a given SiO_2 level at which the number of vertex-linked tetrahedral units is maximized. The simplest explanation is that at this position, about 1:1 Na_2O:B_2O_3, all of the boron-containing units exist as BO_4, spatially connected in three dimensions. At Na_2O:B_2O_3 > 1:1, BO_3 units exist in direct proportion to the amount of Na_2O present. At Na_2O:B_2O_3 > 1:1, non-bridging oxygen exists in increasing quantity as Na_2O increases.

Good agreement between first order connectivity theory and experimental measurements of chemical durability is obtained in Na_2O-B_2O_3-SiO_2 glasses. The correlation is between structure invariant parameters calculated for an idealized network and isodurs. Each contour line of structure invariance is a line of variable composition. These theoretical lines correspond to experimentally determined lines of equivalent durabilities (isodurs).

There are a wide variety of glasses containing B_2O_3 which are of commercial interest (Figure 8). In many cases, isodurs might be used as an empirical tool to predict changes in their durability.

Additional work should be done. NMR should be capable of directly measuring RO_4 units in a ternary system. Better definition of the isodurs can be achieved with additional measurements. The shape of isodurs obtained in acid and base solutions will aid in understanding structure since different mechanisms of dissolution will prevail. The kinetics of dissolution at various points in the ternary will help to quantify interpretations. Composition variables, such as the K_2O and Li_2O substitutes for Na_2O in borosilicate systems and the addition of alkaline earths should be studied. This work appears to be a fruitful avenue to pursue in light of the results discussed above.

Finally, it seems apparent that knowledge gained about the Na_2O-B_2O_3 system is useful in understanding the Na_2O-B_2O_3-SiO_2 system; some recent studies referenced by Griscom[1] suggest that a connectivity maximum occurs in the binary also at about 1:1=Na_2O:B_2O_3.

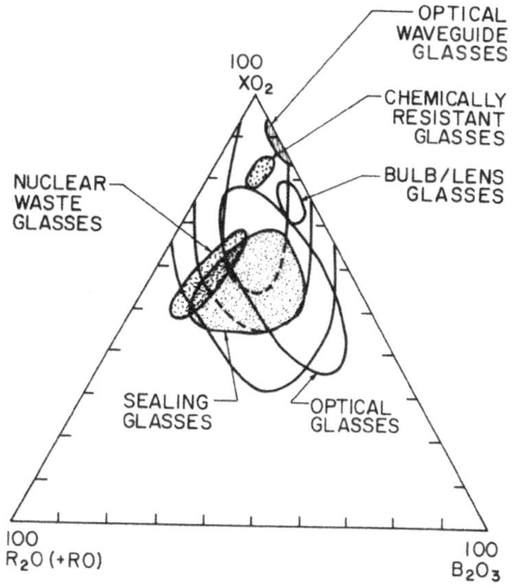

Figure 8. Commercial Glass Composition Areas.

ACKNOWLEDGEMENTS

Mr. J. K. O'Bryan and Ms. M. L. Nelson conducted the chemical durability tests, Mr. G. B. Carrier did the electron microscopy and Dr. R. A. Burdo and Mr. O'Bryan made the chemical analyses.

REFERENCES

[1]D. L. Griscom, "Borate Glass Structure". pp 1-128, this volume.

[2]M. B. Volf, Technical Glasses (Sir Isaac Putnam and Sons, Ltd., London, 1961).

[3]ASTM Designation: C-225-58, 1973 Annual Book of Standards (Am. Soc. Testing Mats., Philadelphia, 1973).

PROPERTIES OF SILICA GLASSES CONTAINING SMALL AMOUNTS OF B_2O_3

William C. LaCourse and Harrie J. Stevens

New York State College of Ceramics

Alfred, New York

INTRODUCTION

In the classic mixed alkali effect a maximum occurs in the electrical resistivity and activation energy for conductivity (ΔE_A) when one progressively substitutes one alkali (R_1) with another (R_2). This effect is illustrated in Figure 1. LaCourse,[1] using the electrodynamic model of Hendrickson and Bray,[2, 3] has predicted that in certain glasses containing a single alkali ion type, but with two dissimilar bonding sites for the alkali ion, an effect similar to the mixed alkali effect may be observed (Figure 2). Taylor and Rindone[4] have postulated a similar effect to account for effects of water on the internal friction in glasses.

The requirements for observing a "mixed site effect" are

1. The glass must contain two or more dissimilar bonding sites for alkali ions.

2. The vibrational frequencies of alkali ions on these sites must be different.

3. Both sites must be occupied.

4. The sites must be nearest neighbors.

Amorphously phase separated glasses may not satisfy the fourth criterion since sites of one kind are generally segregated.

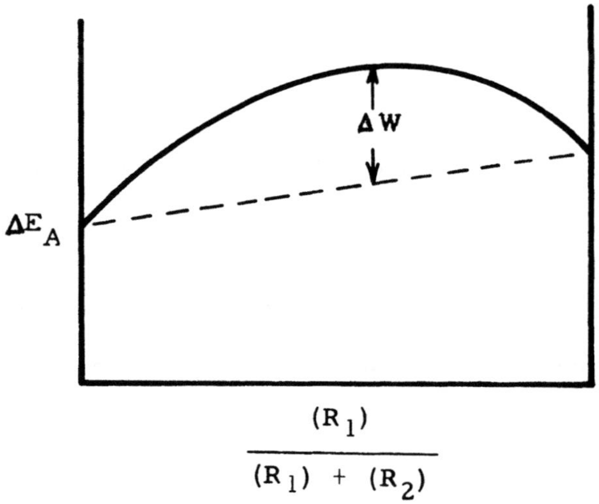

Figure 1. Classical mixed alkali effect.

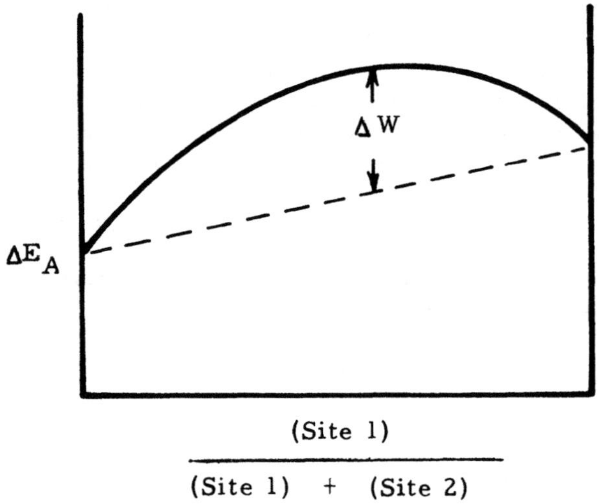

Figure 2. Mixed site effect.

According to Hendrickson and Bray,[2,3] in the normal mixed alkali effect the increase in activation energy, which occurs when one alkali is partially substituted for another, is caused by the mutual interaction of electric fields produced by the oscillating electric dipoles of neighboring alkali ions. The vibrational energy levels of each ion are split by this interaction and, since at low temperatures most ions will exist in the lower energy state, the energy required for moving the ion (ΔE_A) increases. The increase in activation energy for conductivity is essentially equal to the interaction energy between vibrating alkali ions, (ΔW), which in turn is related to the difference in vibrational frequencies of the two ions, ω_1 and ω_2. That is:

$$\Delta W = \frac{-E_1^2 e^2 \left(\omega_2^2 - \omega_1^2\right)}{2\lambda_2\left[\left(\omega_2^2 - \omega_1^2\right)^2 + \omega_1^2 \frac{R_2}{\lambda_2}\right]} \tag{1}$$

Here, E_1 is the electric field of ion 1 seen by ion 2, e is the charge on ion 2, λ_2 is the effective mass of oscillator 2 and R_2 is the damping coefficient of oscillator 2. In the mixed alkali effect the different vibrational frequencies result from the mass difference of the two alkali ions as well as the difference in force constants of the bonds formed by the ions, generally to non-bridging oxygen sites. In the mixed site effect, where only one alkali ion type is present, the different vibrational frequencies result from the mass differences between the anionic sites as well as the difference in force constant of the bonds formed by the alkali at the two different sites.

Alkali-alumino-silicate and alkali-boro-silicate glasses are examples of glasses with two different sites for alkali ions. When B_2O_3 is added to an alkali silicate glass non-bridging oxygen (NBO), sites are converted to four coordinated boron sites as suggested in the following equation:

$$2\left[-\text{Si-O}^{-}\text{Na}^{+}\right] + B_2O \rightarrow 2\left[-\text{Si-O-B-}\right]\text{Na}^{+} \tag{2}$$

Two NBO sites are eliminated and two tetrahedral boron sites are created for each B_2O_3 unit added. Since boron has a +3 charge and, in tetrahedral coordination, shares 4 oxygen ions, the Na^+ previously associated with the NBO site now bonds to the $(BO_4)^-$

site to maintain local electroneutrality. Therefore, in sodium silicate glasses containing boron there are two distinct bonding sites for Na^+, as illustrated in Figure 3 below. This paper investigates the influence of these sites on the properties of glasses in the sodium borosilicate system.

EXPERIMENTAL PROCEDURE

Glasses with the following three general formulas were investigated.

(a) $15\ Na_2O \cdot X\ B_2O_3 \cdot (85 - X)\ SiO_2$ (3)

(b) $25\ Na_2O \cdot X\ B_2O_3 \cdot (75 - X)\ SiO_2$ (4)

(c) $33.3\ Na_2O \cdot X\ B_2O_3 \cdot (66.6 - X)\ SiO_2$ (5)

Values of X were chosen such that in the formulations the ratio of boron to sodium (B/Na) was equal to 0.0, 0.10, 0.15, 0.20 and 0.25. Glasses were prepared from reagent grade or ultrapure materials and melted in a platinum crucible in an electric furnace at temperatures up to 1550°C and for times up to 24 hours, depending on composition. All melts were hand=stirred with a platinum stirrer at least twice. Poured samples were annealed for approximately 1/2 hour and slowly cooled to a temperature below

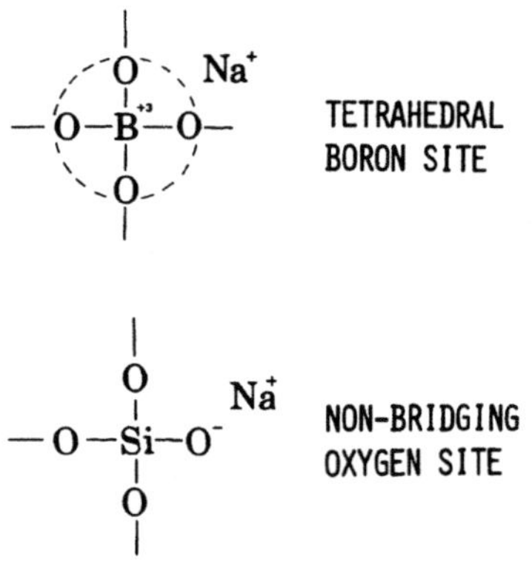

Figure 3. Sodium sites in a sodium borosilicate glass.

the glass transition region. Selected glasses were analyzed and found to be within 2% of the calculated composition. A few key compositions were also investigated using scanning electron microscopy. No evidence of amorphous phase separation was found.

Thermal expansion coefficients between 25°C and 425°C were obtained from an Orton Recording Dilatometer. Annealing points, defined as the temperature at which the viscosity reaches 1.5×10^{13} poises were determined by use of a "Beam Bender" viscometer. Finally, a qualitative measure of the rate of alkali extraction from these glasses in an aqueous media was obtained by monitoring the change in pH of a glass-water slurry with time.

RESULTS AND DISCUSSION

Thermal expansion coefficients and annealing points are summarized in Table I. The thermal expansion coefficient is not strongly influenced by the boron content although a slight decrease is observed in the 25 $Na_2O \cdot X\,B_2O_3 \cdot (75 - X)\,SiO_2$ series. The annealing point is more strongly effected, increasing by up to 40°C. These findings are consistent with the elimination of NBO sites by boron resulting in a "tighter" structure.

Figure 4 illustrates the effect of B_2O_3 on the rate of Na^+ extraction. The values actually plotted indicate the time required for a glass-water solution to reach a pH from a starting pH of 5.8. In all cases the initial boron reduces the rate of alkali extration. The mechanism of this effect is probably twofold. It is well known that the Na^+ extraction results from an ion exchange process with H^+, e.g:

$$\text{Si-O-Na}^+ + \text{HOH} \rightarrow \text{Si-OH} + \text{NaOH} \qquad (6)$$

When the exchange site is Si-O-Na^+ the thermodynamics of the exchange reaction is favorable due to the large strength of the OH bond which forms. However, $(BO_4)^- Na^+$ sites will not take part in the ion exchange process since the negative charge on this site is associated with the tetrahedron rather than a particular oxygen and unpolarizable cations such as H^+ cannot effectively compensate for this charge as well as the Na^+. Therefore, the thermodynamics do not favor H^+-Na^+ exchange at these sites. A second effect is probably related to a decrease in the interdiffusion coefficients for the H^+-Na^+ exchange although this cannot be verified in the present work.

Table I. Thermal Properties

Composition (mole %)			Linear Thermal	Annealing
SiO_2	Na_2O	B_2O_3	Expansion x $10^7/^oC$	Point^oC
85.0	15.0	0	82	497
83.50	15.0	1.5	80	515
82.75	15.0	2.25	84	535
82.00	15.0	3.0	81	531
81.25	15.0	3.75	82	-
75.00	25.0	0	132	471
72.50	25.0	2.5	130	485
71.25	25.0	3.75	-	493
70.00	25.0	5.0	119	490
68.75	25.0	6.25	123	511
66.67	33.3	0	160	455
63.33	33.3	3.33	165	464
61.67	33.3	5.00	160	464
60.00	33.3	6.67	161	-
58.33	33.3	8.33	170	463

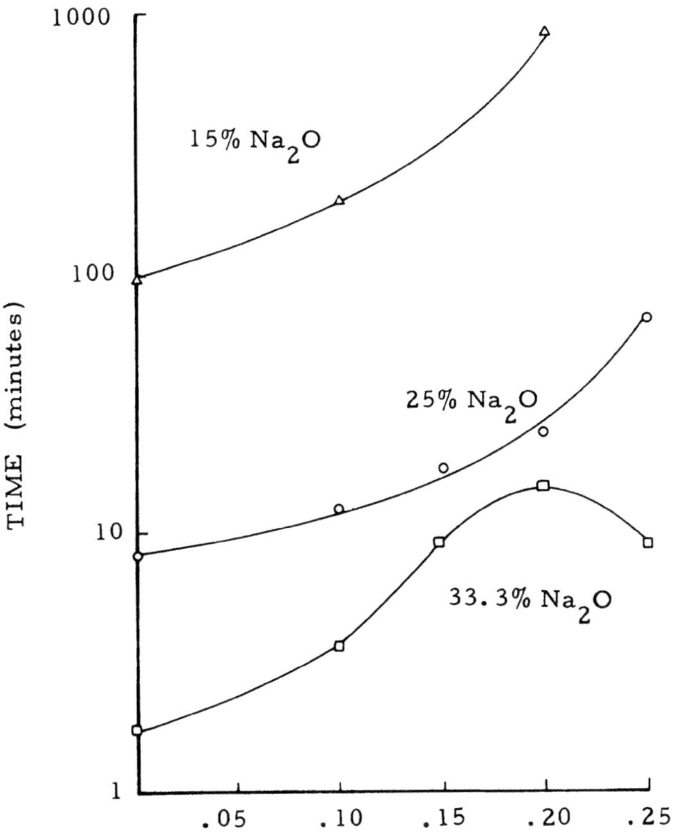

Figure 4. Time to reach pH 9.0 vs. B/Na ratio.

Finally, recent evidence indicates that in alkali borosilicate glasses containing small amounts of boron, the fraction of boron in four coordination (N_4) will be approximately 1 as long as the SiO_2/B_2O_3 ratio is large.[5, 6] The increased rate of alkali extraction in the 33.3 $Na_2O \cdot$ 8.3 $B_2O_3 \cdot$ 58.3 SiO_2 glass is probably related to the formation of BO_3 groups in the glass since the ratio of SiO_2/B_2O_3 in this composition is at the limit for which one expects an N_4 ratio of 1.

Table II summarizes the effect of boron on the electrical properties of these glasses. In the 25% and 33% Na_2O series the conductivity goes through a minimum while the activation energy exhibits a maximum in the same compositional range (B/Na ≈ .15) as shown in Figure 5. In the 15% Na_2O series a continuous decrease in conductivity is observed. These results are strong evidence for a mixed site effect in these glasses.

Table II. Electrical Properties

Composition (mole %)			σ (25°C)	ΔE	σ_0
SiO_2	Na_2O	B_2O_3	$ohm^{-1}cm^{-1}$	kcal/mole	$ohm^{-1}cm^{-1}$
85.00	15.0	0	4.7×10^{-12}	15.3	1.79
83.50	15.0	1.5	2.05×10^{-12}	15.9	1.50
82.75	15.0	2.25	1.4×10^{-12}	16.2	1.25
82.00	15.0	3.0	1.2×10^{-12}	16.6	0.83
81.25	15.0	3.75	8.4×10^{-13}	17.1	0.48
75.00	25.0	0	1.8×10^{-10}	14.8	0.11
72.50	25.0	2.5	1.5×10^{-10}	14.9	0.11
71.25	25.0	3.75	1.5×10^{-10}	15.6	0.035
70.00	25.0	5.0	1.27×10^{-10}	15.3	0.066
68.75	25.0	6.25	8.4×10^{-11}	15.1	0.14
66.67	33.3	0	4.8×10^{-9}	11.2	2.3
63.33	33.3	3.33	2.0×10^{-9}	12.5	0.45
61.67	33.3	5.00	1.5×10^{-9}	14.0	0.05
60.00	33.3	6.67	1.6×10^{-9}	13.6	0.09
58.33	33.3	8.33	1.9×10^{-9}	13.0	0.21

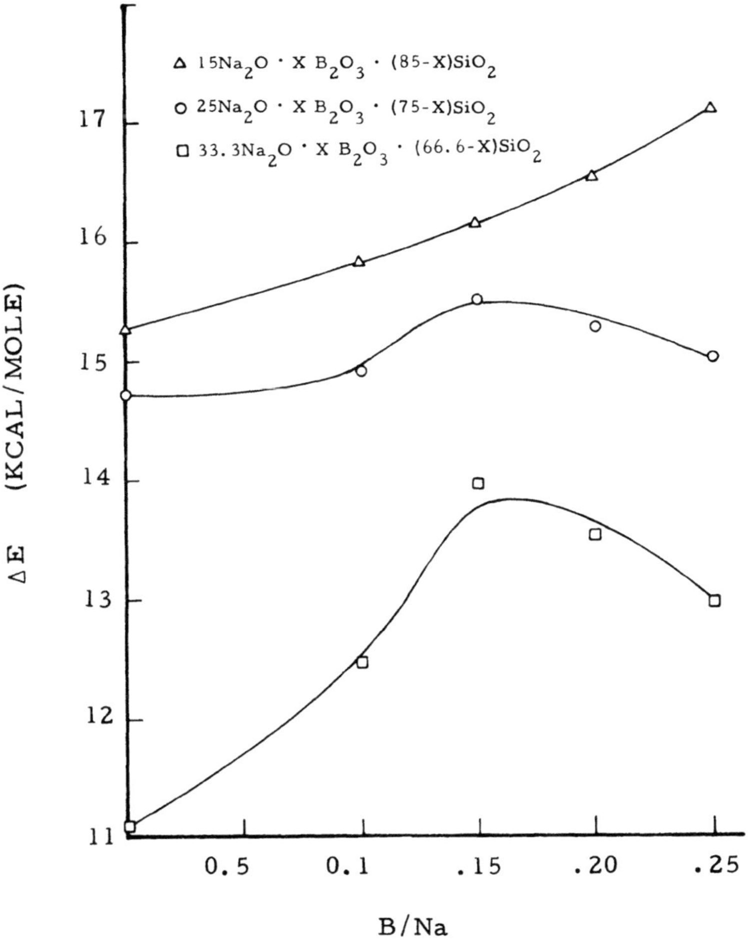

Figure 5. Activation energy vs. B/Na ratio for sodium-borosilicate glasses.

Na^+ will be less tightly bound to a $(BO_4)^-$ site than to a NBO. Therefore, in the absence of a mixed site effect the activation energy for Na^+ diffusion should decrease continuously as the concentration of these sites increases.

The mixed site effect predicts an increase in the activation energy for Na^+ diffusion with the initial B_2O_3. The magnitude of the effect will depend on the difference in vibrational frequencies between the Na^+ on the $(BO_4)^-$ and Si-O^- sites and on the total alkali content (i. e. average Na^+ - Na^+) separation distance. Beyond a certain B/Na ratio the activation energy is predicted to decrease continuously with B_2O_3 additions. The B/Na ratio corresponding to the maximum in activation energy, and the magnitude of the increase cannot be predicted without more information.

The behavior of glasses in the 25% and 33% Na_2O series agrees well with that predicted from the mixed site effect. However, the 15% Na_2O series shows no maximum in the compositional range investigated and the activation energy increases even more than that observed for the 25% series. While it is possible that the maximum occurs at a greater B/Na ratio the magnitude of the activation energy increase is difficult to account for by this model.

The effect of the mixed site effect on other properties has not been considered. It is known, however, that the viscosity of mixed alkali glasses goes through a minimum and the thermal expansion exhibits a weak maximum.[7] These effects are in opposition to those resulting from the structural change occurring in these glasses (Eg. 1) and are probably masked by it. The lack of a strongly dependent thermal expansion for these glasses may be a result of these opposing influences.

REFERENCES

1. W. C. LaCourse, J. Non-Cryst. Solids 21 (1976) 431.
2. J. R. Hendrickson and P. J. Bray, Phys. Chem. Glasses, 13 (1972) 45.
3. J. R. Hendrickson and P. J. Bray, Phys. Chem. Glasses, 13 (1972) 107.
4. T. D. Taylor and G. E. Rindone, J. Non-Cryst. Solids, 14 (1974) 157.
5. M. E. Milberg, J. G. O'Keefe, R. A. Verhelst and H. O. Hooper, Phys. Chem. Glasses, 13 (1972) 79.
6. P. J. Bray, this volume.
7. J. Isard, J. Non-Cryst. Solids, 1 (1969) 235.

OPTICAL PROPERTIES OF THE SODIUM-BORATE GLASS SYSTEM

Karl-Heinz Mader and Thomas J. Loretz

Schott Optical Glass Inc.

Duryea, Pennsylvania 18642

As a direct consequence of the "borate glass anomaly," a phrase describing non-uniform and seemingly inexplicable composition and property behavior of alkali-borate glasses, the borate glass structure has been investigated extensively. D. L. Griscom provided an in-depth review of this experimental and theoretical work.[1]

A great deal of the early papers centered around the coefficients of thermal expansion versus alkali content in the alkaline-borate system. Refractive index, dispersion, UV absorption edge, and density data have also been reported and discussed in the framework of the borate glass structure.

Although individually these physical-optical data mentioned do not provide direct structural insight, in combination their results may do so. The intent of this work was to measure physical-optical properties within a single melt, prepared under optical glass melting conditions and, then; evaluate these data in light of the present understanding of the borate glass structure.

EXPERIMENTAL PROCEDURES: Melts of the binary Na_2O - B_2O_3 alkaline borate glass system in the 0 to 30 mol percent range were made using optical grade H_3BO_3 and Na_2CO_3 raw materials. All melts were prepared in induction heated, open air Platinum pots, using a Platinum stirrer. Melt time, including homogenizing and refining, was approximately five (5) hours.

X-ray fluorescence analysis of the K_α peak of Sodium served to substantiate the actual chemical composition of the melts. Wet chemical analysis, used to standardize the X-ray fluorescence output, were proved by the Analytical Laboratory of the JENAer GLASWERK SCHOTT & GEN., Mainz (Mr. Jenemann).

The codes used in Table I (e.g. NB-X) denote a Sodium (N) - Borate (B) glass of (X) mol percent by batch input. The first glass in the series, B-1, is a pure Boron Oxide composition.

The thermal expansion samples, and those for the Transformation and Littleton Softening Point, were annealed at 40°C per hour. All subsequent physical-optical data were measured on precision-annealed glass (3°C per hour). The purpose of this controlled annealing was to achieve a "zero" strain condition for samples used for measuring the stress-optical coefficient, and to create a uniform reference from which to compare index of refraction, density, and ultraviolet cut-on characteristics.

The thermal expansion coefficient measurements were made dilatometrically, as were the transformation temperatures, (Tg's). The softening point measurements were those of Littleton, following ASTM procedures.

The stress-optical coefficient measurements were made following the National Bureau of Standards approach, using 632.8 (He-Ne Laser) illuminant

Index of refraction measurements were made on a Pulfrich refractometer, using the yellow "d" line of Helium (587.56 nm).

H_2O/OH-concentrations were determined based on the O-H stretching absorption band, 2.8 micron, as measured on a Beckman DK-2A Spectrophotometer.

RESULTS: Tables I and II, and Figures 1 through 10 summarize all results presented in this paper. The physical-optical data measured in the Na_2O - B_2O_3 glass system indicate that certain "unusual" behavior is contained within its makeup, it is not only present in the thermal expansion coefficient data but it is also quite apparent in an evaluation of the refractive index, density, UV cut-on, and stress optical results.

The Coefficient of Thermal Expansion versus composition curve (Fig. 1) exhibits a definite minimum between 16.2 and 17.5 mol percent Na_2O. This phenomenon holds true for all temperature intervals reported between 20°C and 350°C (Table I) but also for the low temperature CTE between -196°C and 25°C (Fig. 2).

The results on the Transformation Temperatures (Viscosity approx. $10^{13.2}$P) and the Littleton Softening Point (Viscosity = $10^{7.6}$P) are summarized in Figure 3. The Transformation Temperature Curve shows non-uniform behavior between 16.2 and 17.4 mol percent Na_2O addition; with a maximum occurring at approx. 20.0 mol percent. The Littleton Softening Point Temperature Curve, very similar to the latter, also exhibits non-uniform behavior in the region between 16.0 and 17.0 mol percent and a maximum at approximately 22.0 mol percent Na_2O addition.

Both the 50 percent and 10 percent Ultraviolet cut-on versus Composition Curves (Fig. 4) exhibit very pronounced changes between 13.0 and 18.0 mol percent Na_2O addition. The turning point in both cases is found to occur at approximately 17.0 mol percent.

The Density versus Composition Curve (Fig. 5) exhibits a subtle non-uniformity between 13.0 and 20.0 mol percent Na_2O addition. The turning point of this disuniformity lies at the composition containing approximately 17.0 mol percent Na_2O.

The Refractive Index (n_d) curve (Fig. 6) exhibits a large change between 12.0 and 20.0 mol percent Na_2O addition. The turning point appears to be at approximately 16.0 mol percent. An analogous and complimentary profile is found for the Stress-Optical Coefficient curve (Fig. 8).

DISCUSSION OF RESULTS: Since historically, the coefficients of thermal expansion (CTE's) of the binary Na_2O - B_2O_3 glass system established the so-called "borate glass anomaly",[3] any discussion of physical-optical properties of this glass system has to start with these data. Remeasuring the CTE's also allows for easy cross-referencing with previously published results. This is of importance, since most of these published data are based on batch input compositions, melted under a wide variety of melting conditions, water contents, and other parameters of structural and compositional influence.

The "borate glass anomaly" had been suggested by the observed existence of abrupt property versus composition changes in binary alkaline borate glasses. Biscoe and Warren[4] introduced the coordination change of three coordinated (BO_3) triangles to four coordinated (BO_4) tetrahedral groups with increasing Sodium-Oxide content in the Na_2O - B_2O_3 system; concluding that the (BO_4) population peaks with the introduction of non-bridging oxygen at approximately 16 mol percent Na_2O. They explain the measured decrease of the coefficient of expansion with the progressive change of the essentially, two-dimensional (BO_3) triangular structure into the three-dimensional tetrahedral (BO_4) structure.

Table I

Chemical Composition and Experimental Results for the Na_2O-B_2O_3 System Studied

PROPERTY	CONDITION	MEAS. ERROR	CODE													
Composition	Mol% Na_2O	±0.1%	B-1 0.01	NB-5 4.4	NB-10 8.7	NB12.5 11.5	NB-15 13.7	NB-16.8 15.8	NB-17.5 16.2	NB-18.8 17.4	NB-20 18.4	NB-21.5 19.6	NB-22.5 20.0	NB-23.8 22.5	NB-25 23.6	NB-30 28.9
Thermal Expansion Coefficient ($\alpha \times 10^7/°C$)	-196.25°C	$\pm 1.5 \times 10^{-7}/°C$	140.0	94.6	N.M.	71.5	69.3	67.4	65.9	N.M.	69.1	N.M.	N.M.	71.9	N.M.	81.4
	20.50°C	$\pm 0.8 \times 10^{-7}/°C$	149.3	103.0	98.8	88.7	87.5	80.7	86.0	85.6	86.2	86.8	87.6	90.4	90.4	102.1
	20.100°C		154.3	107.2	99.0	92.6	89.8	84.7	86.7	86.9	86.9	88.3	89.4	91.9	93.1	102.3
	20.150°C		149.0	109.9	100.5	94.9	92.3	87.8	87.7	89.1	89.2	91.2	91.6	94.7	96.7	107.4
	20.200°C		151.1	113.0	101.8	96.1	94.5	90.4	89.5	90.4	92.3	93.1	95.1	97.1	98.2	110.3
	20.250°C		—	116.00	105.3	97.9	90.9	93.3	90.9	92.4	94.1	95.3	97.6	98.7	101.2	112.8
	20.300°C		—	—	—	101.1	93.7	94.9	93.7	94.3	95.0	97.2	99.1	101.4	104.3	114.9
	20.350°C		—	—	—	—	—	97.9	96.9	96.3	96.2	99.6	101.3	104.0	108.5	117.1
Transformation Temperature (°C)	Dilatometer	±3.0°C	257	301	340	378	387	414	429	448	453	462	469	469	467	464
Softening Point (°C)	Littleton	±3.0°C	364	411	456	480	495	514	525	539	543	546	551	550	549	542
Stress-Opt. Coeff ($K \times 10^6$ nm/cm/dy/cm²)	3°C/hr. anneal 25°C	$\pm 5 \times 10^{-8}$K	9.32	6.40	5.17	4.92	4.76	4.70	4.62	4.45	4.33	4.20	4.05	3.74	3.63	3.19
Density (gm/cm³)		±0.001	1.852	1.972	2.060	2.093	2.121	2.148	2.150	2.178	2.189	2.215	2.221	2.257	2.271	2.345
Refractive Index (n_d)		$\pm 5 \times 10^{-5}$	1.46536	1.48387	1.49442	1.49662	1.49641	1.50024	1.49984	1.50155	1.50210	1.50468	1.50500	1.50806	1.50979	1.51611
Molar Refractivity (M_R) (CM³)		±0.02	10.40	10.05	9.75	9.60	9.48	9.37	9.35	9.24	9.20	9.12	9.09	8.97	8.93	8.68
50% UV Cut-on (nm)		±2 nm	254	251	259	261	263	269	270	276	281	285	286	294	295	303
10% UV Cut-on (nm)		±2 nm	192	196	199	203	206	213	215	223	229	235	234	238	238	243

*N.M. — Not Measured

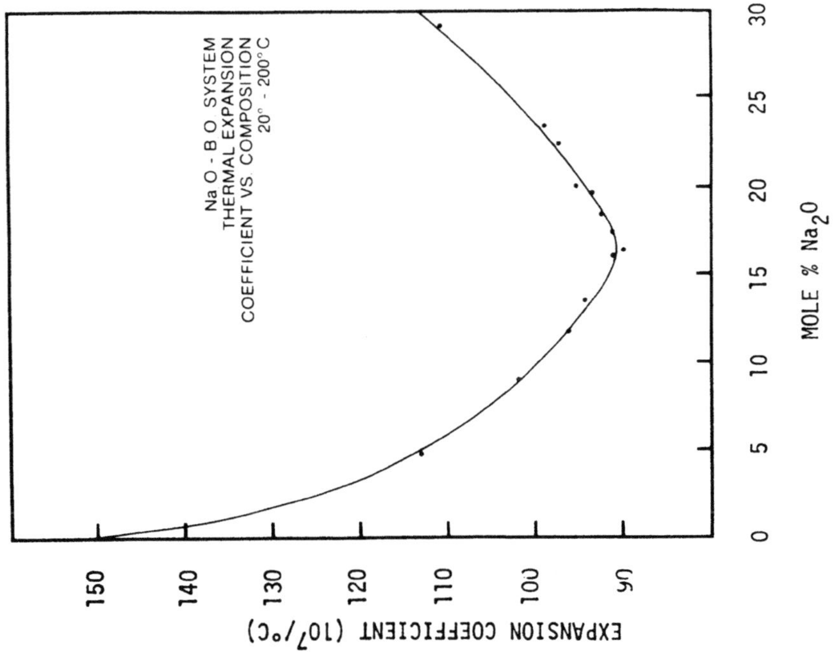

FIGURE 1

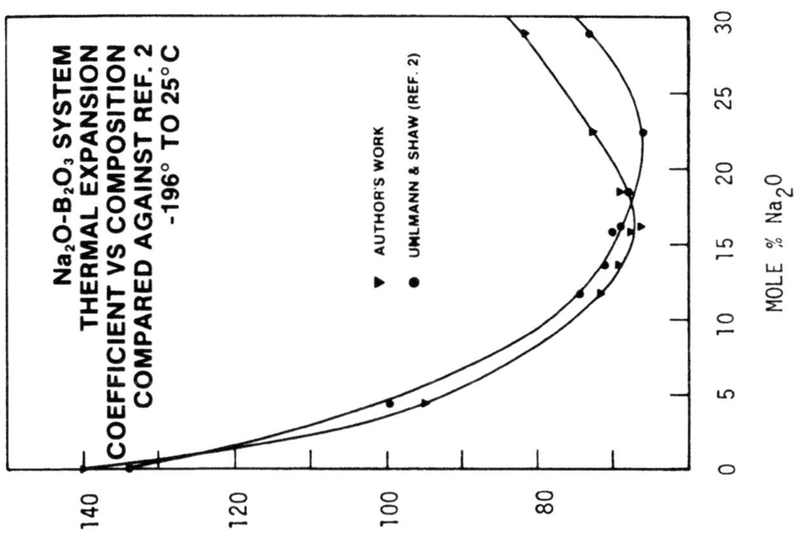

FIGURE 2

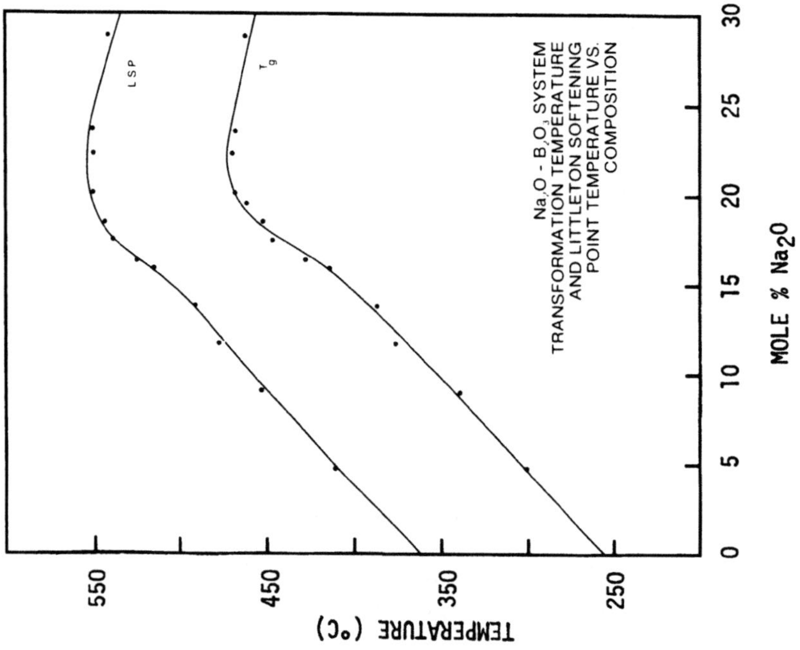

FIGURE 3

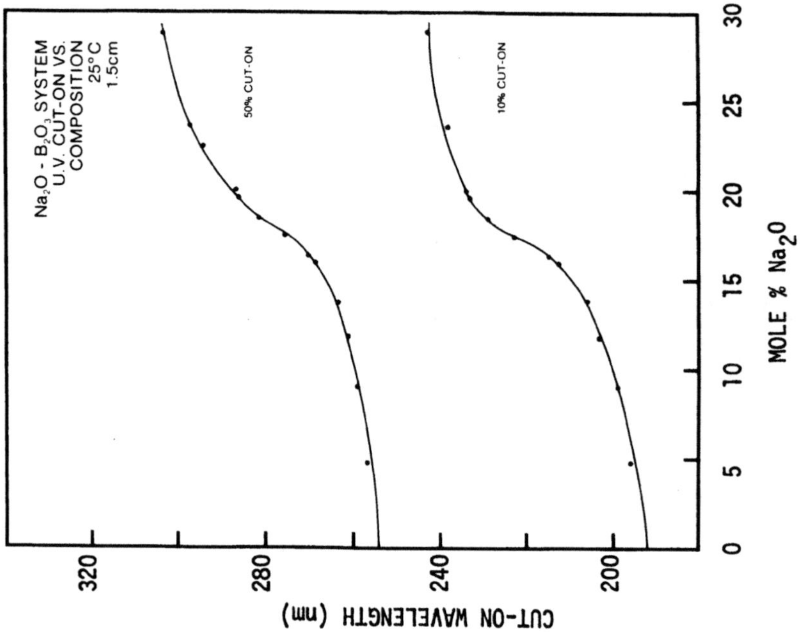

FIGURE 4

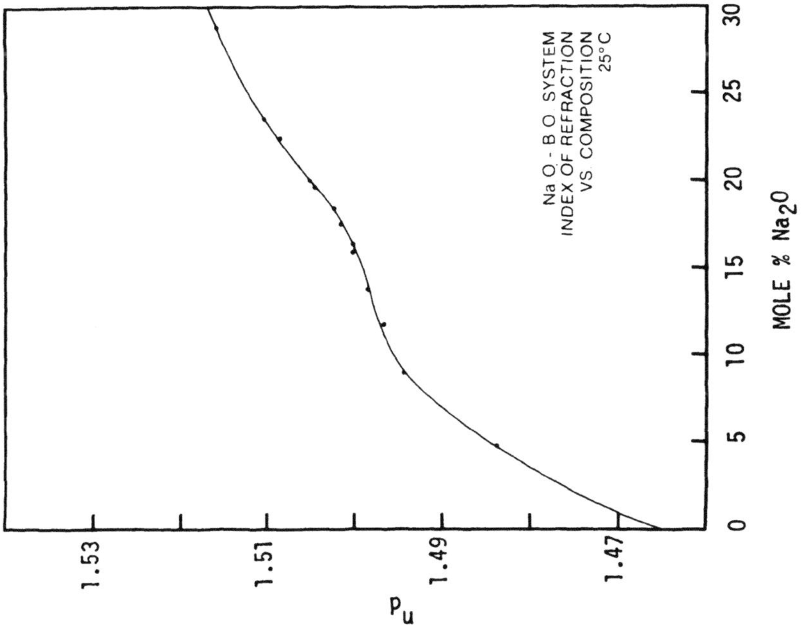

FIGURE 6

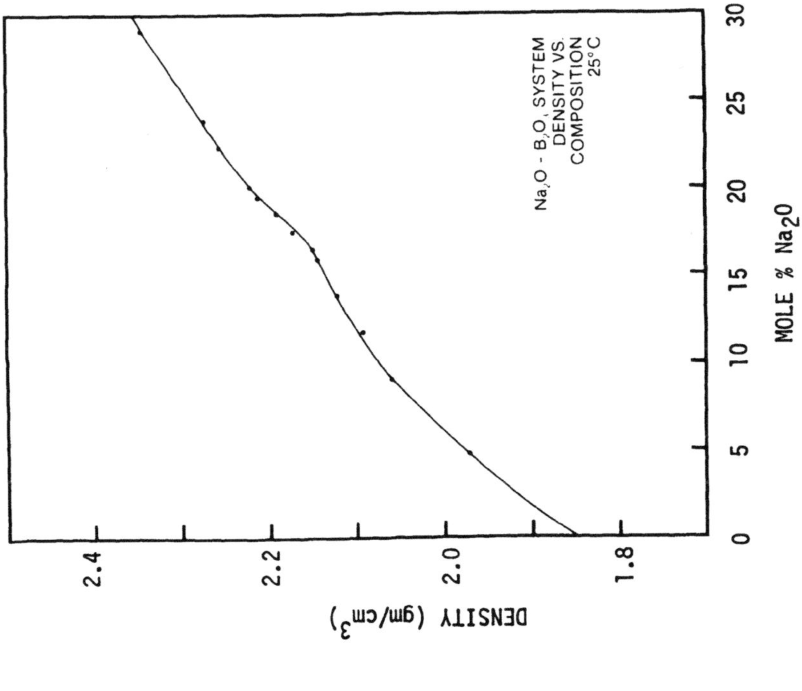

FIGURE 5

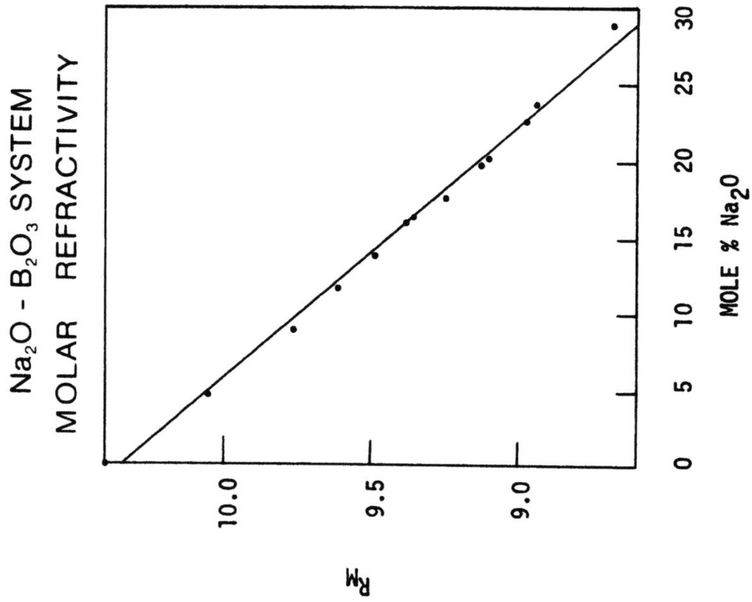

FIGURE 7

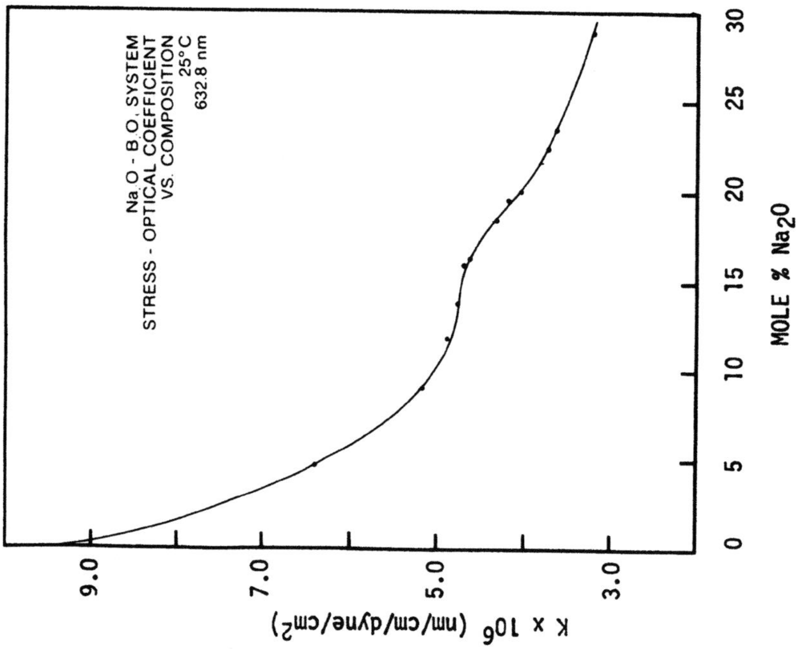

FIGURE 8

The observed minimum, at approximately 16 mol percent Na_2O, "is interpreted as the loosening of the structure by an increase in the number of singly bonded oxygens." The roll-off of the density curve near the same composition is similarly explained. Work by Krogh-Moe et.al.[5] was in agreement with Warren's picture of the $(BO_3)/(BO_4)$ coordination change, but Bray and O'Keefe's[6] NMR data resulted in a direct measurement of the four coordinated (BO_4) concentration and indicated the (BO_4) to increase in number up to a 30 mol percent Na_2O introduction. With these results it seemed the "borate glass anomaly" could no longer be interpreted by means of simple boron coordination changes alone.

Uhlman and Shaw[2] have reviewed the experimental work concerning the expansion coefficient measurements in the binary alkaline-borate glasses, adding low temperature CTE - measurements of their own. They dismiss the idea of a true "borate glass anomaly," based on their broad, almost non-existent, flat minima in the five alkaline-borate systems (Li through Cs) they investigated. They contend that the results "reflect a competition between two processes: the formation of BO_4 tetrahedrea, tending to decrease the thermal expansion coefficient and the introduction of modifying cations, tending to increase it."

Figure 2 presents the work of Uhlmann and Shaw on the Sodium-Borate glass system. The points taken from their paper are plotted together with those of these investigators, for the same temperature range. Since the curves of Uhlmann and Shaw are referenced to batch input, it becomes necessary; based on the vaporization characteristics of the binary alkaline-borate glass melt, as observed in work reported here, to shift all points approximately by one to two mol percent Na_2O to lower values. Assuming that this correction is applicable to the melting conditions used by Uhlmann and Shaw, it becomes evident that their results closely parallel the measurements presented here. In fact, a true minima seems to exist in the low temperature (-196°C to 25°C) thermal expansion coefficient between 16.0 and 18.0 mol percent Na_2O; as well as in all other high temperature CTE-ranges (see Table I and Figure 1).

Before elaborating on this point, more of the physical-optical data of the Na_2O - B_2O_3 system have to be presented. Figure 3 shows the Littleton Softening Points (lg viscosity = 7.6 P) and Transformation Temperatures (lg viscosity approximately 13.2 P) of the Na_2O - B_2O_3 glass system, between 0 and 30 percent mol Na_2O. These results are in qualitative agreement with Tg-temperature measurements in binary alkaline-borate systems by Karsch,[7] viscosity measurements of molten alkali borates by Shartis, Capps, and Spinner,[8] and softening point measurements by Eversteijn, Stevels, and Waterman,[9] to mention only a few. These papers document a viscosity increase with increasing introduction of Na_2O into the Borate structure at viscosities higher than 10^4 poise and explain the viscosity increase with the transformation of two-dimensional (BO_3) boroxol groups into the three-dimensional (BO_4) tetra-borate structure. The change of slope observed in our measurements, between 16 and 17.5 mol percent Na_2O, and the obvious maximum at approximately 20 mol percent Na_2O, suggest in qualitative terms that not just two processes -- as suggested by Uhlmann and Shaw in discussing the CTE data -- but, rather, three (or more) assert themselves on the structure of the alkaline-borate glass system in the compositional range investigated. Though the observed change in the viscosity versus composition curves is only small, it suggests an underlying structural change not explainable with the structural information on hand.

Krogh-Moe suggested that the concept of variable boron coordination has to be more than the structural change from two- into three-dimensional network structures. He explained the viscosity and compressibility characteristics of vitreous B_2O_3 through low activation energy coordination changes, postulating that a certain degree of (BO_4) coordination exists in pure vitreous B_2O_3, or can easily be induced by external forces ("mechanical" coordination). Alkaline oxides would diminish this low activation energy, coordination change, and, through a "chemical" coordination influence, create a viscosity increase. With increasing alkaline-oxide concentration, this network modifying influence becomes more important; the latter influence predominates at Na_2O concentrations higher than 20.0 mol percent.

In order to gain insight into the presence or non-presence of singly-bonded or non-bridging oxygens, the UV cut-on characteristics in the binary Na_2O - B_2O_3 glass system were investigated.

McSwain, Borelli, and Su[10] have measured the UV-absorption characteristics of binary sodium borate glasses based on the generally accepted qualitative understanding that the UV-absorption edge in glasses is determined by the oxygen bond strength in the glass forming network. The absorption edge corresponds to the transition of an electron belonging to an oxygen atom into an excited state. Any change of the oxygen bond in the glass network, for instance, the formation of non-bridging oxygen through the introduction of alkaline oxide, changes the absorption characteristics to longer wavelengths, e.g. lower energies.

McSwain et. al. report that between 0 an 15 mol percent Na_2O, the UV-absorption edge shifts only by 6 nm, but that a sharp increase in absorption occurs between the 15 and 20 mol percent Na_2O - B_2O_3 binaries, resulting in a UV-absorption edge shift of 27 nm. Between 20 and 33 mol percent Na_2O, the shift decreases to only 15 nm. The above investigators explain their observed UV-absorption shift with a sharp increase of non-bridging oxygen between 15 an 20 mol percent Na_2O followed by a gradual further increase with increasing Na_2O content above 20 mol percent Na_2O.

The results reviewed agree very well with the measurements presented in this paper (Table I and Figure 4), keeping in mind that a direct quantitative comparision must take into account sample thickness. Between 0 and 14 mol percent, we observed cut-on changes of 9 nm for the 50 percent cut-on or 14 nm for the 10 percent cut-on points. Between 14 and 20 mol percent of Na_2O the UV-absorption edge changes abruptly by 23 nm for the 50 percent cut-on points or 28 nm for the 10 percent cut-on points, decreasing between 20 and 30 mol percent Na_2O to 17 or 9 nm respectively.

McSwain's work, which pointed out that the shift of the UV-edge between 15 and 20 mol percent Na_2O compositions could be explained by the appearance of non-bridging oxygen in the borate glass structure, is not supported by the work of Krogh-Moe or Bray and O'Keefe, who predicted a smooth increase in four coordinated boron (BO_4) in this concentration range. The relatively abrupt change in UV-absorption characteristics, if associated with the absorption characteristics of oxygen in (BO_4) tetrahedral coordination, would not be in accordance with the simple linear (BO_4) increase as measured earlier by NMR techniques. This either leaves us with the appearance of non-bridging oxygen as the most plausible explanation for the UV-absorption shift or we will have to assume a more complicated model, which would have to explain the UV-cut-on shift by other changes in the glass matrix. It would also have to support the cation size dependency which together with similar observations of UV-cut-on shifts in the alkaline silicate systems lend further support to the non-bridging oxygen reasoning.

If we have to include the appearance of substantial amounts of non-bridging or singly-bonded oxygens, at Na_2O concentration between 15 and 20 mol percent, overlapping with the smooth transition of triangular (BO_3) groups into three-dimensional tetrahedral (BO_4) coordinated groups, we would expect a deviation in viscosity from the almost linear increase observed up to 15 mol percent Na_2O, to lower viscosities above the 15 percent mark. Our measurements indicate that the opposite takes place! The viscosity increases up to 20 mol percent Na_2O and then begins to drop at Na_2O concentrations above 23 to 25 mol percent.

Measurements of the near infrared absorption of glasses can provide structural information specifically about the amount of water contained in the glass structure. In the binary alkaline borates studies, the probability of a large number of (OH) groups being present is very high, especially when the glass has been melted under open atmosphere conditions.

Work done by Wulff and Majumder,[11] as well as, Everstein, Stevels, and Waterman previously cited, indicated the presence of (OH) groups in the B_2O_3 - glass structure and their influence on the physical-optical data as measured. Refractive index, viscosity, density, coefficient of thermal expansion, and other glass properties depend very much on water content, which itself is highly batch material and melting condition dependent.

Knowing the sensitivity of the Borate glass structure to the presence of OH groups, the band position and peak height of the 2.8 micron O-H stretching mode were determined. The results for the binary glass melts discussed are summarized in Table II and Figure 9.

With increasing introduction of Na_2O into the B_2O_3 structure, the 2.8 micron absorption decreases, with the absorption band shifting parallel to larger wavelength. Figure 9 shows that the

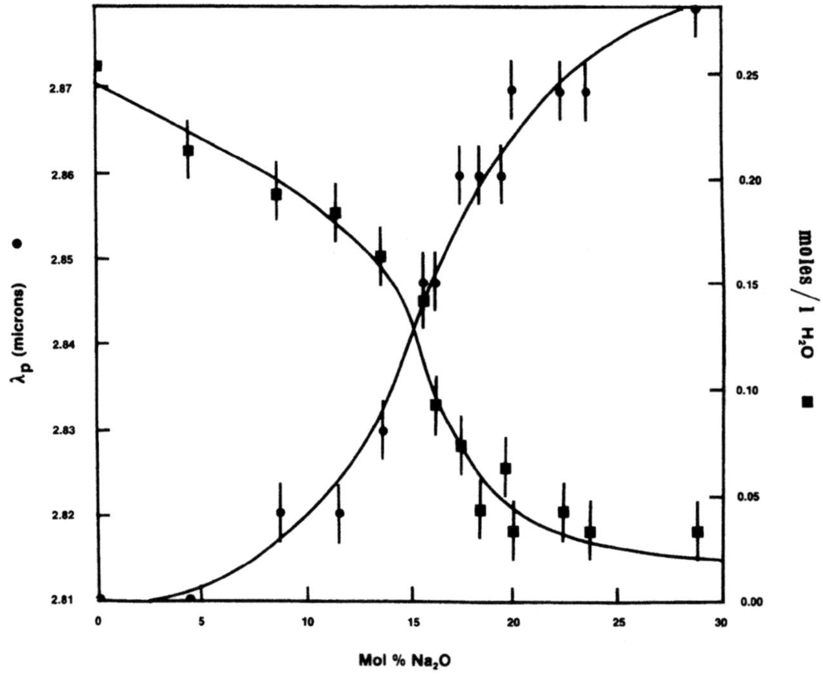

FIGURE 9. Near-Infrared band, O-H determination for the Sodium Borate glass system studied.

main shift occurs between 13 and 18 mol percent Na_2O. It cannot be overlooked that, again, this "sudden" property change occurs in a composition range formerly associated with the "borate glass anomaly."

These results are in agreement with Scholze,[12] who, in his work on water in glass, had already investigated the binary Lithium, Sodium, and Potassium - Oxide Boron-Oxide glass systems. He concluded, that, with the introduction of Na_2O into the B_2O_3 structure, the (BO_3) coordination changes to (BO_4) symmetry, and the alkaline cations occupy the voids of the glass network. The appearance of non-bridging oxygen with further increasing Na_2O content stops the shift of the OH-stretching band, since -- as Scholze argues -- from that point on, bonded OH groups are being incorporated into the network. The latter should have an absorption band in the 3.5 micron region, which Scholze, in his work, did not study.

It should be noted, that all compositions had to be measured after water-free polishing; since, especially with the glass compositions below 15 mol percent, H_2O surface attack led to a rapid deterioration of surface quality. Scholze, Borelli, and coworkers, and more recently Franz[13] reported an OH associated band at 3.15 microns identifying it with the formation of H_3BO_3 on the glass surface. This band, like the 2.8 micron OH - stretching band, has not been detected in pure B_2O_3 glasses formed under water-free conditions. (Another absorption was found in the series of compositions discussed between 3.35 and 3.45 microns, apparently not associated with surface water attack, exhibiting absorption decreases with increasing Na_2O content, thereby, suggesting a structural relationship with the observed 2.8 micron OH band phenomena.)

In interpreting our results on the 2.8 micron band, we have to agree with Scholze that with increasing alkaline content, fewer OH groups are maintained in the binary alkaline borate structure. In the composition range between 14 and 20 mol percent Na_2O, the water content (based on the absorption intensity of the 2.8 micron peak only) decreases by a factor of 4. We calculated the OH concentration in vitreous B_2O_3 using the absorption coefficient, ε, of 141 $mole^{-1}\ cm^{-1}\ l$, as cited in the work of Franz. The water concentration of 0.25 - 0.31 mole/liter corresponds to approximately 0.24 percent by weight as reported by other authors for similar compositions and melting conditions.[14,15]

With the noticeable exception of the thermal expansion coefficients, all other properties seem to follow the properties changes expected in accordance with the decreasing OH concentration, with increasing Na_2O content in the range between 14 and 17 mol percent

TABLE II. H_2O / OH Absorption results for the Sodium Borate Glass system studied.

CODE	COMPOSITION (Mol % Na_2O)	WAVELENGTH (λ_p microns)	ABSORPTION (A)	H_2O CONTENT (Moles/l)
B - 1	< 0.01	2.81	3.52	0.25 *
NB - 5	4.4	2.81	2.96	0.21
NB - 10	8.7	2.82	2.68	0.19
NB - 12.5	11.5	2.82	2.54	0.18
NB - 15	13.7	2.83	2.26	0.16
NB - 16.8	15.8	2.85	1.97	0.14
NB - 17.5	16.2	2.85	1.13	0.08
NB - 18.8	17.4	2.86	0.85	0.06
NB - 20	18.4	2.86	0.56	0.04
NB - 21.5	19.6	2.86	0.85	0.06
NB - 22.5	20.0	2.87	0.42	0.03
NB - 23.8	22.5	2.87	0.56	0.04
NB - 25	23.6	2.87	0.42	0.03
NB - 30	28.9	2.88	0.42	0.03

$t = 0.10$ cm

$\varepsilon = 141$ mole^{-1} cm^{-1} l

* = 0.24% H_2O by weight

Na_2O: Densities and refractive indices decrease, relatively speaking, and viscosity (but not the thermal expansion) increases.

The densities measured for the binary Na_2O - B_2O_3 system are shown graphically in Figure 5. The densities increase with changing slope in the region of 15 to 20 mol percent Na_2O (referred to as the "roll-off" of the density curve) in agreement with measurements by Coenen[16], Eversteijn[9], Shartis[8], and Bresker[17]. The density increase - corresponding to a decrease of the Molar volume - is generally explained as the impact of the (BO_3) to (BO_4) change. with the alkaline occupying the voids of the glass network. With further increasing sodium content, the sodium modifies the structure resulting in a loosening of the packing density. The latter effect has been found to depend very much on the size and the full strength of the alkaline ion. As indicated earlier, Krogh-Moe emphasized the concept of variable coordination in B_2O_3, with the alkaline oxide increasing the activation energy necessary for the (BO_3) to (BO_4) coordination change.

The refractive indices of the Na_2O - B_2O_3 system (Figure 6) follow the structural changes indicated by the dnesity (ρ). Both are directly related through the Molar Refractivity (equation 1),

$$R = \frac{n^2 - 1}{n^2 + 2}\frac{M}{\rho} = \frac{4 \pi N_L}{3} \alpha \qquad (1)$$

which is a measure of the polarizability (α) as determined primarily by the oxygen atoms in the borate glass structure. The refractive index then depends on the polarizability and the Molar volume of the glass $(\frac{M}{\rho})$; it increases with α and decreases with $\frac{M}{\rho}$.

In case of the Na_2O - B_2O_3 system reported, the Molar refractivity (Figure 7) decreases uniformly with increasing Na_2O content: Therefore, if any major structural change occurs in the composition range investigated, it does not result in a sudden change of polarizability. It seems safe to assume that the structural changes as evident in the properties measured are smoothly overlapping, leaving little room for any abrupt anomaly.

The Stress-Optical Coefficient corelates the response in optical retardation to an applied uniaxial stress:

$$K = \frac{\varepsilon_o}{\sigma} = \frac{n_\perp - n_{11}}{\frac{n_o}{\sigma}} = \frac{\Delta n}{\frac{n_o}{\sigma}} \qquad (2)$$

Where: ε_o = optical strain, σ = applied uniaxial stress, $n_\perp$ = index in $\perp$ direction, n_{11} = index in 11 direction, n_o = unstressed index.

The measurements of the Stress-Optical Coefficient, (K), again reflect the changes typical for the boron glass structure. The shape of the Stress-Optical Coefficient versus composition curve--specifically the nonuniformity in the curve--is the response to the "n_0" term of equation (2) and follows, therefore, the characteristics of the density and refractive index curves. The measured K value of 9.32 x 10^{-6} dyne/nm/cm/cm^2 for vitreous B_2O_3 is predictably high compared to that for common SiO_2 or borosilicate glasses (app. 3.5 to 2.7 x 10^{-6}). It is a measure of the high compressibility of the B_2O_3 glass.

Explaining the high compressibility and the specific flow viscosity characteristics of the B_2O_3 structure in comparison to vitreous SiO_2, Krogh-Moe has pointed out that only low activation energy is required for coordination changes in the pure vitreous boron oxide. Any introdution of alkaline atoms, then, should inhibit these low energy boron coordination changes, resulting in the drop of compressibility and Stress-Optical Coefficient with alkaline introduction -- as observed.

The qualitative model[1] (Figure 10), based on Krogh-Moe's melting point depression studies describes -- supported by NMR, IR, Raman-scattering, and other data -- the binary alkaline borate glasses with 0 to 20 mol percent Na_2O as consisting of boroxol, diborate, and tetraborate structures, partially dissociated. Tetraborate structures dominate in alkali-borate glasses of 20 mol percent alkali oxide with no boroxol groups present.

The specific make-up of the binary Na_2O - B_2O_3 system manifests itself in the cumulative physical-optical data as measured. Without being able to pinpoint the influence of any single structural component, we tend to conclude that the structural change associated with the introduction of Na_2O into the B_2O_3 network results in "chemically" induced coordination change from (BO_3) to (BO_4) coordination with the alkaline oxide occupying the voids of the vitreous B_2O_3 structure.

With higher concentrations of alkaline oxide acting as modifying cations, non-bridging oxygen is introduced, resulting in a structural change which decreases the (OH) content in the glass matrix.

In the concentration range between 14 and 17 percent, all of these influences overlap and result in producing some non-linear property changes ("anomalies") in the physical-optical properties -- as discussed.

At Na_2O concentration above 20 mol percent, the modifying cation influence seems to become the dominating factor and determines the change of physical-optical properties with composition.

The actual structural change cannot be deducted from the data presented here. A more detailed analysis has to include the structural data as summarized in the Krogh-Moe model for the alkaline borate glasses.

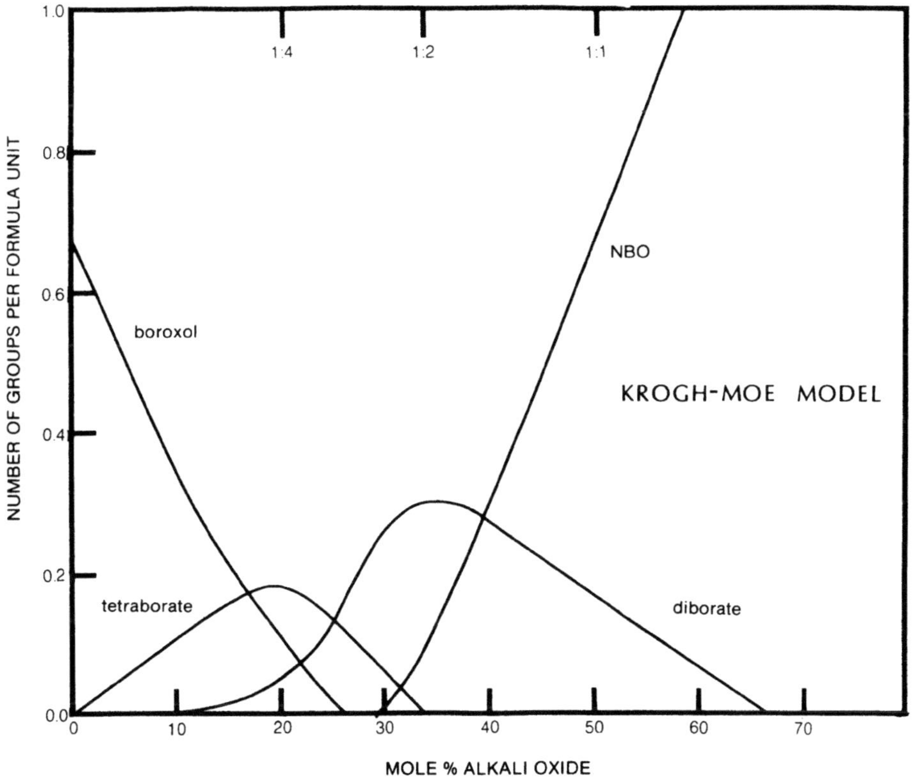

Figure 10. The Krogh-Moe structural interpretation model for the binary alkali-borate system.

REFERENCES

1. Griscom, D.L., "Pre-Conference Manuscript on Borate Glass Structure, " Alfred, New York (1977).

2. Uhlmann, D.R. and Shaw, R.R., "J. Non-Crystal Solids," 1, (1969) 347.

3. Gooding, E.J., and Turner, W.E.S., "J. Soc. Glass Tech.," 18, (1934), 32.

4. Biscoe, J. and Warren, B.E., "J. Am. Ceram. Soc.," 21, (1938), 287. Warren, B.E., "J. Appl. Phys.," 13, (1942), 602.

5. Krogh-Moe, J., "Phys. Chem. Glasses," 1 (1960), 26. Svanson, S.E., Forslind, E., Krogh-Moe, J., "Phys. Chem.," 66 (1962), 174.

6. Bray, P.J. and O'Keefe, J.G., "Phys. Chem. Glasses," 4, (1963) 37.

7. Karsch, K.-H., "Glastech. Ber.," 35, (1962), 234.

8. Shartis, L., Capps, W., and Spinner, S., "J. Amer. Cer. Soc.," 36, (1953), 35.

9. Eversteijn, F.C., Stevels, J.M., and Waterman, H.I., "Phys. Chem. Glass," 1, (1960), 123.

10. McSwain, B.D., Borelli, N.F., and Su, G-J., "Phys. Chem. Glass," 4, (1963), 1.

11. Wulff, P. and Majumder, S.K., "Z. Phys. Chem.," 31B, (1963), 319.

12. Scholze, H., "Glastech. Ber.," 32, (1959), 142.

13. Franz, H., "Glastech. Ber.," 38, (1965), 54.

14. Brueckner, R., "Glastech. Ber.," 37, (1964), 413.

15. Poch, W., "Glastech. Ber.," 37, (1964), 533.

16. Coenen, M., "Glastech. Ber.," 35, (1962), 16.

17. Bresker, R.I., Russian Dissertation.

EFFECT OF WATER CONTENT ON DENSITY, REFRACTIVE INDEX, AND TRANSFORMATION TEMPERATURE OF ALKALI BORATE GLASSES

HELMUT FRANZ

PPG INDUSTRIES, Inc., Glass Research Center

P.O. Box 11472, Pittsburgh, PA 15238

1. INTRODUCTION

The solubility of water vapor in boric oxide [1] and binary alkali borate melts [2] has been found to be proportional to the square root of the partial pressure P_{H_2O}, indicating that water vapor introduced at $P_{H_2O} < 1$ atm is dissolved in the form of hydroxyl groups (-OH), in analogy to the incorporation of water in silicate melts [3-6]. Correlations between the bond state of dissolved hydroxyl groups and the structure of alkali borate [2] and alkali silicate [7] glasses have been derived from solubility data and infrared absorption spectra. The effect of water content on some properties of silicate glasses have been discussed previously [8]. This study will present experimental data on the effect of the hydroxyl concentration on density, refractive index, and transformation temperature of binary lithium, sodium, and potassium borate glasses containing 0 to 33 mole percent R_2O. The corresponding properties of water-free alkali borate glass systems have been extrapolated. Results are discussed in terms of a simplified structural model interpreting the effects of incorporated hydroxyl groups on structure sensitive properties of alkali borate glasses.

2. EXPERIMENTAL

The binary alkali borate glasses were melted from reagent grade boric acid and alkali carbonates. Different levels of water content were prepared by bubbling the molten glass at 900°C with nitrogen containing controlled partial pressures of water vapor. After saturation of a melt, samples for determination of properties by various methods were quenched rapidly.

Alkali concentrations were determined by acidimetric titration. Hydroxyl concentrations were calculated from the infrared absorption spectra of thin polished disks, using the absorption coefficients of the OH absorption band at 2.8 to 2.9 μ from Ref. [2].

Density (ρ) measurements were made on annealed, bubble-free granules of glass by the sink-float method described in [8].

Refractive index (n) was measured with an ABBE-refractometer.

Transformation temperatures (T_G) were determined from the dilatometer curves of annealed rods.

Changes of ρ, n, and T_G due to alkali evaporation during saturation with water vapor were compensated for by interpolation of the properties of water-free glasses at each constant level of alkali concentration (8, 16, 24, or 32 mole % R_2O).

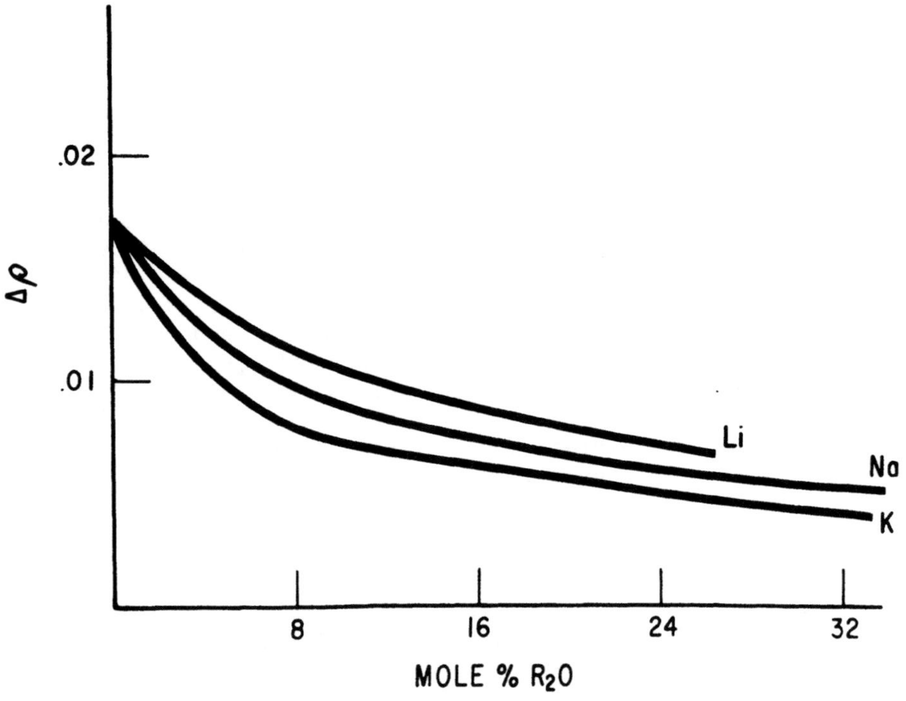

Fig. 1. Δρ per mole % H_2O in alkali borate glasses.

3. RESULTS

3.1. Density

The density ρ of the three alkali borate glass systems increases with the water content. The change in density, $\Delta\rho$ per mole % H_2O, increases in the order K-borate < Na-borate < Li-borate < B_2O_3 glass, but in all three alkali borate system, $\Delta\rho$ becomes smaller with increasing alkali content (Fig. 1).

The density of the water-free glasses containing 0 to 11 mole % R_2O increases in the order ρ Li < ρ Na < ρ K. Above 11 mole % R_2O, the order changes of ρ Li < ρ K < ρ Na. (Fig. 2)

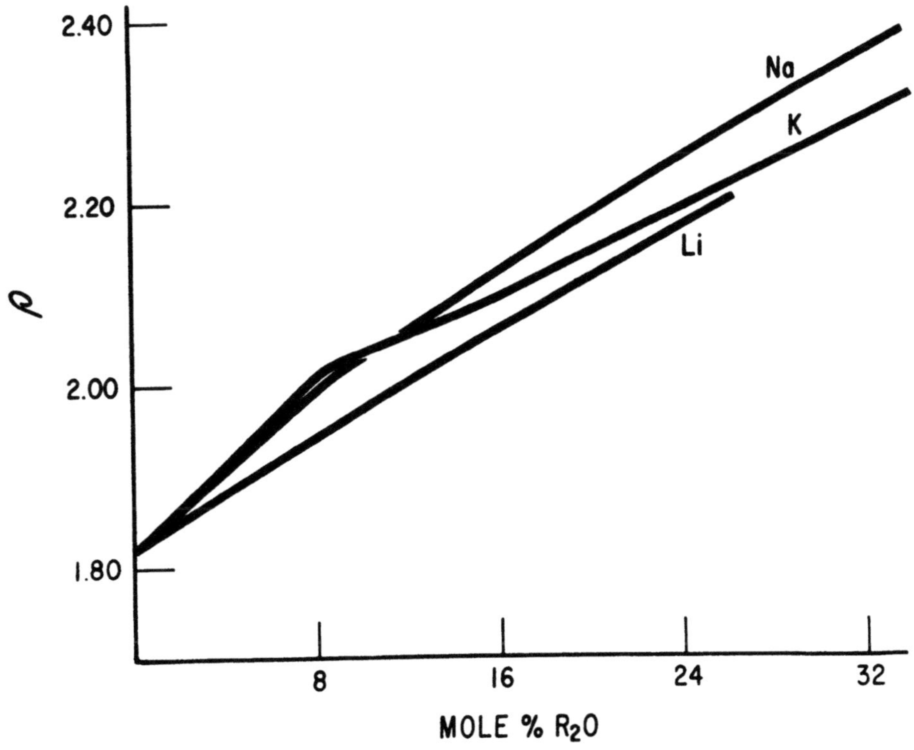

Fig. 2. Density ρ of water-free alkali borate glasses.

3.2 Refractive Index

The refractive index n of the three alkali borate glass systems increases with the water content. The change in refractive index, Δn per mole % H_2O, is almost identical in Li- and Na-borate glasses, but slightly higher in K-borate compositions. With increasing alkali content, n first decreases, reaching a minimum at 24 mole % R_2O, then slightly increases. (Fig. 3)

The refractive index n of water-free Li-, Na-, and K-borate glasses, at constant alkali content, was found to be identical in the range of 0 to 7 mole % R_2O. At 7 mole % R_2O, sharp breaks occur in the n vs. composition curves of K- and Na-borate glasses. At higher alkali contents, n increases in the order $n_K < n_{Na} < n_{Li}$. (Fig. 4)

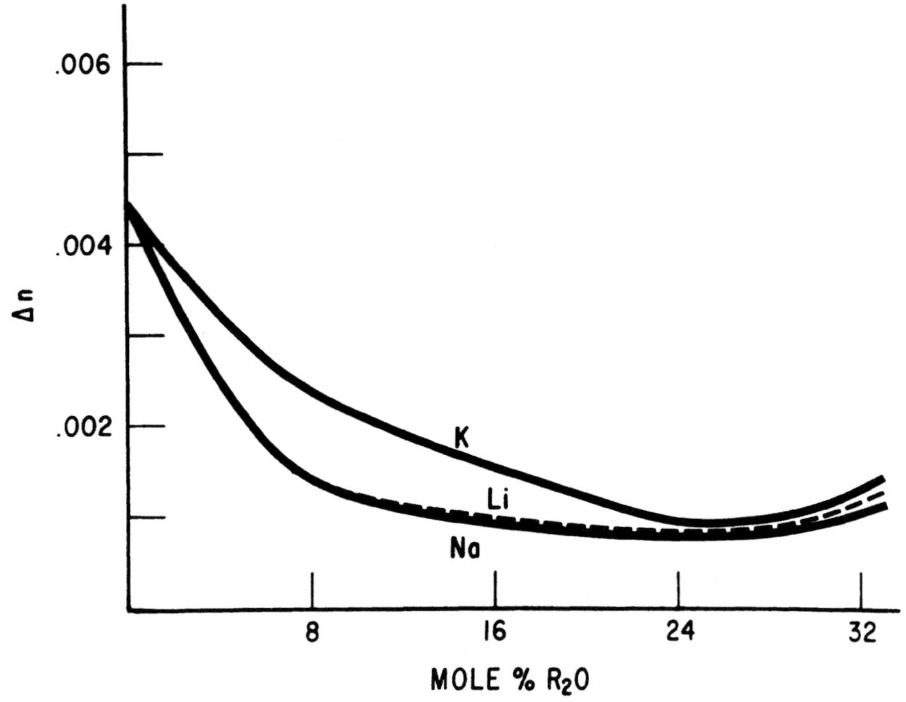

Fig. 3. Δn per mole % H_2O in alkali borate glasses.

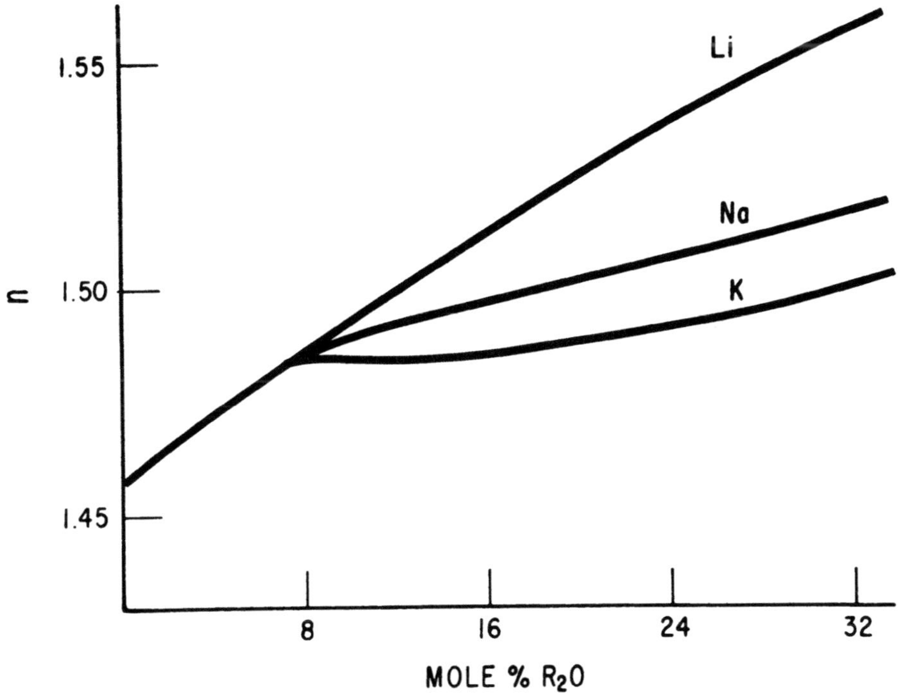

Fig. 4. Refractive index n of water-free alkali borate glasses

3.3. Transformation Temperature

The transformation temperature T_G of the three alkali borate glass systems decreases with the water content.

The change in transformation temperature, $-\Delta T_G$ per mole % H_2O, increases in the order Li-borate < Na-borate < K-borate glasses. (Fig. 5)

The transformation temperature T_G of water-free Li-, Na-, and K-borate glasses, at constant alkali content, was found to be identical up to 8 mole % R_2O. At higher alkali contents, the T_G vs. composition curve of K-borate glasses branches off at 8 mole % K_2O, that of Na-borate glasses at 16 mole % Na_2O, resulting in an increasing order $T_{G_K} < T_{G_{Na}} < T_{G_{Li}}$. (Fig. 6)

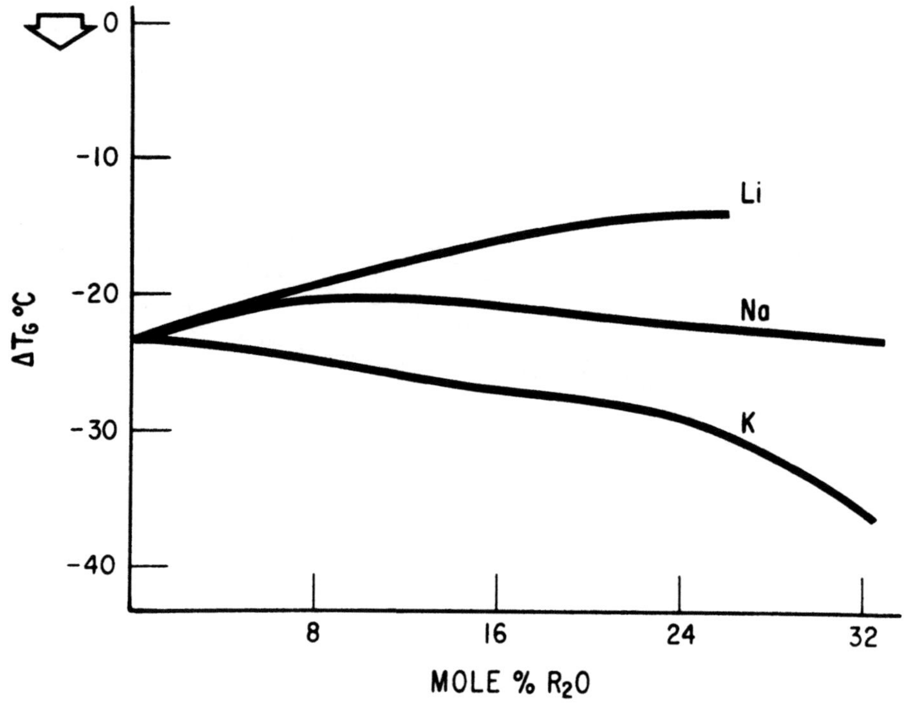

Fig. 5. ΔT_G per mole % H_2O in alkali borate glasses

4. CONCLUSIONS

The structure of borate glasses, including binary alkali-borates, has been reviewed and discussed extensively in the pre-conference manuscript [9]. The structural analysis is based mainly on the evaluation of structure-sensitive experimental methods such as x-ray diffraction, NMR and ESR, Raman and IR-spectroscopy, and some thermodynamic data of vitreous and crystalline borate systems.

In this study, some structure-sensitive physical properties such as density, refractive index, and transformation temperature, and the effect of incorporated hydroxyl groups on these properties, have been chosen as indicators for some structural changes in borate glass systems as a function of composition.

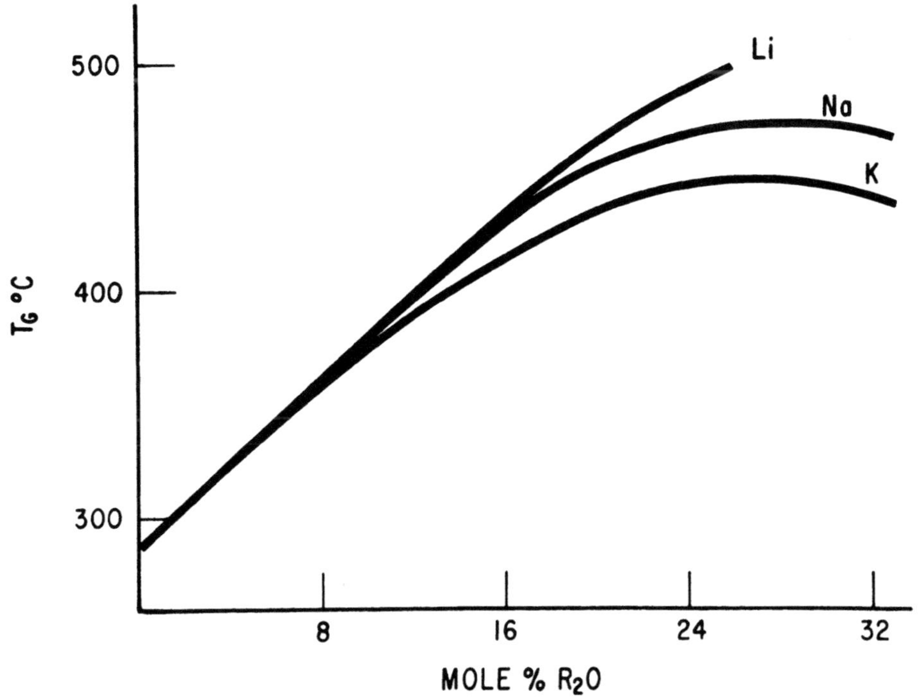

Fig. 6. Transformation temperature T_G of water-free alkali borate glasses

The general statement of Krogh-Moe's structural theory, as discussed in [9], describes the structure of alkali borate glasses as a disordered framework of some well-defined polyborate groupings. It follows that the alkali cations will occupy interstitial spaces within the polyborate groupings or in the framework formed by these segments.

Starting from the pure B_2O_3-glass composition, the linear plot of density, refractive index, and transformation temperature vs. composition indicates that up to approximately 7 mole % R_2O, all Li^+, Na^+, and K^+ cations will fit into these interstitial spaces without expanding or distorting the framework. Density increases are in relation to the mass added to the borate framework by the "cation stuffing" in the order Li < Na < K, (Fig. 2), whereas refractive index and transformation temperature, properties related mainly to the structure of the framework, remain identical. (Figs. 4, 6)

At higher alkali concentrations, only Li^+ cations will be accommodated in interstitial spaces without distortion, whereas the larger K^+ and, to a lesser extent, Na^+ cations will expand and distort the borate framework, as indicated by breaks in the curves for n and ρ (Figs. 2, 4). Stretching of the framework will also weaken some network bonds and reduce their dissociation energy, thereby lowering T_G. This effect becomes significant above 8 mole % K_2O or 16 mole % Na_2O. (Fig. 6)

The simplified structural model developed from the ρ , n, and T_G vs. composition pattern of water-free alkali borate glasses also offers an interpretation for the effect of incorporated hydroxyl groups on these properties. (Figs. 1, 3, 5)

The density ρ , as mentioned above, reflects the degree of "stuffing" of interstitial spaces with cations and, at least in K- and Na-borate systems, the expansion of the framework at higher alkali concentrations. Incorporation of OH-groups will add less to the density in K- and Na-borate glasses where the interstitial spaces are already filled and the framework may be expanded already, than in Li-borate glasses where hydroxyl groups still can occupy open space and may contract the structure slightly by formation of weak hydrogen bonds to adjacent bridging oxygens in the borate network. With increasing alkali content, the effect of hydroxyl groups on the density becomes less significant. (Fig. 1)

The changes in refractive index n and transformation temperature T_G due to dissolved water can be attributed to changes in the borate framework structure by the conversion of =B-O-B= bridges into pairs of =B-OH groups according to equation (1):

$$=B-O-B= + H_2O \longrightarrow 2 =B-OH. \qquad (1)$$

The effect of hydroxyl groups on the refractive index n becomes smaller as the alkali content increases, but apparently it is more pronounced in the expanded K-borate structure than in the less- or unstretched framework of Na- or Li-borate glasses, respectively. (Fig. 3)

This structural interpretation is manifested by the effect of dissolved water on the transformation temperature T_G. (Fig. 5) In Li-borate glasses, $-\Delta T_G$ per mole % H_2O decreases with the alkali content, as it is to be expected from the incorporation of a small increment of network modifier H_2O to a system containing already increasing amounts of network modified R_2O. In Na-borate glasses, however, this effect is counteracted, and at $Na_2O > 8$ mole % overcompensated by the relaxation of the slightly expanded borate framework due to the rupture of oxygen bridges according to equation (1).

In K-borate glasses with even more expanded framework structures, the latter effect is dominating throughout the compositional range of 0 to 33 mole % K_2O, resulting in increasing $-\Delta T_G$ per mole % H_2O with increasing alkali concentration.

REFERENCES

[1] H. Franz, Glastechn. Ber. 38 (1965), 54-59.
[2] H. Franz, J. Amer. Ceram. Soc. 49 (1966), 473-477.
[3] H. Scholze, H. O. Mulfinger, and H. Franz, Advances in Glass Technology, Plenum Press, Inc., New York (1962), 230-248.
[4] H. Franz and H. Scholze, Glastechn. Ber. 36 (1963), 347-356.
[5] J. M. Uys and T. B. King, Trans. AIME 227 (1963), 492-500.
[6] C. R. Kurkjian and L. E. Russell, J. Soc. Glass Technol. 42 (1958), 130-144T.
[7] H. Franz and T. Kelen, Glastechn. Ber. 40 (1967), 141-148.
[8] H. Scholze, H. Franz, and L. Merker, Glastechn. Ber. 32 (1959), 421-426.
[9] D. L. Griscom, "Borate Glass Structure", pp. 1-128, this volume.

ACOUSTIC SPECTRA OF GLASSES IN THE SYSTEM $Na_2O-B_2O_3$

J. T. Krause and C. R. Kurkjian

Bell Laboratories

Murray Hill, New Jersey 07974

Introduction

The subject of the structure of vitreous B_2O_3 and $Na_2O-B_2O_3$ glasses has been a controversial one for many years. Initially Warren[1] proposed a theory, based mainly on x-ray studies, that vitreous B_2O_3 is composed of BO_3 triangles, with each oxygen shared between two borons. The addition of Na_2O up to 15 mole % introduces extra oxygen into the network, converting some of these triangles to BO_4 tetrahedra. The extra oxygens are doubly-bonded to boron and thereby strengthen the network. At Na_2O contents greater than 15%, no more tetrahedra are formed, rather singly-bonded oxygens are produced in the same manner as they are in Na_2O-SiO_2 glasses. This results in broken B-O-B linkages with a consequent weakening of the network. On the basis of NMR investigations, however, it appears that BO_4 tetrahedra are produced continuously up to about 30-35 mole % Na_2O. An extension of these concepts is made through the work and theory of Krogh-Moe[3] which states that the BO_3 triangles and BO_4 tetrahedra are not merely joined at corners in a random network, but that the structure contains well-defined borate groups randomly distributed.

Results in the literature on both high[4] and intermediate[5,6] frequency dielectric studies, and high[7] and low[8] frequency mechanical studies, indicate a rather complicated absorption spectrum for glasses in this system. As indicated in a previous paper[9] a technique developed for making acoustic measurements at 20 MHz from 1.4°K to 1000°K permits a study of most relaxations of interest in glasses. In contrast to earlier work, several relaxation mechanisms can be studied in a single apparatus and on the same glass sample.

The purpose of this work is to study the acoustic behavior of B_2O_3 and $Na_2O-B_2O_3$ glasses over as wide a temperature range as possible to elucidate both the structure and acoustic behavior in this system.

Experimental

The glasses were melted from analytical grade boric acid and sodium carbonate in a platinum boat at approximately 950°C in air and annealed to remove strains. Because of previous findings several samples were melted at 1200°C and bubbled with dry nitrogen to remove water. The higher temperature was used for easier bubbling and is expected to affect only the water content. Sample sizes varied slightly, but were approximately 1/2 inch square by 2 inches long for the high temperature measurements. The 1 inch long samples were polished optically flat and parallel to 10 seconds of arc on the two opposite faces normal to the one inch length. The 2 inch samples were prepared in the form of bars with a 1/2 inch step cut in one end. 20 MHz x-cut or y-cut quartz resonators were bonded to the samples with either Dow Corning 200 silicone oil, viscosity 12,500 c.s. or Dow Resin 276-V9. The pulse echo, pulse superposition, and differential path methods, described in reference (9), were used for the determination of loss and velocity of propagation.

Results and Discussion

Elastic Modulus

Figure 1 shows the data for the room temperature longitudinal velocity ($V_L = \sqrt{\lambda+2\mu/\rho}$) of the wet and dry $Na_2O-B_2O_3$ glasses and Fig. 2 shows the shear velocity ($V_s = \sqrt{\mu/\rho}$) of the wet glasses. Since the moduli are proportional to the square of the velocity, a plot of velocity vs. composition is indicative of relative network strength. Also shown in these figures are data for Na_2O-GeO_2[9] glasses and Na_2O-SiO_2 glasses.[10] In silicate glasses the velocity decreases continuously as a network modifier (NWM) is added. This is what one would expect on the basis of the classical random network model; e.g., as Na_2O is added and doubly-bonded oxygen are replaced by singly-bonded oxygen, the stiffness of the network would be expected to decrease. Both the borate and germanate glasses, however, show a maximum in acoustic velocity as NWM is added. If one assumes with Warren[1], and Bray[2] that B_2O_3 consists of BO_3 triangles, and that BO_4 tetrahedra are formed as Na_2O is added, the velocity would be expected to increase since the formation of tetrahedra is essentially a cross-linking mechanism with doubly-bonded oxygens resulting in a strengthened network. As pointed out in reference (9) the same interpretation also seems to hold in the case of the Na_2O-GeO_2 glasses. In the

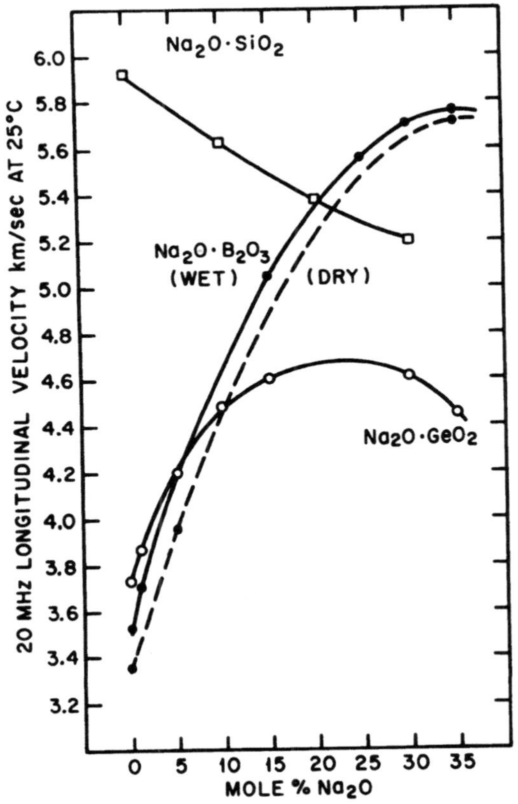

Fig. 1 Longitudinal Velocity vs. Composition for Sodium Silicate, Germanate, and Borate Glasses

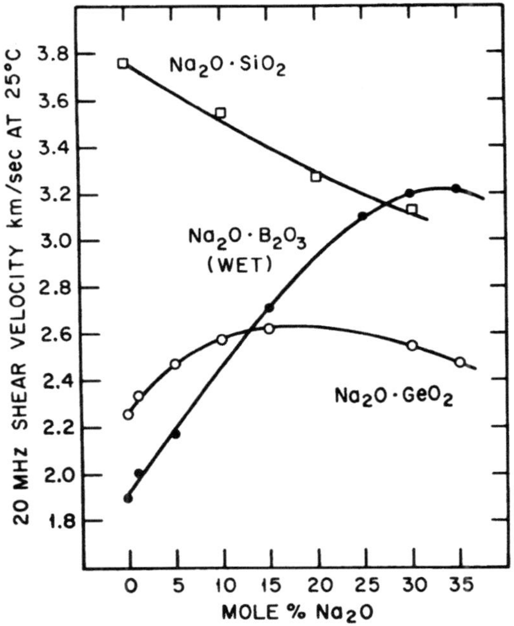

Fig. 2 Shear Velocity vs. Composition for Sodium Silicate, Germanate, and Borate Glasses

germanate case the evidence supports a proposed change of Ge coordination from 4 to 6 with a similar crosslinking up to 15 mole % Na_2O. In both the germanate and borate glasses, as NWM is added beyond certain concentrations, the cessation of the coordination change and subsequent production of singly-bonded oxygens or destruction of the higher coordination state results in a weakening of the network and a consequent decrease in the velocity.

Acoustic Loss

As indicated in a previous note,[11] the presence of water or OH in the borate structure has an extremely large effect on the acoustic loss spectrum and appears to give rise to an additional absorption at 300°K having an activation energy of 6.3 Kcal/mole. This value agrees well with that found dielectrically and also attributed to water.[6] The dry B_2O_3 spectrum shows only one peak at 60°K which is more intense than in the wet glass. This data is reproduced in Fig. 3. The amount of OH has since been determined

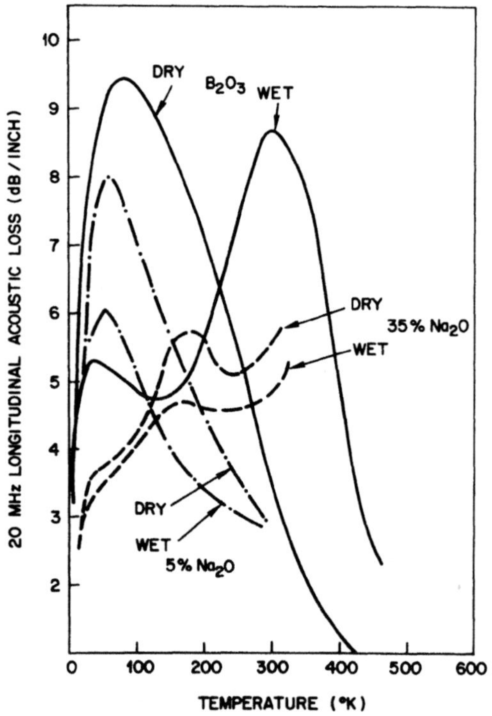

Fig. 3 Acoustic Loss vs. Temperature for Wet and Dry B_2O_3, $5NaO \cdot 95B_2O_3$, and $35Na_2O \cdot 65B_2O_3$.

to be 1.1% by wt. for the wet glass and 500 ppm for the dry. Also included in Fig. 3 are data for wet and dry 5% and 35% soda glasses. Note that the 35% soda glass shows little difference between wet and dry loss at 60°K. This is in keeping with the known decreased water solubility at the higher soda concentration.(12) A more complete loss spectrum of the wet Na_2O-B_2O_3 system is shown in Fig. 4. As soda is added to wet B_2O_3, the low temperature peak now at 45°K first increases in amplitude and temperature position up to about 5% soda and then decreases rapidly at greater concentrations. However, as soda is added to dry B_2O_3, there occurs just a continuous decrease in peak amplitude as well as peak temperature position. Thus the presence of OH must be taken into account in comparison of systematic additions of Na_2O to the borate structure. Figures 3 and 4 indicate that a new absorption peak at approximately

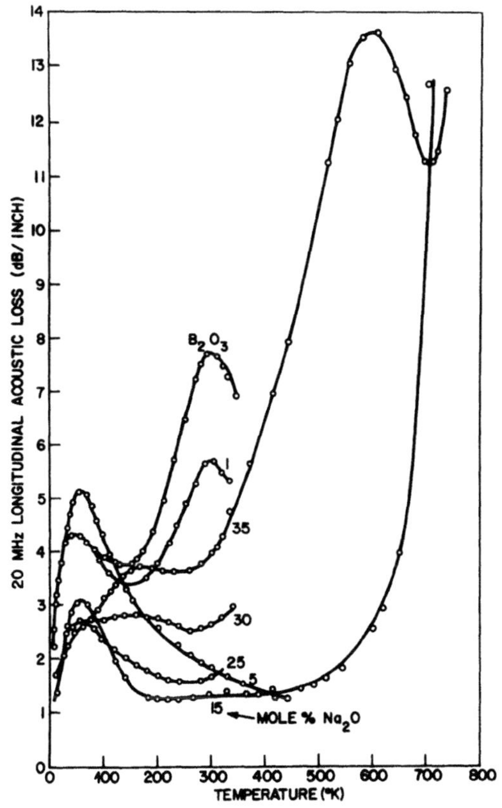

Fig. 4 Acoustic Loss vs. Temperature for Wet Sodium Borate Glasses.

160°K appears at about 25 % Na_2O which increases in amplitude with increasing soda content. The mechanism of these absorptions is believed to be the transverse vibration of the three and four coordinated oxygens. In B_2O_3 the 60°K loss peak is close to the temperature at which a peak is found in fused SiO_2. In their study of this glass, Anderson and Bömmel[13] have associated this acoustic absorption with transverse vibrations of doubly-bonded (bridging) oxygen ions tetrahedrally coordinating silicon. As Na_2O is added to B_2O_3, the intensity of the 60°K peak would be expected to decrease since the number of 3-fold coordinated borons is decreasing. The peak which appears at 160°K may then be associated with vibrations of oxygen atoms which are in 4-fold coordination with boron. The intensity of this peak would be expected to increase with soda content as is found experimentally.

The mechanism for the water peak at 300°K is left unresolved. This peak has no parallel in other glass systems.

Sodium Ion Loss

For the glass containing 35 mole % Na_2O, a large peak is resolved at about 575°K which is presumably due to sodium ion diffusion. The activation energy for this absorption is 23 Kcal/mole. Previous low frequency measurements by Rotger[14] on a similar glass gave an activation energy of 12.2 Kcal/mole. The discrepancy here is quite large and unaccounted for.

Temperature Dependence of Velocity

The acoustic velocity vs. temperature curves are shown in Figs. 5-7. These data are included mainly for completeness. They do, however, suggest a change in coordination number with soda addition to B_2O_3. The temperature dependence becomes less negative with soda additions up to about 15%. At concentrations above this the dependence becomes again increasingly negative. This trend towards a positive dependence suggests the formation of BO_4 species from analogy with the strong positive temperature dependence of velocity of the glasses SiO_2, GeO_2, and $Zn(PO_3)_2$ having structures with four coordinated bonding.[15]

Summary

The sound velocity increase with sodium concentration in borate and germanate glasses is similar and opposite to that shown by silicate glasses. This is interpreted as a network stiffening in the germanate and borate glasses resulting from an increase in coordination number. In the borate case this is associated with production of BO_4 tetrahedra up to about 35% soda. The acoustic loss spectrum of B_2O_3 shows two peaks one of which at 300°K is due

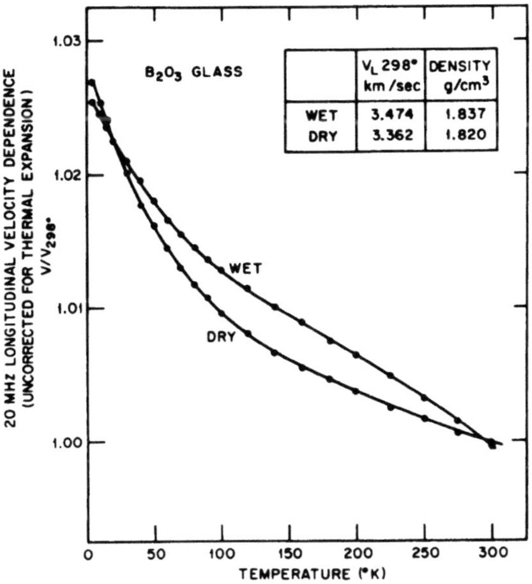

Fig. 5 Longitudinal Velocity vs. Temperature for Wet and Dry B_2O_3 Glasses.

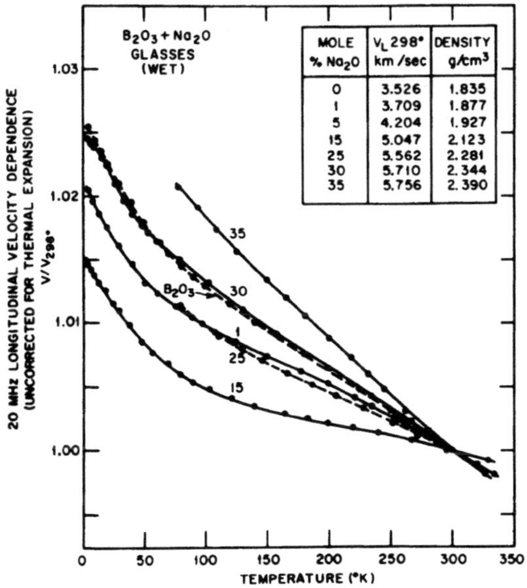

Fig. 6 Longitudinal Velocity vs. Temperature for Wet Sodium Borate Glasses.

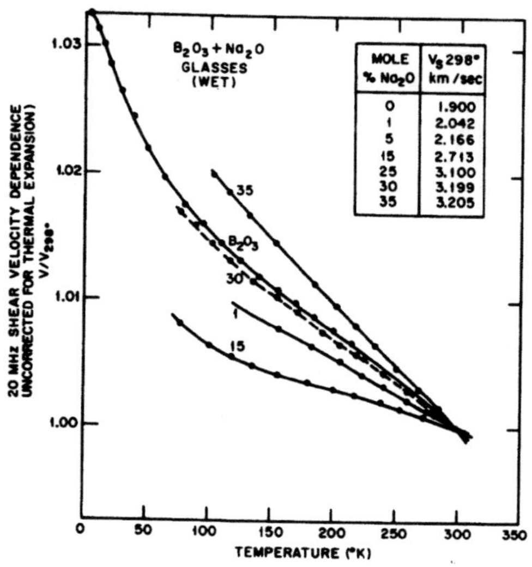

Fig. 7 Shear Velocity vs. Temperature for Wet Sodium Borate Glasses.

to water or OH ion. This peak is entirely absent in dry B_2O_3 where only the low temperature peak at 60°K remains. This peak, associated with boron in triangular coordination decreases in amplitude with soda addition. With continued addition of soda a new peak appears at about 160°K which is thought due to the production of four coordinated boron. At high soda concentration (35%) a peak at about 575°K is assigned to migration type relaxation due to sodium ion diffusion.

REFERENCES

1. J. Biscoe and B. E. Warren, "X-ray Diffraction Study of Soda-Boric Oxide Glasses", J. Am. Ceram. Soc. 21 289-93 (1938).
2. P. J. Bray and J. G. O'Keefe, "Nuclear Magnetic Resonance Investigations of the Structure of Alkali Borate Glasses", Phys. Chem. Glasses 4 37-46 (1963).
3. J. Krogh-Moe, Phys. Chem. Glasses 3 101 (1962).
4. V. A. Ioffe, "Dielectric Losses in Boron-Alkali Glasses at Low Temperatures", J. Tech. Phys. 1 498-505 (1956).
5. J. Volger and J. M. Stevels, "Further Experimental Investigations of the Dielectric Losses of Various Glasses at Low Temperatures", Philips Res. Reps. 11 452-470 (1956).

6. W. J. Th. Van Gemert, "Relaxation Phenomena in Borate and Phosphate Glasses", Doctorate Thesis, The Technische Hogeschool, Eindhoven, June 28, 1977.
7. R. E. Strakna and A. T. Savage, "Ultrasonic Relaxation Loss in SiO_2, GeO_2, B_2O_3, and As_2O_3 Glasses", J. Appl. Phys. 35 1445-50 (1964).
8. K. H. Karsch and E. Jenckel, "Thermische und Mechanishe Intersuchungen on Alkalabortglas", Glastech. Ber. 34 397-400 (1961).
9. C. R. Kurkjian and J. T. Krause, "Acoustic Spectra of Na_2O-GeO_2 Glasses", J. Am. Ceram. Soc. 49 134-38 (1966).
10. A. V. Gladkov and V. V. Tarasov, "Investigation of the Polymeric Structure of Inorganic Glasses", The Structure of Glass, Vol. 2, Consultants Bureau, Inc., New York (1960).
11. C. R. Kurkjian and J. T. Krause, "Effect of OH Content on the Acoustic Spectra of B_2O_3 Glass", J. Am. Ceram. Soc. 49 171-172 (1966).
12. Helmut Franz, "Solubility of Water Vapor in Alkali Borate Melts", J. Am. Ceram. Soc. 49 473-77 (1966).
13. O. L. Anderson and H. E. Bömmel, "Ultrasonic Absorption in Fused Silica at Low Temperatures and High Frequencies", J. Am. Ceram. Soc. 38 125-131 (1955).
14. H. Rotger, "Über das Elastische Relaxationsverhalten von Einfachen und Gemischten Alkali-Silikaten und von Borax", Glartech. Ber. 31 54-60 (1958).
15. J. T. Krause and C. R. Kurkjian, "Vibrational Anomalies in Inorganic Glass Formers", J. Am. Ceram. Soc. 51 226-227 (1968).

INDUSTRIAL BORATE GLASSES

Emil W. Deeg

American Optical Corporation

Southbridge, MA

1. INTRODUCTION

One of the first factory produced boro-silicate glasses was made by Maës in Clichy, France around 1850. It was a glass one would call today a "boro-silicate crown" and its composition, according to an analysis reported by E. Zschimmer [1], was 56% silica, 7% boric acid, 17% potassia, 2% calcia, 14% zinc oxide, 4% lead oxide. Articles made of this glass included hollow ware and most likely optical elements. Although there appears to be no direct connection between this early boro-silicate glass and today's wide variety of boron containing glasses, optical elements and hollow ware for special applications are still articles produced commonly from them.

Systematic, empirical studies relating glass composition to optical properties, carried out by Harcourt and Stokes in England during the middle and by O. Schott and Abbe in Germany in the last quarter of the 19th century included boric oxide as glass constituent. The breakthrough permitting large scale production of boro-silicate glasses with properties most desirable for laboratory ware and chemical equipment was achieved by Sullivan and Taylor in this country in the first quarter of the 20th century [2].

2. LABORATORY GLASSES

A number of boro-silicate glasses with high chemical durability, good thermal shock resistance, high upper use temperature and reasonably low working point became known as laboratory glasses. Some of them show low thermal after effect in addition to these properties. Articles manufactured from these glasses include laboratory ware, containers for pharmaceutical and medical products, pipe lines and equipment for chemical and food processing plants, transparent cooking and oven ware, windows for high temperature environment and even mirror blanks for optical elements in reflective systems and tubing, bulbs and lenses for illuminators.

Although there are differences in the degree of corrosion resistance of different glass compositions within the group of boro-silicate laboratory glasses, these differences are small. As a group these glasses have the highest chemical durability. Whereas the degree of attack by most inorganic and organic liquids is negligible, the glasses are severely attacked by hydrofluoric acid and by fluorides. Basic solutions and phosphoric acid attack these glasses slightly (see Table 1 for examples).

Thermal shock resistance of a (homogeneous!) glass article is determined by its shape, size, its thermal, mechanical and chemical history and by material constants of which the coefficient of thermal expansion is the most important one. Low thermal expansion results in high thermal shock resistance.

A. Matthiesen discovered in 1866 that the coefficient of thermal expansion of glass can vary with time, depending on the thermal history of the sample [4]. This "thermal after effect" influences the precision of glass thermometers as discussed e.g. in [5]. R. Weber found in 1879 that certain glass compositions show a reduced thermal after effect, and are therefore more suitable for manufacture of thermometers [6]. Because these instruments are in direct contact with solutions at different temperatures, chemical durability and high thermal shock resistance are requirements which have to be met also. Boro-silicate glasses

TABLE 1: Examples of corrosion of durable glass compositions. For more precise data the catalogs of suppliers should be consulted. See also [3].

Solution (aquoeus)	Temperature in ^{o}C	Exposure in hrs.	Approximate Weight Loss in mg/cm^2
5% Na_2CO_3	150	6	10.0
5% NaOH	80	6	0.3
3% NH_4OH	80	100	0.3
85% H_3PO_4	100	24	0.01

provide the means to solve this set of problems.

Today it is common knowledge that most boro-silicate glasses show a tendency towards phase separation. The discovery of this phenomenon by Hood and Nordberg in 1934 is connected with, although not limited to, a new manufacturing process for producing high silica laboratory ware [7]. Already in the original patent the potential of utilizing the phase separation process for glass purification is mentioned:

> "Oxides, such as the oxides of iron and cobalt, have been found to concentrate mainly in the soluble phase during the heat treatment and the separation of phases which takes place therewith. Such oxides are therefore practically entirely removed during the acid leaching step of the process. Therefore proper glasses treated in accordance with our process have a high transmission for ultra-violet radiation, even though prepared from iron-bearing batch materials."

Those developments resulted in the highly useful glass compositions listed in Table 2. They also required studies and process development leading to improved refractories, tank design and forming techniques. The improved chemical and physical properties of the industrial boro-silicate glasses made it necessary to refine test methods which provided a stimulus for scientific metrology. The discovery of phase separation finally lead to a re-assessment of our opinions on the validity of certain glass structure models. It is difficult to imagine that all this was accomplished because the glass industry attempted to provide the scientific community with better test tubes, flasks and thermometers.

3. OPTICAL GLASSES

Boron containing optical glasses are used chiefly for lenses in high quality instruments, for light filters with specific spectral transmission, and for ophthalmic photochromic lenses. The index of refraction of glasses is within the visible spectrum a monotonic decreasing function of the wavelength of light. The focal length of a lens therefore is different for different wavelengths. Using the thin lens formula

$$\frac{1}{f} = (n-1) \cdot \left(\frac{1}{R_1} - \frac{1}{R_2}\right)$$

this difference can be expressed as

TABLE 2: Examples of boron containing laboratory glass compositions. Concentration in round numbers in wt.%, ingredients amounting to less than 1% are omitted. - Linear thermal expansion (20-300°C) in 10^7 per deg. C; working point (10^4 poise) in °C; density in g/cm^3. - Glasses 9 and 13 are fiber glasses, 12 and 14 thermometer glasses, 8 and 11 pharmaceutical glasses, 4,5 and 10 sealing glasses

Glass No.	1	2	3	4	5	6	7	8	9	10	11	12	13	14
Therm. Exp.	8	32	40	41	42	42	49	49	50	52	55	64	80	90
Workg. pt.	>1800	1260	1195	1140	1095	1190	1190	1165	-	1040	1140	1125	-	1000
Density	2.2	2.2	2.3	2.3	2.3	2.5	2.4	2.4	2.5	2.3	2.5	2.4	2.5	2.6
SiO_2	96	81	76	71	75	62	76	74	53	67	71	68	65	67
B_2O_3	3	12	11	16	18	5	8	10	8	22	8	8	5	2
Al_2O_3	-	2	3	-	-	17	4	5	15	3	4	9	4	3
MgO	-	-	-	1	-	7	-	-	3	-	-	-	3	-
CaO	-	-	1	-	-	8	1	-	19	-	-	6	14	7
BaO	-	-	4	-	-	-	4	4	-	-	4	-	-	-
ZnO	-	-	-	-	-	-	-	-	-	-	-	-	-	7
PbO	-	-	-	4	-	-	-	-	-	-	-	-	-	-
Na_2O	-	3	4	2	6	1	6	6	-	3	7	9	9	14
K_2O	-	1	1	5	-	-	-	-	-	4	-	-	-	-
Fe_2O_3	-	-	-	-	-	-	-	-	-	-	2	-	-	-
MnO	-	-	-	-	-	-	-	-	-	-	3	-	-	-

$$\Delta f = - \frac{\Delta n}{(n-1)} \cdot f \qquad (1)$$

where

f = focal length,

n = index of refraction,

R_1 and R_2 = radii of curvature of the two lens surfaces.

The quantity $\frac{\Delta n}{n-1}$ is usually expressed in terms of specific wavelengths (see also Table 3), such as:

587.56 nm (= helium d-line)
486.13 nm (= hydrogen F-line)
656.27 nm (= hydrogen C-line)

It is abbreviated by

$$\nu_d = \frac{n_d - 1}{n_F - n_C} = \frac{n_d - 1}{\Delta n_{F,C}} .$$

TABLE 3: Spectral lines frequently chosen for designation of optical glass properties.

Wavelength in nm	"Fraunhofer" symbols	Light source	Color
365.01	i	Hg	UV
404.66	h	Hg	violet
435.84	g	Hg	blue
479.99	F'	Cd	blue
486.13	F	H	blue
546.07	e	Hg	green
587.56	d	He	yellow
589.3 1)	D	Na	yellow
643.85	C'	Cd	red
656.27	C	H	red
706.52	r	He	red
768.2 1)	A'	K	red
852.11	s	Cs	IR
1013.98	t	Hg	IR

1) Mean of doublet.

ν_d is called the "Abbe number" or the "reciprocal dispersion" of the glass for the d-line. The difference

$$\Delta n_{F,C} = n_F - n_C$$

is the dispersion of the glass between the blue F-line and the red C-line. According to equation (1) the difference of the focal length for blue and red light of a flint glass lens with Abbe number 35 would be

$$\Delta f = -\frac{1}{\nu_d} \cdot f = -\frac{1}{35} f \quad - 0.03 f,$$

or approximately 3% of the focal length f.
This "chromatic aberration" must be corrected for. It can be done by combining two lenses made of glasses with different ν_d-values. The result is a lens doublet with identical focal length for the F- and C- line. Between and beyond these two wavelengths a difference in focal length for different wavelengths still exists. To correct for this "secondary spectrum" additional criteria for the second derivatives of the refractive indices must be met. Optical glasses therefore are described not only by their refractive index at the d-line and their ν_d-value but also by a set of partial dispersion ratios, such as e.g. $(n_h-n_g)/(n_F-n_C)$, or by partial dispersion values such as (n_h-n_g), (n_g-n_F), $(n_{F'}-n_e)$, etc. The introduction of boron oxide into optical glass melts resulted in a wide variety of lenses permitting first as well as second order correction of chromatic aberration. Some of the glasses are listed in Table 4. An analysis of approximately 250 optical glasses shows that 55% of them contain more than 5 weight % B_2O_3 and only 27%, most of them in the flint and heavy flint group, are boron free. The prime reason for the presence of B_2O_3 in optical glasses is its influence on the dispersion in the short wavelength range of the visible spectrum. Other reasons are improved chemical durability, reduced devitrification tendency and ease of melting of some of the glasses in question. Not all boron containing optical glasses are boro-silicates. Some of them can be classified as rare earth borates, lead borates, barium borates, fluoro-boro-silicates and antimony-boro-silicates.

Among approximately 110 analyzed filter glasses, including heat screen glasses, 75% contained boron oxide, 39% of them more than 5 weight %. The highest B_2O_3 concentrations (approximately 20 weight %) are found in neutral gray and in gold ruby glasses. Base glasses are mostly boro-silicates with the exception of infrared absorbing (heat screen) glasses which are essentially calcium-aluminum-phosphates containing approximately 4 weight % B_2O_3 and up to 2% iron oxide. Because of the influence of melting

TABLE 4: Examples of clear optical glasses containing boron oxide. Concentration in round numbers in wt. %, ingredients amounting to less than 1% are omitted. Optical data are included for illustration. It cannot be expected that the compositions listed will yield precisely these values. Glass 22 is a Faraday effect glass according to [8].

Glass No.	15	16	17	18	19	20	21	22
n_d	1.5331	1.5182	1.6073	1.5512	1.5687	1.7130	1.5578	high
ν_d	58.1	65.2	59.5	49.6	63.1	53.9	53.8	Verdet
$(n_e-n_C)/(n_F-n_C)$	0.5425	0.5459	0.3903	0.5392	0.5450	0.5426	0.5434	Constant
$(n_F-n_e)/(n_F-n_C)$	0.4575	0.4541	0.4574	0.4608	0.4550	0.4574	0.4566	
$(n_g-n_F)/(n_F-n_C)$	0.5474	0.5321	0.5421	0.5558	0.5383	0.5442	0.5434	
$(n_h-n_g)/(n_{F'}-n_{C'})$	0.4519	0.4363	0.4466	0.4648	0.4415	0.4499	0.4512	
SiO_2	55	68	36	46	44	2	5	6
ZrO_2	-	-	-	-	-	4	-	4
B_2O_3	7	13	13	14	19	40	53	40
Al_2O_3	-	-	5	3	1	-	10	-
Sb_2O_3	-	-	-	21	-	-	-	-
La_2O_3	-	-	-	-	-	41	-	3
Tb_4O_7	-	-	-	-	-	-	-	35
WO_3	-	-	-	-	-	-	-	3
CaO	-	-	-	-	-	6	-	9
BaO	-	1	46	-	29	-	-	-
ZnO	20	-	-	4	-	6	17	-
PbO	-	-	-	-	-	-	10	-
Na_2O	1	9	-	-	3	-	-	-
K_2O	16	0	-	12	4	-	5	-

TABLE 5: Examples of boron containing optical filter glasses. Concentration in round numbers in wt. %, ingredients amounting to less than 1% are omitted. Details on preparing glass 27 are described in [10].

Glass No.	23	24	25	26	27	28	29	30	31
Characteristics	green	orange	green	blue	photo-chromic	purple	gray	red	fluores-cent
SiO_2	1	52	62	52	57	57	60	49	63
ZrO_2	-	-	-	-	2	-	-	-	-
P_2O_5	70	-	-	-	-	-	-	-	-
B_2O_3	4	4	10	12	16	19	19	20	16
Al_2O_3	7	-	-	6	9	1	1	2	-
Sb_2O_3	-	-	-	-	-	-	-	20	-
MgO	7	-	-	-	-	-	-	-	-
CaO	8	-	-	-	-	-	-	-	-
BaO	-	-	-	-	7	-	-	-	-
ZnO	-	19	-	6	-	4	4	-	-
PbO	-	-	10	-	5	-	-	-	-
Li_2O	-	-	-	-	2	-	-	-	-
Na_2O	-	-	9	-	2	14	-	2	-
K_2O	-	22	8	23	-	-	11	6	18
KF	-	-	-	-	-	-	4	-	3
Colorants	Fe	Cd-sulfo selenide	Cu/Cr	Cu/Co	Ag-hal. Cu	Nd/Pr	Fe/Co	Au	U

conditions and subsequent heat treatment on the spectral transmission of optical filter glasses only examples of base glass compositions are given in Table 5. Information on colorants is reported in a qualitative form. One of the early publications related to the effect of boron oxide on the color of chromium containing glasses is cited under [9].

4. CONCLUSION

The importance of the work of the borate glass pioneers mentioned above can hardly be fully appreciated by us who take for granted the availability of high quality optical instruments and cameras, of thermal shock resistant and chemically inert laboratory ware, and of high purity pharmaceutical and medical products. Progress made during the past 50 years in the chemical, pharmaceutical, and food processing industries was strongly influenced, if not determined, by the availability of borosilicate glasses. Fiber enforced plastics rely on the advantages of boro-silicate glass fibers. High and medium quality lens systems include usually at least one element made of a boron containing glass. Sealing and solder glasses used in the electronics industry utilize boron oxide.

Search for and discovery of new uses of known borate glasses as well as development of technically useful new compositions is still on-going. Questions calling for scientific explanation of the properties, particularly of the more complex borate glasses, are still unanswered.

REFERENCES

[1] Zschimmer, E.: Die Glasindustrie in Jena, Eugen Diederichs Verl., Jena, 1912.

[2] Sullivan, E.C. and Taylor, W.C.: U.S. Pat. 1,304,622 and 1,304,623 (1919).

[3] Deeg, E. and Richter, H.: Glas im Laboratorium, Editio Cantor, Aulendorf, 1965.

[4] Matthiesen, A.: Phil, Mag. (4), 31 (1866), p. 149.

[5] Deeg, E.: Glastechn. Ber. 31 (1958), p. 8.

[6] Weber, R.: Ann. Phys. 6 (1879), p. 431.

[7] Hood, H.P. and Nordberg, M.E.: U.S. Pat. 2,106,744 (1938) and 2,221,709 (1940).

[8] Deeg, E.W., Krohn, D. A.and Graf, R.E.: U.S. Pat. 3,935,020 (1976).

[9] Zsigmondy, R.: Ann. Phys. 4 (1901), p. 60.

[10] Armistead, W.H. and Stookey S.D.: U.S. Pat. 3,208,860 (1965).

APPLICATION OF BORATE GLASSES AND VARIOUS BORON BEARING GLASSES TO THE MANAGEMENT OF FRENCH RADIOACTIVE WASTES

R. BONNIAUD - A. REDON - C. SOMBRET -

COMMISSARIAT A L'ENERGIE ATOMIQUE - FRANCE -

MARCOULE - B.POSTALE 170 - 30200 BAGNOLS S/CEZE -

INTRODUCTION AND HISTORICAL BACKGROUND

The storage of fission products solutions generated by the reprocessing of spent nuclear fuels, has been operated until now in stainless steel tanks. This management does not give rise to any danger provided that suitable devices are able to withdraw the released heat and to make harmless possible leakages. Nevertheless the increase of nuclear energy in the world, emphasizes the need to use safer methods. The main one is founded upon the solidification of the radioactive liquids.

French works about the solidification of high level radioactive liquid wastes, started up in 1957.

In order to set up a solid putting up good chemical properties, it was attempted at first to make synthetic crystals like micas [1]. This gave successful results when a single fission product element was taken into account. But it appeared by using multiple fission products micas, that only one element could take place in the lattice, the others remaining between the layers, and escaping easily under leaching conditions.

To face up this feature, it was decided to make fission products glasses. For 20 years, improvement of the glass quality, determination of physical and chemical properties and the perfection of the technological equipments have been achieved [2 to 10].

CHARACTERISTICS OF THE RADIOACTIVE SOLUTIONS

All the solutions are made of nitric acid but the kind and the quantity of the dissolved elements is sharply subjected to important items which are :

TABLE I : CHARACTERISTICS OF HIGH LEVEL WASTE SOLUTIONS -

Reactor System	Type of fuel	Burn up	Concentration rate	Acidity (N)	APPROXIMATIVE CHEMICAL COMPOSITION ($g.l^{-1}$)										
					Al	Na	Mg	Fe	Ni	Cr	F	P	Gd	Actinides	F.P. oxides
MTR	U.Al/ Pu.Al	< 500 $MWj.kg^{-1}$	12 $m^3.t^{-1}$	~ 1.8	81	2/3		1/2			10/12				light
GRAPHITE/GAS	Sicral	1 000 $MWj.t^{-1}$	30 $l.t^{-1}$	1.5/2	30/35	19/23	4/5	15/17	1/2	2/3	8	1		3	15
	(Natural Uranium)	4 000 $MWj.t^{-1}$	100 $l.t^{-1}$	1.5	10/2	3/6	1/3	4/8	1/2	2/3	5	1		1/2	45
LWR	UO_2	33 000 $MWj.t^{-1}$	500 $l.t^{-1}$	1.5	< 1	1		1/40	< 1	< 1			15/24 or 0	4/6	75
FBR	UO_2/ PuO_2	60 000 $MWj.t^{-1}$ (mean value)	2 000 $l.t^{-1}$	1/1.5		20		15/20	1/2	2/3			20/30	2/3	25

- the type of the nucleary fuel,
- the rate of burn up,
- the possibility to declad or not the spent fuel prior to reprocess,
- in the latter case, the components of the cladding hulls,
- chemicals added in the course of the reprocessing,
- the final concentration,
- the duration of the interim storage.

Table 1 shows the composition of the solutions stored or to be stored in France. The main one on account of the french nucleary policy is at the moment that one generated by the reprocessing of LWR spent fuels. See table II.

CHARACTERISTICS OF THE GLASSES

Through a general acceptance now, the fate of the glasses is, at first, after making, a storage under cooling to dissipate the releasing heat due to the power of the radionuclides. This stage ought to be achieved for some years in an engineered facility. Then, the glasses will be transfered to an ultimate repository which could be a geological disposal.
Hence, owing to the thermal gradient and self irradiation, a change of the physicochemical properties is bound to happen.
So it is important to elaborate a product subjected to the smallest change or the change of which generating minor effects upon the properties. Besides, the glasses must comply with other requirements, chiefly :

- compatibility with the liquid composition,
- low leaching rate for aqueous solutions,
- easy making through an industrial process,
- large volume reduction factor.

The lattest item might be limited nevertheless for some reasons, mainly the problem related to the release of a very high specific heat.

TABLE II : EXPECTED REPROCESSING NEEDS IN FRANCE

(in tons of fuel)

Reactor system	1980	1985	1990
LWR	650	1 350	1 750
FBR	10	35	90

The range of the compositions of the various fission products solutions does not allow to achieve a glass suitable enough to the vitrification of the whole. So, as a matter of fact, a single type of glass had to be considered in connection with one type of liquid.

Silicates, as the most stable glasses, have been choosen to be basic products. The silicates become boroaluminosilicates for MTR and GG wastes. Boron is introduced as a fluxing agent and aluminium is used to make allowance of the quantity of Al in the solution.
The cases of FBR and LWR wastes are a little bit different. The corresponding glasses are borosilicates. Most of the fission products oxides have a tendancy to decrease the liquid temperature and boron is used mainly to loosen the network and to allow an entire vitrification. For instance let us consider compositions of a ternary diagram SiO_2 / Na_2O / FP oxides which contain an additional amount of boric oxide. With 10 % B_2O_3 content, there are 3 different areas in the operating zone at 1 200°C, one corresponding to unmelted products, another one, two phases products and the third one to pure glasses. See figure 1.
The more B_2O_3 content increases, the more the 2 phases products area tends to vanish. With 18 % B_2O_3 it is indeed smaller. See figure 2.

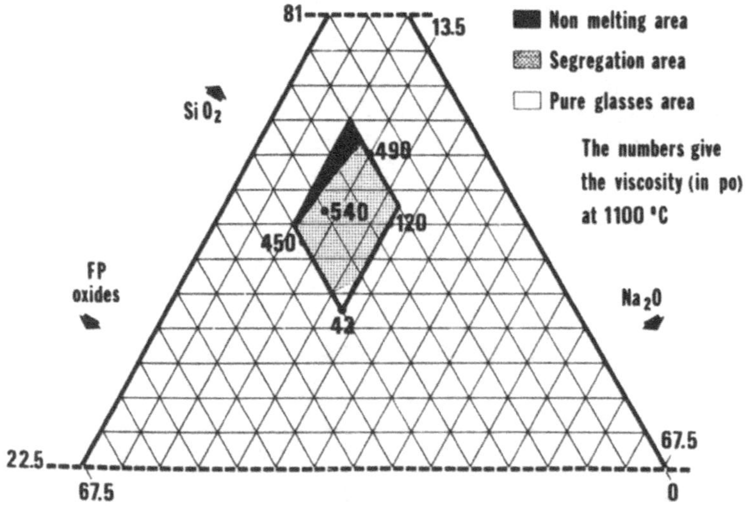

Figure 1 : L.W.R.GLASSES DIAGRAM WITH 10% B_2O_3

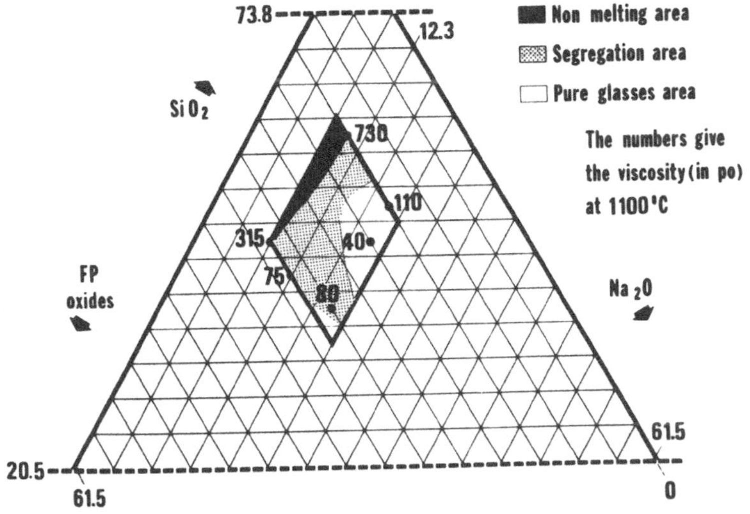

Figure 2 : L.W.R GLASSES DIAGRAM WITH 18% B_2O_3

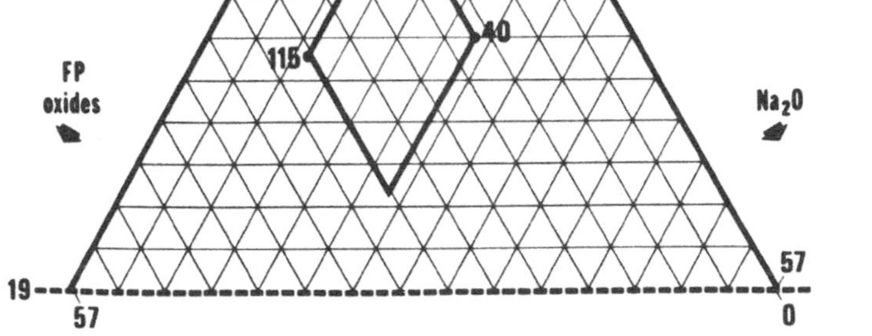

Figure 3 : L.W.R. GLASSES DIAGRAM WITH 24% B_2O_3

TABLE III : EXAMPLES OF GLASS COMPOSITIONS -

TYPE OF REACTOR		GLASS COMPOSITION (WEIGHT %)								Reduct. Volume coeffic.	Glass volume per ton of spent fuel
		SiO_2	Na_2O	B_2O_3	Al_2O_3	Fe_2O_3	MgO	NiO $+Cr_2O_3$	P.F oxides		
GRAPHITE/GAZ	Military	48.8	15.0	14.2	8.4	2.6	6.3	0.2	4.5	5.4	5.6 liters
	Commercial	42.7	14.2	17.8	8.6	1.6	1.0	1.4	12.7	7.0	14.0 liters
MTR		35.0	19.4	14.4	23.2	1.7	-	-	1.3	3.5	3.4 m^3
LWR		49.0	8.2	13.3	5.0	-	-	-	24.5	7.5	67 liters
FBR *		40.0	18.2	18.2	13.9	0.7	-	0.3	7.8	7.5	160 liters

*: This example is especially devoted to a former spent fuel batch from Phenix reactor :

- mean burn up (core + blanket) = 32 000 $MWj.t^{-1}$
- concentration rate = 1 200 $l.t^{-1}$

With 24 % B_2O_3 or over it does not exist any more. See figure 3. X ray diffraction pattern pointed out the non vitreous phase as a sodium molybdate. The microprobe showed in addition to Na and Mo, Cs and Sr elements.
Table III shows some examples of glass compositions.
The viscosity of the LWR glasses is rather low. The viscosity of the others in higher. The glasses the viscosity of which is in excess of 600 poises at 1 100°C, must be ruled out on account of the industrial techniques.
The well-known opposed effects of aluminium and boric oxide are checked for all compositions. In LWR glasses the part of the fission products oxides conduces to decrease the viscosity.

PROPERTIES OF THE GLASSES

The main properties investigated are :

- the behaviour of the products submitted to thermal treatment,
- the behaviour under β irradiation,
- the loss of activity by aqueous leaching,
- the effect of α radionuclides.

Two types of thermal treatment are performed. The aim of the first one is to measure the speed of the growth of the crystals in order to evaluate the respective tendancy to devitrify. The second one is a storage simulation. The glasses are kept during a long time at 500°C or 600°C and examined at stated intervals.
In the aluminoborosilicates, the different phases, chiefly in the case of fission products high content, are molybdates and chromites, and nepheline in alumina high content glasses. See figure 4. Uranium oxides and molybdates are frequently found in borosilicate glasses. See figure 5.

The distribution of the fission products has not been yet wholly established, but it appears till now that strontium and rare earths settle as molybdates.
An irradiation test was carried out in a accelerator under the following conditions :

- beam energy : 3 MeV
- beam electrical intensity : 400 A
- irradiation power : 1.35 KW
- electron flux : 1.75×10^{13} $cm^{-2}.s^{-1}$
- effective duration : 288 h
- steady temperature : 445°C
- integrated dose : 10^{11} rads.

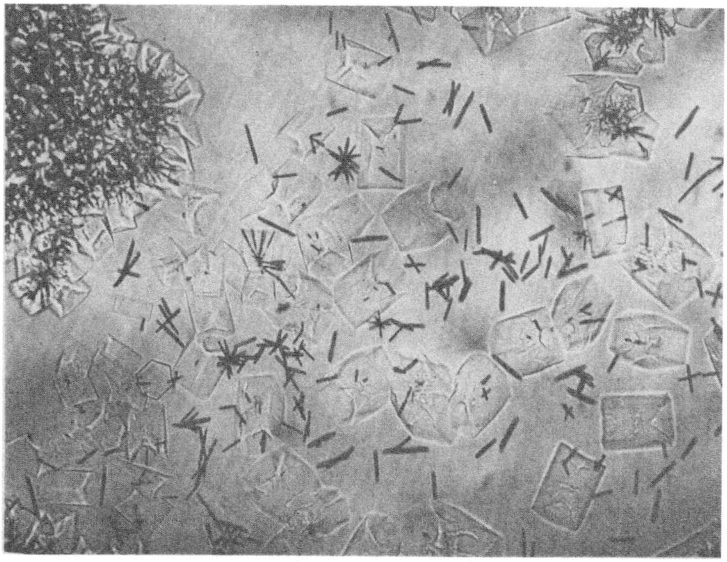

Figure 4 : DEVITRIFICATION IN A Na-FISSION PRODUCTS ALUMINOBOROSILICATES (MAGNIFICATION 195)

Figure 5 : DEVITRIFICATION IN A Na-FISSION PRODUCTS BOROSILICATES (MAGNIFICATION 195)

No difference could be observed between standard and irradiated samples about the structure by means of optical and scanning microscope or infrared absorption, neither about the chemical properties (leaching tests) nor stored energy (DTA, differential calorimetry).
Thus, the results allow to be hopeful but only to a certain extend for the dose rate is not related to an infinite storage but to 70% of the cumulated energy released by MTR glasses stored indefinitly or only 10 % of the energy released in 1 000 years by LWR glasses.

The leaching of the glasses is made by tap-water at room temperature. The glasses are in cylindrical form of diameter 100 mm and height 100 mm. Their weight is in the range of 2 000 g.
The glass blocks and the leaching are performed in a shielded cell (Vulcain, previously described) [11] [12]. The leaching is operated under flowing and the water is renewed every days.

The leaching liquor is then subjected to a gross β counting and radiochemical analysis to measure the quantity of Cs, Sr, Ce and Ru (when possible). The leaching rate is expressed as :

$$e = \frac{a}{A} \times \frac{P}{S}$$

with :

a : radioactivity of the liquor
A : initial radioactivity of the glass
P : weight (of the glass block in grams)
S : surface of the glass block in cm2.

After a 20 to 30 days period, the leaching rate reaches a constant value. This one is, for the gross β activity, in the range of 10^{-7}, whatever the composition of the glass may be. No difference has been noticed between one and two phases LWR glasses.
The leaching of devitrified glasses is under achievement. The releasing energy due to the α emitters plays a leading part after a 1 000 years storage. See table IV.

TABLE IV : COMPARISON OF RELEASED ENERGIES FROM α AND F.P. ELEMENTS

Spent time from discharge (years)	W_A Actinides thermal power (W/t)	W_{FP} FP thermal power (W/t)	$\frac{W_A}{W_{FP}+W_A}$ X 100
1	315	10 000	3.1
10	70.60	1 030	6.4
100	10.10	104	8.9
1 000	2.27	1.77 10-2	99.2

RADIONUCLIDES		GLASS COMPOSITION (WEIGHT %)											Weight of the sample (g)	α (Ci)	β (Ci)	Released energy in 1 year kWh
		SiO_2	F.P.Ox	Na_2O	B_2O_3	Al_2O_3	Fe_2O_3	MgO	NiO + Cr_2O_3	AmO_2	PuO_2	CmO_2				
MIXED Pu	WEIGH %	45.4	23.1	9.4	19.3	-	-	-	-	-	2.8	-	1700	209 (April 75)	357 (April 75)	60.4
238	29.4															
239	42.6															
240	15.9															
* 241	7.6	47.1	19.6	9.8	20.2	-	-	-	-	-	3.3	-	1924	278 (April 75)	474 (April 75)	80.2
242	4.5															
^{241}Am		50.8	24.7	8.3	13.4	-	-	-	-	2.8	-	-	1736	150	-	68.5
^{244}Cm		39.7	-	18.1	18.1	14.4	5.0	2.5	0.9	-	-	1.3	50	52 (June 75)	-	810.0

* : β emitter.

TABLE V : CHARACTERISTICS OF SOME α BEARING GLASSES -

So it is important, owing to the harmful complexion of actinides to determine their effect upon the glassy material. α bearing glasses were made in Vulcain cell.
The characteristics of some of them are plotted in table V. The neutron flux generated by α-n reactions is extremely low. The rate is 1 to 2 neutrons for 10^6 emissions (making allowance of spontaneous fission). Most of them are high energy particules. See figure 6. The composition of the glasses, although the boric oxide content varied from 14 % to 24 % in weight, has no relationship with the value of the neutron flux. Many samples are stored and retriewed at stated intervals to be examined. The scheduled investigations are :

- evaluation of nuclides distribution
- leaching rate of the nuclides
- behavior of mechanical properties
- behavior of the structure
- measurement of released He
- measurement of occluded He.

All the works are being carried out now. The first results related to occluded He show that approximately 85 % of the α particules produce helium in the case of ^{241}Am and 60 to 70 % in the case of ^{238}Pu. The leaching rate of americium is 10 times lower than the mean value of the one of the β emitters. It is also at least 10 times lower for plutonium.

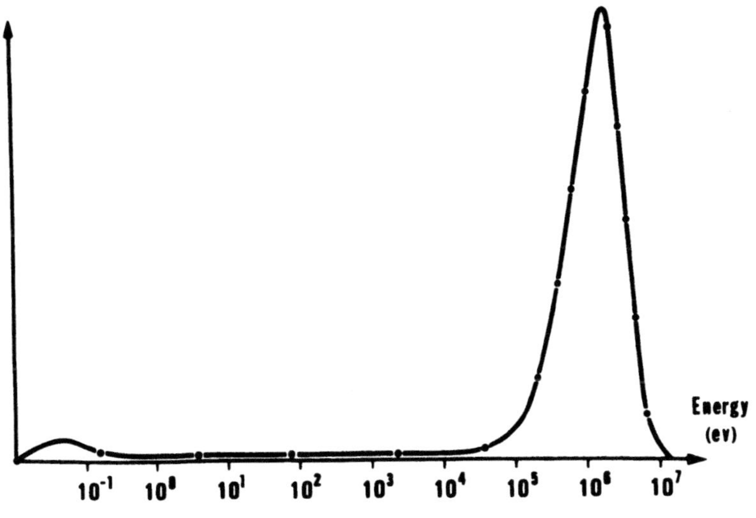

Figure 6 : NEUTRON FLUX GENERATED BY α-n REACTION IN A L.W.R. GLASS

TECHNOLOGICAL RESEARCH AND DEVELOPMENT

Two processes, a batch one and a continuous one have been developped. The batch process which is a pot vitrification method was the first in operation. See figure 7. A pot made of inconel and heated by induction at 10 000 cycles is fed with the fission products solution and the raw materials. Theses ones are in the form of a glass crushed and mixed with water to be introduced as a slurry. The feeding is ensured at a constant flow rate under moderate temperature in order to realise gently drying and calcination until the pot is 75 % filled. The feeding is then stopped and the temperature increased to 1 180° to allow the melting. After refining, the glass is pourred through a pipe fitted at the bottom of the pot and heated by induction. This pipe, kept cool during making, is somewhat a thermal valve.

The gas off are processed with a condensor, a ruthenium trap, washing columns (water and caustic) and filters. A ruthenium trap is used owing to the volatilisation of Ru which occurs during the various step of the making, and in addition to its radioactive complexion, could initiate some plugging in the ducts.
The results given by the prototype were so attractive that a fully radioactive pilot plant was built in Marcoule and have been operated from 1969 to 1973, in processing waste solutions issued of natural uranium spent fuels (SICRAL).
12 t of glass of a maximum specific activity of 3 000 Ci/l were produced in this plant called PIVER. During each run, using a pot 250 mm diameter and 2.5 m high, 200 liters of solution were vitrified, generating 90 kg of glass.

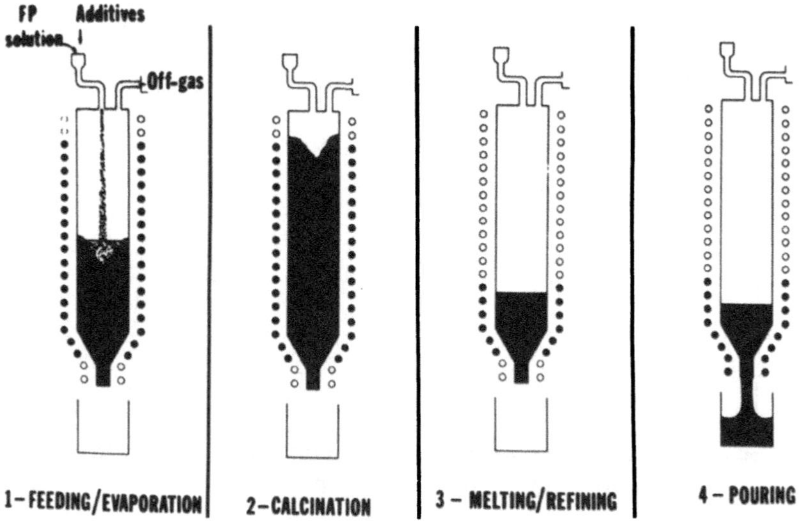

Figure 7 : PRINCIPLE OF POT-VITRIFICATION TECHNIQUE

The glass disposed in stainless steel containers is stored in an experimental storage. This facility comprises 32 pits, 10 meters deep, enclosed in a concrete structure which is equipped with a forced air cooling. Each pit can contain 10 containers or 3 used pots. The pots were removed every 30 runs, not owing to corrosion matter but to a possible buckling. 164 containers are presently stored in this disposal.

This technique is perfectly satisfactory but unfortunately it does not meet the needs due to the growth of commercial nuclear energy. Besides, it shows some disadvantages comparing to a continuous process. See table VI.

TABLE VI : COMPARISON OF CONTINUOUS AND DISCONTINUOUS PROCESSES

	POT VITRIFICATION	CONTINUOUS VITRIFICATION
TECHNOLOGY	Simple mechanical device	More elaborate apparatus
WORKING FACILITIES	Many remotely handling Diff. to automatize	Easy
CAPACITY	Low subjected to salt amount	Non sensitive to salt amount
FINAL PRODUCT	Good	Good
EFFECT OF AL IN WASTE	Wrong	Rather good
EFFECT OF ALKALINE IN WASTE	Slight casual foaming	Caking possibility
ABILITY IN GLASS CHOICE	Limited	Large scale
MELTING TEMPERATURE	Limited (metallic pot)	No compulsion with inside induction
DUST ESCAPE	Low	Higher
INVESTMENT	25 % more	
WORKING COST	25 % more	

Nevertheless it will be resumed a little to vitrify some cubic meters of high level wastes generated by the reprocessing of Phoenix spent fuels (FBR) in order to make glass blocks of a very high specific radioactivity.
The continuous technique is composed of a rotary calciner compled to a melting furnace. An outlined flow diagram of the process is shown in figure 8.

The calciner is a rotating tube, 27 cm. outside diameter, 12 cm. thick, and 3.6 m long, set in order to have a slight slope. A loose bar is located inside the tube to prevent possible caking. An end-fitting is set a each end of the tube. The upper one allows the liquid feeding and the off-gas exhaust. The lower one is devoted to glass additives feeding, calcined product draining, and connection with the melting unit. Air tightness is ensured by graphite rings locked on each end-fitting and that slide on a flange borne by the tube. The rotating speed is 30 rpm.

The tube is heated by a 70 KW electrical resistance-heated furnace. This furnace is composed of four independent heating zones.The four electric power consumptions are recorded.

The solution is fed through the upper end-fitting, evaporated and dried, and then partially calcined along. The maximum feed rate of such an apparatus is 40 liters/hr.
The solution is lifted by a double air lift that overflows into a small vessel where it is mixed with a recycling solution coming from the first off-gas treatment apparatus and with an organic additive that improves the granulometric distribution of the calcined product.

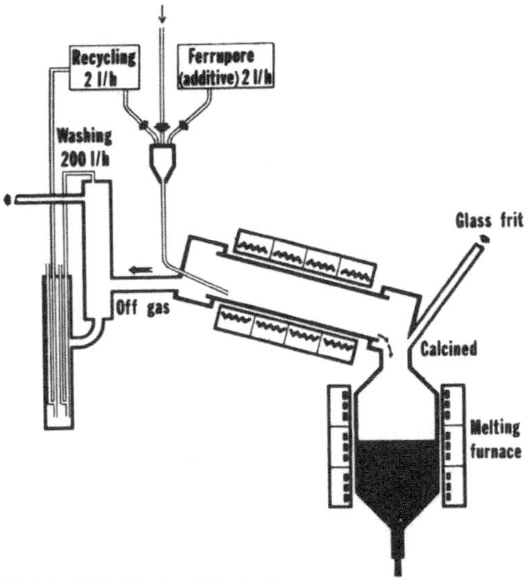

Figure 8 : FLOW DIAGRAM OF THE CONTINUOUS VITRIFICATION PROCESS

Vitrification is made in a melting furnace composed of a metallic pot heated by 10 KHz frequency induction. The pot is a 1.7 m. high, 350 mm. inside diameter, and 6.35 mm. thick cylinder.The bottom is made up with a cone and a drain tube. There are three independent cylindrical induction coils, and one conical and one small coil related to the drain tube, which is used as a thermal valve. The additives are a primary glass introduced via a lock chamber and through the lower end-fitting of the calciner in the form of small flakes (frit). This glass falls in the melting furnace simultaneously with the calcined product. They react with the melted glass and make the glass level rise. When the level reaches a given upper limit, input power is supplied on thermal valve coil and the glass drains off. A siphon is fitted on the top of the drain tube in order to keep some glass in the bottom to resume glassmaking and to ensure air-tightness.During the pouring of the glass, additives and solution feedings are not interrupted.

The melting pot vacuum is about 10 cm. H_2O. Melting temperature, which is measured with thermocouples, is approximatively 1,150°C.
Total decomposition of the nitrates occurs in the melting pot. The maximum capacity is 18 kg/h of glass.

The off-gas, escaping from the melting furnace and the calciner, contains water, nitrogen oxides, some volatile products (boron, cesium, ruthenium, and fluorine compounds), and dusts. These dusts are trapped by a wet scubber, which sprays the stream with acid solution. This solution is drained back to the calciner at a steady flow rate.
The consequent lack of liquid is balanced with water injection controlled by a level gauge. The gas is then condensed and the continuation of the treatment that follows is conventional (absorption column, absolute filters, and so on). A gas pump located at the end of the treatment ensures a gas correct vacuum.

A first prototype equipment has been operated successfully for over 8 000 h with various simulated high level wastes (MTR, LWR, natural U). From this prototype, the Marcoule vitrification Plant (AVM) was designed and built, scale to scale.
In this plant, previously described [8] the glass is poured in 50 cm. diameter and 1 m. high metallic containers. The containers are weighed during pouring. After filling, they are closed with a lid, welded by means of a plasma torch, and decontaminated. Then, they are transferred to the interim storage facility where they are put in vaults equipped with deep cylindrical sleeves under forced-air cooling. The maximum temperature of exhausted air which is filtered is 100°C. When natural convection cooling is sufficient forced circulation is shut off.

The storage facility is made up of three artificial vaults. The structure, which is reinforced concrete, is totaly underground except for the cover blocks. A mild-steel impervious jacket is countersunk in the structure. 220 vertical sleeves made of metallic tubes are stacked by means of cross pieces. Each tube, 0.6 m. diameter and 10 m. deep, can hold ten containers. The cover blocks are metallic caissons filled with 1.5 in. thick concrete.
The air for the induced draft cooling is supplied through suitable flues and flows in the annular space formed by the containers and the sleeves. The hot exhausted air is filtered and pulled by fans to a stack. The structure is capped with a hangar, protecting a 350 kN bridge-crane. This bridge-crane carries the containers in shielded casks from the vitrification facility to the storage area. The descending motion in the tubes is operated by hoist.

The maximum radioactivity of one container will be, at the beginning of storage time, approximately 6.75/Kcal/hr.(7.8 KW). Taking radioactive decay, load input, and aged high level waste solutions presently stored into account, the maximum heat power that will evolved when the vault is totaly filled is about 1 MW.

AVM has been designed to treat the daily Marcoule production and to resorb the liquid presently stored within 10 years. The storage facility is large enough for this purpose and more over could be extended to meet further requirments. There has been no decision as yet in France about the institution of a national depository, but if such a management policy were set up, it would be possible to retrieve wholly the containers. A general section of the plant is shwon in Figure 9 and an extended view in Figure 10.

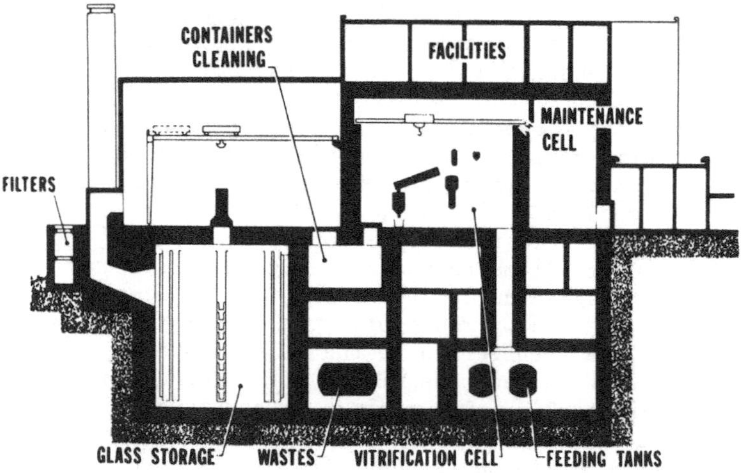

Figure 9 : GENERAL SECTION OF A.V.M.

Figure 10 : A.V.M (MARCOULE VITRIFICATION PLANT)

The first non-radioactive run took place in February 1977, the radioactive operation is expected in the next half year.

PROSPECTS

First AVM working time is expected to give much worthy informations usable for La Hague vitrification. But a lot of LWR runs operated in a nonradioactive prototype predicts that this way will be a successful one.
In La Hague case (800 tons/yr.of LWR spent fuels), it will be necessary to use a melting furnace with a higher capacity.

To this purpose, a prototype in glass induction furnace has been developped in Marcoule. Its frequency is in the range of 100 to 200 kHz. The walls of the chamber are made of refractory ceramics cooled with a water jacket [13].

This process would be suitable also to vitrify high level radioactives wastes issued from a 1500 ton/yr reprocessing plant under the following conditions : high level waste partially denitrated and concentrated with a factor 2, two calciners, and two high frequency furnaces. See figure 11.

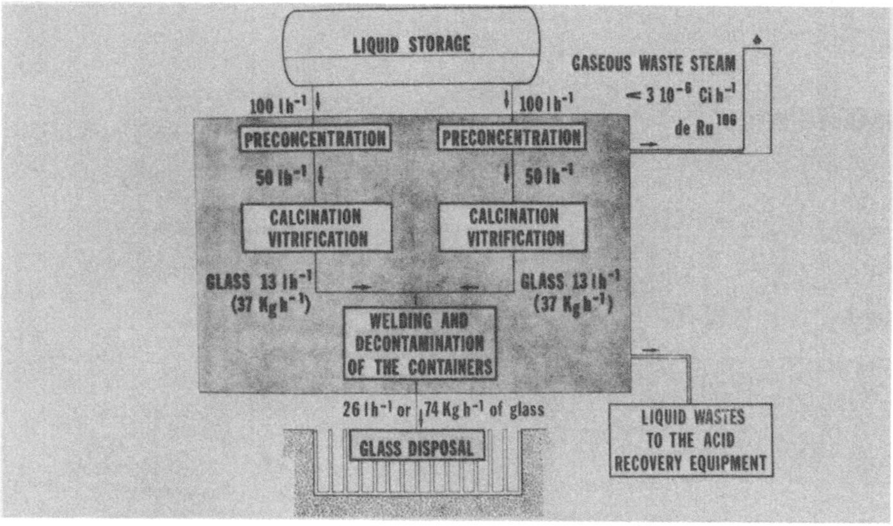

Figure 11 : DIAGRAM FEATURING THE VITRIFICATION OF THE WASTES GENERATED BY A 1500 t/y REPROCESSING PLANT

REFERENCES

1. Essais d'incorporation de solutions concentrées de produits de fission dans des verres et des micas. R. BONNIAUD, P. COHEN, C. SOMBRET. 2ème conférence internationale des Nations Unies sur l'utilisation de l'énergie atomique à des fins pacifiques. A/CONF.15 P/1176. GENEVE Mai 1958.

2. Vitrification of concentrated solutions of fission products : Study of glasses and their characteristics. R. BONNIAUD, C. SOMBRET, F. LAUDE. Symposium on treatment and storage of high level radioactive wastes. IAEA VIENNA - October 1962.

3. Vitrification of concentrated solutions of fission products: Technological studies. R. BONNIAUD, C. SOMBRET, D. RANCON. Symposium on treatment and storage of high level radioactive wastes. IAEA VIENNA - October 1962.

4. La vitrification dans le domaine des effluents radioactifs. R. BONNIAUD, C. SOMBRET. Verres et Réfractaires. Vol. 25 n° 2 - 1971.

5. Experience acquise en France dans le traitement par la vitrification des solutions concentrées de produits de fission. R. BONNIAUD, F. LAUDE, C. SOMBRET. Symposium on the managment of radioactive wastes from fuel reprocessing. OECD. PARIS - November 1972.

6. Continuous vitrification of concentrated solutions of fission products. A. JOUAN, C. SOMBRET. 1st conference of the european nuclear society " Nuclear Energy Maturity " PARIS. April 1975.

7. Statement of research in the field of solidification of high level radioactive wastes in France. R. BONNIAUD, C. SOMBRET. 77th annual meeting of the American Ceramic Society. WASHINGTON DC. May 1975.

8. Franch industrial plant " AVM " for continuous vitrification of high level radioactive wastes. R. BONNIAUD, C. SOMBRET, A. BARBE. AICHE meeting. LOS ANGELES - November 1975.

9. Developpement actuel du procédé de vitrification continue des solutions de produits de fission.
R. BONNIAUD, R. CARTIER, A. JOUAN, C. SOMBRET.
International symposium on the management of radioactive wastes from the nuclear fuel cycle IAEA/OECD. VIENNA - March 1976. (IAEA.SM.207/27).

10. Confinement de la radioactivité dans les verres.
R. BONNIAUD, F. LAUDE, G. RABOT, C. SOMBRET.
International symposium on the management of radioactive wastes from the nuclear fuel cycle. IAEA/OECD. VIENNA - March 1976. (IAEA.SM 207/36).

11. Some bearings about the behaviour of ultimate stored radionuclides in the form of glass or glassy matrix product.
R. BONNIAUD, F. PACAUD, C. SOMBRET.
Congrès de la Société Française de Radioprotection. VERSAILLES - Mai 1974 - VII. SFRP/I -

12. Equipement intérieur de la chaîne VULCAIN.
Note C.E.A. n° 1419 p. 397. R. BONNIAUD-A.JOUAN-C.SOMBRET

13. Nouveaux développements dans la fusion électrique des verres réfractaires.
R. BONNIAUD, J. MOULIN, J. REBOUX, C. SOMBRET.
Verres et Réfractaires - Vol. 26 n° 4- 5 - 1972.

APPLICATION OF BORATE GLASSES IN ELECTRONICS

R. R. Tummala

IBM System Products Division, East Fishkill

Hopewell Junction, New York 12533

INTRODUCTION

Glasses are widely used in the electronics industry. Such applications as sealing to form hermetic packages, bonding ferrites to form storage heads, depositing coatings on silicon for passivation, forming dielectrics for high speed transmission of signals, and fabrication of displays which may require as many as six different glasses, all require glasses with significant amounts of B_2O_3 in their composition. While B_2O_3, by itself, does not have much application, primarily because of its poor chemical durability, glasses containing B_2O_3 content in the range of 5 to 40% do offer technological properties, one application that takes advantage of the low-melting properties of B_2O_3 is in a plasma display panel; a schematic of which is shown in Fig. 1. This display which consists of as many as six different glasses to fabricate the assembly, requires approximately 1 mil thick bubble-free coating of a dielectric glass on commercial soda-lime-silica substrates. Since typical soda-lime-silica substrates of any significant size cannot withstand processing temperatures in excess of about 600°C, dielectric glasses with very steep viscosity as a function of temperature had to be developed. Since the sealing of the panel is done to dielectric surfaces, it meant that further low-melting yet thermal expansion matching the dielectric had to be developed. Both these glasses were developed in the lead borate system with modifications to obtain steep viscosity for the dielectric and low-softening and flow out temperatures for the seal.

DIELECTRIC GLASSES

The requirements of an ideal dielectric for the plasma display application are as follows:

1) 1 mil-thick bubble-free coating as fired at about 600°C or below.
2) Thermal expansion matching commercial soda-lime-silica substrates.
3) Good dielectric properties under electric field and high secondary emission.
4) Very steep viscosity so as form bubble-free film and still allow sealing without structural deformation of the dielectric during sealing operation.
5) High transparency.

Examples of dielectric glasses explained in the PbO-B_2O_3 system are shown in Table 1. The first glass in this table is a typical lead glass that is most commonly used for solder glass sealing applications. This glass did fire bubble-free at about 600°C but its deformation temperature (found to correspond to log viscosity = 10.3 poises) was only 430°C, which did not allow sealing even with the lowest melting seal glass. The lowest melting seal glass required about 485°C. The dielectric deformation temperature was raised by the incorporation of MgO, but the magnitude of the increase required was not possible with MgO alone. The glasses with MgO contents in excess of 3% began to crystallize. The addition of CaO to glasses containing MgO raised the deformation temperature to the required level without proportionately increasing the high temperature viscosity. Na_2O was added to stabilize dielectric-to-soda-lime-silica reactions. The viscosity of the dielectric glass No. V shown in Fig. 2 as compared with the viscosity of pure B_2O_3 and Pyrex glass indicates the extent steepness of viscosity achieved as a function of temperature.

Even with the high steep viscosity achieved for the dielectric glasses, it still had not been possible to fire the dielectrics below about 600°C, glasses II to V requiring firing temperatures in excess of 650°C. It was found, however, that most of these glasses responded to firing in wet atmospheres. For example, high B_2O_3 glasses such as glass IV although required 650°C in dry air, could be fired substantially bubble-free at 610°C. The lowering of viscosity achieved in wet and dry atmos-

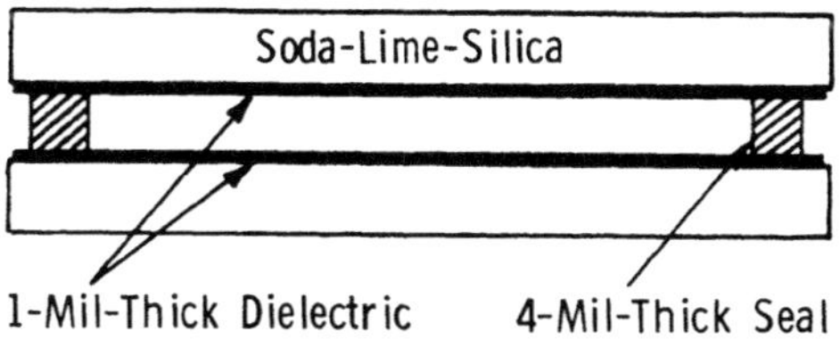

Fig. 1. Schematic of plasma display.

Table 1. Dielectric glasses (lead).

PbO :	73.5	69.6	63.4	56.0	56.0
B_2O_3 :	12.7	13.6	17.0	21.5	21.5
SiO_2 :	13.6	13.6	12.0	12.5	12.5
Al_2O_3 :	0.2	0.2	0.2	1.0	1.0
CaO :	0	0	4.4	5.0	5.0
MgO :	0	3.0	3.0	2.0	2.0
Na_2O :	0	0	0	2.0	2.0
Ts (^oC)	480	506	540	535	570
Flow in air (^oC)	600	680	650	650	720
Flow in wet air (^oC)	560	620	610	610	700
Temperature @ log η = 10.3	430	463	490	490	500

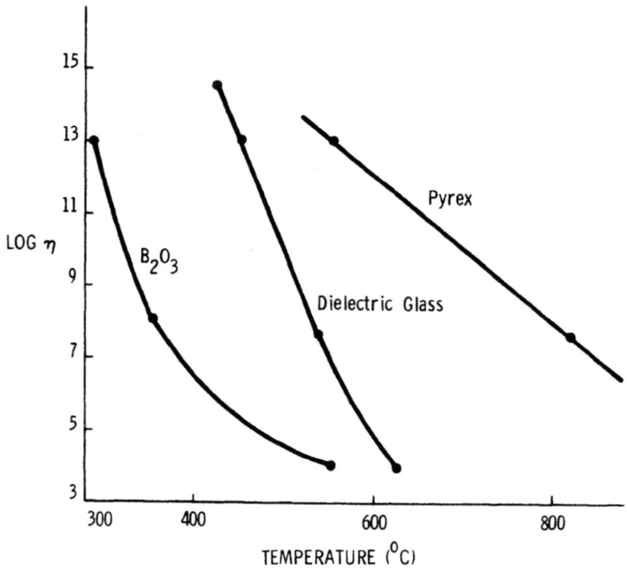

Fig. 2. Steep viscosity of dielectric glass compared with B_2O_3 and Pyrex glasses.

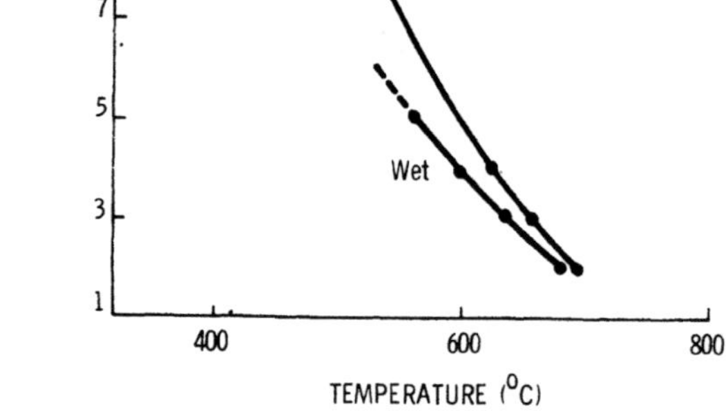

Fig. 3. Viscosity of dielectric glass in wet and dry atmospheres.

pheres for this glass is shown in Fig. 3, and is in general agreement with the reductions reported previously (1). The incorporation of OH ions either as free, bonded, or both (2) in the glassy network apparently caused the viscosity reduction.

SEAL GLASSES

The requirements of seal glass are as follows:

1) Thermal expansion matching the substrate and dielectric (about 80×10^{7}/°C.
2) Softening temperature as low as possible but no higher than 415°C so sealing is performed below the 485°C deformation temperature.
3) Hermetically seal the panel to provide long life.
4) Vitreous in nature so preforms can be drawn.

The summary of all commercial glasses by Cox (2) indicates that no vitreous glass, with a thermal expansion coefficeint of about 80×10^{7}/°C and softening of 415°C is available. Typical commercial glasses have softening temperature around 440°C. The seal glass developed to meet the requirements is shown in Table 2. This glass is a modification of lead-zinc-borate glass with small amounts of SiO_2, Al_2O_3, CuO and Bi_2O_3 to impart low softening and low thermal expansion. The strength of this glass as increased by dil. HNO_3 etching and the stress corrosion in humid atmospheres has been reported previously (3, 4).

SUMMARY

Dielectric and seal glasses in the PbO-B_2O_3 system with B_2O_3 in the range of 14 to 21.5% have been developed to meet the requirements of a gas panel. Dielectric glasses in the PbO-B_2O_3 and SiO_2 system with additions of CaO and MgO produced the required steep viscosity, and this behavior combined with higher responsiveness of higher B_2O_3 (compared with SiO_2) glasses to wet firing resulted in the desired dielectric films. The seal glasses with modifications of PbO-ZnO-B_2O_3 system by additions of CuO, Al_2O_3, Bi_2O_3 and SiO_2 have resulted in vitreous seal glass with the required low-melting properties.

Table 2. Seal glass composition.

PbO	: 66	
B_2O_3	: 14	
SiO_2	: 2	
Al_2O_3	: 3.5	$TCE = 82 \times 10^{-7}/^{0}C$
CuO	: 2.0	$Ts = 413^{0}C$
Bi_2O_3	: 2.0	
ZnO	: 10.5	

ACKNOWLEDGMENTS

The author wishes to acknowledge the contributions of T. A. Sherk, J. M. Brownlow, R. C. Buchanan and B. J. Foster.

REFERENCES

(1) H. Scholze, Glass Ind. 47 (1966) 670.
(2) S. M. Cox, Survey of Glass Materials in Micro-Electronics, Ministry of Technology, London, England (Sept. 1968).
(3) R. R. Tummala and B. J. Foster, J. Am. Ceram. Soc. 60 (1977) 170.
(4) R. R. Tummala, J. Non. Cryst. Solids, 19 (1975) 263.

REACTION CURED BOROSILICATE GLASS COATING FOR LOW-DENSITY FIBROUS SILICA INSULATION

Howard E. Goldstein, D. B. Leiser, and V. Katvala

Ames Research Center, NASA

Moffett Field, California

INTRODUCTION

The Space Shuttle Orbiter, shown in Fig. 1 will be the world's first reusable space vehicle. Basic to the development of such a vehicle is the requirement for a heat shield that can survive multiple reentries into the earth's atmosphere at temperatures up to 1400°C. For the majority of the orbiter's surface, a material was required that had good insulative properties, could survive temperatures up to 1260°C, and would be extremely light weight. It would also be required to survive large temperature gradients (>1000°C/cm) and severe thermal shock resulting from the entry environment (Fig. 2). The convective heating environment which occurs during the vehicle's entry is unique in that it can result in chemical interactions between the high-temperature gases and the solid surface. These interactions cause an abnormally high vaporization rate to occur among the less refractory compounds present in a heat-shield surface (1). Therefore, the stability of the heat shield in this type of environment is critical to the success of the vehicle.

A family of materials called reusable surface insulation (RSI) was developed for this application in the late 1960's. These materials were low density, rigidized ceramic fiber materials that have high emittance and water-impervious coatings (2). As a result of extensive testing, a silica RSI that has a density of 0.144 gm/cc (9 lb/ft^3) was chosen (3) for most of the Space Shuttle surface. The two major advantages of the material were its high resistance to thermal shock and its good insulative efficiency (4). The thermal shock resistance of silica RSI is a result of its extremely low thermal expansion coefficient ($\sim 4 \times 10^{-7}$ cm/cm- °C) and its

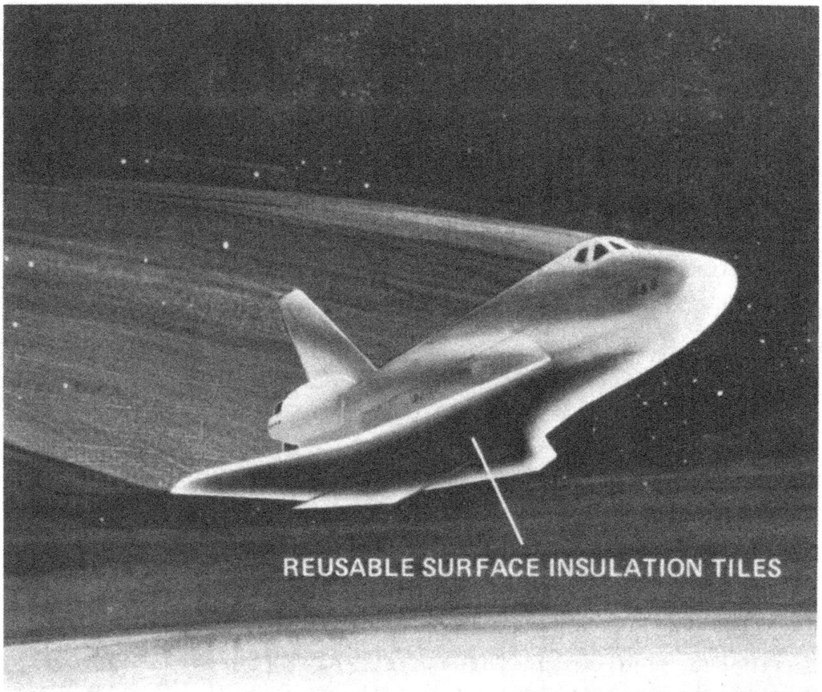

Figure 1. Space Shuttle orbiter during atmospheric entry.

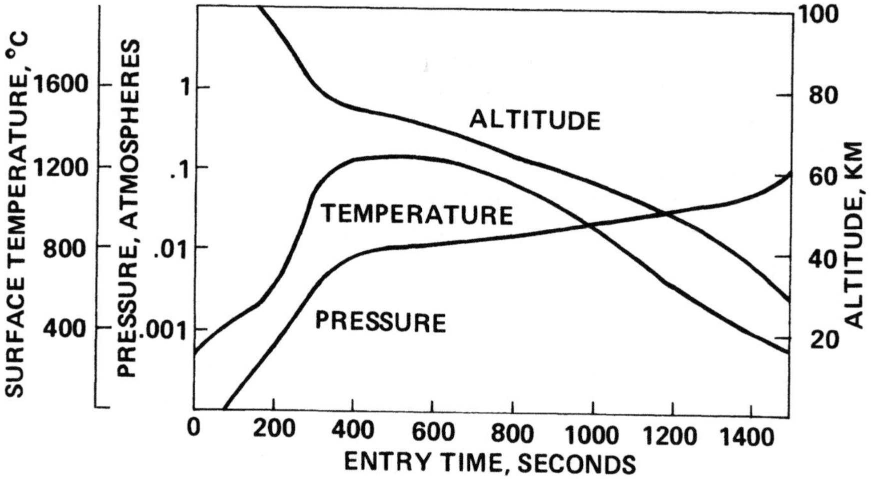

Figure 2. Shuttle entry environment.

high strain to failure (0.1 to 0.2%) attributable to its fibrous nature. The insulative efficiency is provided by the very small fiber diameter (1 to 2 μm) and the low bulk density of the insulation (2).

In 1973 tests demonstrated that the state-of-the-art coatings for the silica did not meet the requirements for the Space Shuttle orbiter (5). These requirements were: the coating must be chemically and physically compatible with the substrate such that no devitrification or thermal stress cracking would occur during the Shuttle lifetime of approximately 24 hours at 1260°C; it must not change dimensions because of foaming* or become porous or suffer an unacceptable reduction in emittance during its planned lifetime; it must remain waterproof as well; the coated tile must be manufactured to tolerances of ±0.06 cm and coating weights must be as low as possible to minimize total heat shield weight. The coatings used in 1973 did change dimensions because of foaming, and did lose their high emittance in the aeroconvective heating environment. In addition, the coatings devitrified and, as a result, cracked after 10 to 30 simulated reentries. These coating deficiencies led to studies to develop a coating that could meet the previously described requirements.

COATING DEVELOPMENT

In order to meet the Shuttle orbiter service life requirements, a glass coating material has to have an extremely low thermal expansion coefficient that closely approaches that of silica. It must remain stable against devitrification for at least 24 hours at 1260°C, even after exposure to severe sea salt contamination. It must have a total emittance that is greater than 0.8 and that does not change as a result of exposure to a convective heating environment. It must glaze at a low enough temperature to minimize the distortion and shrinkage of the low-density silica substrate. A number of systems were studied to meet these requirements; they included low expansion coefficient ceramics such as lithium aluminum silicates. These did not meet the requirements because of the volatility of lithium oxide at high temperature and low pressures and because they were mineralizers or contaminants to the silica substrate.

Ultralow-expansion glasses** were among the other candidate systems that were evaluated. These glasses rapidly devitrified

---

*Foaming is defined as a porous surface layer that is formed as the direct result of volatilization of parts of the coating.

**ULE - Corning Glass Works.

at temperatures above 1200°C; they severely contaminated the silica substrate causing rapid devitrification at 1260°C; they also required very high glazing temperatures.

Borosilicate glasses were found to be the most promising candidates because of their low thermal expansion coefficients (Fig. 3) and good thermal stability. In order to make a coating that met the requirements discussed previously, a number of borosilicate glass variations were evaluated. At first, very high purity prereacted borosilicates with low alkali contents (<0.1%) were studied. These glasses, even with 15% boron oxide content, would not glaze satisfactorily in the temperature range (below 1260°C) required for compatibility with the silica substrate. They also were not as stable against devitrification as was desired. In order to obtain the lower glazing temperature, several fluxes that included alkali borosilicate glasses and extremely high boron-oxide content frits (up to 30% boron oxide) were tried. None of these was successful; some had poor stability against devitrification and some required too high a glazing temperature. Another type of composition considered was a two-phase frit formed by boration of a powdered silica, a high silica borosilicate glass† and a porous high silica borosilicate glass.‡ The resulting borated frits were then calcined at low enough temperatures (<1100°C) so that the frit particles were not fused but the boron oxide was sintered to the individual frit particles. The borated silica was found to be less stable against devitrification than the original silica material and its fusibility had not been significantly enhanced. A 2% or more addition of boron oxide was found to stabilize the basically unstable high silica borosilicate frits§ for the required 24 hours at 1260°C as shown in Fig. 4. The fusibility (reactivity) of the porous glass version was significantly improved, whereas that of the nonporous glass was only slightly increased. Further testing demonstrated that the borated (3% additional boron oxide) porous high silica borosilicate glass, when fused, was stable for at least 800 hours at 1260°C and was fully compatible with the silica substrate, without causing devitrification for the anticipated Shuttle orbiter lifetime.

Another approach studied to increase the basic glass stability was the use of a reducing agent to decrease the oxygen content of the glass, since it had been found (6) that oxygen-reduced silica

---

†Vycor 7913 - Corning Glass Works.

‡Vycor 7930 - Corning Glass Works.

§Devitrification of High Silica (>90%) Borosilicate Glasses. Presentation by D. B. Leiser at the 77th Annual Meeting of the American Ceramic Society in Washington, D.C., May 1975 (unpublished).

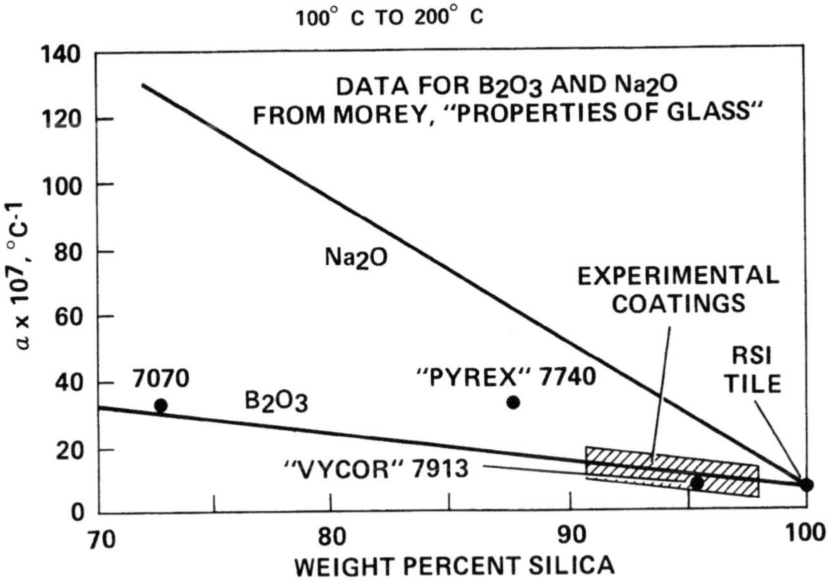

Figure 3. Linear thermal expansion coefficient of high silica glasses and fibrous silica insulation.

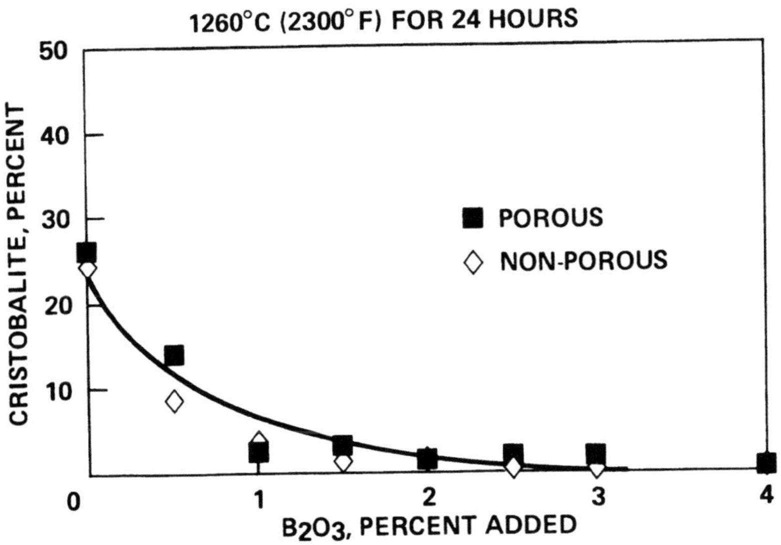

Figure 4. Effect of boron oxide on porous and non-porous high-silica borosilicate glass (Vycor) devitrification.

glasses were more stable than stoichiometric silica. Among the reducing agents tried were carbon, silicon, boron, tetraboron silicide and hexaboron silicide. The carbon and silicon proved to be unacceptable because (a) they did not further stabilize the glass, and (b) the carbon formed a gaseous oxide which tended to foam the glass. The boron and boron silicides, on the other hand, formed glasses, which, when fused with the two-phase borosilicate frit, were not only more stable against devitrification than before, but were either white, grey, or black, depending on the fusion temperature and concentration of components. The stabilizing mechanism appeared to be the formation of a boron-oxide-rich borosilicate glass phase between the silica-rich borosilicate glass particles rather than the oxygen depletion theorized in Ref. 6.

In order to meet the requirement for a high emittance (>0.80) at 1260°C, the borosilicate glass matrix must be opacified, since these glasses are quite transparent by themselves between the wavelengths 0.3 to 4.0 μm. A number of emittance agents were studied for this purpose, including oxides such as iron, chromium, hafnium, and cobalt. All of these oxides were either found to be contaminants to the silica substrate or unstable in the convective heating environment. They caused rapid devitrification, particularly at the interface between the silica insulation and the glass coating. Carbides were also tried but they generally formed carbon monoxide and dioxide when exposed to high temperatures in the borosilicate matrix. These gases would then act as blowing agents, foaming the coating when it became plastic at 1200°C or above.

The boron and boron silicides were subsequently reexamined for use as possible emittance agents, since they were only partially oxidized when fused with the reactive borosilicate glass frit (as was discovered in earlier studies) and they did appear to have a high emittance. Further studies showed that the most effective of these materials was commercially designated as tetraboron silicide.¶ It consisted of a mixture of several boron silicides. Several compositional variations of the glaze were subsequently tried. They included variations in both the tetraboron silicide content and the boron-oxide content of the reactive glass frit. The final composition, described in the following section, was selected to give a low glazing temperature consistent with the required long-term stability and high emittance at 1260°C, while retaining a minimum thermal expansion coefficient.

¶Cerac Corporation.

Coating Composition and Processing

The reaction-cured borosilicate glass is prepared by mixing a reactive high-silica borosilicate glass frit and the bimetallic alloy tetraboron silicide. The frit is prepared by mixing a commercially available porous high-silica borosilicate glass (Vycor 7930♠), with boric anhydride (boron oxide) in hot water at 85°C. The mixture contains sufficient dissolved boric anhydride to adjust the total boron-oxide content of the resultant frit to 5.75% by weight. The solution is continuously stirred while the temperature is maintained at 85°C, eliminating the excess water by evaporation until the resultant slurry is too thick to mix further. The slurry is then dried for about 16 hours at 70°C until the majority of the water is removed. The dried material is screened (16 mesh) before sintering at 1090°C for at least 1 hour. After sintering, the glass is crushed and passed through a 325-mesh screen. The frit "as prepared" is insoluble in water and consists of two phases which are a high-silica borosilicate glass interior and a very reactive exterior of high-boron borosilicate glass. The final composition is prepared by mixing the previously described reactive frit with 2.5% by weight tetraboron silicide (B_4Si). This mixture is ball-milled for approximately 3 hours with a methylcellulose prebinder and ethyl alcohol. The alcohol is the carrier during the subsequent spray application of the coating material to the low density RSI. After spraying, the article is dried for approximately 5 hours at 70°C. The coating is then glazed (reaction cured) by inserting it in a furnace at 1200°C. After 90 minutes, the coating is rapidly cooled to room temperature. The rapid cooling after curing is required to minimize the "as processed" residual tensile strain on the coating.

Strain reduction can be explained by two mechanisms:

(a) The coating effectively deforms the "soft" low-density silica substrate because the substrate remains above its strain temperature as the coating's temperature rapidly cools below its strain point. The substrate therefore takes up some of the normal strain differential through deformation if the coating is cooled rapidly enough.

(b) The low-expansion-coefficient silica substrate remains hot, and therefore fully expanded, when the higher expansion coefficient coating is cooled below its strain point. The substrate then contracts upon cooling and reduces the residual strain of the higher expansion coating. This process is essentially identical to tempering.

♠Corning Glass Works

Figure 5 shows a schematic of the reaction curing process. The reaction curing process is produced by the partial oxidation of tetraboron silicide to form a reactive borosilicate glass. This glass fuses with the boron-rich borosilicate exterior of the "as prepared" frit particle, as shown in the figure. The oxidation of tetraboron silicide particles is not uniform. The tetraboron silicide particles on the coating surface are completely oxidized because the external surfaces of the coating are exposed to more oxygen than the interior of the coating. Figure 6 shows the reaction cured glass structure after curing on the low-density silica substrate. In the optical micrograph, the coating appears multilayered. This is caused by the relative transparency of the exterior surfaces compared with the interior. The transparency indicates no remaining tetraboron silicide particles while opacity indicates their presence. Figure 6 also shows a scanning electron micrograph of the coating that indicates the coating's nonporous nature after curing. The nature of the reaction curing process is further illustrated in Fig. 7. Two exotherms are shown in a Differential Thermal Analysis (DTA) of a representative coating mix. One occurs at 300°C when the organic prebinder methycellulose is

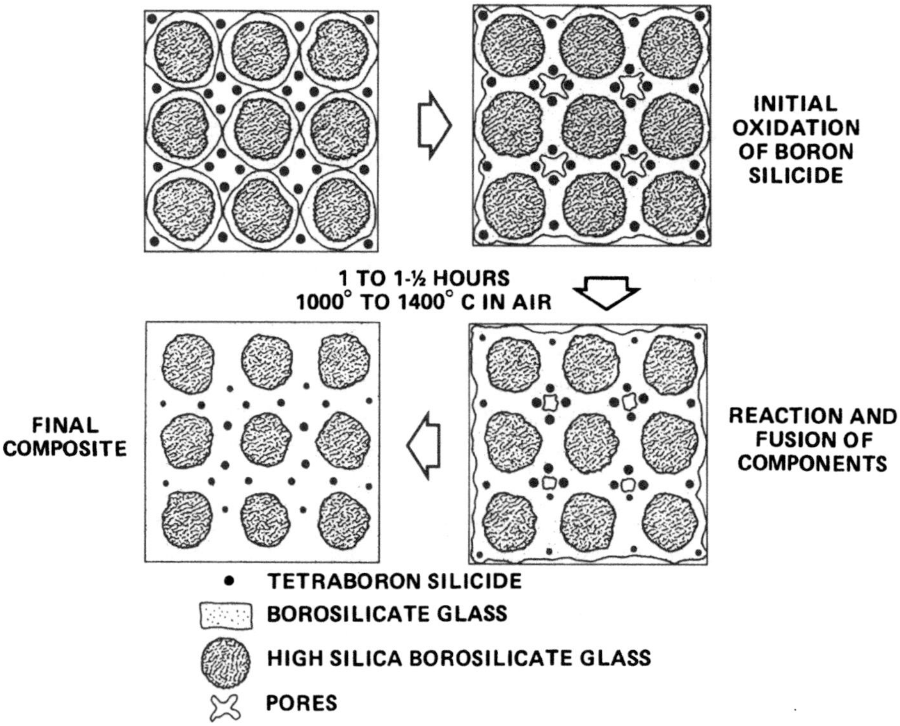

Figure 5. Schematic of reaction curing process.

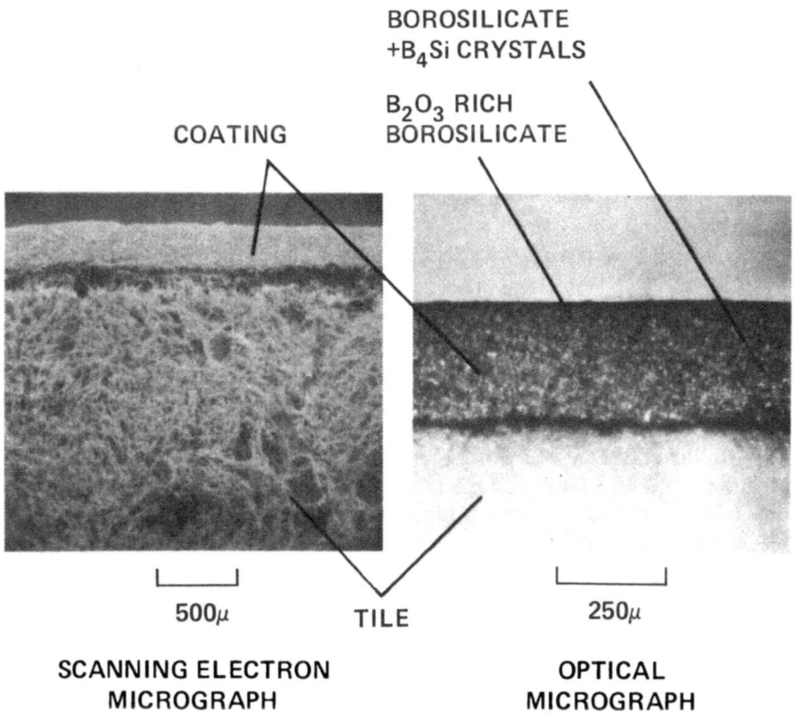

Figure 6. RCG glass coating structure.

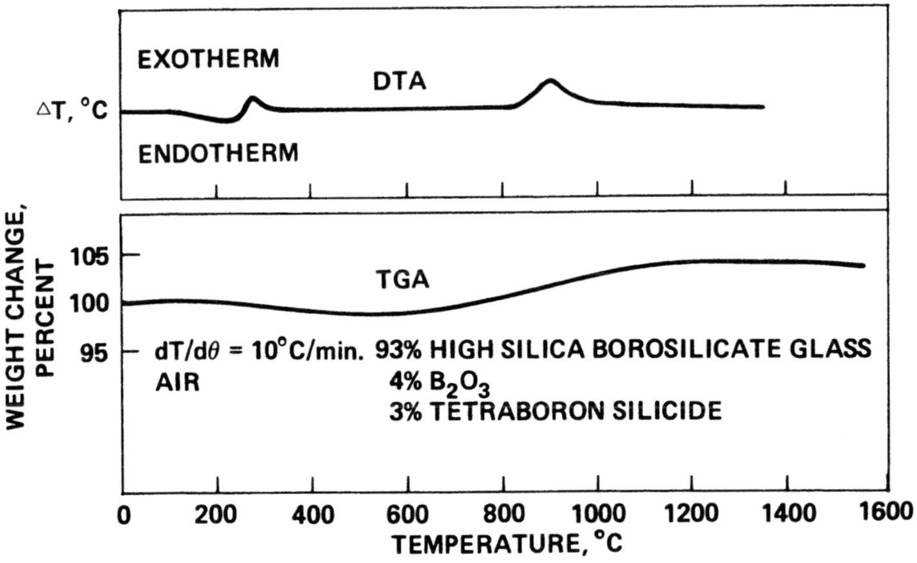

Figure 7. Thermal analysis of typical reaction cured glass.

oxidized and the other occurs at 900°C when the tetraboron silicide is partially oxidized. The simultaneous weight gain of approximately 3% observed in the TGA is caused by the retention of tetraboron silicide oxidation products. During the glazing process, the coating mixture is brought up to 1200°C nearly instantly, which causes the reaction and fusion processes to occur simultaneously. The tetraboron silicide particles are then protected from complete oxidation by the viscous borosilicate glass that surrounds them.

One of the more important properties of a glaze or coating which will be cycled from low to high temperatures and back is its thermal expansion properties relative to those of its substrate. Thermal expansion versus temperature for the reaction-cured glass (RCG) coating and the silica substrate is illustrated in Fig. 8. The figure shows that both the bulk RCG glass and the thin (0.030 cm) coating with equivalent compositions have essentially the same expansion characteristics. Additionally, the figure shows the thermal expansion characteristics of the unborated glass frit (used to form the reactive glass frit) and that its expansion increases 10% when borated with 2.5% boron oxide. The addition of 2.5% tetraboron silicide to form the "basic" RCG coating increases the thermal expansion an additional 30%. The figure also shows that there is a 200 to 300 μcm/cm difference in expansion between the coating and the silica substrate at 800°C. This 200 to 300 μcm/cm difference in expansion could cause cracking in glasses that have the usual modulus of 7×10^9 N/m^2 or greater. Fortunately, the silica RSI

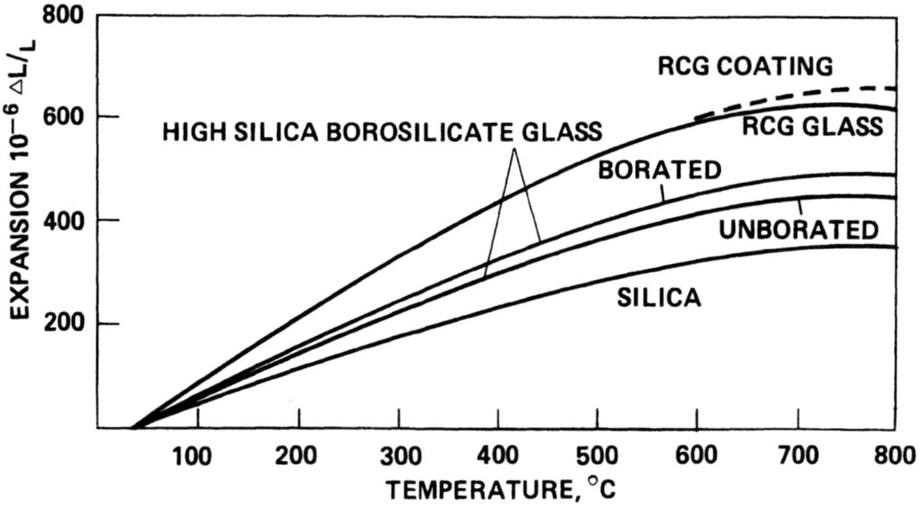

Figure 8. Linear thermal expansion of silica and high silica borosilicate glasses.

substrate modulus is approximately 3.5×10^7 N/m^2, and therefore will absorb much of that residual strain by deformation. The residual strain is further reduced by the tempering or rapid cooling treatment used in processing. As a result of these factors, the residual strain measured in the coating varies from 40 to about 100 μcm/cm, as fabricated, rather than the 200 to 300 μcm/cm that would be expected because of thermal expansion differences only.

Summary of Coating Performance

Reaction-cured glass coatings have been applied in thicknesses from 0.025 to 0.050 cm to both rigid fibrous reusable surface insulation and foamed silica insulating brick densities varying from 0.12 to 1.0 gm/cc, and glazed successfully. The coatings and bulk RCG glass are highly resistant to thermal shock, and are chemically and morphologically stable to temperatures of 1260°C and above for extended periods of time at atmospheric or reduced pressure.

The reaction-cured glass coating has undergone extensive testing on the low-density silica RSI to evaluate its overall performance for planned Space Shuttle missions. These tests have shown that the basic coating does not devitrify after 800 hours at 1260°C. In addition, they have demonstrated that the coating, even if exposed to a normal Cape Kennedy launch environment with the equivalent of 5 years sea salt deposited on its outer surfaces, does not devitrify in 24 hours at 1260°C. Careful examination of the interface between the RCG coating and the silica RSI substrate have further shown that no devitrification or other morphological change occurs at the interface after the exposure to 1260°C for 24 hours. Some of these overall performance tests have been directed at evaluating the reaction cured glass on RSI tiles in simulated Space Shuttle entry environments. These included multiple tests on coated 15 by 15 by 5 cm tiles, which were performed using radiant test facilities to simulate 100 Shuttle entry flights for the environment shown in Fig. 2 to determine if any degradation of the coating occurred. No degradation of the coating's optical, physical, or morphological properties was noted. A very limited number of coating microcracks that resulted from inclusions or flaws in the coating or mechanical damage were observed. In another test series, 100 simulated flights in a convective heating environment were completed. Forerunners to the reaction cured glass coating all failed this test series. The reaction cured glass coating passed the tests with only a slight, but acceptable, degradation of the coating's optical properties. Specifically, its emittance decreased from 0.9 to 0.8 (Requirement: $\varepsilon \geq 0.8$) with no other significant changes occurring.

REFERENCES

1. Stewart, David A. and Leiser, Daniel B., "Effect of Radiant and Convective Heating on the Optical and Thermochemical Properties of Reusable Surface Insulation," AIAA Paper No. 76044, presented at AIAA 11th Thermophysics Conf., San Diego, Calif., July 14-16, 1976.

2. Goldstein, Howard E., Buckley, John D., King, Marshal H., Probst, Hubert B. and Spiker, Ike K., NASA TM X-2570.

3. Beasley, R. M., Izu, Y. D., Nakano, Ozolin A. A. and Pechman, A., "Fabrication and Improvement of LMSC's All-Silica RSI," NASA TM X-2719, Sept. 1973, pp. 1-17.

4. Pigg, O. E., "Results of RSI Thermal-Structure Analysis," NASA TM X-2721, Sept. 1973, pp. 1227-1269.

5. Ransom, P. O. and Dicus, D. L., "Effects of Sea Salts on the Physical Characteristics of Reusable Surface Insulation," NASA TM X-2720, Sept. 1973, pp. 765-792.

6. U.S. Patent 3,378,431, April 16, 1968. Method of Making Carbon-Containing Glass and Product Thereof, by Smith, Carlyle F., Crandall, William B.

INDEX

GPSR Compliance

The European Union's (EU) General Product Safety Regulation (GPSR) is a set of rules that requires consumer products to be safe and our obligations to ensure this.

If you have any concerns about our products, you can contact us on ProductSafety@springernature.com

In case Publisher is established outside the EU, the EU authorized representative is:

Springer Nature Customer Service Center GmbH
Europaplatz 3
69115 Heidelberg, Germany

Batch number: 10838062

Printed by Printforce, the Netherlands